# PLATELET PROTEOMICS

## WILEY-INTERSCIENCE SERIES IN MASS SPECTROMETRY

**Series Editors**

Dominic M. Desiderio
*Departments of Neurology and Biochemistry*
*University of Tennessee Health Science Center*

Nico M. M. Nibbering
*Vrije Universiteit Amsterdam, The Netherlands*

John R. de Laeter • *Applications of Inorganic Mass Spectrometry*
Michael Kinter and Nicholas E. Sherman • *Protein Sequencing and Identification Using Tandem Mass Spectrometry*
Chhabil Dass • *Principles and Practice of Biological Mass Spectrometry*
Mike S. Lee • *LC/MS Applications in Drug Development*
Jerzy Silberring and Rolf Eckman • *Mass Spectrometry and Hyphenated Techniques in Neuropeptide Research*
J. Wayne Rabalais • *Principles and Applications of Ion Scattering Spectrometry: Surface Chemical and Structural Analysis*
Mahmoud Hamdan and Pier Giorgio Righetti • *Proteomics Today: Protein Assessment and Biomarkers Using Mass Spectrometry, 2D Electrophoresis, and Microarray Technology*
Igor A. Kaltashov and Stephen J. Eyles • *Mass Spectrometry in Biophysics: Confirmation and Dynamics of Biomolecules*
Isabella Dalle-Donne, Andrea Scaloni, and D. Allan Butterfield • *Redox Proteomics: From Protein Modifications to Cellular Dysfunction and Diseases*
Silas G. Villas-Boas, Ute Roessner, Michael A.E. Hansen, Jorn Smedsgaard, and Jens Nielsen • *Metabolome Analysis: An Introduction*
Mahmoud H. Hamdan • *Cancer Biomarkers: Analytical Techniques for Discovery*
Chabbil Dass • *Fundamentals of Contemporary Mass Spectrometry*
Kevin M. Downard (Editor) • *Mass Spectrometry of Protein Interactions*
Nobuhiro Takahashi and Toshiaki Isobe • *Proteomic Biology Using LC-MS: Large Scale Analysis of Cellular Dynamics and Function*
Agnieszka Kraj and Jerzy Silberring (Editors) • *Proteomics: Introduction to Methods and Applications*
Ganesh Kumar Agrawal and Randeep Rakwal (Editors) • *Plant Proteomics: Technologies, Strategies, and Applications*
Rolf Ekman, Jerzy Silberring, Ann M. Westman-Brinkmalm, and Agnieszka Kraj (Editors) • *Mass Spectrometry: Instrumentation, Interpretation, and Applications*
Christoph A. Schalley and Andreas Springer • *Mass Spectrometry and Gas-Phase Chemistry of Non-Covalent Complexes*
Riccardo Flamini and Pietro Traldi • *Mass Spectrometry in Grape and Wine Chemistry*
Mario Thevis • *Mass Spectrometry in Sports Drug Testing: Characterization of Prohibited Substances and Doping Control Analytical Assays*
Sara Castiglioni, Ettore Zuccato, and Roberto Fanelli • *Illicit Drugs in the Environment: Occurrence, Analysis, and Fate Using Mass Spectrometry*
Ángel Garciá and Yotis A. Senis (Editors) • *Platelet Proteomics: Principles, Analysis, and Applications*

# PLATELET PROTEOMICS

## Principles, Analysis, and Applications

Edited By

ÁNGEL GARCÍA
YOTIS A. SENIS

A JOHN WILEY & SONS, INC., PUBLICATION

Published by John Wiley & Sons, Inc., Hoboken, New Jersey
Published simultaneously in Canada



***Library of Congress Cataloging-in-Publication Data:***

Platelet proteomics : principles, analysis, and applications / edited by Ángel García, Yotis A. Senis.
p. ; cm.—(Wiley-Interscience series on mass spectrometry)
Includes bibliographical references and index.
ISBN 978-0-470-46337-6 (cloth)
1. Blood platelets. 2. Proteomics. I. García, Ángel, 1968- II. Senis, Yotis, 1970- III. Series: Wiley-Interscience series on mass spectrometry.
[DNLM: 1. Blood Platelets–physiology. 2. Cardiovascular Diseases–physiopathology. 3. Proteomics–methods. 4. Thrombosis–physiopathology. WH 300]
QP97.P555 2011
612.1′17–dc22

2010033323

Printed in the United States of America

oBook ISBN: 978-0-470-94029-7
ePDF ISBN: 978-0-470-94028-0
ePub ISBN: 978-1-118-00207-0

10 9 8 7 6 5 4 3 2 1

# CONTENTS

# CONTRIBUTORS

**Jan-Willem N. Akkerman, PhD**, Department of Clinical Chemistry and Haematology, University Medical Center Utrecht, Utrecht, The Netherlands

**Gloria Álvarez-Llamas, PhD**, Department of Immunology, Fundación Jiménez Diaz, Madrid, Spain

**Wadie F. Bahou, MD**, Department of Medicine, State University of New York at Stony Brook, Stony Brook, New York

**María G. Barderas, PhD**, Department of Vascular Physiopathology, Hospital Nacional de Parapléjicos, Toledo, Spain

**Natasha E. Barrett, PhD**, Institute for Cardiovascular and Metabolic Research, School of Biological Sciences, University of Reading, Reading, United Kingdom

**Katleen Broos, PhD**, Laboratory for Thrombosis Research, KU Leuven Campus Kortrijk, Kortrijk, Belgium

**Julia Maria Burkhart**, Leibniz Institut für Analytische, Wissenschaften ISAS e.V., Dortmund, Germany

**Niklaas Colaert**, VIB Department of Medical Protein Research and UGent Department of Biochemistry, Ghent University, Ghent, Belgium

**Bernard de Bono, MD, PhD**, European Bioinformatics Institute, EMBL Cambridge, United Kingdom

**Hans Deckmyn, PhD**, Laboratory for Thrombosis Research, KU Leuven Campus Kortrijk, Kortrijk, Belgium

**Fernando de la Cuesta, PhD**, Department of Immunology, Fundación Jiménez Diaz, Madrid, Spain

**Ángel García, PhD**, Department of Pharmacology, Facultade de Farmacia, Universidade de Santiago de Compostela, Santiago de Compostela, Spain

**Kris Gevaert, PhD**, VIB Department of Medical Protein Research and UGent Department of Biochemistry, Ghent University, Ghent, Belgium

**Jonathan M. Gibbins, PhD**, Institute for Cardiovascular and Metabolic Research, School of Biological Sciences, University of Reading, Reading, United Kingdom

**Dmitri V. Gnatenko, PhD**, Department of Medicine, State University of New York at Stony Brook, Stony Brook, New York

**Andreas Greinacher, MD**, Institute for Immunology and Transfusion Medicine, Ernst-Moritz-Arndt University of Greifswald, Greifswald, Germany

**Kenny Helsens, PhD**, VIB Department of Medical Protein Research and UGent Department of Biochemistry, Ghent University, Ghent, Belgium

**Francis Impens, PhD**, VIB Department of Medical Protein Research and UGent Department of Biochemistry, Ghent University, Ghent, Belgium

**Lennart Martens, PhD**, VIB Department of Medical Protein Research and UGent Department of Biochemistry, Ghent University, Ghent, Belgium

**James P. McRedmond, PhD**, UCD Conway Institute, University College Dublin, Belfield, Dublin, Ireland

**Marie N. O'Connor, PhD**, MRC Laboratory for Molecular Cell Biology, University College London, London, United Kingdom

**Rudolf Oehler, PhD**, Department of Surgery, Center of Translational Research, Medical University of Vienna, Vienna, Austria

**Isabelle I. Salles, PhD**, Department of Haematology, Imperial College London, London, United Kingdom

**Yotis A. Senis, PhD**, Centre for Cardiovascular Sciences, Institute of Biomedical Research, School of Clinical and Experimental Medicine, College of Medical and Dental Sciences, University of Birmingham, Birmingham, United Kingdom

**Albert Sickmann, PhD**, Leibniz Institut für Analytische, Wissenschaften ISAS e.V., Dortmund, Germany

**David M. Smalley, PhD**, Mass Spectrometry Laboratory, Maine Institute for Human Genetics and Health, Bangor, Maine

**Ronald G. Stanley, PhD**, Institute for Cardiovascular and Metabolic Research, School of Biological Sciences, University of Reading, Reading, United Kingdom

**Leif Steil, PhD**, Interfaculty Institute for Genetics and Functional Genomics, Department of Functional Genomics, Ernst-Moritz-Arndt University of Greifswald, Greifswald, Germany

**Thomas Thiele, MD**, Institute for Immunology and Transfusion Medicine, Ernst-Moritz-Arndt University of Greifswald, Greifswald, Germany

**Daphne C. Thijssen-Timmer, PhD**, Department of Experimental Immunohaematology, Sanquin Research, and Landsteiner Laboratory, AMC, University of Amsterdam, Amsterdam, The Netherlands

**Michael G. Tomlinson, DPhil**, School of Biosciences, College of Life and Environmental Sciences, University of Birmingham, Birmingham, United Kingdom

**Katherine L. Tucker, PhD**, Institute for Cardiovascular and Metabolic Research, School of Biological Sciences, University of Reading, Reading, United Kingdom

**Joël Vandekerckhove, PhD**, VIB Department of Medical Protein Research and UGent Department of Biochemistry, Ghent University, Ghent, Belgium

**Fernando Vivanco, PhD**, Department of Immunology, Fundación Jiménez Diaz, Madrid, Spain, and Department of Biochemistry and Molecular Biology, Facultad de Química, Universidad Complutense, Madrid, Spain

**Uwe Völker, PhD**, Interfaculty Institute for Genetics and Functional Genomics, Department of Functional Genomics, Ernst-Moritz-Arndt University of Greifswald, Greifswald, Germany

**Maria Zellner, PhD**, Institute of Physiology, Center for Physiology and Pharmacology, Medical University Vienna, Vienna, Austria

**Irene Zubiri**, Department of Immunology, Fundación Jiménez Diaz, Madrid, Spain

# FOREWORD

Major progress in science often occurs through an unexpected observation or the advancement in technologies that enables new insights into recognized problems. In my lifetime in platelet research, I believe that the most important technological development is that of proteomics. The platelet field has lagged behind many other fields in hematology because of the absence of a nucleus and the difficulties in working with the platelet "mother cell," the megakaryocyte, which is present at low levels in bone marrow. Thus, routine molecular biology procedures are not readily applicable to the study of platelet function and, while platelets do retain low levels of messenger RNA (mRNA), contamination is always a concern. Indeed, such challenges contributed to the delay in identifying the major platelet receptor for ADP, the $P2Y_{12}$ receptor (the target of the second most widely selling drug in terms of world sales, clopidogrel), which was achieved only in 2001.

The opportunity provided by proteomics at the turn of the millennium of generating a comprehensive list of platelet proteins and providing information on their posttranslational modifications was therefore the start of a new era in platelet research. For the first time, it was possible to have the potential of a comprehensive and quantitative list of proteins in platelets along with vital key information on their posttranslational modifications. Such information is critical for advancements in platelet research. Almost immediately, new platelet proteins, and new platelet protein families, were identified including signaling molecules such as the Dok and RGS families and novel receptors, including CLEC-2 and G6b-B. This paved the way for new insights into platelet function, such as the realization that CLEC-2 and platelets play a critical role in lymphatic development. Thus, no longer would graduate students need to spend their time in the coldroom trying to purify a platelet protein that may not even be present.

Of course, with any new technology, there were many key issues to overcome at the beginning, most notably concerning sensitivity given the remarkable range of expression levels of proteins in platelets. Thus the initial hype and expectation would never be able to achieve the success that was demanded. However, platelet proteomics has now matured such that we have a comprehensive list of proteins in platelets and are beginning to make inroads into the descriptions of posttranslational modifications in resting and activated cells and thus are gaining a dynamic insight into the mechanisms that control platelet function. The mapping of protein phosphorylation in resting and activated platelets is particularly exciting as this paves the way for a more complete understanding of platelet signaling networks. We are nevertheless far from reaching the potential of proteomics, most notably with respect to clinical applications. It is in this area where the field has the greatest potential to make a contribution that is over and above that of genomics, as platelets are controlled by proteins.

The field is now over 10 years old and is approaching the challenging teenage years that occur before the full potential is realized. The present book is very timely in that it consolidates the information that has accumulated up to now and predicts the future. It will therefore be of considerable value to those already working in the field and to those who are considering entering into this exciting area of research.

The book is divided into parts dealing with methodological issues, analysis of the platelet proteome, and integrated proteomics in health and disease, with each chapter written by experts in the field, and edited by two rising investigators. Platelet proteomics has come of age in a very short time. If teenagers would take the time to read relevant material, they might mature more quickly. For those interested in platelet proteomics, this book is a must.

Steve P. Watson

*Centre for Cardiovascular Sciences*
*Institute of Biomedical Research*
*College of Medical and Dental Sciences*
*University of Birmingham, UK*
*June 2010*

# PREFACE

The goal of this book is to provide a comprehensive source of information about platelet proteomics. *Platelet Proteomics: Principles, Analysis, and Applications* introduces the reader to the basic principles of modern proteomics technology and its application to platelet research.

*Platelets* are small anucleate cells that circulate in the blood and play a fundamental role in hemostasis. From a pharmacological perspective, unwanted platelet activation is related to thrombotic and cardiovascular disease, presently recognized as the leading cause of death in the Western world. Since platelets do not contain a nucleus, analysis of the proteome is an ideal way to study mechanistically how platelets function. It is only in the past decade (i.e., since 1999) that platelet proteomics research started, thanks mainly to the recent developments in proteomics instrumentation, especially in the mass spectrometry field. Since then, several research groups worldwide have focused on this emerging field and made important progress on the analysis of the proteome of basal and activating platelets. This has led in some cases to the discovery of new proteins that could eventually become drug targets for platelet-based diseases. The hope is that in the near future, platelet proteomics can be widely applied to translational research related to such disease conditions.

We believe that this is an opportune time for an inaugural book on platelet proteomics that could serve as a reference for those interested in the field. The intended audience for the book includes people involved in platelet research at either basic or translational level. These include hematologists, cardiologists, blood bankers, and researchers in thrombosis and haemostasis, as well as students and fellows in these fields. From a technological perspective, our hope is that this book will be of interest to general proteomics and trascriptomics researchers, particularly those involved in platelet and cardiovascular research.

The authors of each of the 14 chapters are world leaders in the field. The book is organized into three parts with specific chapters serving as links between them; for instance, Chapter 4 provides an overview of two-dimensional gel-electrophoresis-based proteomics and at the same time explains its application to platelet signaling studies. Part I is a general overview of platelets, sample preparation, and mass-spectrometry-based proteomics. Part II presents analyses of the platelet proteome, including discussions of global approaches and subproteomes. Part III focuses on integrated "-omics" and application to disease.

Finally, we would like to thank Prof. Dominic M. Desiderio, editor of the Wiley-Interscience Series on Mass Spectrometry, where this volume is included, for motivating us to take this adventure ahead. This book would not have been created without his perseverance.

Ángel García
Yotis A. Senis

# ACRONYMS*

| | |
|---|---|
| ABCC | ATP binding cassette transporter |
| ACE | angiotensin converting enzyme |
| ARP | actin-related protein |
| BAMBI | bone (morphogenic protein) and activin membrane-bound inhibitor |
| BC | buffy coat |
| BLAST | basic alignment search tool |
| BLOC | biogenesis of lysosome-related organelle complexes |
| BMEC | bone marrow endothelial cell |
| BMP | bone morphogenic protein |
| CAD | collision-activated dissociation; coronary artery disease |
| CDC | cell division control |
| CDG | congenital disorder(s) of glycosylation |
| CIB | calcium and integrin binding |
| CID | collision-induced dissociation |
| CLP | common lymphoid progenitor |
| CMP | common myeloid progenitor |
| COFRADIC | combined fractional diagonal chromatography |
| CRP | collagen-related peptide |
| CV | coefficient of variation |

*Partial list only. Common terms (LDL, MALDI, NMR, PCR, TOF, UV, etc.), chemical compound abbreviations (AP, SDS, etc.), nongeneric acronyms (FDA, WHO, etc.), and most of the exclusively medical terms found in Chapter 14 (CVD, MI, etc.) are omitted here.

| | |
|---|---|
| DGE | digital gene expression |
| DIGE | difference/differential in-gel electrophoresis |
| DTP | direct tissue proteomics |
| ECD | electron capture dissociation (ETD—electron transfer dissociation) |
| ECM | extracellular matrix |
| EDG | endothelial differentiation gene |
| EGFR | epidermal growth factor receptor |
| ERLIC | electrostatic repulsion (hydrophilic interaction) liquid chromatography |
| ES | embryonic system |
| ESI | electrospray ionization |
| EST | expressed sequence tag |
| FAK | focal adhesion kinase |
| FFE | freeflowing electrophoresis |
| FPR | false-positive rate |
| GAP | GTPase activating protein |
| GEF | guanylate exchange factor |
| GFP | gel-filtered platelet |
| GRB | growth (factor) receptor binding |
| HILIC | hydrophobic interaction liquid chromatography |
| HST | hematopoietic stem cell |
| ICAT | isotope-coded affinity tag |
| ICPL | isotope-coded protein label |
| ICR | ion cyclotron resonance |
| IEF | isoelectric focusing |
| ILK | integrin-linked kinase |
| IMAC | immobilized metal affinity chromatography |
| IMP | integral membrane protein |
| IPG | immobilized pH gradient |
| IRS | insulin receptor substrate; internal reflection spectroscopy |
| ITAM | immunoreceptor Tyr-based activation motif (ITIM—immunoreceptor Tyr-based inhibitory motif) |
| iTRAQ | isotope tagging and relative and absolute quantitation |
| JAM | junctional adhesion molecule |
| LAT | linker for activation of T cells |
| LCM | laser capture microdissection |
| LMD | laser microdissection |
| LSPAD | localized statistics of protein abundance distribution |
| MFI | mean fluorescent intensity |
| MMP | matrix metalloprotease |
| MO | morpholinooligonucleotide (*not* molecular orbital in this ms.) |
| MPP | multipotent progenitor |
| MPV | mean platelet volume |
| MRM | multireaction monitoring |

| | |
|---|---|
| MudPIT | multidimensional protein identification technology |
| NLS | neutral loss scanning |
| OCS | open canalicular system |
| ORF | open reading frame |
| PACAP | pituitary adenylcyclase activating polypeptide |
| PAF | platelet activating factor |
| PAI | protein abundance index |
| PAR | protease-activated receptor |
| PC | platelet concentrate |
| PEAR | platelet endothelial aggregation receptor |
| PECAM | platelet endothelial cell adhesion molecule |
| PMF | peptide mass fingerprinting |
| PMP | peripheral membrane protein |
| PPI | protein-protein interaction |
| PPAR | peroxisome proliferator-activated receptor |
| PPP | platelet-poor plasma (PRP—platelet-rich plasma) |
| PRT | pathogen reduction technology |
| PSL | platelet storage lesion |
| PTM | posttranslational modification |
| RACK | receptor for activated C kinase |
| RGS | regulator of G-protein signaling |
| ROS | reactive oxygen species |
| RT | reactive thrombocytosis; reverse transcriptase/transcription (as in RT-PCR); room temperature |
| SAC | serial affinity chromatography |
| SAGE | serial analysis of gene expression |
| SCX | strong cation exchange |
| SELDI | surface-enhanced laser desorption/ionization |
| SERCA | sarcoplasmic/endoplasmic reticulum calcium |
| SILAC | stable isotope labeling by amino acids in cell culture |
| SLAM | signaling lymphocyte activation molecule |
| SMART | switching mechanism at 5′ end of RNA template(s) |
| SNAP | synaptosome-associated protein |
| SNARE | soluble NSF (*N*-ethylmaleimide-sensitive factor) attachment (protein) receptors |
| SPD | storage pool deficiency |
| SRM | selective reaction monitoring |
| TAP | tandem affinity protein |
| TAR | thrombocytopenia and absent radius |
| TILLING | targeting-induced local lesion(s) in genomes |
| TMA | tissue microarray |
| TMD | transmembrane domain |
| TMT | tandem mass tag |
| TRAIL | TNF (tumor necrosis factor)-related apoptosis inducing ligand |

| | |
|---|---|
| TRAP | thrombin receptor activating peptide |
| TSA | thrombus size area |
| 2DGE | two-dimensional gel electrophoresis (*not* 2-DE); 1DGE (*not* 1-DE) |
| VAMP | vesicle-associated membrane protein |
| VASP | vasodilator-stimulated phosphoprotein |
| vWF | von Willebrand factor |
| WASP | Wiskott-Aldrich syndrome protein |
| WBC | whole-body cooling; white blood cell |
| WGA | wheatgerm agglutinin |
| Y2H | yeast two-hybrid |

# PART I

# GENERAL OVERVIEW: PLATELETS, SAMPLE PREPARATION, AND MASS SPECTROMETRY-BASED PROTEOMICS

# 1

# PLATELETS AND THEIR ROLE IN THROMBOTIC AND CARDIOVASCULAR DISEASE: THE IMPACT OF PROTEOMIC ANALYSIS

Ronald G. Stanley, Katherine L. Tucker,
Natasha E. Barrett, and Jonathan M. Gibbins

**Abstract**

This chapter provides an overview of how proteomics research impacts on our understanding of platelets. In addition to their role in hemostasis, inappropriate platelet activation is strongly related to the leading cause of death in Western societies: thrombotic cardiovascular disease. The known processes of platelet activation and the signaling mechanisms that regulate these are detailed here, but this knowledge is incomplete. Mass-spectrometry-based proteomics has already contributed to a growth in the understanding of platelets and presents itself as a tool that can unravel the details of the control of platelet function in health and disease through continuing refinements in technology and experimental design toward the development of diagnostic tools and antithrombotic drugs.

*Platelet Proteomics: Principles, Analysis and Applications*, First Edition.
Edited by Ángel García and Yotis A. Senis.

## 1.1 INTRODUCTION

*Platelets* are anucleate blood cells derived from megakaryocytes that perform a pivotal role in the regulation of hemostasis, a physiologic response to injury that prevents excessive bleeding at sites of injury. Inappropriate platelet activation, however, can lead to the pathological condition *arterial thrombosis*, the formation of a blood clot within a blood vessel resulting in occlusion of bloodflow. This is a critical event that occurs at the site of lipid-rich atherosclerotic plaques to trigger both heart attack [myocardial infarction (MI)] and stroke.

Cardiovascular disease (CVD) is the main cause of death in Westernized societies, and the World Health Organization (WHO) estimated that by 2010 CVD will also be the leading cause of death in developing countries [1]. Similar rates and trends of heart disease are seen across northern Europe and North America so that in both the United States and the United Kingdom CVD is responsible for 35% of deaths each year, with about half of these deaths attributed to coronary heart disease (CHD) and about a quarter to stroke [2,3]. With changes in lifestyle and improvements in pharmaceutical and surgical intervention, mortality rates have been falling since the early 1970s, with reduction rates slowest in younger individuals and fastest in those over 55 years old. This reduction in death rate is confounded by reported increases (6.0–7.4% in men, 4.1–4.5% in women) in the incidence and prevalence of cardiovascular disease and stroke across most age ranges [3]. This increase is most consistently found in people over 75 years old and may reflect the fact that more people in developed countries are living longer [4].

Investigations using basic methods of cell biology and targeting specific signaling molecules or pathways have led to a deeper understanding of platelet biology and the mechanisms that regulate platelet activity. This has resulted in the development of safer and more efficacious antithrombotic strategies of medication [5–7] and to the identification of a number of platelet proteins as potential therapeutic targets [8,9]. Mass-spectrometry (MS)-based methods of proteomics have brought additional tools to the study of cells and tissues: direct measurement and characterization of proteins and peptides, providing information such as identity and *de novo* amino acid sequence. Hence, in the absence of appreciable levels of regulation at the genome level (although, curiously, protein synthesis by platelets has been reported [10,11]), where the analysis of platelet biology is not complicated by significantly changing levels of total protein, proteomics methods are particularly suitable. The principal regulation of platelets is achieved by changes in protein interactions, translocation within the cell, and posttranslational modifications. Hence, the unparalleled levels of sensitivity inherent in proteomic methods of analysis that enable sequence isoform and post-translational modification of low abundance proteins and of protein complexes enables the examination of both normal and disease-induced changes in platelet proteins. This level of information can provide insights not easily gained by alternative methods.

### 1.1.1 Regulation of Platelet Function

Knowledge of the normal regulation of platelet function may establish new mechanisms by which platelet activity in diseased blood vessels can be controlled and the risk of atherosclerosis be reduced. By examining platelet signaling processes and understanding the molecular interactions that are articulated as platelet function, it becomes possible to identify molecules or groups of molecules as potential targets for therapeutic treatment or for use as biomarkers—diagnostic molecular markers of disease. The intention of the following section is to outline the molecular processes involved in platelet activation.

The process of platelet activation and thrombus formation, controlled by ligand–receptor interactions and intracellular signaling events, is outlined in Figure 1.1, while the principal platelet receptors, ligands, and the key signaling events associated with them are described below and illustrated in Figure 1.2.

The generation of a thrombus involves the initial formation of a platelet plug followed by stabilization of this plug through fibrin deposition (coagulation). The

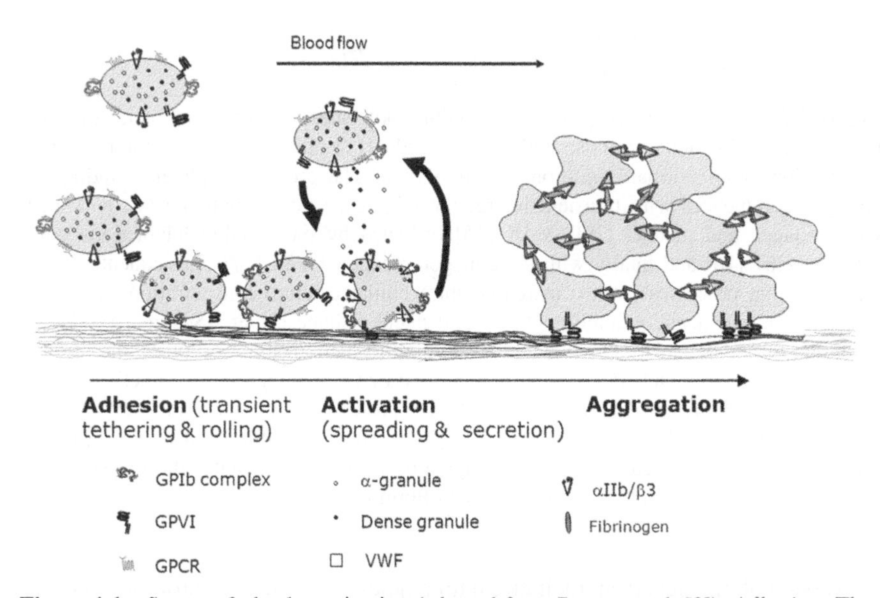


**Figure 1.1** Stages of platelet activation (adapted from Barrett et al. [8]). *Adhesion:* The glycoprotein von Willebrand factor (vWF) binds to exposed collagen and under conditions of high blood shear a transient and unstable interaction with the platelet GPIb-V-IX receptor complex slows the movement of the platelets, allowing other interactions to occur. More stable and direct interactions between collagen and the platelet integrin $\alpha2\beta1$ [16] enable binding to the collagen receptor GPVI. *Activation:* GPVI molecules, complexed with the Fc receptor (FcR) $\gamma$-chain, cluster and stimulate signaling that results in rapid shape change so that platelets spread to cover the damaged endothelium. The secretion of positive-feedback signals from $\alpha$-granules and dense granules attract more platelets to the growing thrombus and activate them. *Aggregation:* Platelet activation increases the affinity of integrin $\alpha IIb\beta3$ for its plasma fibrinogen ligand, leading to aggregation [17–22].

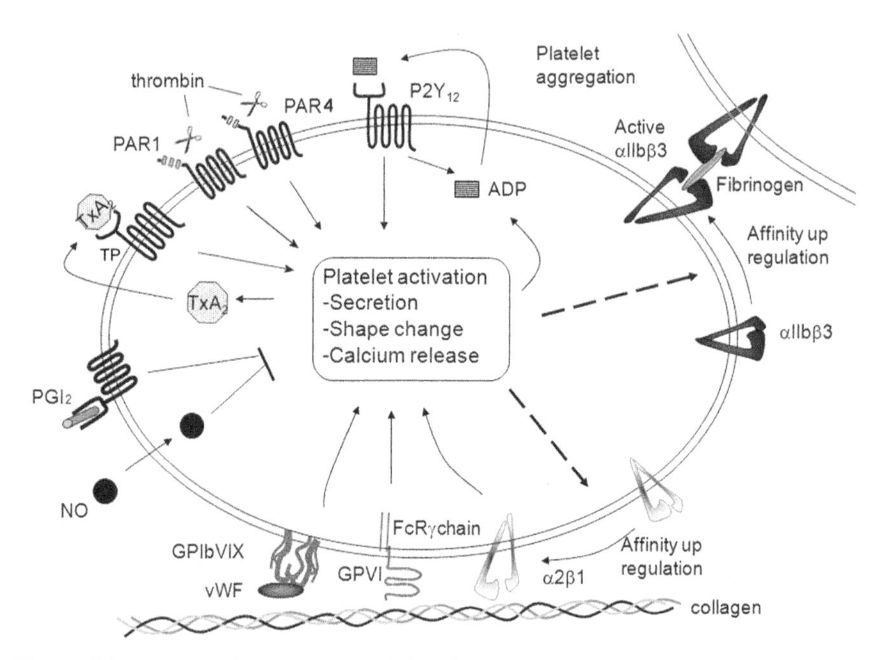


**Figure 1.2** Key platelet receptors and signaling molecules. The principal platelet receptors and their ligands are illustrated. Nitric oxide [12] and $PGI_2$ [13] inhibit platelet aggregation under normal conditions of circulation. The process of platelet binding to collagen commences with transient interaction of plasma von Willebrand factor (vWF) with collagen and platelet GPIb-V-IX [15] followed by more stable interactions with integrin α2β1 [16] and finally with the collagen receptor GPVI [17]. Subsequent activatory signaling processes lead to increased intracellular calcium concentration, secretion of α-granule and dense granule contents, and platelet shape change. Secreted products, including thromboxane $A_2$ ($TXA_2$) and ADP, bind to their specific receptors, the thromboxane (TP) receptor and ADP receptors $P2Y_1$ and $P2Y_{12}$, adding to platelet activation by positive feedback [47–53]. Thrombin activation of protease-activated receptors (PAR1 and PAR4) stimulates further signaling activities [54,56,163]. The affinity of integrin αIIbβ3 for fibrinogen is increased via *inside-out* signaling [17]. This facilitates the formation of platelet aggregates through cross-binding with fibrinogen.

conversion of platelets from their circulating quiescent form to a thrombus may be characterized in three distinct phases: adhesion, activation, and aggregation [8].

***1.1.1.1 Adhesion*** Under normal conditions of blood circulation, molecules such as nitric oxide (NO) [12] and prostaglandin $I_2$ ($PGI_2$) [13] released from healthy endothelial cells inhibit platelet activation. When the endothelium is damaged, however, this antiactivatory signaling is disrupted and the proactivatory subendothelial matrix is exposed, initiating a series of events that initially cause platelet binding to the damaged surface. The multimeric plasma glycoprotein von Willebrand factor (vWF) is released from endothelial cells into the plasma at high concentrations [14]. This vWF binds to exposed collagen and, under conditions of high shear, undergoes conformational changes that enable interaction with the glycoprotein (GP)Ib component of the platelet GPIb-V-IX receptor complex [15]. This initial binding is transient and unstable, but is sufficient to slow the platelets, enabling direct adhesion to collagen via the integrin $\alpha 2\beta 1$. This stabilizes platelet–collagen interactions [16], allowing interaction of the collagen receptor glycoprotein VI (GPVI).

***1.1.1.2 Activation*** Activation of platelets occurs in two rapid phases, amplified by positive feedback, and results in irreversible aggregation. Phase 1 is initiated by binding of collagen to GPVI, inducing rapid activation of a kinase cascade. This cascade results in multiple signaling events that lead to platelet shape change and secretion of many positive-feedback factors. Phase 2 consists of further activation downstream of these (and other) factors, recruiting more platelets, and propelling platelets into aggregation [17–22].

As a pivotal platelet-specific activatory receptor, GPVI and its signaling pathway has been the subject of much research and is described in detail in many articles and reviews [22–25]. Platelets deficient in GPVI are unable to aggregate on collagen stimulation, yet have no major bleeding defect [25,26], thus highlighting the possible benefit of therapeutically targeting GPVI or its signaling pathway components. The binding of collagen to GPVI results in receptor clustering and subsequent tyrosine phosphorylation of the noncovalently associated Fc receptor (FcR) $\gamma$-chain. Two conserved tyrosine residues found within the immunoreceptor tyrosine-based activatory motif (ITAM) of the FcR $\gamma$-chain are phosphorylated by the Src family kinases Fyn and Lyn [27]. This enables docking of the tyrosine kinase Syk via its two Src homology 2 (SH2) domains, and thus the kinase cascade is initiated [18–21]. Syk becomes autophosphorylated on several tyrosine residues and induces tyrosine phosphorylation of residues in the adaptor protein linker for activation of T cells (LAT) [26–28]. LAT acts as a scaffold for signaling molecules such as phospholipase C$\gamma$2 (PLC$\gamma$2) and phosphatidylinositol 3-kinase (PI3K). PLC$\gamma$2 catalyzes the conversion of phosphatidylinositol(4,5)bisphosphate to inositol(1,4,5)trisphosphate and diacylglycerol, triggering a rise in intracellular calcium. The increased intracellular concentration of calcium initiates secretion of both $\alpha$-granules and dense granules [31–33], the contents of which include fibrinogen, vWF, coagulation factors V and XIII, ADP, and serotonin, and act in an autocrine and paracrine fashion to further stimulate platelets [17,22,34–38]. PI3K catalyzes the conversion of phosphatidylinositol (4,5) bisphosphate to phosphatidylinositol (3,4,5) trisphosphate. This phosphorylated lipid recruits Pleckstrin homology (PH) domain containing molecules to the cell membrane, such as protein-dependent kinases (PDKs) and protein kinase B (PKB; also known as Akt) [17,39–42], where they become activated and induce further signaling, ultimately leading to upregulation in the affinity of the fibrinogen receptor integrin $\alpha IIb\beta 3$ [38,43–45].

Factors secreted from platelets act as positive-feedback signals to attract more platelets to join the growing thrombus and to activate them. These factors include locally high concentrations of secondary agonists such as

ADP, adrenaline, 5-hydroxytryptamine (5HT) and thromboxane $A_2$ ($TXA_2$), each of which binds to specific receptors on the platelet plasma membrane. Activation of phospholipase $A_2$ results in the liberation of arachidonic acid from membranes, which is, in turn, converted to thromboxane $A_2$ ($TXA_2$) via the actions of cyclooxygenase (COX) and thromboxane synthase [44]. Liberated $TXA_2$ then binds to thromboxane–prostaglandin (TP) receptors, contributing to positive-feedback activation of platelets [47,48].

Platelets possess two receptors for ADP: $P2Y_1$ and $P2Y_{12}$, both of which are G-protein-coupled receptors (GPCRs). $P2Y_1$ is essential for platelet activation, while $P2Y_{12}$ amplifies and sustains $P2Y_1$-initiated signaling [49–53]. The serine protease thrombin is generated through activation of the coagulation pathways and acts as a powerful platelet agonist by cleaving the *N* terminus of the protease-activated receptors PAR1 and PAR4. The uncleaved part of the receptor forms a "tethered ligand" that interacts with extracellular loops of the receptor to stimulate intracellular signaling [54–57].

***1.1.1.3 Aggregation (Thrombus Propagation)*** The final step in platelet activation is platelet crosslinking via fibrinogen and its receptor integrin αIIbβ3. Many independent signaling pathways (e.g., those stimulated by collagen, thrombin, ADP, and thromboxane $A_2$ ($TXA_2$) ultimately lead to an increased affinity of αIIbβ3 for its ligand through a process known as *inside-out signaling* [17]. High-affinity αIIbβ3 mediates platelet–platelet adhesion through bivalent interaction with fibrinogen or with vWF and an aggregate begins to form. In addition to linking platelets together, the binding of the receptor to its ligand results in outside-in signaling that further amplifies platelet activation [58,59]. Therapeutic targeting of a single pathway involved in the upregulation of αIIbβ3 should, therefore, reduce platelet activation while leaving other activatory pathways intact, thus maintaining hemostasis [8].

Outside-in signaling leads to the remodeling of the actin cytoskeleton and subsequent platelet shape change, including the formation of fillopodia, lamellipodia, and platelet spreading [59,60]. Processes involved in clot retraction, thrombus stabilization, and wound repair are also receiving attention as an emerging concept of sustained signaling within the thrombus [59,61–63]. This research focuses on the roles of contact-dependent ligand–receptor signaling involving a number of platelet surface molecules. These include the junctional adhesion molecules (JAM-A and JAM-B), Eph receptor tyrosine kinases and their ephrin ligands, Sema4D and its platelet receptors CD72 and plexin-B1, Gas6 and its interactions with the Axl, Tyro3, and Mer tyrosine kinase receptors [63].

### 1.1.2 Cardiovascular Disease and Platelets

In the majority of instances, the development of a platelet-rich thrombus at the site of an atherosclerotic plaque is the underlying cause of acute cardiovascular events, including coronary heart disease (myocardial infarction) and ischemic stroke [64]. Atherosclerosis is a highly complex chronic disease that is strongly linked with dyslipidemia, hypercholesterolemia, and inflammation, and which requires decades to progress from its initiation to the formation of pathogenic atherosclerotic plaques [65]. Vulnerable plaques that are structurally weak and rupture or erode [64,66,67] can lead to occlusion of blood vessels. Depending on the location of the plaque, tissues (cardiomyocytes in the case of heart attack and cerebral neurons with ischemic stroke) are deprived of oxygen and die.

Investigations into the potential roles of platelets in atherosclerosis have given rise to an increased awareness that they may be significant factors in the initiation, progression and outcome of this disease, for example [66–68]. Consistent with this, antiplatelet drugs have been found to be beneficial in reducing the incidence of nonfatal events in clinical trials [71].

***1.1.2.1 Atherosclerosis*** The processes involved in atherosclerotic plaque formation are complex and multifactorial. While still not fully understood, atherogenesis essentially follows a sequence of initiation, fatty streak formation, mature complex plaque formation, and finally atherothrombosis—the acute pathological complication of thrombus formation on plaque lesions.

*Initiation* Atherosclerosis commonly occurs at bends, branches and bifurcations of the aorta and its subsidiaries such as the coronary and cerebral arteries [73,74]. There, laminar blood flow is disturbed and turbulent eddies of recirculating blood are formed [74,75] allowing increased endothelium–blood particle contact and suppressing endothelial cell expression of platelet adhesion inhibiting nitric oxide (NO) [12]. *In vivo* models have also been used to show that modified blood shear rate is related to thrombus formation [76].

The retention [77] and partial oxidation of low-density lipoprotein (LDL) molecules [78,79] within the intima of the arterial wall creates a proinflammatory environment where the expression of endothelial adhesion molecules leads to the recruitment of monocytes from the circulation [80].

*Fatty Streak Formation* Within the subendothelium, monocyte-derived macrophages release inflammatory cytokines and growth factors [81,82], while expressing LDL binding scavenger receptors [83,84]. LDL molecules are phagocytized and oxidized to form "foamy" cholesteryl ester-rich lipid droplets. The macrophage foam cells subsequently die and their lipid contents accumulate to form a necrotic lesion core in the developing atherosclerotic plaque [85].

*Mature, Complex Plaques* Vascular smooth muscle cells migrate and proliferate around and above the lipid core where they secrete extracellular matrix proteins so that the mature plaque is overlaid by a collagen-rich fibrous cap and a monolayer of endothelial cells. The growth of plaques into the arterial lumen causes partial occlusion, but this is not pathogenic in itself [64].

*Atherothrombosis* Rupture or erosion of the plaque exposes collagen and possibly also oxidized lipids [86] to circulating platelets leading to thrombus formation at the site of injury. More recently there has been an increasing awareness that the pathogenesis of atherosclerosis is not dependent on plaque size, but on the likelihood of disruption to plaques and the subsequent nature of overlaying thrombi i.e. a large lipid core, thin fibrous cap with few smooth muscle cells, and an abundance of proteases such as matrix metalloproteases (MMPs), cathepsins, and collagenases [64,87].

***1.1.2.2 Platelet Involvement in Atherosclerosis*** While platelet involvement in atherosclerosis has traditionally been confined to the final thrombotic stages of the disease, the involvement of platelets in the initiation of atherosclerosis has been suggested by Huo and colleagues [88] in which platelets adhere to undamaged arterial endothelium of the apolipoprotein E-deficient (apo$E^{-/-}$) murine model of atherosclerosis. In this model platelet adhesion leads to the expression of inflammatory molecules and the initiation of atherogenesis. Aggregates of platelets with monocytes and leukocytes have been shown to promote the formation of atherosclerotic lesions in the apo$E^{-/-}$ model [88]. There, activated circulating platelets that express the surface receptor P-selectin were shown to deliver platelet-derived proinflammatory factors to monocytes, leukocytes, and the vessel wall. Platelet-derived chemokines, stored within α-granules and rapidly released on platelet activation, may also play an important role in atherogenesis through the recruitment of monocytes to sites of vascular damage and their differentiation into macrophages [89].

Numerous factors are involved in the likelihood and clinical outcome of atherothrombosis and the direct involvement of platelets in this process. Localized inflammation, bloodflow dynamics and platelet "sensitivity" to activation all appear to be involved. For example, while the proinflammatory cytokine CD40 is expressed by endothelial cells, macrophages, smooth muscle cells, T cells, and platelets, its ligand CD40L (CD154) is released at high levels by platelets after adhesion via αIIbβ3 [90–92]. Ligation of CD40 results in the expression of adhesion molecules, matrix metalloproteases (MMPs), that digest matrix proteins such as collagen fibrils and lead to the development of unstable atherosclerotic lesions [67] and procoagulant tissue factor exposure [93,94].

Dyslipidemia, a major risk factor for atherosclerosis, has been associated with increased platelet reactivity and platelets have been shown to express receptors for both native (n)LDL and oxLDL [95]. Binding of nLDL to its receptor ApoE-R2′ on the platelet surface leads to increased response to platelet agonists through synthesis of thromboxane $A_2$ via stimulation of the p38-mitogen-activated protein kinase (MAPK) pathway [96]. Ligation of scavenger receptors CD36 and SR-A on the platelet surface by oxLDL leads to more sustained activation of p38-MAPK [97] and increased platelet binding to fibrinogen via the integrin αIIbβ3. Platelet CD36 has also been shown by Podrez and colleagues to act as a receptor for oxidised choline glycerophospholipids generated by oxidative stress [98]. This mechanism may be involved in not only increased thrombosis, but also in the generation of foam cells during early atherosclerosis [68,99,100].

### 1.1.3 Platelet Proteomics

Proteomics is a relatively new and rapidly developing approach to acquiring biological information. While initially intended to describe the study of the entire protein content of a biological system or organism, the term is often used to describe the use of technologies such as mass spectrometry (see Chapter 2), together with its associated methods of sample preparation, protein separation (see, e.g., Chapter 4) and subsequent analyses, to perform protein studies that utilize the high levels of sensitivity and high-density throughput within their capability.

Proteomic analysis is a conceptually simple, but powerful tool that can be used to answer specific questions while requiring little prior knowledge about the proteome under examination. It allows the investigation of changes in protein abundance over time in response to stimulus, medication, illness and genetic conditions, or changes in posttranslational modifications such as phosphorylation and glycosylation. The widescale use of robotics and automation in proteomics lends itself to a high level of experimental reproducibility and also provides the potential for high-throughput analysis when required. Hence, in addition to gaining qualitative and quantitative information from an analyte, the interpretation of data from proteomics studies can be applied to the discovery of new drug targets and to diagnostics through the identification of disease biomarkers.

Proteomic analysis has become increasingly sophisticated and has progressed from being reliant on methods such as two-dimensional gel electrophoresis–mass spectrometry [2DGE (also sometimes abbreviated 2-DE)-MS] to being highly inclusive with the incorporation of techniques such as protein array and multidimensional chromatographic procedures in conjunction with one or more mass spectrometric methods (e.g., as described by O'Neill et al. [101] and Lewandrowski et al. [102]). Hence the ability to produce lists of proteins identified from biological samples now occurs alongside the generation of more functionally relevant information. In parallel with technical advances in mass spectrometry, this has been achieved by applying a targeted approach where subsets of proteins are isolated by a common feature such as a shared affinity for a substrate or their subcellular location, as illustrated in Figure 1.3.

These findings focus investigations on areas of particular interest while increasing the likelihood of identifying low-abundance proteins and providing evidence of the function of the identified proteins [102–108]. Although a prior knowledge of the biological system being studied is required, a targeted approach to proteomics deals with simplified systems that can supply additional dimensions of information relevant to answering specific biological and pharmacological questions. Indeed, the number of question-driven publications now exceeds those of global studies.

The proteome of an organism is complex, and there remain technical hurdles that render identification of true and relevant differences between, for example, resting and agonist-stimulated platelet samples, challenging. These include the

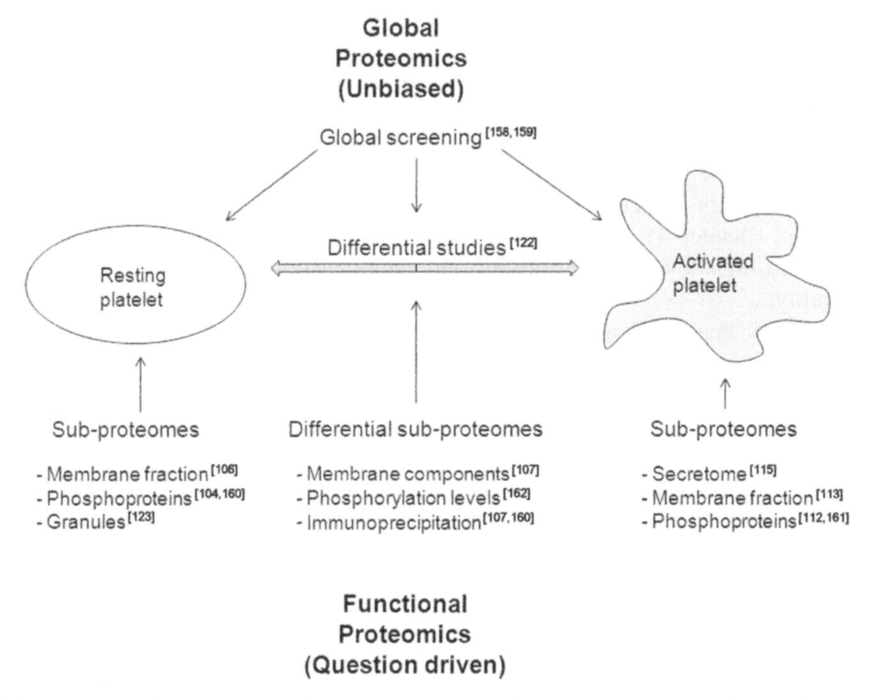


**Figure 1.3** Different approaches to proteomics and mass spectrometry. Schematic representation of different approaches to proteomic analyses of platelets, indicating relevant publications.

difficulties inherent in the large dynamic range in the levels of platelet proteins, where a difference of many orders of magnitude between low- and high-abundance proteins might occur in the same sample. Also, the possible exclusion of hydrophobic, very basic and low- or high-molecular-weight proteins depending on the techniques used. Thus, proteomic studies require careful planning and understanding of the limitations that are inherently present (see Chapter 3).

The ability to identify and characterize proteins is dependent on the availability of genomewide databases. Currently data from mass spectrometry are compared against databases containing protein information that has been theoretically derived from sources such as the human genome project. It has therefore been suggested that a database based on *de novo* sequencing of platelets should be built as a common tool for people working in this field [109]. The mouse genome database is also of great importance as the mouse is the model of choice for investigating protein function in platelets (see Chapter 11).

### 1.1.4 Impact of Proteomics on the Understanding of Platelet Biology

As platelets are anucleate, their activities are controlled predominantly by translocation and posttranslational modification of proteins within the cell. Proteins of

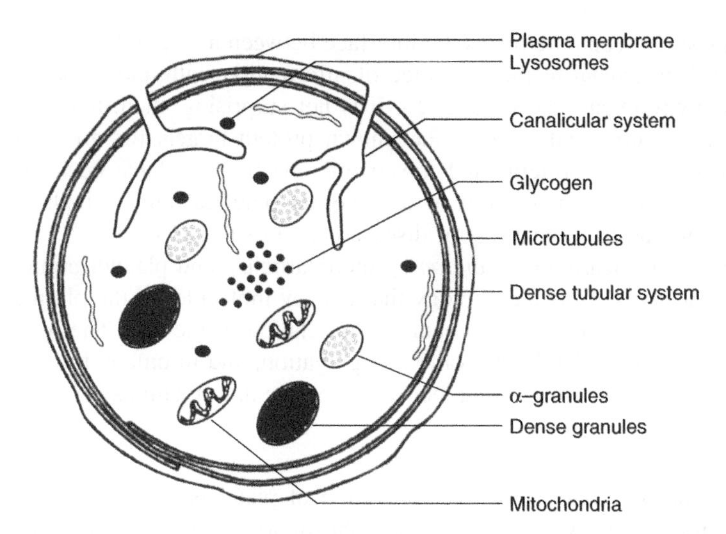


**Figure 1.4** Platelet structure. Schematic cross-sectional drawing of a human platelet indicating major structural features. The *canalicular system* consists of membrane-bound channels that act as a site for granule fusion. The cytoplasm contains a coiled bundle of *microtubules* that maintain the shape of the resting platelet and centrally locate the organelles within the platelet during activation. The cytoplasm contains a number of organelles and secretory granules. *Glycogen* particles and *mitochondria* provide metabolic activity in platelets through glycolysis and oxidative phosphorylation, while small *lysosomes* contain acid phosphatase, arylsulfate, and cathepsin. *α-Granules* contain adhesive proteins, growth factors, and coagulation factors. *Dense granules* are electron-opaque granules that contain a nonmetabolic pool of adenine nucleotides, serotonin, pyrophosphate, and calcium ions. The membrane-bound *dense tubular system* is a sac-like structure containing molecules having enzymatic activity such as peroxidases and dehydrogenases.

importance to platelet function such as signaling proteins, receptors, and ion channels may be present only in low abundance and traditionally have been studied in isolation after purification from complex biological samples. This contrasts with proteomic studies where sample prefractionation techniques have been refined to enable low-abundance proteins to be investigated with a more rapid throughput of information and in the context of global signaling questions, as reviewed, for example, by García [110]. Similarly, methods involving the separation of platelets into sub-cellular components have been key to the successful identification of low-abundance proteins in several platelet proteomic studies [106,111–116].

***1.1.4.1 Protein Localization*** Platelets, like most cells, are highly compartmentalized such that this organization collocalizes proteins with substrates and facilitates interactions in response to stimuli. In the context of signaling proteins, compartmentalization brings molecules together to form the functional units of cell biology. A schematic of the cross section of a platelet, indicating its compartments and organelles, is shown in Figure 1.4.

The plasma membrane forms the interface between a cell and its environment and as such is crucial to the exchange of information with and response to the external environment (see Chapter 5). It is not surprising, then, that membrane proteins make up about 70% of the known protein targets for drugs [117] or that platelet surface proteins and transmembrane proteins have been the focus of several studies with the aim of discovering potential targets both for drug development and as biomarkers of disease [112,113,116,118].

The platelet releasate or *secretome*, microparticles, and platelet granules, are functional aspects of platelet biology that convey molecules to the plasma membrane or to the external environment on stimulation of the cell. The importance of these systems in platelet activation, aggregation, and thrombus formation has made them a focal point of a number of proteomic investigations [115,119–123] (see Chapters 6 and 7).

***1.1.4.2 Phosphoproteomics*** The activation of platelets is controlled by complex signaling pathways in which protein phosphorylation and dephosphorylation play important regulatory roles. It has been estimated that approximately one in three proteins are phosphorylated [124–126] by one or more of the 500+ protein kinases encoded by the human genome [127], while kinases themselves are also regulated by reversible phosphorylation [128]. The phosphoproteome is dynamic, and with most proteins being phosphorylated at multiple sites, the number of reported phosphorylation sites continues to grow. For example, the online database PhosphoSite [129] currently (as of 2010) lists over 63,000 phosphorylation sites from 11,000 proteins. Indeed, kinase inhibitor compounds constitute about 30% of all drug development programs in the pharmaceutical industry [130].

Several different approaches to the analysis of phosphoproteins have been taken. The enrichment of phosphopeptides by immobilized metal affinity chromatography (IMAC) [131] and/or titanium oxide ($TiO_2$) [132,133], often in conjunction with other chromatographic techniques, has become routine in the study of other cell types [134]. These methods provide varying degrees of specificity for the isolation of phosphopeptides from biological samples prior to mass spectrometry, while the immunoprecipitation of platelet proteins with an antiphosphotyrosine antibody [135] has also proved successful.

The application of multiplexed approaches to gain added information to the understanding of protein signaling is illustrated by work from the research group of Lucus Huber [136]. By combining subcellular fractionation with phosphoprotein-specific staining in 2DGE and mass spectrometry, they elucidated the signal transduction of the epidermal growth factor receptor (EGFR) and its influence on cytoskeletal proteins and also on MAPK signaling [137].

The combining of proteomic and microarray analyses of platelet proteins that become phosphorylated on platelet aggregation led to the identification of platelet endothelial aggregation receptor 1 (PEAR1), an EGFR-containing transmembrane receptor, on platelets and endothelial cells [8]. Genotyping of a cohort of patients in conjunction with measurements of platelet activity has subsequently shown PEAR1 to be important in the regulation of platelet activity [139].

***1.1.4.3 Quantitative Proteomics and Platelets*** While it is important to understand where a protein is and whether it has been modified during activation, it is equally important to know how much of that protein or modification is present. Comparative quantitation has been available for some time for researchers carrying out 2DGE by labeling two or more samples with different dyes in the same gel [differential in-gel electrophoresis (DIGE)], described, for example, by Della Corte and colleagues [140]. In addition to this, there are now several methods that can be used in non-gel-based proteomics. Most software platforms for analysis of data from mass spectrometry now incorporate label-free semiquantitative information of protein abundance utilizing methods such as the exponentially modified protein abundance index (emPAI) [141]. The relative contributions of proteins under different conditions can also be achieved by labeling with stable isotopes.

Stable isotope labeling by amino acids in cell culture (SILAC) [142] is a highly popular labeling process, but is considered unsuitable for use with platelets as the labeling is carried out during cell culture. However, studies of Kindlin-3 in red blood cells [143] showed that it is possible to SILAC-label whole mice via SILAC, and this approach could be applied to platelet research in the future. An alternative labeling method—isotope tagging for relative and absolute quantitation (iTRAQ)—does not rely on cell culture or *in vivo* methods and has been applied to the study of stored platelets [144].

***1.1.4.4 Diagnostics*** Since the late 1990s, substantial progress has been made in understanding the regulation of platelet-function, including the characterization of new ligands, platelet specific receptors, and cell signaling pathways. Because of the asymptomatic nature of many of the stages of atherogenesis, diagnosis of early stage cardiovascular problems and the likelihood of thrombotic complications is difficult. Thus, in addition to the need for greater understanding of the biological processes involved, there is also a need for sensitive and effective predictive molecular markers of disease—biomarkers—to be used as diagnostic tools [145].

Biological fluids and in particular plasma are easy to obtain and would therefore provide an ideal medium for diagnosis. However, it can be difficult to gain distinct diagnostic information from plasma because of the complexity of its proteome [146] and the biological variation between individuals in the number and reactivity of candidate biomarker proteins [147,148]. These problems may be overcome, however by an increasing number of novel low-abundance proteins being discovered within subproteomes [149] and the possibility that a multiplex of biomarkers rather than a single diagnostic molecule could provide accurate and reliable early diagnosis [147]. Many datasets from platelet proteomic experiments are now available and bioinformatic analytic methods, combined with functional information on candidate proteins, may be applied to the search for combinations of molecules that would yield suitable diagnostic tools.

*1.1.4.5* ***Platelets as Antithrombotic Targets*** The participation of platelets in thrombosis and atherosclerosis (first proposed by Ross in 1976) [150–152] is well established, and the platelet has become a key target in therapies to combat cardiovascular disease. While antiplatelet therapies are used widely, current approaches lack efficacy, lead to drug resistance, or are associated with side effects, including problem bleeding [5,153,154].

Here, proteomic technologies have provided new insights into platelet signaling and have identified several candidate proteins as suitable targets for antithrombotic medication. For example, proteomics screening has led to the identification of two homophilic adhesion receptors in platelets, CD84 and CD150, signaling lymphocyte activation molecules (SLAM) [104,138]. Similarly, mass spectrometry and proteomics methods are being applied to the understanding of atherosclerosis as a whole and from the numerous disciplines involved in its studies [155–157] (see Chapter 14).

## 1.2 CONCLUDING REMARKS

Since the 1990s, proteomic studies of platelets have ranged from global approaches to changes within whole-platelet samples in response to stimuli [158,159]; to the analysis of very specific subproteomes, including the phosphoproteome of resting [104,106] and stimulated [104,160] platelets; studies on the secretome [122], the membrane fraction of unstimulated platelets [112,113], and proteins upregulated in surface fractions on stimulation [111,114,116,161]; and immunoprecipitation experiments [162]. Each of these studies has identified proteins not previously identified in platelets and has produced new and valuable information. Proteins new to platelets continue to be discovered through the use of proteomic methods. The evolution of mass spectrometry and associated technologies provides improved speed, accuracy and ease of use. Analytical software, which has often been a bottleneck in the experimental process, is becoming more integrated with the emergence of proteomics platforms and pipelines that can readily convert raw data into information with greater confidence and reduced numbers of false-positive identifications. In turn, the growing experience of expert scientists in platelet proteomics is indicated in the development of more question-oriented experiments and a trend toward integration between proteomics with functional genomics, transcriptomics, informaticists, and mathematicians to gain greater depth and relevance of information.

Chapters 2–14 in this book have been written by key scientists in the fields of platelet proteomics, transcriptomics, and functional genomics, with a special focus on the technology and the application to platelet research and platelet-related diseases.

## REFERENCES


1. World Health Organisation (WHO), *Strategic Priorities of the WHO Cardiovascular Disease Programme*, available at http://www.who.int/cardiovascular_diseases/priorities/en/index.html (accessed 12/03/09).
2. American Heart Association; *Heart Disease and Stroke Statistics—2009 Update*, available at http://www.americanheart.org (accessed 12/03/09).
3. British Heart Foundation Statistics Website, http://www.heartstats.org/homepage.asp (accessed 12/03/09).
4. United Nations Development Programme, *UN Human Development Report 2005*, available at http://hdr.undp.org/en/reports/global/hdr2005/ (accessed 12/03/09).
5. Meadows TA, Bhatt DL, Clinical aspects of platelet inhibitors and thrombus formation, *Circ. Res.* **100**(9):1261–1275 (2007).
6. Garg R, Uretsky BF, Lev EI, Anti-platelet and anti-thrombotic approaches in patients undergoing percutaneous coronary intervention, *Cath. Cardiovasc. Intervent.* **70**(3):388–406 (2007).
7. Steinhubl SR, Badimon JJ, Bhatt DL, Herbert JM, Luscher TF, Clinical evidence for anti-inflammatory effects of antiplatelet therapy in patients with atherothrombotic disease, *Vasc. Med.* **12**(2):113–22 (2007).
8. Barrett NE, Holbrook L, Jones S, Kaiser WJ, Moraes LA, Rana R, et al., Future innovations in anti-platelet therapies, *Br. J. Pharmacol.* **154**(5):918–939 (2008).
9. Bhatt DL, Topol EJ, Scientific and therapeutic advances in antiplatelet therapy, *Nat. Rev. Drug Discov.* **2**(1):15–28 (2003).
10. Harrison P, Goodall AH, "Message in the platelet"—more than just vestigial mRNA! *Platelets* **19**(6):395–404 (2008).
11. Zimmerman GA, Weyrich AS, Signal-dependent protein synthesis by activated platelets: New pathways to altered phenotype and function, *Arterioscler. Thromb. Vasc. Biol.* **28**(3):s17–s24 (2008).
12. Ware JA, Heistad DD, Seminars in medicine of the Beth Israel Hospital, Boston. Platelet-endothelium interactions, *N Engl J Med* **328**(9):628–635 (1993).
13. Bourgain RH, Inhibition of PGI2 (prostacyclin) synthesis in the arterial wall enhances the formation of white platelet thrombi *in vivo*, *Hemostasis* **7**(4):252–255 (1978).
14. Perutelli P, Molinari AC, Von Willebrand factor, von Willebrand factor-cleaving protease, and shear stress, *Cardiovasc. Hematol. Agents Med. Chem.* **5**(4):305–310 (2007).
15. Reininger AJ, VWF attributes—impact on thrombus formation, *Thromb. Res.* **122** (Suppl. 4):S9–S13 (2008).
16. Savage B, Saldivar E, Ruggeri ZM, Initiation of platelet adhesion by arrest onto fibrinogen or translocation on von Willebrand factor, *Cell* **84**(2):289–297 (1996).
17. Gibbins JM, Platelet adhesion signaling and the regulation of thrombus formation, *J. Cell. Sci.* **117**(Pt. 16):3415–3425 (2004).
18. Gibbins J, Asselin J, Farndale R, Barnes M, Law CL, Watson SP, Tyrosine phosphorylation of the Fc receptor gamma-chain in collagen-stimulated platelets, *J. Biol. Chem.* **271**(30):18095–18099 (1996).

19. Gibbins JM, Okuma M, Farndale R, Barnes M, Watson SP, Glycoprotein VI is the collagen receptor in platelets which underlies tyrosine phosphorylation of the Fc receptor gamma-chain, *FEBS Lett*. **413**(2):255–259 (1997).
20. Poole A, Gibbins JM, Turner M, van Vugt MJ, van de Winkel JG, Saito T, et al., The Fc receptor gamma-chain and the tyrosine kinase Syk are essential for activation of mouse platelets by collagen, *EMBO J* **16**(9):2333–2341 (1997).
21. Tsuji M, Ezumi Y, Arai M, Takayama H, A novel association of Fc receptor gamma-chain with glycoprotein VI and their co-expression as a collagen receptor in human platelets, *J. Biol. Chem*. **272**(38):23528–23531 (1997).
22. Nieswandt B, Watson SP, Platelet-collagen interaction: Is GPVI the central receptor? *Blood* **102**(2):449–461 (2003).
23. Varga-Szabo D, Pleines I, Nieswandt B, Cell adhesion molecules in platelets, *Arterioscler. Thromb. Vasc. Biol*. **28**;403–412 (2008).
24. Watson SP, Auger JM, McCarty OJ, Pearce AC, GPVI and integrin alphaIIb beta3 signaling in platelets, *J. Thromb. Haemost*. **3**(8):1752–1762 (2005).
25. Surin RW, Barthwal MK, Dikshit M, Platelet collagen receptors, signaling and antagonism: Emerging approaches for the prevention of intravascular thrombosis, *Thromb. Res*. **122**(6):786–803 (2008).
26. Kato K, Kanaji T, Russell S, Kunicki TJ, Furihata K, Kanaji S, et al., The contribution of glycoprotein VI to stable platelet adhesion and thrombus formation illustrated by targeted gene deletion. *Blood* **102**(5):1701–1707 (2003).
27. Lockyer S, Okuyama K, Begum S, Le S, Sun B, Watanabe T, et al., GPVI-deficient mice lack collagen responses and are protected against experimentally induced pulmonary thromboembolism, *Thromb. Res*. **118**(3):371–380 (2006).
28. Briddon SJ, Watson SP, Evidence for the involvement of p59fyn and p53/56lyn in collagen receptor signaling in human platelets, *Biochem. J*. **338** (Pt. 1):203–209 (1999).
29. Pasquet JM, Gross B, Quek L, Asazuma N, Zhang W, Sommers CL, et al., LAT is required for tyrosine phosphorylation of phospholipase cgamma2 and platelet activation by the collagen receptor GPVI, *Mol. Cell. Biol*. **19**(12):8326–8334 (1999).
30. Asazuma N, Wilde JI, Berlanga O, Leduc M, Leo A, Schweighoffer E, et al., Interaction of linker for activation of T cells with multiple adapter proteins in platelets activated by the glycoprotein VI-selective ligand, convulxin, *J. Biol. Chem*. **275**(43):33427–33434 (2000).
31. Ragab A, Severin S, Gratacap MP, Aguado E, Malissen M, Jandrot-Perrus M, et al., Roles of the C-terminal tyrosine residues of LAT in GPVI-induced platelet activation: Insights into the mechanism of PLC gamma 2 activation, *Blood* **110**(7):2466–2474 (2007).
32. Rink TJ, Smith SW, Tsien RY, Cytoplasmic free Ca2+ in human platelets: Ca2+ thresholds and Ca-independent activation for shape-change and secretion, *FEBS Lett*. **148**(1):21–26 (1982).
33. Knight DE, Hallam TJ, Scrutton MC, Agonist selectivity and second messenger concentration in Ca2+-mediated secretion, *Nature* **296**(5854):256–257 (1982).
34. Walker TR, Watson SP, Synergy between Ca2+ and protein kinase C is the major factor in determining the level of secretion from human platelets, *Biochem. J*. **289** (Pt. 1):277–282 (1993).

35. Heemskerk JW, Siljander PR, Bevers EM, Farndale RW, Lindhout T, Receptors and signaling mechanisms in the procoagulant response of platelets, *Platelets* **11**(6):301–306 (2000).
36. Leo A, Schraven B, Networks in signal transduction: the role of adaptor proteins in platelet activation, *Platelets* **11**(8):429–445 (2000).
37. Jackson SP, Nesbitt WS, Kulkarni S, Signaling events underlying thrombus formation, *J. Thromb. Haemost.* **1**(7):1602–1612 (2003).
38. Moroi M, Jung SM, Platelet glycoprotein VI: Its structure and function, *Thromb. Res.* **114**(4):221–233 (2004).
39. Barry FA, Gibbins JM, Protein kinase B is regulated in platelets by the collagen receptor glycoprotein VI, *J. Biol. Chem.* **277**(15):12874–12878 (2002).
40. Kroner C, Eybrechts K, Akkerman JW, Dual regulation of platelet protein kinase B, *J. Biol. Chem.* **275**(36):27790–27798 (2000).
41. Woulfe D, Jiang H, Morgans A, Monks R, Birnbaum M, Brass LF, Defects in secretion, aggregation, and thrombus formation in platelets from mice lacking Akt2, *J. Clin. Invest.* **113**(3):441–450 (2004).
42. Morello F, Perino A, Hirsch E, Phosphoinositide 3-kinase signaling in the vascular system, *Cardiovasc. Res.* **82**(2):261–271 (2009).
43. Calderwood DA, Integrin activation, *J. Cell. Sci.* **117**(Pt. 5):657–666 (2004).
44. Liddington RC, Ginsberg MH, Integrin activation takes shape, *J. Cell. Biol.* **158**(5):833–839 (2002).
45. Shattil SJ, Kashiwagi H, Pampori N, Integrin signaling: The platelet paradigm, *Blood* **91**(8):2645–2657 (1998).
46. Siess W, Siegel FL, Lapetina EG, Arachidonic acid stimulates the formation of 1,2-diacylglycerol and phosphatidic acid in human platelets. Degree of phospholipase C activation correlates with protein phosphorylation, platelet shape change, serotonin release, and aggregation, *J. Biol. Chem.* **258**(18):11236–11242 (1983).
47. Armstrong RA, Platelet prostanoid receptors, *Pharmacol. Ther.* **72**(3):171–191 (1996).
48. Halushka PV, Mais DE, Morinelli TA, Thromboxane and prostacyclin receptors, *Prog. Clin. Biol. Res.* **301**:21–28 (1989).
49. Offermanns S, Toombs CF, Hu YH, Simon MI, Defective platelet activation in G alpha(q)-deficient mice, *Nature* **389**(6647):183–186 (1997).
50. Hollopeter G, Jantzen HM, Vincent D, Li G, England L, Ramakrishnan V, et al., Identification of the platelet ADP receptor targeted by antithrombotic drugs, *Nature* **409**(6817):202–207 (2001).
51. Tolhurst G, Vial C, Leon C, Gachet C, Evans RJ, Mahaut-Smith MP, Interplay between P2Y(1), P2Y(12), and P2X(1) receptors in the activation of megakaryocyte cation influx currents by ADP: Evidence that the primary megakaryocyte represents a fully functional model of platelet P2 receptor signaling, *Blood* **106**(5):1644–1651 (2005).
52. Gachet C. Regulation of platelet functions by P2 receptors, *Annu. Rev. Pharmacol. Toxicol.* **46**:277–300 (2006).
53. Andre P, Delaney SM, LaRocca T, Vincent D, DeGuzman F, Jurek M, et al., P2Y12 regulates platelet adhesion/activation, thrombus growth, and thrombus stability in injured arteries, *J. Clin. Invest.* **112**(3):398–406 (2003).

54. Kahn ML, Zheng YW, Huang W, Bigornia V, Zeng D, Moff S, et al., A dual thrombin receptor system for platelet activation, *Nature* **394**(6694):690–694 (1998).

55. Kahn ML, Nakanishi-Matsui M, Shapiro MJ, Ishihara H, Coughlin SR, Protease-activated receptors 1 and 4 mediate activation of human platelets by thrombin, *J. Clin. Invest*. **103**(6):879–887 (1999).

56. Coughlin SR, Thrombin signaling and protease-activated receptors, *Nature* **407**(6801):258–264 (2000).

57. Leger AJ, Jacques SL, Badar J, Kaneider NC, Derian CK, Andrade-Gordon P, et al., Blocking the protease-activated receptor 1–4 heterodimer in platelet-mediated thrombosis, *Circulation* **113**(9):1244–1254 (2006).

58. Calvete JJ, Structures of integrin domains and concerted conformational changes in the bidirectional signaling mechanism of alphaIIbbeta3, *Exp. Biol. Med*. (Maywood) **229**(8):732–744 (2004).

59. Shattil SJ, Newman PJ, Integrins: Dynamic scaffolds for adhesion and signaling in platelets, *Blood* **104**(6):1606–1615 (2004).

60. Boylan B, Gao C, Rathore V, Gill JC, Newman DK, Newman PJ, Identification of FcgammaRIIa as the ITAM-bearing receptor mediating alphaIIbbeta3 outside-in integrin signaling in human platelets, *Blood* **112**(7):2780–2786 (2008).

61. Zhu L, Bergmeier W, Wu J, Jiang H, Stalker TJ, Cieslak M, et al., Regulated surface expression and shedding support a dual role for semaphorin 4D in platelet responses to vascular injury, *Proc. Natl. Acad. Sci. USA* **104**(5):1621–1626 (2007).

62. Prevost N, Woulfe DS, Jiang H, Stalker TJ, Marchese P, Ruggeri ZM, et al., Eph kinases and ephrins support thrombus growth and stability by regulating integrin outside-in signaling in platelets, *Proc. Natl. Acad. Sci. USA* **102**(28):9820–9825 (2005).

63. Brass LF, Jiang H, Wu J, Stalker TJ, Zhu L, Contact-dependent signaling events that promote thrombus formation, *Blood Cells Mol. Dis*. **36**(2):157–161 (2006).

64. Shah PK, Inflammation and plaque vulnerability, *Cardiovasc. Drugs Ther*. **23**(1):31–40 (2009).

65. Lusis AJ, Atherosclerosis, *Nature* **407**(6801):233–241 (2000).

66. Libby P, Molecular and cellular mechanisms of the thrombotic complications of atherosclerosis, *J. Lipid Res*. **50** (Suppl):S352–S357 (2009).

67. Halvorsen B, Otterdal K, Dahl TB, Skjelland M, Gullestad L, Oie E, et al., Atherosclerotic plaque stability—what determines the fate of a plaque? *Prog. Cardiovasc. Dis*. **51**(3):183–194 (2008).

68. Weyrich A, Cipollone F, Mezzetti A, Zimmerman G, Platelets in atherothrombosis: new and evolving roles, *Curr. Pharm. Des*. 2007; **13**(16):1685–1691 (2007).

69. Kraaijeveld AO, de Jager SC, van Berkel TJ, Biessen EA, Jukema JW, Chemokines and atherosclerotic plaque progression: towards therapeutic targeting? *Curr. Pharm. Des*. **13**(10):1039–1052 (2007).

70. Akkerman JW, From low-density lipoprotein to platelet activation, *Int. J. Biochem. Cell. Biol*. **40**(11):2374–2378 (2008).

71. Badimon L, Vilahur G, Platelets, arterial thrombosis and cerebral ischemia, *Cerebrovasc. Dis*. **24**(Suppl. 1):30–39 (2007).

72. Friedman MH, Hutchins GM, Bargeron CB, Deters OJ, Mark FF, Correlation between intimal thickness and fluid shear in human arteries, *Atherosclerosis* **39**(3):425–436 (1981).

73. Glagov S, Zarins C, Giddens DP, Ku DN, Hemodynamics and atherosclerosis. Insights and perspectives gained from studies of human arteries, *Arch. Pathol. Lab. Med.* **112**(10):1018–1031 (1988).

74. Cooke JP, Flow, NO, and atherogenesis, *Proc. Natl. Acad. Sci. USA.* **100**(3):768–770 (2003).

75. Nesbitt WS, Mangin P, Salem HH, Jackson SP, The impact of blood rheology on the molecular and cellular events underlying arterial thrombosis, *J. Mol. Med.* **84**(12):989–995 (2006).

76. Nesbitt WS, Westein E, Tovar-Lopez FJ, Tolouei E, Mitchell A, Fu J, et al., A shear gradient-dependent platelet aggregation mechanism drives thrombus formation, *Nat. Med.* **15**(6):665–673 (2009).

77. Williams KJ, Tabas I, The response-to-retention hypothesis of early atherogenesis, *Arterioscler. Thromb. Vasc. Biol.* **15**(5):551–561 (1995).

78. Cyrus T, Witztum JL, Rader DJ, Tangirala R, Fazio S, Linton MF, et al., Disruption of the 12/15-lipoxygenase gene diminishes atherosclerosis in apo E-deficient mice, *J. Clin. Invest.* **103**(11):1597–1604 (1999).

79. Watson AD, Leitinger N, Navab M, Faull KF, Horkko S, Witztum JL, et al., Structural identification by mass spectrometry of oxidized phospholipids in minimally oxidized low density lipoprotein that induce monocyte/endothelial interactions and evidence for their presence *in vivo*, *J. Biol. Chem.* **272**(21):13597–13607 (1997).

80. Mestas J, Ley K, Monocyte-endothelial cell interactions in the development of atherosclerosis, *Trends Cardiovasc. Med.* **18**(6):228–232 (2008)

81. Hamilton JA, Byrne R, Jessup W, Kanagasundaram V, Whitty G, Comparison of macrophage responses to oxidized low-density lipoprotein and macrophage colony-stimulating factor (M-CSF or CSF–1), *Biochem. J.* **354**(Pt. 1):179–187 (2001).

82. Nelken NA, Coughlin SR, Gordon D, Wilcox JN, Monocyte chemoattractant protein-1 in human atheromatous plaques, *J. Clin. Invest.* **88**(4):1121–1127 (1991).

83. Collot-Teixeira S, Martin J, McDermott-Roe C, Poston R, McGregor JL, CD36 and macrophages in atherosclerosis, *Cardiovasc. Res.* **75**(3):468–477 (2007).

84. Suzuki H, Kurihara Y, Takeya M, Kamada N, Kataoka M, Jishage K, et al., A role for macrophage scavenger receptors in atherosclerosis and susceptibility to infection, *Nature* **386**(6622):292–296 (1997).

85. Osterud B, Bjorklid E, Role of monocytes in atherogenesis, *Physiol. Rev.* **83**(4):1069–1112 (2003).

86. Shaw PX, Horkko S, Tsimikas S, Chang MK, Palinski W, Silverman GJ, et al., Human-derived anti-oxidized LDL autoantibody blocks uptake of oxidized LDL by macrophages and localizes to atherosclerotic lesions *in vivo*, *Arterioscler. Thromb. Vasc. Biol.* **21**(8):1333–1339 (2001).

87. Libby P, The molecular mechanisms of the thrombotic complications of atherosclerosis, *J. Intern. Med.* **263**(5):517–527 (2008).

88. Huo Y, Schober A, Forlow SB, Smith DF, Hyman MC, Jung S, et al., Circulating activated platelets exacerbate atherosclerosis in mice deficient in apolipoprotein E, *Nat. Med.* **9**(1):61–67 (2003).

89. Gleissner CA, von Hundelshausen P, Ley K, Platelet chemokines in vascular disease, *Arterioscler. Thromb. Vasc. Biol*. **28**(11):1920–1927 (2008).

90. May AE, Kalsch T, Massberg S, Herouy Y, Schmidt R, Gawaz M, Engagement of glycoprotein IIb/IIIa (alpha(IIb)beta3) on platelets upregulates CD40L and triggers CD40L-dependent matrix degradation by endothelial cells, *Circulation* **106**(16):2111–2117 (2002).

91. Andre P, Nannizzi-Alaimo L, Prasad SK, Phillips DR, Platelet-derived CD40L: The switch-hitting player of cardiovascular disease, *Circulation* **106**(8):896–899 (2002).

92. Chakrabarti S, Blair P, Freedman JE, CD40–40L Signaling in vascular inflammation, *J. Biol. Chem*. **282**(25):18307–18317 (2007).

93. Orfeo T, Butenas S, Brummel-Ziedins KE, Mann KG, The tissue factor requirement in blood coagulation, *J. Biol. Chem*. **280**(52):42887–42896 (2005).

94. Breitenstein A, Camici GG, Tanner FC, Tissue factor: beyond coagulation in the cardiovascular system, *Clin. Sci*. (Lond.) **118**(3):159–172 (2009).

95. Korporaal SJ, Relou IA, van Eck M, Strasser V, Bezemer M, Gorter G, et al., Binding of low density lipoprotein to platelet apolipoprotein E receptor 2′ results in phosphorylation of p38MAPK, *J. Biol. Chem*. **279**(50):52526–52534 (2004).

96. Kramer RM, Roberts EF, Um SL, Borsch-Haubold AG, Watson SP, Fisher MJ, et al., p38 Mitogen-activated protein kinase phosphorylates cytosolic phospholipase A2 (cPLA2) in thrombin-stimulated platelets. Evidence that proline-directed phosphorylation is not required for mobilization of arachidonic acid by cPLA2, *J. Biol. Chem*. **271**(44):27723–27729 (1996).

97. Korporaal SJ, Van Eck M, Adelmeijer J, Ijsseldijk M, Out R, Lisman T, et al., Platelet activation by oxidized low density lipoprotein is mediated by CD36 and scavenger receptor-A, *Arterioscler. Thromb. Vasc. Biol*. **27**(11):2476–2483 (2007).

98. Podrez EA, Byzova TV, Febbraio M, Salomon RG, Ma Y, Valiyaveettil M, et al., Platelet CD36 links hyperlipidemia, oxidant stress and a prothrombotic phenotype, *Nat. Med*. **13**(9):1086–1095 (2007).

99. Siegel-Axel D, Daub K, Seizer P, Lindemann S, Gawaz M, Platelet lipoprotein interplay: Trigger of foam cell formation and driver of atherosclerosis, *Cardiovasc. Res*. **78**(1):8–17 (2008).

100. Massberg S, Brand K, Gruner S, Page S, Muller E, Muller I, et al., A critical role of platelet adhesion in the initiation of atherosclerotic lesion formation, *J. Exp. Med*. **196**(7):887–896 (2002).

101. O'Neill EE, Brock CJ, von Kriegsheim AF, Pearce AC, Dwek RA, Watson SP, et al., Towards complete analysis of the platelet proteome, *Proteomics* **2**(3):288–305 (2002).

102. Lewandrowski U, Wortelkamp S, Lohrig K, Zahedi RP, Wolters DA, Walter U, et al., Platelet membrane proteomics: A novel repository for functional research, *Blood* **114**(1):e10-9 (2009).

103. Maguire PB, Foy M, Fitzgerald DJ, Using proteomics to identify potential therapeutic targets in platelets, *Biochem. Soc. Trans*. **33**(Pt. 2):409–412 (2005).

104. Nanda N, Bao M, Lin H, Clauser K, Komuves L, Quertermous T, et al., Platelet endothelial aggregation receptor 1 (PEAR1), a novel epidermal growth factor repeat-containing transmembrane receptor, participates in platelet contact-induced activation, *J. Biol. Chem*. **280**(26):24680–24689 (2005).

105. Foy M, Harney DF, Wynne K, Maguire PB, Enrichment of phosphotyrosine proteome of human platelets by immunoprecipitation, *Methods. Mol. Biol.* **357**:313–318 (2007).

106. Zahedi RP, Begonja AJ, Gambaryan S, Sickmann A, Phosphoproteomics of human platelets: A quest for novel activation pathways, *Biochim. Biophys. Acta* **1764**(12):1963–1976 (2006).

107. García A, Prabhakar S, Hughan S, Anderson TW, Brock CJ, Pearce AC, et al., Differential proteome analysis of TRAP-activated platelets: Involvement of DOK-2 and phosphorylation of RGS proteins, *Blood* **103**(6):2088–2095 (2004).

108. Pixton KL, Is it functional? Report on the first BSPR/EBI meeting on functional proteomics, *Proteomics* **4**(12):3762–3764 (2004).

109. Cagney G, McRedmond J, A central resource for platelet proteomics, *Arterioscler. Thromb. Vasc. Biol.* **28**(7):1214–1215 (2008).

110. García A, Proteome analysis of signaling cascades in human platelets, *Blood Cells. Mol. Dis.* **36**(2):152–156 (2006).

111. García A, Prabhakar S, Brock CJ, Pearce AC, Dwek RA, Watson SP, et al., Extensive analysis of the human platelet proteome by two-dimensional gel electrophoresis and mass spectrometry, *Proteomics* **4**(3):656–668 (2004).

112. Moebius J, Zahedi RP, Lewandrowski U, Berger C, Walter U, Sickmann A, The human platelet membrane proteome reveals several new potential membrane proteins, *Mol. Cell. Proteom.* **4**(11):1754–17561 (2005).

113. Senis YA, Tomlinson MG, García A, Dumon S, Heath VL, Herbert J, et al., A comprehensive proteomics and genomics analysis reveals novel transmembrane proteins in human platelets and mouse megakaryocytes including G6b-B, a novel immunoreceptor tyrosine-based inhibitory motif protein, *Mol. Cell. Proteom.* **6**(3):548–564 (2007).

114. Kaiser WJ, Holbrook LM, Tucker KL, Stanley RG, Gibbins JM, A functional proteomic method for the enrichment of peripheral membrane proteins reveals the collagen binding protein Hsp47 is exposed on the surface of activated human platelets, *J. Proteome Res.* **8**(6):2903–2914 (2009).

115. Piersma SR, Broxterman HJ, Kapci M, de Haas RR, Hoekman K, Verheul HM, et al., Proteomics of the TRAP-induced platelet releasate, *J. Proteom.* **72**(1):91–109 (2009).

116. Tucker KL, Kaiser WJ, Bergeron AL, Hu H, Dong JF, Tan TH, et al., Proteomic analysis of resting and thrombin-stimulated platelets reveals the translocation and functional relevance of HIP-55 in platelets, *Proteomics* **9**(18):4340–4354 (2009).

117. Hopkins AL, Groom CR, Target analysis: a priori assessment of druggability, *Ernst Schering Res. Found. Workshop* **2003**(42):11–17 (2003).

118. Senis YA, Tomlinson MG, Ellison S, Mazharian A, Lim J, Zhao Y, et al., The tyrosine phosphatase CD148 is an essential positive regulator of platelet activation and thrombosis, *Blood* **113**(20):4942–4954 (2009).

119. García BA, Smalley DM, Cho H, Shabanowitz J, Ley K, Hunt DF, The platelet microparticle proteome, *J. Proteome Res.* **4**(5):1516–1521 (2005).

120. Coppinger J, Fitzgerald DJ, Maguire PB, Isolation of the platelet releasate, *Meth. Mol. Biol.* **357**:307–311 (2007).

121. Ruiz FA, Lea CR, Oldfield E, Docampo R, Human platelet dense granules contain polyphosphate and are similar to acidocalcisomes of bacteria and unicellular eukaryotes, *J. Biol. Chem.* **279**(43):44250–44257 (2004).
122. Maynard DM, Heijnen HF, Horne MK, White JG, Gahl WA, Proteomic analysis of platelet alpha-granules using mass spectrometry, *J. Thromb. Haemost.* **5**(9):1945–1955 (2007).
123. Hernandez-Ruiz L, Valverde F, Jimenez-Nunez MD, Ocana E, Saez-Benito A, Rodriguez-Martorell J, et al., Organellar proteomics of human platelet dense granules reveals that 14–3-3zeta is a granule protein related to atherosclerosis, *J. Proteome Res.* **6**(11):4449–4457 (2007).
124. Johnson SA, Hunter T, Kinomics, *Nat. Meth.* **2**(1):17–25 (2005).
125. Cohen P, The regulation of protein function by multisite phosphorylation—a 25 year update, *Trends Biochem. Sci.* **25**:596–601 (2000).
126. Sefton BM and Shenolikar S. 2001. Overview of Protein Phosphorylation. Current Protocols in Protein Science. 13.1.1– 13.1.5. `http://onlinelibrary.wiley.com/doi/10.1002/0471140864.ps1301s00/abstract;jsessionid=32B6B2CA4407A5B0902D8019DEB6D595.d02t02`.
127. Manning G, Whyte DB, Martinez R, Hunter T, Sudarsanam S, The protein kinase complement of the human genome, *Science* **298**(5600):1912–1934 (2002).
128. Gafken PR, An overview of the qualitative analysis of phosphoproteins by mass spectrometry, *Meth. Mol. Biol.* **527**:159–172, ix (2009).
129. PhosphoSitePlus Website, `http://www.phosphosite.org` (accessed 6/18/10).
130. Knight ZA, Shokat KM, Features of selective kinase inhibitors, *Chem. Biol.* **12**(6):621–637 (2005).
131. Sun X, Chiu JF, He QY, Application of immobilized metal affinity chromatography in proteomics, *Exp. Rev. Proteom.* **2**(5):649–657 (2005).
132. Thingholm TE, Larsen MR, The use of titanium dioxide micro-columns to selectively isolate phosphopeptides from proteolytic digests, *Meth. Mol. Biol.* **527**:57–66, xi (2009).
133. Carrascal M, Ovelleiro D, Casas V, Gay M, Abian J, Phosphorylation analysis of primary human T lymphocytes using sequential IMAC and titanium oxide enrichment, *J. Proteome Res.* **7**(12):5167–5176 (2008).
134. Olsen JV, Blagoev B, Gnad F, Macek B, Kumar C, Mortensen P, et al., Global, *in vivo*, and site-specific phosphorylation dynamics in signaling networks, *Cell* **127**(3):635–648 (2006).
135. García A, Senis YA, Antrobus R, Hughes CE, Dwek RA, Watson SP, et al., A global proteomics approach identifies novel phosphorylated signaling proteins in GPVI-activated platelets: Involvement of G6f, a novel platelet Grb2-binding membrane adapter, *Proteomics* **6**(19):5332–5343 (2006).
136. Stasyk T, Schiefermeier N, Skvortsov S, Zwierzina H, Peranen J, Bonn GK, Huber LA, Identification of endosomal epidermal growth factor receptor signaling targets by functional organelle proteomics, *Mol. Cell. Proteom.* **6**(5):908–922 (2007).
137. Taub N, Teis D, Ebner HL, Hess MW, Huber LA, Late endosomal traffic of the epidermal growth factor receptor ensures spatial and temporal fidelity of mitogen-activated protein kinase signaling, *Mol. Biol. Cell* **18**(12):4698–4710 (2007).

138. Nanda N, Andre P, Bao M, Clauser K, Deguzman F, Howie D, et al., Platelet aggregation induces platelet aggregate stability via SLAM family receptor signaling, *Blood* **106**(9):3028–3034 (2005).

139. Jones CI, Bray S, Garner SF, Stephens J, de Bono B, Angenent WG, et al., A functional genomics approach reveals novel quantitative trait loci associated with platelet signaling pathways, *Blood* **114**(7):1405–1416 (2009).

140. Della Corte A, Maugeri N, Pampuch A, Cerletti C, de Gaetano G, Rotilio D, Application of 2-dimensional difference gel electrophoresis (2D-DIGE) to the study of thrombin-activated human platelet secretome, *Platelets* **19**(1):43–50 (2008).

141. Ishihama Y, Oda Y, Tabata T, Sato T, Nagasu T, Rappsilber J, et al., Exponentially modified protein abundance index (emPAI) for estimation of absolute protein amount in proteomics by the number of sequenced peptides per protein, *Mol. Cell. Proteom.* **4**(9):1265–1272 (2005).

142. Ong SE, Mann M, A practical recipe for stable isotope labeling by amino acids in cell culture (SILAC), *Nat. Protoc.* **1**(6):2650–2660 (2006).

143. Kruger M, Moser M, Ussar S, Thievessen I, Luber CA, Forner F, et al., SILAC mouse for quantitative proteomics uncovers kindlin-3 as an essential factor for red blood cell function, *Cell* **134**(2):353–364 (2008).

144. Thon JN, Schubert P, Duguay M, Serrano K, Lin S, Kast J, et al., Comprehensive proteomic analysis of protein changes during platelet storage requires complementary proteomic approaches, *Transfusion* **48**(3):425–435 (2008).

145. Packard RR, Libby P, Inflammation in atherosclerosis: From vascular biology to biomarker discovery and risk prediction, *Clin. Chem.* **54**(1):24–38 (2008).

146. Anderson L, Candidate-based proteomics in the search for biomarkers of cardiovascular disease, *J. Physiol.* **563**(Pt. 1):23–60 (2005).

147. Anderson NL, Anderson NG, The human plasma proteome: History, character, and diagnostic prospects, *Mol. Cell. Proteom.* **1**(11):845–867 (2002).

148. Winkler W, Zellner M, Diestinger M, Babeluk R, Marchetti M, Goll A, et al., Biological variation of the platelet proteome in the elderly population and its implication for biomarker research, *Mol. Cell. Proteom.* **7**(1):193–203 (2008).

149. Alvarez-Llamas G, de la Cuesta F, Barderas ME, Darde V, Padial LR, Vivanco F, Recent advances in atherosclerosis-based proteomics: New biomarkers and a future perspective, *Expert Rev. Proteom.* **5**(5):679–691 (2008).

150. Ross R, Harker L, Hyperlipidemia and atherosclerosis, *Science* **193**(4258): 1094–1100 (1976).

151. Rossi EC, Platelets and thrombosis, *J. Chron. Dis.* **29**(4):215–219 (1976).

152. Ross R, Atherosclerosis: The role of endothelial injury, smooth muscle proliferation and platelet factors, *Triangle* **15**(2–3):45–51 (1976).

153. Ferguson JJ, The role of oral antiplatelet agents in atherothrombotic disease, *Am. J. Cardiovasc. Drugs* **6**(3):149–157 (2006).

154. Mayr FB, Jilma B, Current developments in anti-platelet therapy, *Wien Med. Wochenschr.* **156**(17–18):472–480 (2006).

155. Didangelos A, Simper D, Monaco C, Mayr M, Proteomics of acute coronary syndromes, *Curr. Atheroscler. Rep.* **11**(3):188–195 (2009).

156. Lepedda AJ, Cigliano A, Cherchi GM, Spirito R, Maggioni M, Carta F, et al., A proteomic approach to differentiate histologically classified stable and unstable plaques from human carotid arteries, *Atherosclerosis* **203**(1):112–118 (2009).
157. de la Cuesta F, Alvarez-Llamas G, Gil-Dones F, Martin-Rojas T, Zubiri I, Pastor C, et al., Tissue proteomics in atherosclerosis: elucidating the molecular mechanisms of cardiovascular diseases, *Expert Rev. Proteom.* **6**(4):395–409 (2009).
158. Dittrich M, Birschmann I, Stuhlfelder C, Sickmann A, Herterich S, Nieswandt B, et al., Understanding platelets. Lessons from proteomics, genomics and promises from network analysis, *Thromb. Haemost.* **94**(5):916–925 (2005).
159. Maguire PB, Fitzgerald DJ, Platelet proteomics, *J. Thromb. Haemost.* **1**(7):1593–1601 (2003).
160. Maguire PB, Wynne KJ, Harney DF, O'Donoghue NM, Stephens G, Fitzgerald DJ, Identification of the phosphotyrosine proteome from thrombin activated platelets, *Proteomics* **2**(6):642–648 (2002).
161. Marcus K, Moebius J, Meyer HE, Differential analysis of phosphorylated proteins in resting and thrombin-stimulated human platelets, *Anal. Bioanal. Chem.* **376**(7):973–993 (2003).
162. Foy M, Harney DF, Wynne K, Maguire PB, Enrichment of phosphotyrosine proteome of human platelets by immunoprecipitation, in Vivanco F, ed., *Cardiovascular Proteomics, Methods and Protocols* (Series in Methods in Molecular Biology, Vol. 357), Springer/Humana Press, 2007; pp. 313–318.
163. Leger AJ, Covic L, Kuliopulos A, Protease-activated receptors in cardiovascular diseases, *Circulation* **114**(10):1070–1077 (2006).

# 2

# MASS-SPECTROMETRY-BASED PROTEOMICS: GENERAL OVERVIEW AND POSTTRANSLATIONAL MODIFICATION ANALYSIS IN THE CONTEXT OF PLATELET RESEARCH

Julia Maria Burkhart and Albert Sickmann

**Abstract**

In recent years, mass spectrometry-based proteomics has become an indispensable tool allowing the analysis of protein patterns and sensitive changes of post-translational modifications (PTMs). Mass spectrometry has been increasingly applied for platelet research, thereby enabling the analysis of functionally relevant processes on a molecular level, especially with regard to PTM-regulated activation/inhibition. In this chapter a general overview including the analysis of PTM is presented and respective platelet studies are discussed.

## 2.1 INTRODUCTION: MULTIDIMENSIONAL ANALYSIS OF COMPLEX PROTEIN SAMPLES

The *proteome* can be defined as the entire set of proteins being translated within a cell at a given timepoint under defined conditions. Next to a variety of thousands of proteins, there are isoforms and posttranslationally modified species that increase the complexity of the cell enormously and consequently also the

*Platelet Proteomics: Principles, Analysis and Applications*, First Edition.
Edited by Ángel García and Yotis A. Senis.

requirements for the analytical techniques used to detect them. Beside sample complexity, another major issue is the so-called dynamic range, which relates to the vast differences in the concentration of proteins or isoforms within a cell.

Mass spectrometry has become the method of choice for the analysis of complex protein samples. Implying high selectivity and sensitivity, mass spectrometry allows the elucidation of primary sequences, including posttranslational modifications of peptides and proteins; moreover, mass spectrometry can be applied to identify protein–protein interactions.

Mass spectrometry-based proteomics has been introduced into the field of platelet research. Because of their anucleate nature, platelets show very limited *de novo* protein synthesis and for this reason are well-suited model systems for proteome analysis. Originally, platelet proteomics studies focused on two-dimensional gel electrophoresis in combination with matrix-assisted laser desorption/ionization–mass spectrometry (MALDI-MS), yet with the introduction of new and improved methods, a variety of proteomic strategies have been applied to address distinct biological questions. In this chapter we will summarize the field of platelet proteomics with a special emphasis on posttranslational modifications.

### 2.1.1 Separation Techniques

In order to reduce sample complexity, a prerequisite in proteome research is prefractionation. Currently, there have been two major groups of separation techniques applied in proteome research: gel-based and non-gel-based. Whereas both have advantages as well as disadvantages, more recently a clear trend toward non-gel-based methods can be seen.

***2.1.1.1 Gel-Based Separation*** In 1975, two-dimensional gel electrophoresis (2DGE; also abbreviated 2-DE) was independently invented by O'Farrell [1] and Klose [2]. This technique is based on the subsequent separation of proteins referring to two independent physicochemical properties. In the first dimension, proteins are separated according to their net charge in an isoelectric focusing (IEF) step, whereas in the second dimension the proteins are separated according to their molecular weight (MW) in a common sodium dodecylsulfate polyacrylamide gel electrophoresis (SDS-PAGE). Since the isoelectric point (pI) and MW are two independent protein properties, the resolution is amplified, leading to a high separation capacity of up to 10,000 protein spots on a single 2DGE gel. For a more detailed overview of 2DGE in platelet proteome and signaling analysis, the reader is referred to Chapter 4.

***2.1.1.2 Chromatographic Separation*** While in case of 2DGE the primary separation occurs on the protein level with subsequent proteolytic digestion, non-gel-based methods focus mainly on peptide-centric approaches. Here, the proteome is digested by a specific protease, most commonly trypsin, and then the generated peptides are subjected to separation. In this context high-performance liquid chromatography (HPLC) is the method of choice in terms of automation, reproducibility, resolution, and sensitivity. Since the digest of whole proteomes leads to a large number of different peptides, there are two main approaches to deal with those complex samples: (1) the mixture can be separated by a combination of orthogonal parameters—by multidimensional liquid chromatography (MDLC) prior to mass spectrometry—and (2) specific classes of peptides can be isolated prior to further analysis, for instance, by using combined fractional diagonal chromatography (COFRADIC) [3]. Although both techniques allow the reduction of sample complexity, offline or online MDLC, as well as COFRADIC, they both have a general limitation in MS, so-called undersampling. For acquisition, mass spectrometers randomly chose the most abundant ions occurring at a distinct timepoint for fragmentation analysis. Because of the enormous complexity of common biological samples, only a subset of peptides is analyzed; as a consequence, to cope with undersampling, the researcher may repeat the analysis several times with the help of inclusion/exclusion lists or also gas-phase separation [4]. Alternatively, research groups and HPLC as well as MS vendors focus on specific aspects to overcome this phenomenon. They aim to reduce the large number of analytes by improving the separation of peptides prior to MS and the duty cycle, mass accuracy, sensitivity, and dynamic range of MS systems. Another promising approach is the combination of peptide IEF and HPLC [5]. Additionally, specific analytical tasks are increasingly being addressed by using targeted approaches discussed below.

***2.1.1.3 Cofradic*** A novel and powerful approach, combined fractional diagonal chromatography (COFRADIC), was introduced by Gevaert et al. in 2002 [6]. COFRADIC consists essentially of two identical separation steps with a chemical or enzymatic reaction in between allowing for the isolation of specific subsets of peptides by the induction of retention timeshifts between the first and second separations. The derivatization step alters the chromatographic properties of peptides containing the respective residues. So far, COFRADIC has been utilized to target Met- [6] as well as Cys- [7] containing peptides, glycopeptides [8], phosphopeptides [9], and *N*-terminal peptides [3]. General aspects concerning COFRADIC and comparison of the original *N*-terminal protocol to more recent improved protocols are described in Chapter 8.

### 2.1.2 Mass Spectrometry of Peptides

***2.1.2.1 Instrumentation*** *Mass spectrometry* can be defined as the determination of mass-to-charge ratios ($m/z$) of ions in gas phase. MS instrumentation consists of the three main components: an ion source, a mass analyzer, and a detection system that is capable of detecting even single ions and amplifying the signal. Finally, the multitude of MS-generated information has to be evaluated during data processing in order to illustrate biochemical status. Essential mass spectrometric elements are discussed in the following outline, including (1) ionization and (2) mass-to-charge analysis with the respective mass analyzers.

1. *Ionization.* Prior to separation, molecules need to be ionized and transferred into gas phase using different techniques. Ionization techniques used initially were not convenient for generating stable ions when applied to large biomolecules such as proteins and peptides. Only the so-called soft ionization techniques turned out to be suitable for conversion of large molecules into gas phase without affecting integrity. Today the most important ionization methods for proteins and peptides are electrospray ionization (ESI) and matrix-assisted laser desorption/ionization (MALDI), which were developed in the late 1980s [10]. While in ESI analytes are ionized out of a solution to render it compatible with liquid chromatography, in MALDI simple peptide mixtures are ionized out of crystalline matrix with the help of a short laser pulse. The reduction of HPLC solvent flow rates from several microliters to a few nanoliters per minute (nano-HPLC) resulted in a 100–1000-fold increase in sensitivity in ESI-MS [11]; nevertheless, to cope with issues such as coeluting analytes, a highly sensitive and reproducible nano-HPLC-MS/MS setup requires the optimal adjustment of HPLC separation, sample volumes, and preparation (e.g., preconcentration and desalting). In contrast to ESI, MALDI is utilized predominantly to analyze samples containing only a few analytes. Consequently, MALDI is less suited for complex samples; for this purpose, the technique that can be used is a LC-MALDI setup where chromatographic elution fractions are automatically spotted to the matrix containing MALDI target via a robotic system. The crystallized spots can be analyzed following LC separation. Although it is considered to be a time-saving technique, depending on instrumentation and applied gradients, the LC-MALDI setup actually takes almost as much time as do comparable approaches based on ESI-MS, such as, for instance, the analysis of samples from an ion exchange chromatography prefractionation. Once crystallized on a MALDI target, the samples can be reanalyzed at a later timepoint provided that the target has been stored at optimal conditions.
2. *Mass-to-Charge Analysis.* The ionization techniques ESI and MALDI can be combined with different mass analyzers. A variety of commonly used mass analyzers are summarized in the following paragraphs, including (a) time-of-flight (TOF) analyzers, (b) quadrupole-based analyzers, (c) ion trap analyzers, (d) ion cyclotron resonance analyzers, and (e) orbitraps.
   a. *TOF Analyzers.* Time-of-flight (TOF) analyzers accelerate ions in an electric field (~30 kV) and measure the drift time needed to fly through a field-free flight tube of 0.5–4 m length until reaching the detector. The entire system is under high vacuum, and as each ion is accelerated to virtually the same kinetic energy, ions with different masses have different velocities and therefore can be separated. Since the applied voltage and the length of the field-free drift zone are known, $m/z$ ratios can be calculated from the measured flight time. TOF analyzers generally have a high resolution (10,000–40,000), and as a discontinuous measuring method, TOF is well suited to the pulsed-ion generation in

MALDI sources. To compensate for a reduction in resolution due to spatial and kinetic energy distribution, a variable delay of a few hundred nanoseconds is applied and the acceleration voltage is switched on with a fast pulse (delayed extraction/pulsed-ion extraction) [12]. Resolution can be enhanced by employing an electrostatic mirror deflecting the ions' trajectories, referred to as *reflectron*. Thus, the drifting zone of the ions is prolonged and small kinetic energy differences in a given ion population can be compensated, since faster ions penetrate deeper into the electrostatic mirror before deflection.

b. *Quadrupole Analyzers.* In quadrupole instruments, ions are separated in an electric field created by two pairs of parallel rods. Opposite rods are connected electronically, and two voltages—direct current (DC) and radiofrequent (RF) alternating-current (AC)—are supplied. In this electric oscillating field, ions follow a complex wave-like trajectory. Ions with distinct characteristics are able to transverse the quadrupole on a stabile trajectory and reach the adjacent quadrupole or detector, respectively. Varying the voltages while the DC/AC ratio remains constant allows the selection of ions with different $m/z$ ratios. In theory, stable trajectories may be calculated by solution of the Mathieu equation. Depending on the analytical purposes, quadrupoles can be run in different modes and circuitries. A widespread setup is the triple-quadrupole mass analyzer, which can be used as $m/z$ filters and therefore can be applied for specific scanning techniques, such as precursor ion scanning [13], neutral loss scanning [14], and selected reaction monitoring [15].

c. *Ion Trap Analyzers.* Ion trap mass analyzers utilize a three-electrode configuration to trap ions in a small volume. Thus, the applied electric field allows retention of ions during classic operations in MS (i.e., full scan, fragmentation, product ion scan). Moreover, the setup of ion selection and fragmentation as well as analysis of resulting fragments can be repeated ($MS^n$), thereby providing additional information. Because of their ability to trap and accumulate ions, ion trap analyzers increase the signal-to-noise-ratio, yielding a high sensitivity. Ion trap analyzers include linear and three-dimensional (3D) mass spectrometers, which differ in their geometries. While 3D or spherical ion traps consist of a ring electrode separating two hemispherical electrodes, linear ion traps trap the ions in a cavity built up by a quadrupole in RF mode and locking lenses in DC voltage at both sides of the quadrupole.

d. *Ion Cyclotron Resonance Analyzers.* The highest mass resolution and mass accuracy is provided by Fourier transform ion cyclotron resonance (FTICR) mass analyzers. Ions are injected into a cavity of typically cubical or cylindrical geometry. In the presence of a strong magnetic field, ions experience a perpendicular force known as *Lorentz force*. As a consequence, ions travel in an orbital trajectory called *cyclotron motion* with harmonic oscillation. On excitation, ions can be detected by current images whereby, according to Fourier transformation, oscillating frequencies correlate with $m/z$ ratios of the ions and the applied magnetic strength.

e. *Orbitrap Analyzers.* Orbitrap mass analyzers have gained attention since the newly introduced instrumentation detects at high resolution and allows mass accuracies in the sub-part-per-million (<ppm) range [16]. In addition, orbitrap analyzers have higher sensitivity and scan speed than does FTICR instrumentation. Compared to the aforementioned mass analyzers, orbitraps do not use RF or magnetic fields to retain ions in the mass analyzer cavity, but trap them by electric fields. Orbitraps consist of two electrodes: an axial electrode called a "spindle" and an outer electrode allowing for interplay between electrostatic attraction and centrifugal forces by the ions themselves. This balance induces the ions to travel in harmonic oscillation along the axial electrode. In order to characterize ions with respect to $m/z$ ratios, researchers calculate oscillation frequencies of the ions using Fourier transformation, as in the case of FTICR. For this purpose, the axial component of their oscillating trajectory is detected as an image current.

The abovementioned mass analyzers show characteristics that qualify them for particular applications. To combine the strengths of different analyzers, hybrid mass spectrometers were developed comprising two different types of mass analyzers within a single instrument, for example, the hybrid triple-quadrupole/linear ion trap enables applications such as selected reaction monitoring (SRM), precursor ion scanning (PIS), and neutral loss scanning (NLS) in combination with the high sensitivity and $MS^n$ capability of ion trap instruments.

***2.1.2.2 Peptide Sequencing by MS/MS*** In order to elucidate the amino acid sequence of a given peptide, molecular ions have to be activated to undergo fragmentation for conducting MS/MS analysis [17]. Decomposition depends on a variety of factors, including the primary amino acid sequence, the amount of internal energy, charge state, or the mode of activation. Different modes of energy transfer are used for peptide fragmentation, including collision-induced dissociation (CID), electron capture dissociation (ECD) [18], or electron transfer dissociation (ETD) [19], infrared multiphoton dissociation (IRMPD), and post-source decay (PSD) [20], thereby resulting in different and characteristic patterns of fragments. Those fragments can be detected by mass spectrometry only if they carry a charge. If this charge is retained on the $N$ terminus, fragments are classified—depending on the cleavage site at the peptide backbone—as $a$, $b$, or $c$ ions; if the charge is conserved on the $C$ terminus, fragments are assigned as $x$, $y$, or $z$ ions. On CID activation, mainly $b$ and $y$ ions are generated whereas ETD preferentially yields $c$ and $z$ ions as depicted in Figure 2.1. The given nomenclature was first introduced by Roepstorff and Fohlmann in 1984 [21] and further modified by Johnson et al. [22].

With the advent of sequenced genomes from a variety of organisms, nowadays identification of peptides and therefore proteins is performed mostly with the help

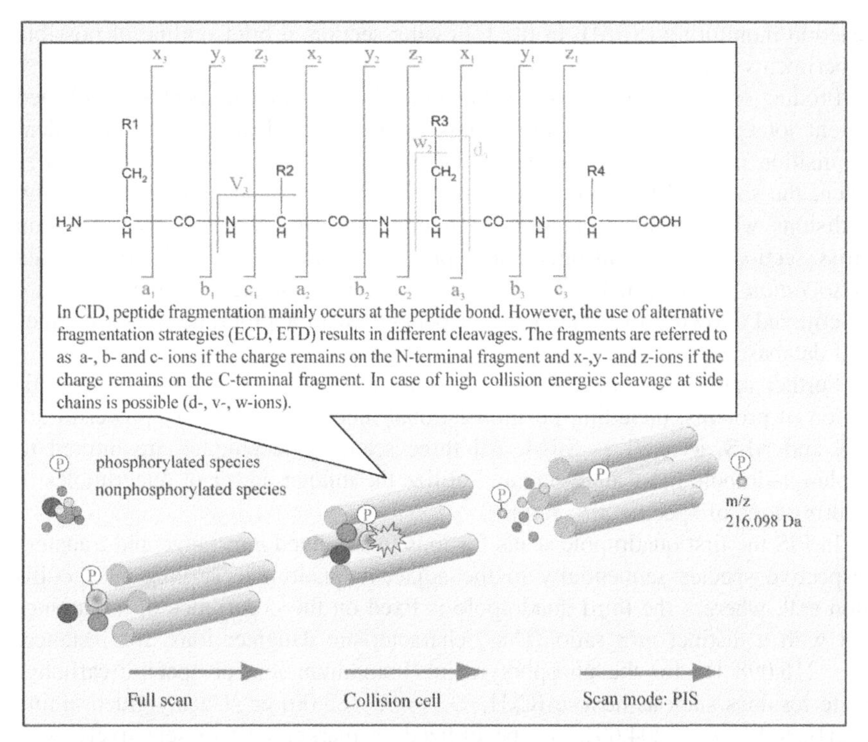


**Figure 2.1** Schematic illustration of a triple-quadrupole mass analyzer operating in precursor ion scanning (PIS) mode. Here quadrupole 3 (Q3) monitors for a characteristic daughter ion that occurs on fragmentation. In this case, Q3 scans for the phosphotyrosine immonium ion with *m/z* 216.098 Da.

of search engines in conjunction with protein databases. Search algorithms such as Sequest [23], Mascot [24], Phenyx [25], or Omssa [26] use fragmentation data to explore diverse protein databases (Swissprot, NCBI, IPI) in either a probabilistic or heuristic manner [27]. Experimental MS/MS spectra are usually matched to theoretical MS/MS spectra of all peptides generated by *in silico* digestion of protein databases. Positive matches correspond to the experimental parent ion mass. Thus peptide ions can be sequenced and hence proteins identified.

In contrast to peptide sequencing using MS/MS, in peptide mass fingerprinting (PMF), usually conducted following MALDI-TOF analysis, tryptic peptide masses in the mass spectrum are matched to respective *in silico* digests within a database search. This technique is mostly applied to 2DGE analysis. For unambiguous identification, high-purity spots are required and the assignment of unique peptide masses is essential.

***2.1.2.3 Scanning Principles*** Specific scan functions utilized in protein mass spectrometry with the different types of mass analyzers are product ion scan (PS), precursor ion scan (PIS), neutral loss scan (NLS), and selective reaction monitoring (SRM). In the following section, a brief outline of possible experiments is provided.

Product ion scan enables the acquisition of fragment ion spectra of selected parent ions, either based on presettings or in so-called information-dependent acquisition mode. PS starts with (1) the isolation of an ion with selected $m/z$. Then, the so-called (2) parent ion undergoes fragmentation, induced mostly by collisions with an inert gas such as helium or a gas with a larger collision cross section such as nitrogen, in a process referred to as *collision-induced dissociation* (CID). Finally, (3) $m/z$ values of the generated fragment ions are determined allowing for sequencing of peptides with appropriate search engines and databases.

Further applications that are in a way restrictive, but—in terms of identification of proteins, including posttranslational modifications—very powerful are PIS and NLS, as well as SRM. All three scanning techniques are limited to triple-quadrupole mass analyzers and utilize the unique power of quadrupoles in filtering ions of specific $m/z$ ratios.

In PIS the first quadrupole scans for ions in a defined $m/z$ range and transfers respective species sequentially to the adjacent quadrupole serving as a collision cell, whereas the third quadrupole is fixed on the occurrence of a daughter ion with a distinct $m/z$ ratio. Thus, characteristic daughter ions, for instance, $m/z$ 216.098 Da for the phosphotyrosine immonium ion, or specific carbohydrate residues such as hexose ($C_6H_{11}O_5^+$, $m/z$ 163.06) or $N$-acetylglucosamine ($C_8H_{14}NO_5^+$, $m/z$ 204.09), can be utilized as marker ions to selectively scan solely for parent ions containing the corresponding modification. Therefore, for unambiguous alignment the use of high-resolution instruments is recommended.

Another useful approach for the elucidation of posttranslational modifications is the NLS. In this approach, an offset of a distinct neutral loss is implemented between the first and the third quadrupole, such as characteristic losses of 98 Da ($H_3PO_4$) in case of phosphopeptides or to a lesser extent 80 Da ($HPO_3$). Selective reaction monitoring (SRM) is even more specific enabling monitoring the occurrence of a distinct and pre-defined pair of parent ion and fragment ion $m/z$ ratios, referred to as *transition*, whereas PIS and NLS allow for analyzing an entire subset of peptides with defined properties. Since only defined transitions are analyzed, SRM is highly selective as well as sensitive and can be used for quantification.

## 2.2 GLOBAL ANALYSIS OF THE PLATELET PROTEOME

In 1990s, the analysis of the platelet proteome has focused on the global ensemble of proteins as well as on biological function. Hence, the entity of proteins has been converged in global approaches, subcellular analysis, and differential analysis on stimulation.

In the initial platelet proteome analyses, experiments were conducted using classic gel-based approaches allowing the visualization of a high number of proteins from complex mixtures. These studies [28–33] are summarized in Chapter 4. In 2005, by applying the gel-free COFRADIC technology (discussed in Chapter 8) to platelets [34], Martens et al. were able to extend the number of identified proteins. Using this approach, a share of 14% was classified as transmembrane proteins; thus a major drawback of 2DGE, the separation of membrane proteins, could be circumvented.

## 2.3 ANALYSIS OF SUBCELLULAR COMPONENTS

Proteome research generally has to deal with highly complex samples and also with a large dynamic range, where proteins with a few copies per cell coexist with proteins of several millions and even more copies per cell. One strategy to cope with this important issue is the characterization of subcellular components, dramatically reducing sample complexity on the one hand and additionally gaining important information about protein localization on the other hand.

In the following chapters of this book emphasis is placed on sample preparation and further proteomic analysis of subcellular compartments such as the platelet plasma membrane (Chapter 5), granules (Chapter 6), functional organelles (Chapter 6), and microparticles (Chapter 7).

## 2.4 POSTTRANSLATIONAL MODIFICATIONS

Many proteins undergo co- and posttranslational modifications during or after their translation at cytoplasmic ribosomes, which are not readily predictable from the genome. These temporospatial (time- and space-dependent) alterations of proteins often occur in a transient and substoichiometric manner allowing for the delicate regulation and coordination of various inter- and intracellular processes such as enzyme regulation, signal transduction, protein localization, and protein–protein interactions [35]. Consequently, the complexity and dynamics of the proteome increases enormously.

Currently there are more than 300 posttranslational modifications (PTMs) known to occur physiologically [36]; however, proteomic methods for the analysis of phosphorylation and glycosylation, which are among the most abundant PTMs, are the best established. Nevertheless, important cellular functions are regulated by acetylation, lipidation, methylation, nitrosylation, oxidation, sulfation, or ubiquitination and have thus far been underrepresented in proteomic research; however, they are accessible to mass spectrometry. Concerning platelets, several studies dealing with protein phosphorylation and glycosylation using proteomic approaches have been published. Phosphorylation in particular, which is essential for platelet activation [37], plays a pivotal role in physiologic and pathologic processes of homeostasis, inflammation, and/or coronary heart disease [38]. In the following sections proteomic analysis of protein phosphorylation and glycosylation is introduced with a detailed focus on the enrichment of modified proteins/peptides and further mass spectrometric requirements for MS.

### 2.4.1 Phosphorylation

Phosphorylation of proteins and peptides is probably the most abundant form of PTM. The majority of proteins are at least transiently phosphorylated during their lifetime. A diverse concert of protein phosphatases and protein kinases defines the transient character of this modification, which contributes to the regulatory mechanism of numerous biological and biochemical processes, including protein folding, function, activity, interaction, location, and degradation [39].

In general, there are four types of protein phosphorylations [40]: (1) $O$-phosphates, (2) $N$-phosphates, (3) acylphosphates, and (4) $S$-phosphates. The $O$-phosphates are the most abundant, occuring on serine (pSer), threonine (pThr), or tyrosine residues (pTyr). Rarely is this modification found on hydroxyproline and hydroxylysine. While $O$-phosphates are stable under acidic conditions, $N$-phosphates attached to arginine, histidine, and lysine residues show better stability under alkaline conditions. Acylphosphates occur on carboxylic groups of aspartic acid and glutamatic acid and $S$-phosphates, on cysteine residues. Acylphosphates exhibit a high reactivity and are therefore labile under acidic as well as alkaline conditions. $S$-phosphates offer a moderate stability under acidic and alkaline conditions [40].

The transient character and low stoichiometry of phosphorylation renders the analysis of phosphorylation sites a demanding and challenging task.

***2.4.1.1 Phosphorylation in Platelets*** Because of their anucleate character, platelets exhibit a limited capability of *de novo* protein synthesis [41]. Furthermore, the sensitive regulation between inhibition and activation must respond to sudden stimuli within seconds. Accordingly, platelets are regulated by PTMs that control both protein activity and interaction. In this context phosphorylation plays a critical role not only in the initial steps of activation and aggregation following exposure to adhesive and activating ligands such as collagen, fibrinogen, vWF, ADP, thrombin, and thromboxane $A_2$, but also in inhibitory pathways that are stimulated by endothelium-derived nitric oxide (NO), prostacyclin ($PGI_2$), and immunoreceptor tyrosine-based inhibitory motif (ITIM)-containing receptors. The initial steps of platelet activation are regulated by a concert of tyrosine and serine/threonine kinases and phosphatases whereby inhibitory events are strongly determined by the cyclic-nucleotide-dependent protein kinases A and G, which are highly expressed in human platelets.

***2.4.1.2 Sample Preparation*** A mandatory and initial step in phosphorylation analysis is the isolation of phosphoproteins/peptides out of an entire cell lysate or a subcellular compartment. In this context the liberation of phosphatases and kinases on cell lysis has to be taken into account. To overcome the enzymatic changes of phosphorylation patterns arising from platelet lysis, the use of protein phosphatase and to a lesser extent protein kinase inhibitors, is an elementary and essential prerequisite beside the usage of protease inhibitor cocktails that limit protein degradation. In order to suppress the activity of a broad range of enzymes, attention should be given to the completeness of inhibitors depending on experimental demands. Furthermore, inhibitors should be added to all used buffers at an early timepoint.

***2.4.1.3 Classical Approaches for Phosphorylation Analysis*** Depending on the complexity of the sample, protein separation can be achieved by one- or two-dimensional polyacrylamide gel electrophoresis (1DGE or 2DGE). Two-dimensional gels allow the separation and detection of protein isoforms and of modified proteins, such as phosphorylated proteins since the introduction of a phosphate group results in an acidic pI shift. Phosphospecific stainings, radiolabeling, and immunodetection can be used for detection of phosphoproteins.

Phosphoproteins can be detected by immunoblotting. In this process Western blots are performed and probed with phosphospecific antibodies. This approach is widely used but depends highly on the efficiency, specificity, and sensitivity of the applied antibody. Moreover, there are only efficient antibodies against phosphotyrosine; antibodies against phosphoserine and phosphothreonine are not reliable in terms of selectivity and sensitivity. General antiphosphotyrosine-specific antibodies have been routinely used to monitor platelet activation and the phosphorylation status of specific signaling proteins following immunoprecipitation. Phosphospecific antibodies are increasingly being used to monitor the phosphorylation status of distinct sites within proteins. However, caution must be taken to validate the specificities of such antibodies. The most sensitive strategy to visualize phosphoproteins is radioactive labeling using $^{32}P/^{33}P$. This procedure can be performed *in vivo* as well as *in vitro*. For *in vivo* labeling, cells are incubated with [$^{32/33}P$]-orthophosphate, whereas *in vitro* labeling is based on the addition of [$\gamma^{32/33}P$]-ATP and a purified protein kinase. Samples are subsequently analyzed by autoradiography, positive spots or gel bands excised and prepared for MS analysis. However, this technique is afflicted with diverse drawbacks, including (1) high concentrations of labeling reagent and protein kinases lead to nonspecific phosphorylation; (2) *in vivo* exposure to [$^{32/33}P$]-orthophosphate causes cellular stress, triggering further nonspecific phosphorylation events, such as DNA damage; and (3) there may be exposure of staff to radioactivity.

Radiolabeling of platelets has been conducted in several published proteomic studies. Marcus et al. [32] monitored tyrosine phosphorylation on thrombin stimulation in platelets radiolabeled with [$^{32}P$]-orthophosphate. Complete cell lysates were visualized by 2DGE using IEF strips in the range of pH 4–7 and pH 6–11; although radiolabeling is a highly sensitive method, detecting phosphotyrosines, which account for less than 1% of all *O*-phosphorylation sites, is challenging in the presence of large amounts of non-phosphorylated species. After tryptic digest of respective protein spots, 55 differential phosphorylated proteins were identified using MALDI-TOF-MS and nano-LC-MS/MS. Diverse comigration events in 2DGE hampered the unambiguous identification of phosphorylation sites for all spots.

***2.4.1.4 Enrichment of Phosphorylated Proteins or Peptides for Mass Spectrometric Analysis*** Because of their low stoichiometry in complex cell digests in combination with their lower ionization efficiency compared to corresponding nonphosphorylated isoforms, the enrichment of phosphorylated peptides prior to analysis is mandatory. To improve signal-to-noise ratios and consequently to address the issues of sensitivity and dynamic range, several phosphoprotein/peptide enrichment strategies have been developed, improved, and successfully applied.

Several studies have focused on the platelet phosphoproteome following distinct stimulation [29,37,42], as well as on the global phosphoproteome of resting platelets [43]. Different activation states of platelets were investigated using (1) immunoprecipitation, (2) strong cationic exchange (SCX) [44], and (3) immobilized ion metal affinity chromatography (IMAC) enrichment [45]. Techniques employed to study the phosphoproteome in other cell and tissue types that could be applied to platelets include (1) metal oxides such as titanium dioxide ($TiO_2$) [46] chromatography and zirconium dioxide ($ZrO_2$) [47], (2) sequential elution from IMAC (SIMAC) [48], (3) hydrophilic interaction liquid chromatography (HILIC) [49], (4) electrostatic repulsion-hydrophilic interaction chromatography (ERLIC) [50], (5) derivatization-based enrichment [51–53], and (9) phosphoamidate chemistry [51,54]. These techniques are discussed in the following paragraphs.

1. *Immunoprecipitation.* In human platelets signaling cascades are both positively and negatively regulated by tyrosine phosphorylation. With a general ratio of pSer : pThr : pTyr = 1800 : 200 : 1, pTyr is relatively rare [55]. However, in contrast to pSer- and pThr-specific antibodies, anti-pTyr-specific antibodies are available, allowing for the selective enrichment of pTyr-containing proteins and peptides [42,56]. In 2002, Maguire et al. analyzed for the first time the tyrosine phosphorylation status of human platelets on thrombin stimulation using immunoprecipitation [42]. They used the monoclonal antibody 4G10 to isolate all dynamic phosphotyrosine events in the thrombin signaling pathway. In this study 67 differentially phosphorylated proteins were reproducibly identified by subsequent 2DGE. In conjunction with further analysis by Western blotting and MALDI-TOF-MS, 10 of these proteins could be positively identified. Tyrosine phosphorylation was characterized in 7 proteins by using MALDI-PMF, among them FAK, Syk, P2X6, and MAPK. For further details, see also Foy et al. [57]. In order to enrich phosphoproteins more efficiently, different techniques can be used in conjunction with immunoprecipitation. This has been demonstrated by combining prefractionation of phosphoproteins using pTyr antibodies prior to proteolytic digest and further enrichment on the peptide level using IMAC [58] or $TiO_2$ [59].

2. *Ion Exchange Chromatography.* In general, ion exchange chromatography has been successfully applied as a means of prefractionating proteins and peptides. However, mainly strong cation exchange (SCX) has been successfully established in widescale phosphoproteome studies. The technique relies on the interaction of positively charged ions with negatively charged functional groups on the SCX solid phase. Thereby selectivity and sensitivity depend highly on the

pH and ionic strength of applied loading and elution buffers. In 2004 Beausoleil et al. introduced SCX at pH 2.7 for the isolation of tryptic phosphopeptides [60]. Under these conditions, tryptic peptides typically bear a net charge of +2, due to the positively charged *N* terminus and *C*-terminal Arg or Lys, whereas acidic residues such as Glu and Asp are mostly protonated. However, phosphopeptides contain an additional negative charge per phosphate group and thus, due to their reduced net charge, exhibit reduced retention during SCX at pH 2.7. Nevertheless, coelution of other peptides, including charge-reduced peptide species such as *N*-terminally acetylated, *C*-terminal, or sialylated peptides, has to be taken into account [61]. Accordingly, multiply phosphorylated peptides yield poor retention and are predominantly detected in the flowthrough or in the early stages of the chromatographic separation. It is highly recommended that enrichment techniques be combined to achieve comprehensive datasets. For example, SCX is widely used as a primary fractionation step followed by IMAC and $TiO_2$-based enrichment of collected fractions.

3. *Immobilized Metal Ion Affinity Chromatography*. Initially introduced for the purpose of affinity purification of proteins [62], immobilized metal ion affinity chromatography (IMAC) was employed for the enrichment of phosphopeptides for the first time in 1986. In this technique, negatively charged phosphopeptides interact with positively charged metal ions ($Al^{3+}$, $Fe^{3+}$, $Ga^{3+}$, $Zr^{4+}$) chelated in a 1 : 1 ratio with nitrilotriacetic acid (NTA), iminodiacetic acid (IDA), or tris(carboxymethylethylendiamine) (TED) on the solid phase. Binding efficiency and enrichment selectivity is highly dependent on the applied conditions. While binding and washing occur in acidic milieu, elution is performed by using salt or pH gradients. Although IMAC has become a widely used technique, it is slightly biased toward multiply phosphorylated peptides and also has an assortment of other associated drawbacks. The major problem is the occasionally high level of non-specific binding due mainly to the copurification of acidic peptides. Additionally, co-purified nonphosphorylated species often show more efficient ionization, so that signals from phosphorylated species are suppressed during MS analysis. To overcome this problem, Ficarro et al. derivatized carboxylic groups of amino acid residues by *O*-methyl esterification using HCl-saturated dried methanol [45]. However, incomplete esterification and possible deamidation of asparagine and glutamine during treatment might increase sample complexity; consequently reaction conditions have to be selected with care. Another approach to reduce non-specific binding is the optimization of loading and washing conditions (e.g. pH and organic solvent) [63]. An applied pH of 2–2.3 results in a high degree of protonation of carboxylic groups on the acidic amino acid residues Asp and Glu [64], thus reducing their interaction with the metal ions. Nevertheless, attention has to be paid to the acidic stability of the solid phase containing metal ions!

4. *Metal Oxides.* Metal oxides were initially used for the enrichment of nucleotides and phospholipids [65]. In 2004 Sano et al. introduced $TiO_2$ as an alternative technique to IMAC for the specific enrichment of phosphopeptides. To date, mainly the metal oxides titanium dioxide ($TiO_2$) [66] and zirconium

dioxide ($ZiO_2$) [47] have been successfully used in phosphoproteomics: alkyl phosphates form monolayers on their surfaces [67]. Compared to silica-based material, metal oxides offer higher stability and can therefore be treated under harsh conditions ($13 < pH < 1$ and $T > 200°C$) [65]. Pinkse et al. established a 2D nano-LC-MS/MS setup that allows the parallel enrichment of phosphorylated peptides on a titanium dioxide precolumn and nonphosphorylated peptides on a reversed-phase precolumn online [66]. A recovery of more than 90% could be achieved for selected peptides. However, the titanium oxide set-up is also prone to copurification of acidic peptides. Larsen et al. improved the selectivity of $TiO_2$ toward phosphopeptides by adding 2,5-dihydroxybenzoic acid (DHB) to the loading buffer. DHB competes with non-phosphorylated peptides for binding sites on the titanium dioxide surface [68].

5. *Sequential Elution from IMAC.* Multiphosphorylated peptide ions are expected to ionize even less efficiently than monophosphorylated peptides do. Furthermore, due to their low intensities, multi-phosphorylated peptides are rarely selected for fragmentation in the presence of mono- and nonphosphorylated peptides. To overcome the limited recovery of multi-phosphorylated peptides in complex samples, a new strategy for affinity enrichment was introduced by Tingholm et al. in 2008 involving the sequential elution of proteins/peptides from IMAC (SIMAC) [48]. Combing the already discussed techniques IMAC and $TiO_2$, SIMAC allows the successive enrichment of mono- and multi-phosphorylated peptides. In this approach, peptide samples are first loaded on IMAC material and flowthroughs from binding and washing steps are consecutively incubated with $TiO_2$. Monophosphorylated peptides are subsequently eluted from the IMAC beads under acidic conditions.

6. *Hydrophilic Interaction Liquid Chromatography.* Hydrophilic interaction liquid chromatography (HILIC) provides characteristics that exploit the increase in hydrophilicity of phosphorylated peptides [69]. Engaged normal-phase chromatography with predominantly organic mobile phase allows retention of polar analytes. Peptides in organic phase are captured on a neutral hydrophilic phase by hydrogen bonding, which is disrupted by decreasing the amount of organic phase. Peptides and phosphorylated peptides elute according to their polarities, giving rise to phosphopeptide-enriched fractions in the later stages of the chromatographic process.

7. *Electrostatic Repulsion Hydrophilic Interaction.* Electrostatic repulsion hydrophilic interaction chromatography (ERLIC) represents a novel approach for peptide-centric enrichment, based on a combination of ion-exchange and hydrophilic interaction [50]. For phosphorylated peptides, a pH sufficiently low to protonate carboxyl groups but not phosphate groups results in an electrostatic interaction with the weak anion exchange stationary phase. Phosphopeptides are eluted by decreasing the organic and/or increasing the salt content.

8. *Derivatization Approaches.* Derivatization approaches allow for enrichment of phosphopeptides and also deal with suppression effects during ionization and the lability of the phosphoester group. Alkali treatment of phosphopeptides induces β-elimination, and highly labile pSer and pThr can be transformed to

dehydroalanine and 2-aminodehydrobutyric acid, respectively. This treatment allows further analysis by mass spectrometry or the addition of nucleophilic functional groups (Michael addition). Phosphopeptides undergoing the latter modification may subsequently be subjected to diagonal chromatography and, due to the added functional groups, separated from nonphosphorylated species. A limitation of β-elimination is the racemization of the β-carbon within the peptide backbone, resulting in peak broadening during reversed-phase (RP) chromatography and consequently in loss of sensitivity [53]. Additionally, in this process the elimination of nonphosphorylated Ser and Thr as well as *O*-glycans may lead to higher false-positive PTM localization. Moreover, the arylphosphate pTyr is not accessible to β-elimination.

9. *Phosphoamidate Chemistry.* Phosphoamidate chemistry (PAC), an alternative derivatization method introduced by Zhou et al. and further modified by Tao et al., allows the general analysis of phosphates including pTyr [51,54]. In contrast to other enrichment strategies, the phosphate group remains intact on the peptide during the four-step procedure. This technique is based on the condensation reaction of phosphopeptides to a polyamine dendrimer catalyzed by carbodiimides and imidazole. Carboxylic groups of the peptides and phosphorylation residues build up amides and phosphoamidates, respectively. On removal of non-phosphorylated peptides using size exclusion chromatography, phosphopeptides are liberated and recovered by acidic hydrolysis and subsequent size exclusion chromatography.

The only study on global phosphorylation events in platelets to date (at the time of this writing) was conducted by Zahedi et al. [43]. In this study, the phosphorylation pattern of "resting" nonstimulated platelets isolated from healthy donors was analyzed. For this purpose, IMAC and SCX enrichment were performed in conjunction with mass spectrometry, to yield 564 phosphorylation sites from 278 unambiguous proteins and 15 ambiguous proteins as a result of peptide redundancies. In this study, 196 phosphorylated peptides were exclusively detected on IMAC enrichment and 187 could be assigned only after SCX separation. Interestingly, applying the two techniques resulted in an overlap of only 111 phosphorylated peptides, which emphasizes the benefit of using different enrichment strategies. For nano-LC-MS/MS, full MS scans were acquired, and the most intensive signals were subjected to MS/MS. To address the issue of low pTyr occurrence, PIS monitoring for the pTyr immonium ion at *m/z* 216.01 was applied.

Phosphoproteins identified in this study could be predominantly classified into proteins participating in signaling activities, reorganization of the actin cytoskeleton, and cell adhesion. In a detailed sequence analysis of the respective phosphorylation sites, particular emphasis was placed on potential substrates of the inhibitory protein kinases A and G that contribute to maintain platelets in a nonactivated state. Both PKA and PKG contain an overlapping consensus sequence (*R*/K-*R*/K-X-*S*/T), which had also been identified in phosphopeptides of 23 other proteins. Of those, several play significant roles in platelet function, such as GPIbα and PDE3A, or have been implicated as important regulators of platelet function, such as G6b. In this study, data interpretation of the relatively low number of phosphorylation sites was performed by manual validation of MS/MS spectra and alternatively by assessing a false discovery rate (FDR) based on the target/decoy search strategy introduced by Gygi and coworkers [70]. In contrast to the tedious manual validation process, the automated FPR search strategy led to the identification of a higher number of phosphorylation sites; these, however, also included hits with (1) incorrect phosphorylation site assignment, (2) poor MS/MS spectrum quality, and (3) false-negative identifications. Therefore, for this still addressable number of phosphopeptides, manual validation was preferred to solely FPR-based data interpretation. Due to the introduction of mass spectrometric instrumentation with high sensitivity and especially high mass accuracy in more recent years, automated data interpretation of large-scale proteome studies gained more and more attention. However, the primary phosphorylation site assignment by search algorithms such as Mascot [24], Omssa [26], and Sequest [23] might be ambiguous depending on the MS/MS spectra. To address this issue, alternative search algorithms with different scorings for phosphopeptides such as MS Quant [71] or Ascore [72] have been developed.

***2.4.1.5 Identification of Phosphorylation Sites by Mass Spectrometry*** Phosphorylation site analysis is performed mostly on the peptide level since analysis on the protein level requires high-sensitivity and cost-intensive analytical instruments such as FTICR-MS. Furthermore, the distinct assignment of phosphorylation sites may be hindered on the protein level. Working on the peptide level in conjunction with specific enrichment techniques allows for the more efficient reduction of sample complexity, thereby compensating for the potential underrepresentation of phosphorylated species. Besides more effective ionization by ESI, phosphorylated peptides may be analyzed in different experimental setups aiming for the distinct identification of phosphopeptides or phosphorylated peptides due to neutral losses during fragmentation. Analyzing phosphorylation sites by mass spectrometry comprises the detection of phosphorylated peptides and the distinct localization of sites of phosphorylation. For this purpose MALDI and ESI ion sources provide different characteristics, both presenting advantages and disadvantages.

In general, labile phosphopeptides are prone to neutral losses on CID, which can be detected by mass spectrometry such as metaphosphoric acid ($-HPO_3$, −80 Da; pSer/Thr/Tyr, less often) and phosphoric acid ($H_3PO_4$, −98 Da; only pSer/Thr, more common).

In MALDI-TOF instruments, phosphorylated peptides will already have undergone metastable decay when they pass the field-free drift, referred to as *postsource decay* (PSD) [73]. Here, the singly charged parent ion dissociates into two parts, the neutral loss molecule and the singly charged parent minus neutral loss ion.The singly charged fragment is deflected in reflector-mode TOF analyzers and finally detected with a shifted $m/z$ value and a weakly resolved isotopic pattern.

Generally, the ionization/detection efficiency of phosphopeptides may be significantly reduced in the presence of nonphosphorylated peptides. To address these issues, the enrichment of phosphorylated species as aforementioned, the usage of additives such as 1% phosphoric acid [74], which reduce adsorption to the metal target and usage of the optimal crystallization matrix [e.g., 2,5-dihydroxybenzoic acid (DHB)], which alters the issue of neutral losses of phosphopeptides compared to α-cyano-4-hydroxycinnamic acid (CHCA), are essential prerequisites.

In contrast, ESI ion sources generate predominantly doubly and triply charged ions in conjunction with tryptic peptides; therefore ESI coupled to an LC-MS/MS system is well suited for the analysis of highly complex sample mixtures in MS/MS mode. Whereas phosphorylation of a peptide results in a mass shift of +80 Da per occurrence, which can be monitored in MS survey scans, in CID-based MS/MS experiments of phosphopeptides the aforementioned neutral losses of phosphoric and metaphosphoric acid can be detected from parent and fragment ions as depicted in Figure 2.2. These neutral losses may severely reduce fragment ion spectrum quality, thus posing one of the major issues in MS analysis of phosphopeptides. The neutral loss from the parent ion in particular in many cases can completely dominate MS/MS spectra; however, this effect is dependent on the type of mass analyzer, peptide sequence [75], and charge. Nevertheless, it can be utilized either for PIS in negative-mode searching for the characteristic −79 *m/z* anion ($PO_3^-$) or for NLS in positive-mode searching for the respective losses from parent ions [76].

Generally, it is significant that *O*-sulfation not only exhibits almost the same mass shift as does *O*-phosphorylation but also demonstrates a similar neutral loss signal of 80 Da on CID. However, since sulfonated peptides undergo a loss of their sulfo-moiety preceding any backbone fragmentation, these isobaric modifications can be differentiated by their fragment ion spectra or by precursor ion scanning [77,78]. While NLS does not allow the analysis of pTyr due to enhanced stability of arylphosphates, PIS for the $PO_3^-$ anion may detect all types of phosphorylation. In PIS mode the first quadrupole scans for the whole mass range, while in the second quadrupole CID induces fragmentation. The third quadrupole monitors the occurrence of the −79 *m/z* $PO_3^-$ anion in negative mode. The occurrence of the respective *m/z* sequencing of the corresponding parent ion is accomplished in positive polarity mode. Sequencing of the peptide backbone is much more efficient in positive polarity mode, and therefore the switch is mandatory even though it is afflicted with spray instabilities. However, this limitation can be partially compensated by the postcolumn addition of isopropanol [79].

### 2.4.2 Glycosylation

Glycosylation represents one of the most abundant posttranslational modifications among cell surface markers and secreted proteins. Approximately 50–60% of all proteins in the human body are considered to be modified by glycosylation [35]. This modification occurs in a wide range of biological processes, influencing protein stability and conformation, cellular signaling, and modulation

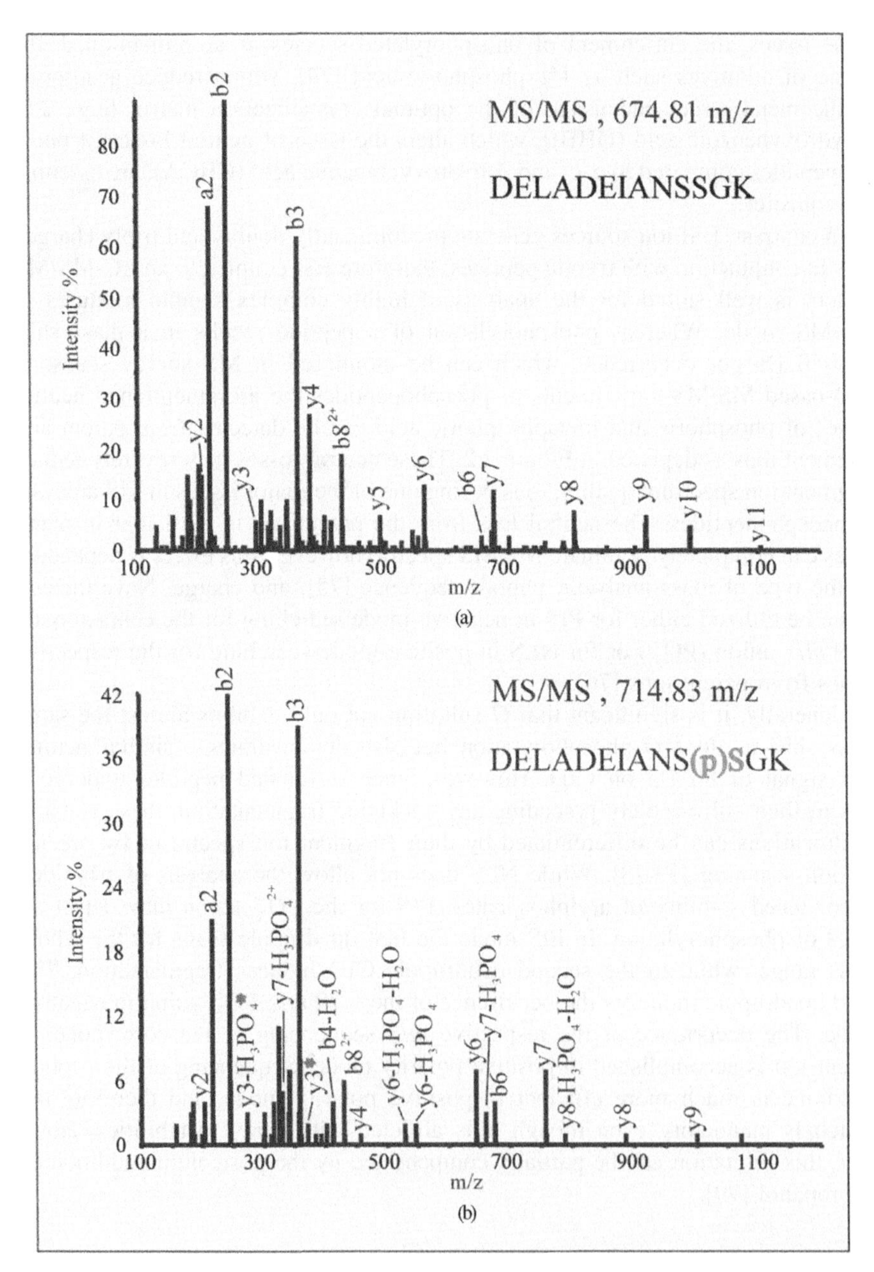


**Figure 2.2** Fragment ion spectra of the nonphosphorylated (a) an phosphorylated (b) versions of the doubly charged peptide DELADEIANS(p)SGK of human myosin-9. On addition of the phosphate moiety (+80 Da) and partial neutral losses of $H_3PO_4$ (−98 Da) from the pSer containing (*b*) and *y* ions and the resulting fragement pattern, the distinct site of phosphorylation can be assigned. Note the shifts of the two $y_3$(*) ions in the phosphorylated compared to the nonphosphorylated version and the difference of 40 *m*/*z* of the doubly charged parent ions.

of binding affinities. Hence, glycosylations are present inside the cell, including the cytoplasm and subcellular organelles, in cell membranes, and in the extracellular matrices. Here, glycoproteins function as enzymes, receptors, antibodies, and plasma proteins as well as structural proteins. To date, only a few proteomic studies have been performed on the platelet glycosylation pattern, although glycoproteins play a crucial role in the process of hemostasis and may also lead to pathologic thrombus formation [80]. Glycoproteins are extensively involved in the entire process of platelet activation.

The inherent heterogeneous and diverse nature of glycoproteins reveals great variability with respect to (1) the carbohydrate–amino acid linkage of the peptide backbone, (2) the variable composition of glycans by 13 monosaccharides [81], (3) the reducing sugar unit, as well as (4) the position and (5) anomeric forms of linkages between respective monosaccharide building blocks. Furthermore, monosaccharides may be modified by phosphorylation, sulfation, methylation, and acetylation. In addition, the complexity of glycosylations is enhanced by possible distributions of glycans to discrete glycosylation sites. While the overall glycan profile of a glycoprotein may stay the same (macroheterogeneity), diverse and differing glycans may be attached to individual sites (microheterogeneity).

Generally, glycosylations linked to asparagine and serine/threonine residues, classified as *N*-glycosylation and *O*-glycosylation, respectively, are among the most prevalent. Moreover, the addition of glycosylphosphatidylinositol (GPI) to protein *C* termini is a well-known modification that mediates anchoring of proteins to the cell membrane. Furthermore, three minor types of glycosylation are differentiated: (1) phosphoglycosylation, characterized by a carbohydrate linkage through a phosphodiester bound to a hydroxyl group containing amino acids such as Ser and Thr; (2) *C*-mannosylation, which attaches a mannose residue to the indole ring of Trp; and (3) ribosylation, where the linkage of the 3-hydroxyl group of ribose to Glu, Asn, Arg, or Cys has been observed.

With the exception of GlcNAc-β-Asn and GPI anchors, in biological processes carbohydrate–amino acid linkage is mediated by direct enzymatic transfer of monosaccharide residues to a distinct amino acid. Oligosaccharides then evolve by successive enzymatic attachment of sugars to the peptide-linked moiety.

*N*-Glycans are connected to the polypeptide backbone in a β-glycosidic manner via *N*-acetyl-D-glucosamine (GlcNAc). In the first instance, a preassembled core structure consisting of the pentasaccharide $(GlcNAc)_2Man_3$ is modified to $Glc_3Man_9(GlcNAc)_2$ and subsequently transferred *en bloc* to the respective consensus sequence Asn-X-Ser/Thr, in which X denotes any amino acid except proline. The cotranslational modification undergoes several processing steps in the endoplasmic reticulum (ER) and the Golgi apparatus and finally results in three basic structural forms with remaining chitobiose core: high-mannose, complex, and hybrid-type glycans, displayed in Figure 2.3. Beyond the relatively simple core structure, the variety of *N*-glycans is amplified by diverse bi-, tri- or tetraantennary branching and modifications such as terminal (2,3)- and (2,6)-linked *N*-acetylneuraminic acid (Neu5Ac), resulting in complete or partial sialylation. Moreover, the introduction of for example, $\alpha_1$-6-linked fucose to the

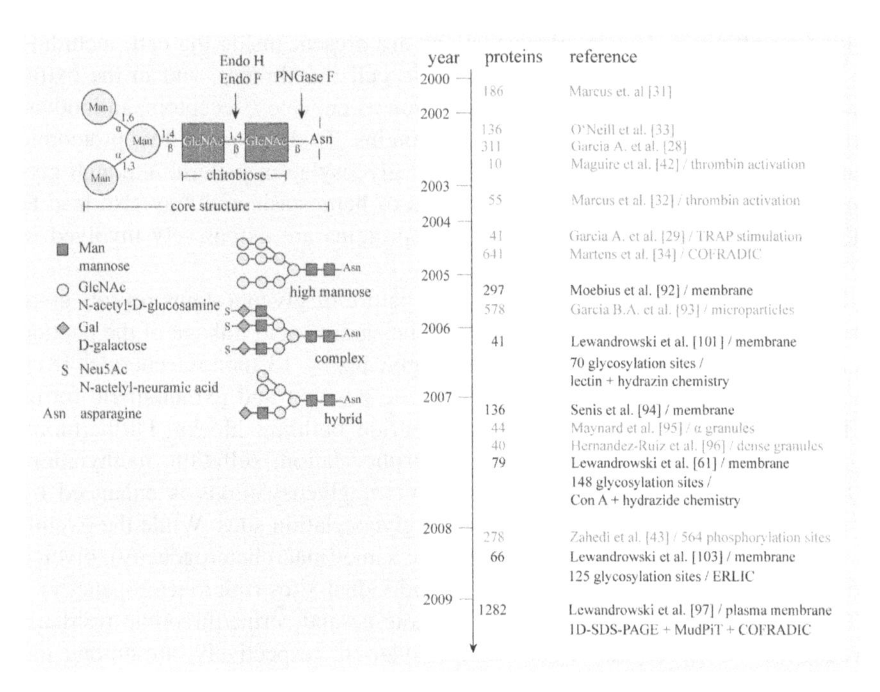


**Figure 2.3** The $N$-glycan core structure $(GlcNAc)_2Man_3$ shown here is part of the three basic structural forms: high mannose, complex, and hybrid types. As depicted in the timescale, recent platelet proteomics focused mainly on membranes and glycoproteins (black) as these components play a crucial role in initial signal transduction. For summarized studies, the number of identified proteins is given.

reducing terminal GlcNac and further the introduction of a bisecting GlcNAc in position 4 of the β-associated core mannose establishes a great diversity of structures.

$O$-Linked glycosylation is generally more heterogeneous than $N$-glycosylation. The end sugar unit linked to hydroxyl-containing amino acids such as serine and threonine and less frequently tyrosine, hydroxylysine, and hydroxyproline is abundantly α 1-bound $N$-acetyl-α-galactosamine (αGalNAc), but may vary [81]. Additionally, respective monosaccharides can be attached in α- or β-anomeric fashion. Moreover, their individual branching structures may differ to a high degree. αGalNAc-based glycans can be classified into eight distinct core structures [82]. While to date no specific consensus sequence for $O$-glycosylation has been described, glycan attachment is found to occur in single or in clustered units on defined proteins stretches, thereby complicating the distinct glycosylation site assignment.

The analysis pathways of $O$-linked glycans and their respective attachment sites are different in comparison to $N$-glycosylation. For instance, in contrast to $N$-glycans, no specific enzyme for the removal of $O$-glycans is available so far. While β-elimination results in $O$-glycan release, degradation of the peptide backbone has to be taken into account, resulting in drawbacks regarding peptide sequencing.

The investigation of *O*-glycosylation remains challenging. Since there have been no studies on *O*-glycosylation in platelets, in the following sections we will discuss the analysis of *N*-glycans and their attachment sites.

***2.4.2.1 Disorders of Platelet Function Depending on Glycosylation Abnormalities*** Congenital disorders of glycosylation (CDG) are inherited metabolic diseases that affect the assembly and the transfer of glycans or their processing. Although current knowledge of CDGs is limited, this is a rapidly growing group of clinical diseases based on the diversity of different types of CDG. Among them, the majority of identified CDGs were reported to include defects of *N*-glycosylation [83].

Concerning platelets, most studies demonstrated glycoprotein mediators of adhesion and aggregation to be involved in CDG. The most common forms of platelet-related CDGs are Glanzmann thrombasthenia (GT) [80] and the Bernard–Soulier syndrome (BSS) [84]. More recently, studies on GT and BSS and further platelet disorders which are not related to impaired glycoproteins, such as familial thrombocytopenias or diverse storage pool diseases, have contributed to our understanding of platelet structure and function. For medical treatment with platelet concentrates, the optimal storage conditions with regard to hemostatic and clearance function must be determined. Unlike other transplantable tissues, platelets are very sensitive to refrigeration and rearrange their surface glycoprotein GPIb to irreversible clusters marking them for phagocytosis [85]. In contrast, *in vitro* assays reported detrimental changes such as the loss of hemostatic function when storing platelets at room temperature. Consequently, its life-saving potential is considerably reduced.

In order to overcome this general bottleneck, Hoffmeister et al. performed *in vitro* galactosylation of murine platelets using endogenous galactosyltransferase activity and UDP-galactose, thereby preventing phagocytosis and clearance [86]. Furthermore, aggregation response from murine platelets to ristocetin of chilled galactosylated platelets did not differ from unmodified ones. However, whereas the murine system showed promising results, a clinical study (phase 1) based on human platelets could not confirm these previous findings [87]. These findings may be ascribed to diverse methodological differences between the studies, while the authors suppose possible differential carbohydrate-mediated mechanisms for clearance of short-term and long-term cold-stored platelets.

***2.4.2.2 Sample Preparation*** For a complete characterization of protein glycosylation, the analysis must be performed on different levels of detail. Although more recent advances in ESI- and MALDI-MS have led to increased sensitivity and mass accuracy, the investigation of intact glycoproteins remains a crucial point with limitations concerning the efficiency of ionization and fragmentation. To identify existing glycosylation sites, possible individual glycoforms and also the composition and structure of the attached glycans, a characterization on the peptide level is often preferable.

In order to investigate the structure of a glycan, it is usually released and separated from the peptide backbone. In case of *N*-glycans, two general procedures are commonly employed: (1) enzymatic digest with endoglycosidases or (2) chemical cleavage. Peptide *N*-glycosidase F (PNGase F) [88] cleaves preferably between the innermost GlcNAc and Asn of *N*-glycopeptides and releases glycans with intact reducing termini; with its additional amidase activity, PNGase F converts the respective asparagine residue to an aspartic acid, resulting in a mass increase of 1 Da. Thereby, on analysis of the peptide moiety, identification of the glycosylation site and the respective protein by mass spectrometry is possible. Additional incubation with $^{18}O/^{16}O$-labeled water during the PNGase F treatment leads to incorporation of respective oxygen isotopes to converted amino acids. Consequently, the MS spectrum features both a characteristic doublet peak and a +3 Da mass shift, allowing the discrimination of common versus artificial deamidation. Alternatively, other glycosidases can be used to mediate cleavage at different sites in the $Man_3GlcNAc_2$ core structure of *N*-glycans, such as endoF, endoH, or PNGase A (see Fig. 2.3).

Conversely, it is also possible to employ chemical cleavage allowing for liberation of both *N*- and *O*-glycans. Chemical cleavage is the only common technique for releasing *O*-glycans since there is no glycosidase available that cleaves between the proximal αGalNAc and the linked amino acid. Because of the harsh reaction conditions, the major drawback of chemical cleavage using hydrazinolysis is the destruction of all peptide bonds, rendering the simultaneous identification of the glycosylation site impossible. Moreover, hydrazine treatment accounts for a couple of modifications on carbohydrates, which, as a consequence, cannot be retraced by MS.

Commonly, *O*-glycans are cleaved by β-elimination under alkaline conditions with NaOH and further addition of $NaBH_4$, thereby converting the reducing terminal sugar to its alditol to prevent ongoing β-elimination and finally destruction of the oligosaccharide (peeling reactions) [89].

For detailed glycan sequencing, exoglycosidases may be employed as single or defined combinations of enzymes. Unfortunately, this experimental approach requires high amounts of purified glycans or glycoproteins. Although there are some enzymes offering a narrow specificity for a distinct carbohydrate as well as linkage and anomericity, the multitude of glycosidases currently does not allow the elucidation of all structural features.

***2.4.2.3 Enrichment of Glycopeptides*** Another challenge regarding the analysis of glycopeptides is the high number of non-glycosylated peptides in a complete cell digest. Already present in relatively low stoichiometry, glycopeptides show lower ionization efficiency and tend not to be selected for MS analysis in a peptide complex mixture. Thus, distinct enrichment of glycopeptides is an essential prerequisite prior to MS analysis. Specific enrichment requires chemical and chromatographic strategies. Methods for enrichment of glycoproteins and/or glycopeptides include (1) lectin affinity chromatography, (2) peroxidase oxidation followed by hydrazide coupling, (3) HILIC, (4) ERLIC, (5) boronic acid chromatography, and (6) glycosylation-specific antibodies.

Techniques used to enrich glycoproteins in the few platelet glycosylation studies to date include lectin affinity chromatography, hydrazine chemistry, and HILIC, which are discussed in the following paragraphs.

1. *Lectin Affinity Enrichment*. Lectins are a class of proteins that specifically interact with carbohydrates. They offer a unique affinity toward specific motives in glycans and are preferentially used to isolate either glycosylated proteins and peptides or to distinguish between glycan residues. Two lectin affinity chromatography–based glycoprotein enrichment techniques have been reported. The multilectin approach introduced by Yang and Hancock [90] relies on a single column that harbors different lectins with broad specificity. The approach offers the global isolation of glycoproteins and glycopeptides from complex mixtures that span a dynamic range of several orders of magnitude. Serial lectin affinity chromatography (SLAC), described by Cummings and Kornfeld in 1984, performs a successive binding of glycoproteins and glycopeptides to lectin columns with different specificities [91]. Employing lectins with broad and narrow affinities (see Table 2.1), SAC promises to be a powerful tool for the elucidation of structural components in glycan residues. However, the results of the experiment are strongly based on the selectivity of lectins and thereby are restricted to a specific subset of glycans. Moreover, the interaction of carbohydrates with lectins is competitive and reversible, so that stringent washing steps may simultaneously copurify also nonmodified peptides.
2. *Hydrazide Chemistry Enrichment*. Initially introduced for the detection of sugar moieties of glycoproteins on SDS-PAGE or on Western blot membranes by reacting oxidized glycoproteins with basic fuchsin [98], in 2003 hydrazide chemistry was used by Aebersold and coworkers for glycoprotein profiling using mass spectrometry [99]. The approach focused on the chemical derivatization of vicinal hydroxyl groups in glycans in order to couple glycosylated proteins to a hydrazine resin. Therefore, the vicinal hydroxyl groups of glycoproteins were initially oxidized by periodate treatment resulting in reactive aldehyde groups. Following this oxidation

**TABLE 2.1 Specific Lectin Carbohydrate Interactions**

| Lectin (Short Form) | Lectin | Origin | Glycan Motif |
|---|---|---|---|
| ConA | Concanavalin A | *Canavalia ensiformis* | α-Mannose |
| WGA | Wheatgerm agglutinin | *Triticum vulgaris* | β-GlcNAc |
| JCA | Jacalin | *Artocarpus integrifolia* | Galactosyl β-1,3-GalNAc |
| UEA1 | *Ulex europaeus* agglutinin 1 | *Ulex europaeus* | Glycosylated species with α-linked fucose |
| SNA | *Sambucus nigra* | *Sambucus nigra* | α-2,6-Bound sialic acid |

step, the applied aldehydes form hydrazones by reaction with hydrazide-derivatized agarose beads via a Schiff-base reaction. The covalent product is stable and allows for stringent washing steps preventing copurification. Elution was performed by PNGase F treatment, releasing only peptides formerly glycosylated. Subsequently, eluted peptides were identified by MS analysis.

3. *HILIC Enrichment*. Alternatively, glycosylated peptides can be enriched using HIILIC. Glycopeptide retention is based mainly on hydrophilic properties and further on the extent of the glycan moiety; therefore, glycopeptides elute in later fractions and can be separated from the non-glycosylated species as was reported in 2004 by Roepstorff and coworkers [100] on the analysis of sialylated glycopeptides. Orthogonal separation by RP-HPLC and HILIC revealed correlation to the numbers of sialylated residues. In current studies HILIC has been utilized for the analysis of glycosylation sites, but did not return information about glycan structure. HILIC was therefore used for desalting, enrichment, or separation of glycopeptides.

With respect to platelets, in 2006 Lewandrowski et al. used concanavalin A (ConA) for the identification of *N*-glycosylated peptides from whole-cell lysates [101]. Essentially the same study design used by Persson et al. was adopted to the platelet system [102]. Two-fold lectin affinity chromatography was performed in order to enrich a broad range of *N*-glycoproteins in the first step. Enriched glycoproteins were tryptically digested following elution from the lectin column. A second lectin affinity step was used to eliminate nonglycosylated peptides from the digest. Following deglycosylation by PNGase F, formerly glycosylated peptides were analyzed by nano-LC-MS/MS. In this study, 47 glycosylation sites on 32 proteins were identified after using this setup. The lectin-based approach was subsequently supplemented with a hydrazide chemistry step as described above. By combining the two enrichment techniques, a total of 70 individual glycosylation sites on 41 different glycoproteins were identified. Interestingly, there was poor overlap between the datasets identified with the applied methods, whereas the majority of sites were attributed to ConA treatment.

In contrast to the protein-based lectin affinity enrichment of glycoproteins, hydrazide chemistry shortens the stepwise procedure described above. Captured glycoproteins on the hydrazide solid support may be enzymatically cleaved, thereby retaining only glycopeptides on the column, while all nonmodified peptides are eliminated.

***2.4.2.4 Separation*** Differential 2DGE may be used to monitor the glycoprotein composition of whole-cell lysates. Glycoproteins often migrate in trains of spots, due to their heterogeneous glycosylation profiles. These trains reflect different isoelectric points and molecular masses. Using a non-gel-based approach, Lewandrowski et al. concentrated on sialylated *N*-glycosylation sites on platelet plasma membrane proteins [61]. A two-step workflow consisting of an aqueous two-phase partitioning system for membrane enrichment and subsequently a specific enrichment of glycopeptides using SCX, referred to as *enhanced N-glycosylation site analysis using strong cation exchange enrichment* (ENSAS), was employed. This technique removed the non-glycosylated background on the peptide level.

To further enrich for glycosylated plasma membrane-derived glycopeptides, ENSAS exploits the high frequency of sialylation on extracellular *N*-glycans introducing additional negative charges. Because of their reduced net charge compared to non-glycosylated tryptic peptides at pH 2.7, sialylated glycopeptides offer less affinity to SCX material and elute at early retention times. This assumption was confirmed by neuraminidase treatment, which removes terminal sialic acid residues. No significant enrichment was achieved when desialylated samples were separated via SCX. Using ENSAS, a total of 148 glycosylation sites distributed on 79 proteins were identified, including G-protein-coupled receptors such as PAR4, PI2R, and CCR4. However, it is limited to the enrichment of sialylated glycopeptides since glycopeptides with neutral glycan moieties do not separate from nonglycosylated peptides on SCX material. Applying ENSAS to glycosylated peptides, the elution of respective peptides is achieved at once and lacks separation. For this reason ERLIC developed for the analysis of phosphorylated species was adopted to glycosylation [103]. Continuing the use of aqueous two-phase partitioning system to initially enrich membrane proteins, ERLIC was used in 2008 by Lewandrowski et al. for the distinct subfractionation of glycopeptides. ERLIC is based on the interaction of negatively charged groups with the weak anion exchange matrix overlaid with hydrophilic effects due to solvent composition. In the respective study glycopeptides were successfully retained and eluted in later fractions, whereas nonglycosylated species were found in flowthrough and early fractions. Subsequent nano-LC-MS/MS analysis revealed 125 glycosylation sites on 66 distinct proteins.

Alternatively, glycosylated peptides can be separated by COFRADIC using common RP-HPLC [8,104]. The setup comprises a PNGase F treatment in between two identical HPLC runs, thereby producing a retention time shift of deglycosylated peptides. However, reflecting the removed glycan nature the shift can be either hydrophilic or hydrophobic.

***2.4.2.5 Mass Spectrometric Analysis*** For the analysis of glycoproteins and glycopeptides, the previously introduced soft ionization methods ESI and MALDI in conjunction with advanced mass spectrometry have been commonly employed. In the late 1980s, those methods displaced fast-atom bombardment (FAB) and led to a significantly better glycopeptide signal response. However, the analysis of complex oligosaccharides and the coupled peptide backbone remains a major challenge. Considering the level of detail, glycosylation analysis comprises the elucidation of both the glycosylation site and glycan structure. Thereby the investigation can be accomplished on the level of their constituents, namely, glycan and peptide, as well as on the level of the entire glycopeptide.

To date, the analysis of glycosylation in platelets has been concentrated on the annotation of *N*-glycosylation sites. In order to enrich for glycosylated species, several approaches based on the characteristic carbohydrate properties such as their interaction with lectins or the reduced positive charge of glycopeptides due to sialylation, were utilized. Nevertheless the involved glycan composition was not determined.

Because of the highly intricate fragmentation spectra that originate from both peptide and glycans residues, demand for strong expertise and detailed introduction is essential. The reader is referred to more detailed reviews in the literature [105–107].

## 2.5 QUANTIFICATION USING MASS SPECTROMETRY

Initially MS was widely used as a qualitative technique to elucidate the presence of proteins, and further PTMs. For instance, MS was used to identify the components of differential spots in 2DGE. More recently, improvements in mass accuracy, sensitivity, selectivity, and speed—under predefined conditions—allow the quantitative assessment of the respective analytes, such as changes in protein expression and changes in PTM patterns between different samples. Typical analytical setups for quantification comprise strategies whereby (1) signal intensities between differentially stable isotope-labeled samples or (2) precursor ions and pairs of precursor and fragment ions, respectively, are compared in so-called label-free or multireaction-monitoring-based approaches.

Several variations for quantification based on stable isotopes have been introduced, including isotope-coded affinity tag (ICAT), isotope-coded protein label (ICPL), and isobaric tag for relative and absolute quantitation (iTRAQ) as chemical labeling strategies; and stable isotope labeling of amino acids in cell culture (SILAC) as a metabolic labeling strategy. To date, these methods have been utilized predominantly for relative quantification but can also be utilized for absolute quantification on employing appropriate standards [108,109]. By introducing stable isotopes [110], such as $^{13}C$, $^{15}N$, and $^{18}O$, the physicochemical properties of heavy and light versions of the same peptide are identical, allowing a common separation on the level of HPLC and a simultaneous detection by MS. Also, since the ionization and detection properties of a given peptide in heavy and light forms with a defined mass shift are the same, comparing the signal intensities of the corresponding $m/z$ values allows for relative quantification. However, the incorporation of deuterium (D) as an isotopic label changes hydrophobicity, resulting in shifts in retention time during RP-LC and is therefore rarely used. Isotopic labeling strategies can be differentiated based on the level of MS quantification. In the case of ICPL and SILAC, this is done on the MS level since light and heavy forms of a given peptide have different masses and yield different precursor ion signals in survey scans. Isobaric labels such as iTRAQ generate a single precursor ion signal for up to eight differentially labeled versions of the same peptide, but are quantified on the MS/MS level on generation of specific reporter ions for each label. Although the use of isobaric labels does not multiply sample complexity on the precursor ion level, quantification on the MS level may result in more accurate quantification, based solely on the higher number of MS compared to MS/MS spectra, which can be used to quantify a single peptide.

Absolute quantification addressing the accurate amount of a defined set of analytes present in a sample has increasingly gained attention, especially with the introduction of the absolute quantification (AQUA) strategy [111]. AQUA is based on the addition of stable isotope-labeled synthetic peptides as internal standards, allowing for absolute quantification of the corresponding native counterparts.

In order to analyze single analytes of interest in complex mixtures, high-resolution and mass accuracy analyzers are highly recommended for both identification and accurate quantification. In general, quantification has to be based on unique peptides that can be assigned to only single proteins. As a result, the number of proteolytic peptides can be limited, especially when monitoring distinct protein isoforms [112]. Furthermore, peptides containing Cys or Met, which are prone to oxidation, should not be considered for quantification.

In principle, quantification of differential samples has to be based on a sufficient number of biological and also technical replicates, to ensure the significance of generated data. Moreover, applying alternative strategies to verify obtained results is desirable. Even when equipped with high-accuracy mass spectrometers, quantification may be limited at a very early and crucial timepoint. A mandatory prerequisite is a reproducible sample preparation that should be limited to a step-reduced workflow containing only essential handlings. Certainly protein digestion as well as desalting should be included in this setup. Even an experienced operator cannot testify on the efficiency of these processes without dedicated controls. Moreover, sample tubes, vials and/or pipette tips should have low protein binding character. Being a highly sensitive instrument quantification with a LC-MS, especially in cases of label-free and SRM quantification, should be carried out successively in order to avoid variations in HPLC separation or spray stability.

In order to analyze single analytes of interest in a complex mixture, high-resolution mass spectrometry is highly recommended for both identification as well as accurate quantification. Even with ppm mass accuracy, the simple comparison of peptide *m/z* ratios may still imply bias due to mass redundancies. Therefore, identification of analytes by MS/MS—also in conjunction with MS/MS quantification as in iTRAQ—is preferable to increase confidence.

Moreover, using stringent data interpretation may validate experimental results. Enclosing only proteotypic peptides that correspond to a single protein for quantitative analysis reduces misinterpretation of the datasets. However, primarily suited peptides may be subjected to missed cleavage or underlie modifications such as oxidation of methionine during sample preparation. As a consequence, respective peptides are excluded from quantification. In order to overcome significant variations, technical and biological replicates are indispensable.

In terms of platelets, stable isotopic labeling is restricted to techniques that chemically fix an isobaric tag to a specific amino acid on protein or peptide level,

ICAT [113], ICPL [114], as well as iTRAQ [115]. In contrast, metabolic labels that incorporate stable isotope-coded amino acids into the primary sequence of proteins, such as SILAC [116] or $^{15}N$ minimal cell media, are not compatible with platelets, which cannot be grown *ex vivo*. Because of their anucleate character and therefore low protein turnover, quantification in platelet research may focus on aspects of variability—for example, between different individuals, healthy versus diseased states—or on PTM-based signaling on stimulation. Monitoring of phosphorylation is of particular importance, as it is tightly connected to platelet activation as well as inhibition.

### 2.5.1 Methods Applied to Platelets

So far quantitative platelet proteome studies were accomplished in a relative way, analyzing differential proteomes from, for example, diseased versus healthy states or nonactivated versus activated platelets using 2DGE. Here, statistical evaluation of up/downregulation is conducted with the help of specific imaging software, while mass spectrometry is used only for qualitative assessment of protein identification within selected regulated spots. An introduction to the basic principles of 2DGE and its application to differential studies, with a primary focus on the analysis of signaling pathways, is covered by Chapter 4. Other examples of 2DGE applications are presented in Chapters 3, 13, and 14.

Several differential platelet proteome studies using 2DGE were published by García et al., O'Neill et al., and Marcus et al. [28,29,31–33,117]. However, to date only a single quantitative study has been conducted by Oehler et al. [118]. The group placed emphasis on the age-dependent variation of platelet protein expression. By factoring in a statistical confidence, the study assessed a total of 900 proteins and found effectively a one- to twofold change between blood donors. In their opinion, in order to detect respective marginal differences, future proteomic studies on platelets will require a highly sensitive setup.

Considering the limitations of 2DGE discussed above, non-gel-based differential studies are now expanding. Since this chapter focuses primarily on HPLC- and MS-based quantification approaches, we refer the interested reader to Chapter 4 and further literature dealing with the experimental and statistical needs of 2DGE-based quantification [119].

### 2.5.2 Quantitative MS Strategies Compatible with Platelets

Stable isotope labeling strategies are based on the incorporation of $^{13}C$, $^{15}N$, and $^{18}O$ and to a lesser extent D tags into proteins or peptides. In case of platelets, only strategies based on chemical attachment of isotopic labeled tags are applicable, and therefore *in vivo*/metabolic labeling is not discussed in this chapter.

In 1999, Gygi et al. introduced ICAT [113], based on a stable isotope label consisting of three functional units: (1) an attaching unit, (2) an isotope containing unit, and (3) an affinity tag. The thiol reactive group can be attached to reduced cysteine sidechains connecting these to the stable isotope coded ($H_8$

and $D_8$) ethylene glycol linker. In turn, the third component is a biotinyl group offering an affinity tag for selective isolation. Using ICAT, only peptides containing a cysteine residue can be identified. Resulting in a great reduction of complexity, this selection excludes the majority of peptides, and often quantification will be based only on a single peptide. This, however, is not recommended as it may lead to distorted conclusions. Another critical point is the large size of the ICAT moiety, which might result in steric hindrance. An alternative to the deuterium-containing label, which induces an alteration of peptide hydrophobicity and therefore a shift in LC separations, a $^{13}C$-based cleavable ICAT (cICAT) label was introduced [120]. In contrast to ICAT, iTRAQ [115] focuses on the entirety of peptides in the sample that are labeled by an isobaric tag. By NHS-ester chemistry, primary amines from *N* termini and ε-amino groups of lysines are labeled. Moreover, this global approach generally enables quantification of PTM peptides, which, in case of highly selective affinity tags as in ICAT, are most probably discarded. Initially introduced as 4-plex, another kit containing eight reagents was subsequently introduced. Generally, iTRAQ consists of a peptide reactive group, a balancer group, and a reporter group. In the following discussion we will refer to the originally introduced 4-plex containing four different and isobaric versions of the label, which then can be used to quantify four samples within a single experiment. Since the labels are isobaric, in contrast to ICAT, quantification is accomplished on the MS/MS level. The four isobaric and differentially labeled versions of a single peptide are detected as a single signal in the survey MS scan, but on fragmentation the isobaric tags dissociate, thus revealing the mass differences of the four different reporter ions. These reporter ions at 114, 115, 116, and 117 *m/z* can be differentiated with respect to incorporated stable isotopes. Differences in their masses are compensated for prior to fragmentation by the adapted stable isotope composition of the respective balancer groups.

Compared to other isotopic labeling strategies, iTRAQ is currently the only one that allows multiplexing of up to eight different samples, thus making it an economical and efficient strategy. However, reliable quantification demands sufficient MS/MS spectra of labeled peptides. In order to acquire the distinct reporter signals in the range of 114–117 Da or 113–121 Da in case of the 8-plex, a mass analyzer with a low mass cutoff and high resolution to enable accurate quantification is mandatory. Depending on the experimental setup iTRAQ implies not only relative but also absolute quantification. To conduct absolute quantification, one of the iTRAQ reagents can be used for labeling synthetic peptides, thus operating as an internal standard of the finally pooled samples. However, this strategy is limited by the fact that labeling efficiency of simple synthetic peptide mixes and complex samples may differ, and therefore absolute quantification may be erroneous.

Alternatively to iTRAQ, a reagent kit called *tandem mass tag* (TMT) based on a similar principle was introduced in 2003 allowing for multiplexing of six samples [121]. Reporter ion signals can be detected in the *m/z* range 126–131. Although quantification strategies based on the incorporation of stable isotopes and stable isotope tags are broadly used, there is increasing focus on alternative strategies for the quantification of peptides and proteins in proteome research, and these strategies are discussed below.

*Selective ion monitoring* (SRM), also referred to as *multireaction monitoring* (MRM), has the power to partially overcome the limitations of the abovementioned quantification strategies and allows for a straightforward and targeted quantification of selected analytes out of highly complex matrices. As previously discussed, SRM combines selectivity and sensitivity and covers a dynamic range of five orders of magnitude, making it suitable for accurate quantification of known proteins and peptides. In order to quantify a specific peptide, predefined transitions—to improve statistical significance at least two transitions per peptide—each comprising a pair of precursor ions and fragment ions, are continuously monitored in repeated cycles. To achieve the utmost sensitivity, transitions must be chosen carefully. The peptides in question must have good MS responses whereby, for unambiguous assignment, their unique character is required. Furthermore, the fragment ions must yield optimal signal intensities. SRM is compatible with both relative and absolute quantification. For quantification, signal intensities of distinct SRM transitions are integrated over time and are consecutively compared between various samples in a so-called label-free setup. Thereby variations in signal intensities depending on the LC-MS system or the sample background are possible, since, due to the lack of stable isotopes, multiplexing is not possible. Consequently, separation and analysis of successive samples have to be highly reproducible. For this reason, in label-free applications, normalization is based on the assumption that the overall protein/peptide concentration in a sample remains unchanged. To overcome variations in signal intensities, the addition of stable isotope-labeled peptides serve as useful tools, also enabling the use of the AQUA strategy.

Alternatively, changes in peptide abundance between different samples may also be addressed by label-free approaches based on LC-MS survey scans. Under these conditions, with the help of appropriate software tools such as OpenMS [122] for all present $m/z$ ratios, changes in signal intensities between different samples are determined on the MS level. Thus, without the use of fragmentation, potentially regulated precursor ions (or, more precisely, $m/z$ values) are determined and can be addressed and identified by following a targeted MS/MS analysis of the respective signals of interest. A major advantage of this approach is that, unlike SRM, it is not restricted to predefined peptides, and the analysis is therefore more global. However, it therefore has to cope with a high background of nonregulated signals, which, in contrast, may serve for normalization. Generally, the introduced quantification strategies may not only be used for addressing protein abundance/expression but may also be applied to analyzing changes in PTMs. The latter is particularly relevant to the investigation of platelet activation. A major issue for quantitative approaches, however, is data interpretation and reproducibility. For reliable results that allow us to draw meaningful and significant conclusions with biological impact, we must focus on (1) the reproducibility of the complete procedure, (2) reliable data interpretation, and consequently (3) conducting an appropriate number of biological as well as technical replicates, which is absolutely imperative.

## 2.6 CONCLUDING REMARKS

Comprehensive proteomic analyses of the platelet proteome in diverse states have presented novel insights into structural and functional components of platelets. Detailed knowledge of signaling pathways in these anucleate cells still remains obscure, thereby missing the essential interfaces linking microcosm and macrocosm. In order to close these gaps of knowledge, the ascending field of proteomics has established a great diversity of tools, which allow the elucidation of a wide dynamic range of proteins and moreover of quantitative profiles. Targeted proteomic approaches of both clinical protein patterns and sensitive changes in PTM on activation or inhibition provide important insights into protein abundance and also PTM stoichiometry and will help us to better understand functionally relevant processes in platelets on a molecular level.

## ACKNOWLEDGMENT

The continuous financial support of the Ministerium für Innovation, Wissenschaft, Forschung und Technologie des Landes Nordrhein-Westfalen, and the Bundesministerium für Bildung und Forschung is gratefully acknowledged.

## REFERENCES

1. O'Farrell PH, High resolution two-dimensional electrophoresis of proteins, *J. Biol. Chem.* **250**(10):4007–4021 (1975).
2. Klose J, Protein mapping by combined isoelectric focusing and electrophoresis of mouse tissues. A novel approach to testing for induced point mutations in mammals, *Humangenetik*. **26**(3):231–243 (1975).
3. Gevaert K, Goethals M, Martens L, Van Damme J, Staes A, Thomas GR, et al., Exploring proteomes and analyzing protein processing by mass spectrometric identification of sorted N-terminal peptides, *Nat. Biotechnol*. **21**(5):566–569 (2003).
4. Spahr CS, Davis MT, McGinley MD, Robinson JH, Bures EJ, Beierle J, et al., Towards defining the urinary proteome using liquid chromatography-tandem mass spectrometry. I. Profiling an unfractionated tryptic digest, *Proteomics* **1**(1):93–107 (2001).
5. Krijgsveld J, Gauci S, Dormeyer W, Heck AJ, In-gel isoelectric focusing of peptides as a tool for improved protein identification, *J. Proteome Res*. **5**(7):1721–1730 (2006).
6. Gevaert K, Van Damme J, Goethals M, Thomas GR, Hoorelbeke B, Demol H, et al., Chromatographic isolation of methionine-containing peptides for gel-free proteome analysis: Identification of more than 800 Escherichia coli proteins, *Mol. Cell. Proteom*. **1**(11):896–903 (2002).

7. Gevaert K, Ghesquiere B, Staes A, Martens L, Van Damme J, Thomas GR, et al., Reversible labeling of cysteine-containing peptides allows their specific chromatographic isolation for non-gel proteome studies, *Proteomics* **4**(4):897–908 (2004).
8. Ghesquiere B, Van Damme J, Martens L, Vandekerckhove J, Gevaert K, Proteome-wide characterization of N-glycosylation events by diagonal chromatography, *J. Proteome Res*. **5**(9):2438–2447 (2006).
9. Gevaert K, Staes A, Van Damme J, De Groot S, Hugelier K, Demol H, et al., Global phosphoproteome analysis on human HepG2 hepatocytes using reversed-phase diagonal LC, *Proteomics* **5**(14):3589–3599 (2005).
10. Karas M, Hillenkamp F, Laser desorption ionization of proteins with molecular masses exceeding 10,000 daltons, *Anal. Chem*. **60**(20):2299–2301 (1988).
11. Wilm M, Shevchenko A, Houthaeve T, Breit S, Schweigerer L, Fotsis T, et al., Femtomole sequencing of proteins from polyacrylamide gels by nano-electrospray mass spectrometry, *Nature* **379**(6564):466–469 (1996).
12. Takach EJ, Hines WM, Patterson DH, Juhasz P, Falick AM, Vestal ML, et al., Accurate mass measurements using MALDI-TOF with delayed extraction, *J. Protein Chem*. **16**(5):363–369 (1997).
13. Steen H, Kuster B, Fernandez M, Pandey A, Mann M, Detection of tyrosine phosphorylated peptides by precursor ion scanning quadrupole TOF mass spectrometry in positive ion mode, *Anal. Chem*. **73**(7):1440–1448 (2001).
14. Resing KA, Johnson RS, Walsh KA, Mass spectrometric analysis of 21 phosphorylation sites in the internal repeat of rat profilaggrin, precursor of an intermediate filament associated protein, *Biochemistry* **34**(29):9477–9487 (1995).
15. Lange V, Picotti P, Domon B, Aebersold R, Selected reaction monitoring for quantitative proteomics: A tutorial, *Mol. Syst. Biol*. **4**:222 (2008).
16. Olsen JV, de Godoy LM, Li G, Macek B, Mortensen P, Pesch R, et al., Parts per million mass accuracy on an Orbitrap mass spectrometer via lock mass injection into a C-trap, *Mol. Cell. Proteom*. **4**(12):2010–2021 (2005).
17. Hunt DF, Yates JR, 3rd, Shabanowitz J, Winston S, Hauer CR, Protein sequencing by tandem mass spectrometry, *Proc. Natl. Acad. Sci. USA* **83**(17):6233–6237 (1986).
18. Zubarev RA, Horn DM, Fridriksson EK, Kelleher NL, Kruger NA, Lewis MA, et al., Electron capture dissociation for structural characterization of multiply charged protein cations, *Anal. Chem*. **72**(3):563–573 (2000).
19. Syka JE, Coon JJ, Schroeder MJ, Shabanowitz J, Hunt DF, Peptide and protein sequence analysis by electron transfer dissociation mass spectrometry, *Proc. Natl. Acad. Sci. USA* **101**(26):9528–9533 (2004).
20. Kaufmann R, Chaurand P, Kirsch D, Spengler B, Post-source decay and delayed extraction in matrix-assisted laser desorption/ionization-reflectron time-of-flight mass spectrometry. Are there trade-offs? *Rapid Commun. Mass Spectrom*. **10**(10):1199–1208 (1996).
21. Roepstorff P, Fohlman J, Proposal for a common nomenclature for sequence ions in mass spectra of peptides, *Biomed. Mass Spectrom*. **11**(11):601 (1984).
22. Johnson RS, Martin SA, Biemann K, Stults JT, Watson JT, Novel fragmentation process of peptides by collision-induced decomposition in a tandem mass spectrometer: Differentiation of leucine and isoleucine, *Anal. Chem*. **59**(21):2621–2625 (1987).

23. Eng JK, McCormack AL, Yates IJR, An approach to correlate tandem mass spectral data of peptides with amino acid sequences in a protein database, *J. Am. Soc. Mass Spectrom*. **5**(11):976–989 (1996).
24. Perkins DN, Pappin DJ, Creasy DM, Cottrell JS, Probability-based protein identification by searching sequence databases using mass spectrometry data, *Electrophoresis* **20**(18):3551–3567 (1999).
25. Colinge J, Masselot A, Giron M, Dessingy T, Magnin J, OLAV: Towards high-throughput tandem mass spectrometry data identification, *Proteomics* **3**(8):1454–1463 (2003).
26. Geer LY, Markey SP, Kowalak JA, Wagner L, Xu M, Maynard DM, et al., Open mass spectrometry search algorithm, *J. Proteome Res*. **3**(5):958–964 (2004).
27. Kapp EA, Schutz F, Connolly LM, Chakel JA, Meza JE, Miller CA, et al., An evaluation, comparison, and accurate benchmarking of several publicly available MS/MS search algorithms: Sensitivity and specificity analysis, *Proteomics* **5**(13):3475–3490 (2005).
28. García A, Prabhakar S, Brock CJ, Pearce AC, Dwek RA, Watson SP, et al., Extensive analysis of the human platelet proteome by two-dimensional gel electrophoresis and mass spectrometry, *Proteomics* **4**(3):656–668 (2004).
29. García A, Prabhakar S, Hughan S, Anderson TW, Brock CJ, Pearce AC, et al., Differential proteome analysis of TRAP-activated platelets: Involvement of DOK-2 and phosphorylation of RGS proteins, *Blood* **103**(6):2088–2095 (2004).
30. Immler D, Gremm D, Kirsch D, Spengler B, Presek P, Meyer HE, Identification of phosphorylated proteins from thrombin-activated human platelets isolated by two-dimensional gel electrophoresis by electrospray ionization-tandem mass spectrometry (ESI-MS/MS) and liquid chromatography-electrospray ionization-mass spectrometry (LC-ESI-MS), *Electrophoresis* **19**(6):1015–1023 (1998).
31. Marcus K, Immler D, Sternberger J, Meyer HE, Identification of platelet proteins separated by two-dimensional gel electrophoresis and analyzed by matrix assisted laser desorption/ionization-time of flight-mass spectrometry and detection of tyrosine-phosphorylated proteins, *Electrophoresis* **21**(13):2622–2636 (2000).
32. Marcus K, Moebius J, Meyer HE, Differential analysis of phosphorylated proteins in resting and thrombin-stimulated human platelets, *Anal. Bioanal. Chem.* **376**(7):973–993 (2003).
33. O'Neill EE, Brock CJ, von Kriegsheim AF, Pearce AC, Dwek RA, Watson SP, et al., Towards complete analysis of the platelet proteome, *Proteomics* **2**(3):288–305 (2002).
34. Martens L, Van Damme P, Van Damme J, Staes A, Timmerman E, Ghesquiere B, et al., The human platelet proteome mapped by peptide-centric proteomics: A functional protein profile, *Proteomics* **5**(12):3193–3204 (2005).
35. Apweiler R, Hermjakob H, Sharon N, On the frequency of protein glycosylation, as deduced from analysis of the SWISS-PROT database, *Biochim. Biophys. Acta* **1473**(1):4–8 (1999).
36. Jensen ON, Modification-specific proteomics: Characterization of post-translational modifications by mass spectrometry, *Curr. Opin. Chem. Biol*. **8**(1):33–41 (2004).

37. García A, Senis YA, Antrobus R, Hughes CE, Dwek RA, Watson SP, et al., A global proteomics approach identifies novel phosphorylated signaling proteins in GPVI-activated platelets: Involvement of G6f, a novel platelet Grb2-binding membrane adapter, *Proteomics* **6**(19):5332–5343 (2006).
38. Wee JL, Jackson DE, Phosphotyrosine signaling in platelets: Lessons for vascular thrombosis, *Curr. Drug Targets* **7**(10):1265–1273 (2006).
39. Reinders J, Sickmann A, State-of-the-art in phosphoproteomics, *Proteomics* **5**(16):4052–4061 (2005).
40. Sickmann A, Meyer HE, Phosphoamino acid analysis, *Proteomics* **1**(2):200–206 (2001).
41. Gnatenko DV, Perrotta PL, Bahou WF, Proteomic approaches to dissect platelet function: Half the story, *Blood* **108**(13):3983–3991 (2006).
42. Maguire PB, Wynne KJ, Harney DF, O'Donoghue NM, Stephens G, Fitzgerald DJ, Identification of the phosphotyrosine proteome from thrombin activated platelets, *Proteomics* **2**(6):642–648 (2002).
43. Zahedi RP, Lewandrowski U, Wiesner J, Wortelkamp S, Moebius J, Schutz C, et al., Phosphoproteome of resting human platelets, *J. Proteome Res*. **7**(2):526–534 (2008).
44. Ballif BA, Villen J, Beausoleil SA, Schwartz D, Gygi SP, Phosphoproteomic analysis of the developing mouse brain, *Mol. Cell. Proteom*. **3**(11):1093–1101 (2004).
45. Ficarro SB, McCleland ML, Stukenberg PT, Burke DJ, Ross MM, Shabanowitz J, et al., Phosphoproteome analysis by mass spectrometry and its application to Saccharomyces cerevisiae, *Nat. Biotechnol*. **20**(3):301–305 (2002).
46. Thingholm TE, Jorgensen TJ, Jensen ON, Larsen MR, Highly selective enrichment of phosphorylated peptides using titanium dioxide, *Nat. Protoc*. **1**(4):1929–1935 (2006).
47. Kweon HK, Hakansson K, Selective zirconium dioxide-based enrichment of phosphorylated peptides for mass spectrometric analysis, *Anal Chem*. **78**(6):1743–1749 (2006).
48. Thingholm TE, Jensen ON, Robinson PJ, Larsen MR, SIMAC (sequential elution from IMAC), a phosphoproteomics strategy for the rapid separation of monophosphorylated from multiply phosphorylated peptides, *Mol. Cell. Proteom*. **7**(4):661–671.
49. McNulty DE, Annan RS, Hydrophilic interaction chromatography reduces the complexity of the phosphoproteome and improves global phosphopeptide isolation and detection, *Mol. Cell. Proteom*. **7**(5):971–980 (2008).
50. Alpert AJ, Electrostatic repulsion hydrophilic interaction chromatography for isocratic separation of charged solutes and selective isolation of phosphopeptides, *Anal. Chem*. **80**(1):62–76 (2008).
51. Zhou H, Watts JD, Aebersold R, A systematic approach to the analysis of protein phosphorylation, *Nat. Biotechnol*. **19**(4):375–378 (2001).
52. Steen H, Mann M. A new derivatization strategy for the analysis of phosphopeptides by precursor ion scanning in positive ion mode, *J. Am. Soc. Mass Spectrom*. **13**(8):996–1003 (2002).
53. Meyer HE, Hoffmann-Posorske E, Korte H, Heilmeyer LM Jr., Sequence analysis of phosphoserine-containing peptides. Modification for picomolar sensitivity, *FEBS Lett*. **204**(1):61–66 (1986).

54. Tao WA, Wollscheid B, O'Brien R, Eng JK, Li XJ, Bodenmiller B, et al., Quantitative phosphoproteome analysis using a dendrimer conjugation chemistry and tandem mass spectrometry, *Nat. Meth.* **2**(8):591–598 (2005).
55. Hunter T, The Croonian Lecture 1997. The phosphorylation of proteins on tyrosine: Its role in cell growth and disease, *Philos. Trans. R. Soc. Lond. B Biol. Sci.* **353**(1368):583–605 (1998).
56. Rush J, Moritz A, Lee KA, Guo A, Goss VL, Spek EJ, et al., Immunoaffinity profiling of tyrosine phosphorylation in cancer cells, *Nat. Biotechnol.* **23**(1):94–101 (2005).
57. Foy M, Harney DF, Wynne K, Maguire PB, Enrichment of phosphotyrosine proteome of human platelets by immunoprecipitation, *Meth. Mol. Biol.* **357**:313–318 (2007).
58. Zheng H, Hu P, Quinn DF, Wang YK, Phosphotyrosine proteomic study of interferon alpha signaling pathway using a combination of immunoprecipitation and immobilized metal affinity chromatography, *Mol. Cell. Proteom.* **4**(6):721–730 (2005).
59. Ficarro S, Chertihin O, Westbrook VA, White F, Jayes F, Kalab P, et al., Phosphoproteome analysis of capacitated human sperm. Evidence of tyrosine phosphorylation of a kinase-anchoring protein 3 and valosin-containing protein/p97 during capacitation, *J. Biol. Chem.* **278**(13):11579–1189 (2003).
60. Beausoleil SA, Jedrychowski M, Schwartz D, Elias JE, Villen J, Li J, et al., Large-scale characterization of HeLa cell nuclear phosphoproteins, *Proc Natl Acad Sci USA* **101**(33):12130–12135 (2004).
61. Lewandrowski U, Zahedi RP, Moebius J, Walter U, Sickmann A, Enhanced N-glycosylation site analysis of sialoglycopeptides by strong cation exchange prefractionation applied to platelet plasma membranes, *Mol. Cell. Proteom.* **6**(11):1933–1941 (2007).
62. Porath J, Carlsson J, Olsson I, Belfrage G, Metal chelate affinity chromatography, a new approach to protein fractionation, *Nature* **258**(5536):598–599 (1975).
63. Posewitz MC, Tempst P, Immobilized gallium(III) affinity chromatography of phosphopeptides, *Anal. Chem.* **71**(14):2883–2892 (1999).
64. Thingholm TE, Jensen ON, Larsen MR, Analytical strategies for phosphoproteomics, *Proteomics* **9**(6):1451–1468 (2009).
65. Ikeguchi Y, Nakamura H, Selective enrichment of phospholipids by titania, *Anal. Sci.* **16**(5):541–543 (2000).
66. Pinkse MW, Uitto PM, Hilhorst MJ, Ooms B, Heck AJ, Selective isolation at the femtomole level of phosphopeptides from proteolytic digests using 2D-NanoLC-ESI-MS/MS and titanium oxide precolumns, *Anal. Chem.* **76**(14):3935–3943 (2004).
67. Hofer R, Textor M, Spencer ND, Alkyl phosphate monolayers, self assembled from aqueous solution onto metal oxide surfaces, *Langmuir* **17**(13):4014–4020 (2001).
68. Larsen MR, Thingholm TE, Jensen ON, Roepstorff P, Jorgensen TJ, Highly selective enrichment of phosphorylated peptides from peptide mixtures using titanium dioxide microcolumns, *Mol. Cell. Proteom.* **4**(7):873–886 (2005).
69. Alpert AJ, Hydrophilic-interaction chromatography for the separation of peptides, nucleic acids and other polar compounds, *J. Chromatogr.* (499):177–196 (1990).
70. Elias JE, Gygi SP, Target-decoy search strategy for increased confidence in large-scale protein identifications by mass spectrometry, *Nat. Meth.* **4**(3):207–214 (2007).

71. Olsen JV, Mann M, Improved peptide identification in proteomics by two consecutive stages of mass spectrometric fragmentation, *Proc. Natl. Acad. Sci. USA* **101**(37):13417–13422 (2004).
72. Beausoleil SA, Villen J, Gerber SA, Rush J, Gygi SP, A probability-based approach for high-throughput protein phosphorylation analysis and site localization, *Nat. Biotechnol*. **24**(10):1285–1292 (2006).
73. Kaufmann R, Spengler B, Lutzenkirchen F, Mass spectrometric sequencing of linear peptides by product-ion analysis in a reflectron time-of-flight mass spectrometer using matrix-assisted laser desorption ionization, *Rapid Commun. Mass Spectrom*. **7**(10):902–910 (1993).
74. Kjellstrom S, Jensen ON, Phosphoric acid as a matrix additive for MALDI MS analysis of phosphopeptides and phosphoproteins, *Anal. Chem*. **76**(17):5109–5117 (2004).
75. Martin DB, Eng JK, Nesvizhskii AI, Gemmill A, Aebersold R, Investigation of neutral loss during collision-induced dissociation of peptide ions, *Anal. Chem*. **77**(15):4870–4882 (2005).
76. Carr SA, Huddleston MJ, Annan RS, Selective detection and sequencing of phosphopeptides at the femtomole level by mass spectrometry, *Anal. Biochem*. **239**(2):180–192 (1996).
77. Rappsilber J, Steen H, Mann M, Labile sulfogroup allows differentiation of sulfotyrosine and phosphotyrosine in peptides, *J. Mass Spectrom*. **36**(7):832–833 (2001).
78. Medzihradszky KF, Darula Z, Perlson E, Fainzilber M, Chalkley RJ, Ball H, et al., O-sulfonation of serine and threonine: Mass spectrometric detection and characterization of a new posttranslational modification in diverse proteins throughout the eukaryotes, *Mol. Cell. Proteom*. **3**(5):429–440 (2004).
79. Williamson BL, Marchese J, Morrice NA, Automated identification and quantification of protein phosphorylation sites by LC/MS on a hybrid triple quadrupole linear ion trap mass spectrometer, *Mol. Cell. Proteom*. **5**(2):337–346 (2006).
80. Varga-Szabo D, Pleines I, Nieswandt B, Cell adhesion mechanisms in platelets, *Arterioscler. Thromb. Vasc. Biol*. **28**(3):403–412 (2008).
81. Spiro RG, Protein glycosylation: Nature, distribution, enzymatic formation, and disease implications of glycopeptide bonds, *Glycobiology* **12**(4):43R–56R (2002).
82. Hounsell EF, Davies MJ, Renouf DV, O-linked protein glycosylation structure and function, *Glycoconj. J*. **13**(1):19–26 (1996).
83. Marquardt T, Denecke J, Congenital disorders of glycosylation: Review of their molecular bases, clinical presentations and specific therapies, *Eur. J. Pediatr*. **162**(6):359–379 (2003).
84. Nurden P, Nurden AT, Congenital disorders associated with platelet dysfunctions, *Thromb. Haemost*. **99**(2):253–263 (2008).
85. Hoffmeister KM, Felbinger TW, Falet H, Denis CV, Bergmeier W, Mayadas TN, et al., The clearance mechanism of chilled blood platelets, *Cell* **112**(1):87–97 (2003).
86. Hoffmeister KM, Josefsson EC, Isaac NA, Clausen H, Hartwig JH, Stossel TP, Glycosylation restores survival of chilled blood platelets, *Science* **301**(5639):1531–1534 (2003).

87. Wandall HH, Hoffmeister KM, Sorensen AL, Rumjantseva V, Clausen H, Hartwig JH, et al., Galactosylation does not prevent the rapid clearance of long-term, 4 degrees C-stored platelets, *Blood* **111**(6):3249–3256 (2008).
88. Tarentino AL, Gomez CM, Plummer TH Jr., Deglycosylation of asparagine-linked glycans by peptide:N-glycosidase F, *Biochemistry* **24**(17):4665–4671 (1985).
89. Sickmann A, Mreyen M, Meyer HE, Mass spectrometry—a key technology in proteome research, *Adv. Biochem. Eng. Biotechnol.* **83**:141–176 (2003).
90. Yang Z, Hancock WS, Approach to the comprehensive analysis of glycoproteins isolated from human serum using a multi-lectin affinity column, *J. Chromatogr. A* **1053**(1–2):79–88 (2004).
91. Cummings RD, Kornfeld S, The distribution of repeating [Gal beta 1,4GlcNAc beta 1,3] sequences in asparagine-linked oligosaccharides of the mouse lymphoma cell lines BW5147 and PHAR 2.1, *J. Biol. Chem.* **259**(10):6253–6660 (1984).
92. Moebius J, Zahedi RP, Lewandrowski U, Berger C, Walter U, Sickmann A, The human platelet membrane proteome reveals several new potential membrane proteins, *Mol. Cell. Proteom.* **4**(11):1754–1761 (2005).
93. García BA, Smalley DM, Cho H, Shabanowitz J, Ley K, Hunt DF, The platelet microparticle proteome, *J. Proteome Res.* **4**(5):1516–1521 (2005).
94. Senis YA, Tomlinson MG, García A, Dumon S, Heath VL, Herbert J, et al., A comprehensive proteomics and genomics analysis reveals novel transmembrane proteins in human platelets and mouse megakaryocytes including G6b-B, a novel immunoreceptor tyrosine-based inhibitory motif protein, *Mol. Cell. Proteom.* **6**(3):548–564 (2007).
95. Maynard DM, Heijnen HF, Horne MK, White JG, Gahl WA, Proteomic analysis of platelet alpha-granules using mass spectrometry, *J. Thromb. Haemost.* **5**(9):1945–1955 (2007).
96. Hernandez-Ruiz L, Valverde F, Jimenez-Nunez MD, Ocana E, Saez-Benito A, Rodriguez-Martorell J, et al., Organellar proteomics of human platelet dense granules reveals that 14-3-3zeta is a granule protein related to atherosclerosis, *J. Proteome Res.* **6**(11):4449–4457 (2007).
97. Lewandrowski U, Wortelkamp S, Lohrig K, Zahedi RP, Wolters DA, Walter U, et al., Platelet membrane proteomics: A novel repository for functional research, *Blood* **14**(1):e10–9 (2009).
98. Eckhardt AE, Hayes CE, Goldstein IJ, A sensitive fluorescent method for the detection of glycoproteins in polyacrylamide gels, *Anal. Biochem.* **73**(1):192–197 (1976).
99. Zhang H, Li XJ, Martin DB, Aebersold R, Identification and quantification of N-linked glycoproteins using hydrazide chemistry, stable isotope labeling and mass spectrometry, *Nat. Biotechnol.* **21**(6):660–666 (2003).
100. Hagglund P, Bunkenborg J, Elortza F, Jensen ON, Roepstorff P, A new strategy for identification of N-glycosylated proteins and unambiguous assignment of their glycosylation sites using HILIC enrichment and partial deglycosylation, *J. Proteome Res.* **3**(3):556–566 (2004).
101. Lewandrowski U, Moebius J, Walter U, Sickmann A, Elucidation of N-glycosylation sites on human platelet proteins: A glycoproteomic approach, *Mol. Cell. Proteom.* **5**(2):226–233 (2006).
102. Persson A, Jergil B, The purification of membranes by affinity partitioning, *FASEB J.* **9**(13):1304–1310 (1995).

103. Lewandrowski U, Lohrig K, Zahedi RP, Wolters D, Sickmann A, Glycosylation site analysis of human platelets by electrostatic repulsion hydrophilic interaction chromatography, *Clin. Proteom.* **4**:25–36 (2008).

104. Gevaert K, Van Damme P, Ghesquiere B, Vandekerckhove J, Protein processing and other modifications analyzed by diagonal peptide chromatography, *Biochim. Biophys. Acta* **1764**(12):1801–1810 (2006).

105. Budnik BA, Lee RS, Steen JA, Global methods for protein glycosylation analysis by mass spectrometry, *Biochim. Biophys. Acta* **1764**(12):1870–1880 (2006).

106. Harvey DJ, Proteomic analysis of glycosylation: structural determination of N- and O-linked glycans by mass spectrometry, *Expert Rev. Proteom.* **2**(1):87–101 (2005).

107. Harvey DJ, Royle L, Radcliffe CM, Rudd PM, Dwek RA, Structural and quantitative analysis of N-linked glycans by matrix-assisted laser desorption ionization and negative ion nanospray mass spectrometry, *Anal. Biochem.* **376**(1):44–60 (2008).

108. Putz S, Reinders J, Reinders Y, Sickmann A, Mass spectrometry-based peptide quantification: Applications and limitations, *Expert Rev. Proteom.* **2**(3):381–392 (2005).

109. Reinders J, Lewandrowski U, Moebius J, Wagner Y, Sickmann A, Challenges in mass spectrometry-based proteomics, *Proteomics* **4**(12):3686–3703 (2004).

110. Gevaert K, Impens F, Ghesquiere B, Van Damme P, Lambrechts A, Vandekerckhove J, Stable isotopic labeling in proteomics, *Proteomics* **8**(23–24):4873–4885 (2008).

111. Gerber SA, Kettenbach AN, Rush J, Gygi SP, The absolute quantification strategy: Application to phosphorylation profiling of human separase serine 1126, *Meth. Mol. Biol.* **359**:71–86 (2007).

112. Mallick P, Schirle M, Chen SS, Flory MR, Lee H, Martin D, et al., Computational prediction of proteotypic peptides for quantitative proteomics, *Nat. Biotechnol.* **25**(1):125–131 (2007).

113. Gygi SP, Rist B, Gerber SA, Turecek F, Gelb MH, Aebersold R, Quantitative analysis of complex protein mixtures using isotope-coded affinity tags, *Nat. Biotechnol.* **17**(10):994–999 (1999).

114. Schmidt A, Kellermann J, Lottspeich F, A novel strategy for quantitative proteomics using isotope-coded protein labels, *Proteomics* **5**(1):4–15 (2005).

115. Ross PL, Huang YN, Marchese JN, Williamson B, Parker K, Hattan S, et al., Multiplexed protein quantitation in Saccharomyces cerevisiae using amine-reactive isobaric tagging reagents, *Mol. Cell. Proteom.* **3**(12):1154–1169 (2004).

116. Ong SE, Blagoev B, Kratchmarova I, Kristensen DB, Steen H, Pandey A, et al., Stable isotope labeling by amino acids in cell culture, SILAC, as a simple and accurate approach to expression proteomics, *Mol. Cell. Proteom.* **1**(5):376–386 (2002).

117. García A, Two-dimensional gel electrophoresis in platelet proteomics research, *Meth. Mol. Med.* **139**:339–353 (2007).

118. Winkler W, Zellner M, Diestinger M, Babeluk R, Marchetti M, Goll A, Oehler R, et al., Biological variation of the platelet proteome in the elderly population and its implication for biomarker research, *Mol. Cell. Proteom.* **7**(1):193–203 (2008).

119. Gorg A, Weiss W, Dunn MJ, Current two-dimensional electrophoresis technology for proteomics, *Proteomics* **4**(12):3665–3685 (2004).

120. Fauq AH, Kache R, Khan MA, Vega IE, Synthesis of acid-cleavable light isotope-coded affinity tags (ICAT-L) for potential use in proteomic expression profiling analysis, *Bioconj. Chem.* **17**(1):248–254 (2006).

121. Thompson A, Schafer J, Kuhn K, Kienle S, Schwarz J, Schmidt G, et al., Tandem mass tags: A novel quantification strategy for comparative analysis of complex protein mixtures by MS/MS, *Anal. Chem*. **75**(8):1895–1904 (2003).

122. Sturm M, Bertsch A, Gropl C, Hildebrandt A, Hussong R, Lange E, et al., OpenMS—an open-source software framework for mass spectrometry, *BMC Bioinformat*. **9**:163 (2008).

# 3

# SAMPLE PREPARATION VARIABLES IN PLATELET PROTEOMICS FOR BIOMARKER RESEARCH

MARIA ZELLNER AND RUDOLF OEHLER

**Abstract**

Platelets are maintained in an inactive state in the peripheral circulation and react rapidly in response to vascular injury. Removing platelets from their physiological environment can also result in platelet activation from a number of stimuli, including agonists generated or released during blood collection, mechanical forces exerted on platelets during blood withdrawal, centrifugation, and platelet washing. Besides maintaining platelets in an inactive state during separation and purification, contaminating plasma proteins and other blood cells must be minimized to avoid spurious proteomics results. The present chapter describes steps involved in collecting and preparing platelets to limit platelet activation and/or contamination and proteome variability resulting from poor sample preparation.

## 3.1 INTRODUCTION

Biomarker research aims at defining a molecular characteristic of a tissue that is indicative for normal biologic processes, pathogenic processes, or pharmacologic responses to a therapeutic intervention. Platelets are an interesting subject for biomarker research as they are involved in a variety of pathophysiological processes. They are essential for primary haemostasis and repair of the

*Platelet Proteomics: Principles, Analysis and Applications*, First Edition.
Edited by Ángel García and Yotis A. Senis.

endothelium following injury. There is also a growing body of evidence that platelets play critical roles in inflammation and immunity, angiogenesis, and metastasis. However, they also play a central role in the development of acute coronary syndromes and contribute to cerebrovascular events. In addition, platelets participate in the process of forming and extending atherosclerotic plaques. Therefore, they are targeted by various classes of antiplatelet drugs. Of particular interest in the field of haemostasis and thrombosis is the lack of a sensitive and specific test for platelet activation that is suitable for routine analysis. Until now, platelet activation is determined by quantifying activation markers on the cell surface using flow cytometry (e.g., P-selectin expression) or by platelet aggregation tests. Such assays can be made only in highly specialized laboratories with the appropriate specialist equipment.

Highly purified platelets are easily accessible from peripheral blood for functional and biochemical analysis. Their lack of a nucleus and only a limited set of mRNAs make them less complex than most other cell types to study. Because of their unique design, the majority of regulatory steps in platelets occur at the level of post-translational modification (PTM). Proteomics is therefore regarded as an ideal method for biomarker research in platelets. PTM-mediated regulation has the advantage that platelets can react quickly to exogenous stimuli, including shear stress [1]. This is, of course, advantageous for the body in case of mechanical injury, where a rapid response prevents excessive blood loss. However, the highly sensitive nature of platelets also renders them hyperresponsive to a number of stimuli in the course of blood collection and preparation. Platelet activation is associated with distinct features of PTMs in the platelet proteome, including protein cleavage and phosphorylation [2,3]. It is therefore important to take all necessary precautions to prevent preparation-induced activation for platelet proteomics studies. This is especially critical in biomarker research, where platelets are prepared from a number of patients and controls. Unwanted platelet activation during preparation could prevail over any disease-related activation of a platelet subpopulation. To ensure that a proteome analysis mirrors the physiological situation in the circulation, the platelet preparation must comply with strict quality control standards. There is ample literature regarding platelet preparation for functional testing of platelets (reviewed in Refs. 4 and 5). The main aim of these protocols is to isolate quiescent platelets, that is, to avoid any preactivation of platelets without irreversibly blocking platelet function. Removal of contaminating blood components (e.g., plasma proteins, other blood cells) is critical for obtaining meaningful proteomics data. Such contamination normally does not interfere with functional tests, but can be easily detected by highly sensitive mass spectrometers. Although caution must be taken to obtain highly pure, resting platelets, it should be kept in mind that mass spectrometers can detect vanishingly low levels of proteins, and a low level of contamination is unavoidable. Therefore, one should not become overly preoccupied with completely eliminating any contaminating proteins or cells, as this is impossible to achieve. Further, potentially "interesting" hits should always be validated by alternative means to confirm platelet expression. This chapter provides an overview of platelet preparation with special emphasis on key aspects of proteome analysis. All preparative steps will be discussed in detail, from patient selection and blood collection to separation of platelets from contaminating plasma proteins and blood cells. Advantages and disadvantages will also be discussed.

## 3.2 SELECTION OF TEST SUBJECTS AND BLOOD COLLECTION

Platelets are influenced by many diseases as well as by different physiological conditions such as age, exercise, nutritional status, and mental stress, as well as the time of day. For example, platelets of elderly individuals display an increase in lipid peroxides and nitric oxide synthase (NOS) activity and a decrease in basal cGMP content, and they have a reduced synthesis of nitric oxide (NO) as well as an impaired response to NO [6]. Platelets are affected not only by thrombotic diseases but also by psychiatric diseases, such as depression or neurodegenerative diseases, such as Alzheimer's disease [7]. Attention should therefore focus on any medications taken by patients, particularly antiplatelet drugs, such as aspirin, clopidogrel, and anti-thrombotic treatments, as they can clearly influence platelet function. Other medications, including antiinflammatory or antidepressive drugs, can also influence platelet reactivity [8].

Platelet reactivity is strongly influenced by the redox status of the platelet. Many nutritional studies demonstrate that supplementing food with antioxidative vitamins E and C significantly reduce platelet activation [9,10]. Polyphenolic anti-oxidants, such as resveratrol, which is present in red wine [11]; lycopene, in tomatoes [12]; or flavanols, in chocolate [13], have an inhibitory influence on platelet reactivity. Regarding the influence of physical stress on platelet function, volunteers arriving at the study center should rest in a supine position approximately 30 min prior to blood sampling. It is also advisable that blood be collected at the same time each day, as circadian rhythms have been reported to alter the aggregability and adhesiveness of platelets [14]. The hypercoagulability of platelets during the morning hours is thought to lead to the peak incidence of myocardial infarction, cerebral infarct, and sudden cardiac death at this time. Most pronounced platelet aggregation was observed at around 11:00 h. Flow cytometric analysis of P-selectin indicate that there is an increase in platelet activation during the period from 06:00 to 09:00 A.M. Platelet activation then decreases gradually during the period from noon to midnight. These changes are accompanied by a similar trend in circulating platelet aggregates [15]. These data indicate that patients as well as the control collective should be selected carefully. However, depending on the patient samples being sought, the researcher may have little control over when a patient is available for sampling.

## 3.3 BLOOD SAMPLING

For platelet proteomics a venipuncture procedure must be used that minimizes spontaneous platelet activation. In addition, blood must be collected into a medium that will not only prevent coagulation but will also preserve the activation status of platelets until the samples can be analyzed. The act of drawing blood can induce platelet activation. Location of an appropriate vein is an essential requisite to ensure successful drawing and quality specimens. The antecubital area is the best suited for phlebotomy, and the vein of choice for venous blood drawing should be the median cubital vein. Veins of the legs, ankles, and feet should preferably not be accessed, as blood in the inferior limb veins can undergo changes in coagulability due to contact with atherosclerotic plaques in arteries [16]. Tourniquets are commonly used prior to routine venipuncture to assist the phlebotomist in locating a suitable vein. It is recommended that tourniquet placement be controlled and light (60 Torr for $< 45$ s) or even that no tourniquet be used and a 21-gauge (or larger) needle be used to puncture the vein. The first 2 mL of blood drawn should be discarded. Such precautions minimize tissue thromboplastin release and red cell hemolysis, both of which can lead to platelet activation.

The blood collection tube may also influence platelet function and proteomics results. It has been shown that spontaneous platelet aggregation is influenced by the way in which blood is drawn into the sampling device. Spontaneous platelet aggregation was significantly lower in samples obtained with a syringe compared to those obtained with Vacutainers, which are commonly used in clinical routine [17]. The clotting times of blood collected into Vacutainer or Vacuette plastic tubes decreases with time, with maximal responses observed 30 min postcollection. In contrast, blood from syringe blood sampling devices displays longer clotting times, which does not decrease with time [18]. The major problem encountered when using small-bore needles as well as evacuated tubes is the large shear force applied to the blood, which can cause hemolysis (i.e., erythrocyte lysis) and increase coagulant activity [16]. Nevertheless, although syringe systems and large needle bore size are preferred over Vacutainer tubes, some clinical samples can be realistically obtained only in the latter. In those cases, the control samples should be collected in the same way. Drawn blood should be mixed gently immediately after drawn into the blood tube containing the appropiate anticoagulant. Blood samples collected into sodium citrate (3.8%) should be processed as soon as possible following collection (ideally within one hour). Blood samples should be stored at room temperature (20–25°C) during this time and not on ice (4°C), as this can induce platelet activation [19]. The storage time can be extended with other anticoagulants, such as a mixture of citrate, theophylline, adenosine, and dipyridamole (CTAD), to 4 h.

### 3.3.1 Anticoagulants

A number of different anticoagulants have been used to measure platelet activation *in vitro*. Sodium citrate (regular concentration 3.8%) is the most frequently used anti-coagulant for platelet analysis, but has limitations due to the difficulty in controlling osmolarity in functional assays [20]. When blood is collected into citrate, there is initially little or no change in platelet shape and volume. However, platelets slowly adopt a spherical shape in citrate and swell progressively over a period of 1–2 h (3–10% increase in volume), which also occurs in the presence of another commonly used anticoagulant, sodium EDTA [21]. For these reasons, citrate is considered unreliable for the measurement of mean platelet volume (MPV). Acid citrate dextrose (ACD) or a combination of ACD and sodium EDTA has been recommended instead, because the MPV values obtained are stable over time [22]. Sodium EDTA is the anti-coagulant recommended by the National Committee for Clinical Laboratory Standards for full blood cell counts and white blood cell differential analysis. Sodium EDTA prevents binding of natural ligands, such as fibrinogen to the integrin $\alpha$IIb$\beta$3 and dissociates the $\alpha$IIb$\beta$3 complex [23]. However, sodium EDTA leads to considerable activation of platelets *in vitro* (Fig. 3.1). Heparin is not used because it activates platelets and affects the staining properties of cells [24]. In our hands, CTAD was the best antiplatelet mixture for preventing platelet activation *in vitro* during the storage of tubes at room temperature for 30 min before analysis (unpublished results). CTAD blood showed a much lower activation-induced platelet factor 4 (PF4) secretion than did EDTA blood or citrate blood (Fig. 3.1). Theophylline and dipyridamole present in CTAD inhibit cAMP phosphodiesterase activity, resulting in elevated intracellular levels of cAMP. Adenosine stimulates membrane adenylyl cyclase, which also elevates intracellular cAMP levels. The resulting increase in platelet cAMP and the inhibition of $Ca^{2+}$-mediated responses lead to a reduction in

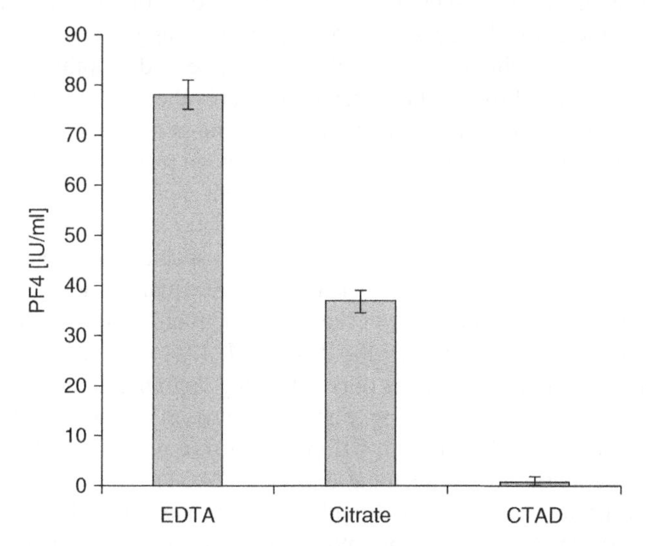


**Figure 3.1** Influence of anticoagulants on platelet activation. Whole blood was taken from the same healthy donors and collected into tubes containing either EDTA, citrate, or CTAD (a mixture of citrate, theophylline, adenosine, and dipyridamole). Samples were centrifuged at 1000*g* for 10 min. The resulting supernatant was centrifuged at 10,000*g* for 10 min to yield platelet-poor plasma (PPP). To monitor platelet activation, the secretion of platelet factor 4 (PF4) was measured in the PPP using a specific ELISA. The graph shows the result of three independent experiments.

platelet activation. Dipyridamole has the disadvantage of being light-sensitive. CTAD anticoagulant tubes should therefore be stored appropriately. Flow cytometric studies of platelet surface antigens have shown that whole-blood samples anticoagulated with CTAD are stable at 20°C under standard laboratory conditions for up to 4 h following blood withdrawal [25].

### 3.3.2 Commonly Used Parameters of Platelet Activation

To control preparation-induced proteome changes, it is highly advisable to monitor the platelet activation status throughout the platelet preparation procedure, at least for initial samples collected as part of any study. This can be done by measuring changes in platelet surface receptor expression, intracellular calcium concentration, cell shape, and other biochemical and functional responses. Early platelet activation induces cytoskeletal reorganization that results in characteristic shape change; platelets lose their discoid shape, adopting a more spherical shape, and begin to develop thin filopodial extensions. At a more advanced stage, they become spiny spheres completely covered with filopodia. Finally, centralization of specific granules and dilation of the open canalicular system occur, as a result of the secretion process.

There are numerous methods for measuring platelet activation that provide valuable information for research or small patient studies. However, none of these tests are currently used to diagnose risk of thrombosis or to guide antiplatelet therapy in routine clinical practice. Light transmission aggregometry is the most commonly used test, which measures the initial rate and amplitude of platelet aggregate formation. Thromboelastography measures the rate and force of blood clot formation. However, these techniques are time-consuming, are not highly standardized and reveal only a very global overview on platelet activation. In contrast, flow cytometry can be used to monitor different aspects of platelet activation and platelet interaction with other cells. Flow cytometry has the added advantages of being specific and sensitive. However, it requires specialist equipment and can be time-consuming and costly. A classical marker for platelet activation by flow cytometry is the surface expression of the α-granule membrane protein P-selectin (also referred to as CD62P, GMP140, PADGEM) [24]. P-selectin is expressed on the surface of activated platelets only following degranulation. Another flow-cytometry-based test for measuring platelet activation involves detecting the active conformation of the integrin αIIbβ3 (also referred to as GPIIb/IIIa and CD41/61) using a specific monoclonal antibody (PAC1). The integrin αIIbβ3 is a receptor for fibrinogen, von Willebrand factor, fibronectin, and vitronectin that is essential for platelet aggregation. Whereas most antibodies directed against the αIIbβ3 complex bind to resting platelets, the monoclonal antibody PACl is directed against the fibrinogen binding site exposed by a conformational change in the integrin αIIbβ3 following platelet activation. PAC1 binds only to activated platelets, not to resting platelets. Fixation with 1% paraformaldehyde before antibody staining is a common preparation step in flow cytometry to prevent artificial activation of the platelets [24]. The advantage of using P-selectin or active αIIbβ3 as markers of activated platelets is that the antibodies are highly specific, the staining protocols and flow cytometry analysis protocols are well established, and the assay is sensitive and accurate.

## 3.4 PLATELET PURIFICATION

The first step in platelet isolation from the whole-blood sample is the separation from other blood cells. This can be most easily done according to their density. Platelets have the lowest density of all cellular particles in the blood (platelet density 1.04–1.06 g/mL), followed by white blood cells (WBC density 1.06–1.09 g/mL), and finally by red blood cells (RBC density 1.09–1.10 g/mL). Mild centrifugation ($< 200g$) of whole blood results in a large RBC pellet accounting for 50–80% of the total volume of the blood in the centrifugation tube. The large interindividual variability in the RBC pellet can be at least partially explained by the sex of the donor, with males having more RBCs than females. The WBCs are visible as a white "buffy coat" at the interface between the RBC pellet and the supernatant on top, termed *platelet-rich plasma* (PRP). The platelets remain in suspension in the PRP, which is straw-colored. The platelet counts in the PRP are normally $3.5–5 \times 10^5\ \mu L^{-1}$, which accounts for the turbid, cloudy appearance of the PRP. The various studies on platelet proteome differ considerably regarding the centrifugal force applied in this first centrifugation step. Our research group centrifuges the anticoagulated whole blood at $50g$ for 20 min [26]. Such a mild centrifugation step minimizes the danger of unwanted platelet activation. However, the WBC layer is less compact and can diffuse slightly into the PRP. As a result, only the top three-fourths (75%) of the PRP is taken for further investigation to prevent contamination with WBC. To avoid the risk of WBC contamination in the PRP (i.e., to obtain a sharper margin between PRP and the cell pellet), many researchers apply higher centrifugal forces. Commonly used centrifugation conditions include $110g$ for 15 min [27], $120g$ for 15 min [28], $150g$ for 10 min [3], $200g$ for 10 min [29], $200g$ for 20 min [30,31], and $300g$ for 20 min [32]. The centrifugation is always performed at room temperature, because cooling induces an additional stress at platelets resulting in activation [33]. Bugert's group investigated the recovery of such a centrifugation in detail [34]. They centrifuged 3 mL citrated whole blood at $150g$ for 20 min. The sample contained $(7.6 \pm 1.3) \times 10^8$ platelets and $(1.4 \pm 0.2) \times 10^7$ WBCs prior to centrifugation. Postcentrifugation, the upper nine-tenths (90%) of the PRP contained $(2.8 \pm 0.5) \times 10^8$ platelets and 215–490 WBCs. This corresponded to a platelet recovery of $36.8\% \pm 3.9\%$ and contamination of 1 WBC $10^{-6}$ platelets (0.0001% WBC contamination). Some researchers use density gradient centrifugation to improve the separation of WBCs from platelets [35]. A good separation of platelets from other blood cell is crucial for successful proteome analysis. Platelets have the lowest volume of all cells in the blood (platelets 4–10 fL, RBCs 81–100 fL, lymphocytes $230 \pm 106$ fL, monocytes $470 \pm 117$ fL, neutrophils $450 \pm 104$ fL). Therefore, contamination of PRP with even a small number of WBCs can result in considerable artifacts of the final proteome analysis. For example, when a platelet preparation shows a purity of 98% with some contaminations with RBC (1.2%) and WBC (0.8%), the platelet contribution to the total cell volume is only 74%. The remaining volume is composed of RBCs (8%) and WBCs (18%). Assuming that the intracellular protein concentration is similar for all three cell types (which it will not be as WBCs contain a nucleus), proteome analysis of a 98% pure platelet preparation will contain a considerable amount of proteins derived from RBCs and WBCs. Platelet purity can subsequently be improved by subjecting the PRP to a second centrifugation step under similar conditions. Separation of platelets from WBCs is especially challenging in samples from patients with strong inflammation or sepsis, as these patients have high levels of leukocytosis (i.e., WBC counts are increased to $>12{,}000\ \mu L^{-1}$, compared to 4000–9000 under normal conditions). WBCs are also activated in these patients, which is associated with changes in cell density [36]. Therefore, repeated centrifugation might not be sufficient for separation of platelets from WBCs. Some researchers pass the PRP through a WBC filter following centrifugation to remove additional WBCs [28,37]. Large WBCs are mechanically trapped in small pores or "dimples" of the polyurethane filter material. These filters have been developed for processing of large blood quantities in blood transfusion centers and are not properly suited for small volumes. A new development in platelet isolation is cell sorting using magnetic microbeads directed against the integrin β3 subunit. However, to date (as of 2010) there is no literature on a proteomic analysis of platelets isolated by this method.

The main purpose of the methods outlined above is to reduce blood cell contamination of platelets. Since cellular contamination can have deleterious effects on proteome analysis, all measures must be taken to minimize this contamination, even at the risk of reduced platelet recovery. Whatever method is applied to separate platelets from other blood cells, it is highly recommended to measure platelet recovery as well as potential contaminations in every PRP sample. Platelet counting can be done manually using a phase contrast microscope. This is a labor-intensive technique, but is highly economical, since only a cell counting chamber and light microscope with phase contrast optics are necessary. Platelet counting can also be carried out using an electronic particle analyser, such as a Coulter particle counter or hematologic analyzer. Cells are detected on the basis of impedance or optical density. Any particle within a defined size range (normally 2–20 fL) is identified as a platelet. In addition, the presence of any amount of WBC and RBC contamination can also be determined. Such an instrument can be used to determine platelet counts in both whole blood and PRP. However, it should be noted that the size and shape of blood cells can vary considerably under certain disease conditions, leading to increased risk of contamination. Similarly, platelet size can also vary in certain disease conditions (e.g., giant platelets in Bernard–Soulier syndrome), which will require altering the centrifugation conditions to maximize platelet recovery.

## 3.5 SEPARATION FROM PLASMA PROTEINS

The second phase of platelet preparation is to wash away plasma proteins from platelets. However, because of platelet physiology and ultrastructure it is not possible to draw a definitive line between the platelet proteome and the plasma proteome. Platelets have the ability to take up proteins from the surrounding plasma via surface receptors, through endocytosis, and also bind plasma proteins on their surface. For example, ApoE containing low-density lipoprotein (LDL) can bind specifically to platelets via the apolipoprotein E receptor 2 (apoER2′) [38]. Correspondingly, apoE has been found in the platelet proteome [39]. The open canalicular system (OCS) constitutes an additional source of platelet-associated plasma proteins. The surface-connected or open canalicular system represents a reservoir of membranes within platelets that opens directly to the extracellular space (i.e., plasma) following activation. The OCS contributes to the high surface area-to-volume ratio of platelets [40]. Each channel joins with other canaliculi of the OCS to form a network of fenestrated conduits spreading throughout the cytoplasm of the platelet. On platelet activation, platelets undergo a dramatic shape change, with α-granules coalescing in the center of the platelet and fusing with OCS. Granule contents are released into the OCS and diffuse out into the extracellular environment [41]. However, protein trafficking is bi-directional through the OCS, with some proteins being released into the plasma by active secretion, whereas other proteins are taken up by the platelets via the OCS [42]. Taken together, these data suggest that there is a constitutive exchange of plasma and platelet proteins. As a result, a platelet proteome represents a snapshot of platelet content, including proteins derived from (1) *de novo* synthesis via translation from platelet mRNA, (2) the megakaryocyte from which the platelet originates, and (3) specific and nonspecific uptake of plasma components. The protein composition of the platelet and the profile of expressed genes are distinct [43]. Factors that influence uptake of proteins and protein binding (e.g., buffer) will alter the composition of the platelet proteome and must be factored in by the investigator. The Platelet Physiology Subcommittee of the International Society on Thrombosis and Haemostasis (ISTH) recommends that information derived from alternative techniques (e.g., Western blotting, flow cytometry) be used to validate proteomics data [43]. The information derived from platelet proteomics studies should therefore be compiled into proteins that are (1) known to have a functional role in platelet regulation, (2) of unknown function that are definitely expressed in platelets, and (3) identified only by proteomic and/or PCR-based techniques that have not been validated by alternative techniques.

### 3.5.1 Centrifugation

Platelets are separated from plasma usually either according to their density by centrifugation or according to their size by gel filtration. Platelet isolation from PRP by centrifugation involves pelleting of platelets and subsequent resuspension in a suitable buffer. In this "Mustard method" (named after Dr. Fraser Mustard, who devised the protocol), much higher centrifugal forces are used to pellet platelets from plasma compared with the initial centrifugation used to separate platelets from other blood cells [44]. More recent studies on the platelet proteome applied 650*g* for 7 min [28], 720*g* for 10 min [3,45], 950*g* for 10 min [37], 1000*g* for 10 min [30,46], or 2000*g* for 10 min [34]. Such high centrifugal forces constitute a remarkable mechanical stress on platelets. Nevertheless, platelets have to be centrifuged a second time to reduce plasma proteins to a negligible level. Using a radioactive inulin–urea assay we showed in a previous study that even after very careful aspiration of supernatant from a cell pellet, $1.9\% \pm 0.46\%$ ($n = 174$) of the total pellet volume derives from extracellular water (i.e., from the supernatant) [47]. Since the total protein content of plasma is 72.8 g/L [48], even a small amount of residual plasma remaining in the platelet pellet would lead to considerable contamination of the platelet proteome. Therefore, it is highly advisable to resuspend platelets in a physiological buffer, such as Tyrode's buffer, and centrifuge them a second time. The resulting "plasma-free" platelets are correspondingly termed "washed platelets" (WPs). To improve the separation of the platelet pellet from the supernatant (i.e., in order to achieve a better platelet recovery without buffer contaminations), some researchers apply a strong centrifugal force (10,000–15,000*g*) at this second centrifugation step [28,49,50]. In that case, however, it is highly recommended to supplement the washing buffer with a potent platelet inhibitor such as 0.1 μmol/L prostaglandin ($PGI_2$), 1 mM EGTA, or another anticoagulant. This largely prevents centrifugation-induced platelet activation. Nevertheless, as shown in Figure 3.2, centrifugation of platelets at 2150*g* induced the expression of various activation markers on platelets, including P-selectin (75%), PAC1 (150%), and CD40L (220%) in the presence of $PGI_2$ compared to platelets purified by gel filtration.

To reduce the degree of platelet activation during centrifugation, the use of a density gradient cushion has been proposed [51]. Platelets are layered onto the surface of a 10–25% metrizamide gradient and pelleted at 1000*g* for 15 min. Metrizamide, a triiodinated benzamido derivative of glucose, is a density gradient material that yields dense aqueous solutions of low viscosity. This should reduce the force to which platelets are exposed during centrifugation. However, a more recent detailed comparative study revealed that this method leads to similar platelet activation as the traditional centrifugation [52].

### 3.5.2 Gel Filtration

A plasma-free platelet suspension may also be prepared by passing PRP through a gel filtration column [53]. The column material most commonly used is either Sepharose 2B or Biogel A-150. These materials have a fractionation range of $10^4$–$10^7$ Da. Platelets have a much higher molecular mass and pass through the material, while plasma proteins are retained in the pores. Gel-filtered platelets (GFPs) are more quiescent and are generally in a better condition than are washed platelets. Figure 3.2 shows that the activation marker CD40L and PAC1 remain unchanged by gel filtration. CD62P is induced by this treatment but to a much lesser extent than by centrifugation. For further avoidance of platelet activation, the column material can be saturated with bovine serum albumin (BSA) before separation. Walkowiak and co-workers proposed to bind BSA covalently to Sepharose 2B [52]. Platelets isolated by this technique showed the same CD62P expression as in the whole-blood sample. Washed platelets, in contrast, showed a

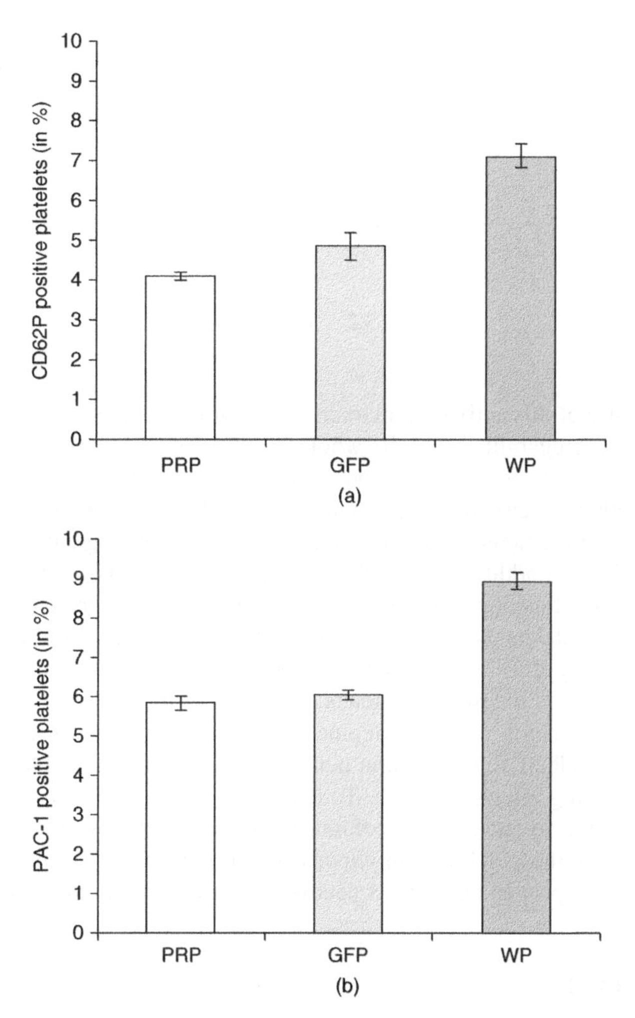


**Figure 3.2** Comparison of the activation state of gel-filtered platelets and washed platelets. Platelet-rich plasma (PRP) was prepared from whole blood by centrifuged at 100g for 20 min. One part of the PRP was passed through a Sepharose 2B column ($10 \times 1$ cm) to obtain gel-filtered platelets (GFPs). Another part of the PRP was centrifuged at 2150*g* for 5 min. Platelets in the pellet were resuspended and washed in Tyrode buffer containing 0.1 $\mu$mol/L $PGI_2$ (pH 6.2) to yield washed platelets (WPs). The graphs indicate the cell surface expression of CD62P (a), PAC1 (b), and CD40L (c) as assessed by flow cytometry ($n = 3$ independent experiments).

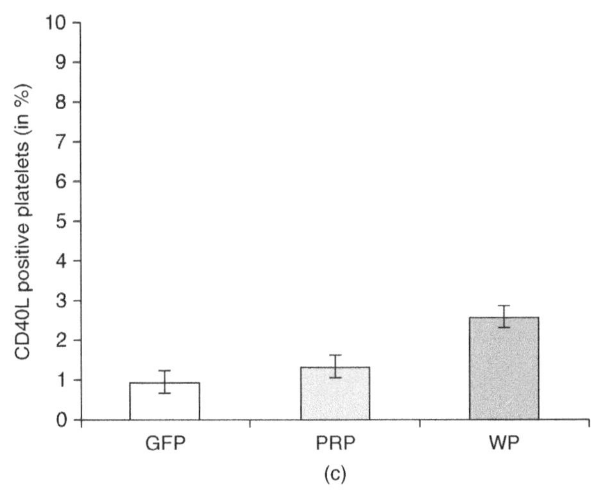


**Figure 3.2** (*Continued*)

twofold increase of this activation marker. In addition, gel-filtered platelets exhibited the same aggregability as in the whole-blood sample. In contrast, in washed platelets the capacity to aggregate was reduced by two-thirds (67%) independently of whether a metrizamide gradient was used. However, it has to be stressed that gel filtration methods are more susceptible to methodological errors than is platelet washing. In addition, the column material can be fissured or too short, leading to an insufficient separation of platelets from plasma proteins. As shown in Figure 3.3, platelets and plasma protein elution are not baseline-separated by gel filtration [Fig. 3.3(a)]. Indeed, plasma proteins elutes from the column together with the last quarter of platelets. Inclusion of these fractions in the GFP results in contamination of the proteome with plasma proteins [Fig. 3.3(c),(d)]. Plasma proteins adhere to the column material and are not removable completely by column washing after separation. To exclude contamination of the next sample, it is advisable to pack a new column with fresh sepharose for each sample [54]. This makes the gel filtration expensive. In addition, a scrupulous quality control of all new packed columns is necessary to guarantee high reproducibility.

## 3.6 CELL LYSIS

The method used to prepare protein samples from plasma-free platelets also affects the results of a platelet proteomic analysis. In many studies washed platelets are centrifuged and the cell pellet is resuspended in a hypotonic buffer containing either nonidet P-40 (NP-40) [3,30,37,45], Triton X-100 [46], or sodium dodecylsulfate (SDS) [32]. The cell membrane becomes leaky and membrane proteins are partially solubilized by this treatment. Cell debris are then removed by centrifugation at high centrifugal force ($>10{,}000g$). The remaining supernatant

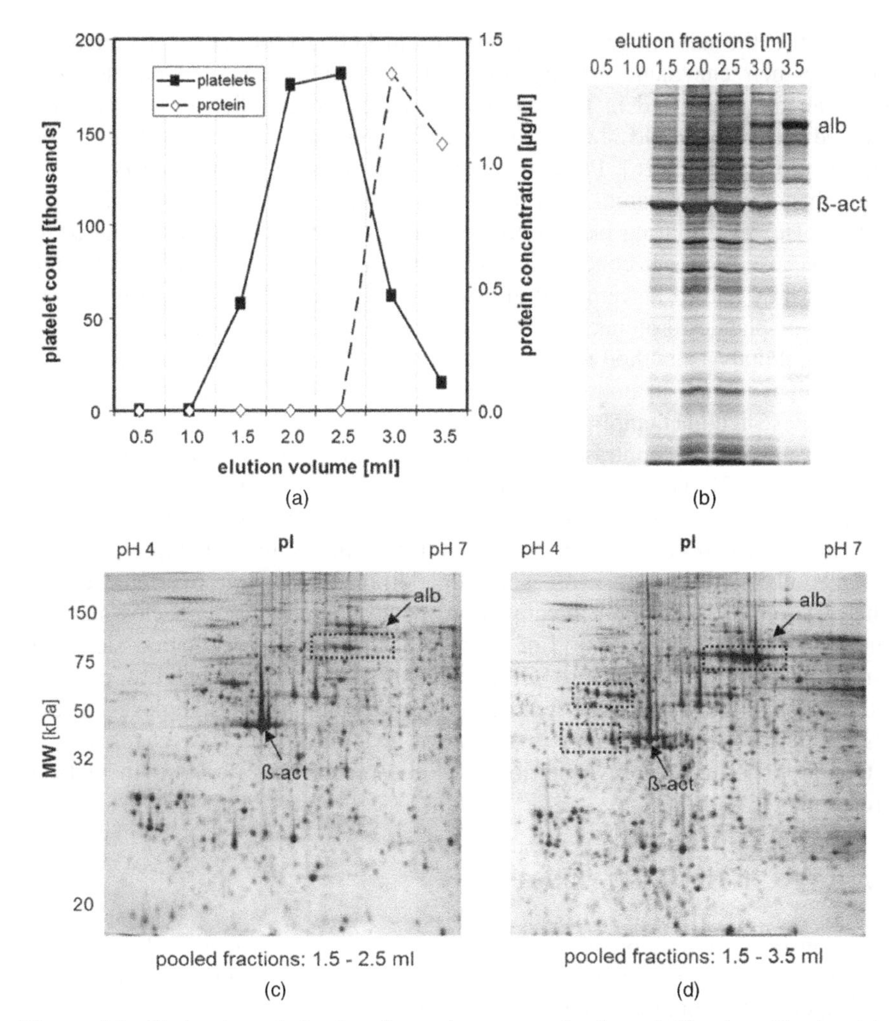


**Figure 3.3** Separation of platelets from plasma proteins by gel filtration. Platelet-rich plasma was passed through a Sepharose 2B column ($11 \times 1$ cm). The column was then washed with phosphate-buffered saline, and fractions of 0.5 mL were collected. (a) Elution profile indicating the platelet count (straight line) and the protein content (broken line) of each fraction; (b) silver-stained SDS-PAGE gel visualizing proteins obtained from each gel filtration fraction; (c, d) silver-stained 2D gels corresponding to different pools of fractions. The dashed-line frames indicate the position of plasma proteins (alb = plasma albumin; β-act = platelet β-actin).

can be used for proteome analysis. It has to be noted that intracellular proteases may still be active after such a treatment, and the researcher should provide sufficient protease inhibitors. Other protocols precipitate platelet proteins from a suspension of washed platelets [2,49] or of gel filtered platelets [54]. Platelets are instantly disintegrated by this method, and proteins become insoluble. This method has the advantage that all proteins including proteases, are denatured and lose their enzymatic activity. In addition, after precipitation platelet proteins can be solubilized in the buffer and in the concentration of choice. In a detailed study we could show that proteomics findings vary according to the precipitation method [26]. Protein precipitation from GFP with TCA/acetone or with ethanol was compared. Both methods resulted in a similar protein yield per platelet. However, during protein precipitation with ethanol salts precipitated as well. Because high salt concentrations are not compatible with most proteomic methods they had to be removed afterward by dialysis. Comparing a proteome analysis of platelet proteins prepared either by the TCA/acetone method or by the ethanol/dialysis method revealed that about 90% of all proteins were present in a similar amount in both 2D gels. However, about 9% of all proteins were underrepresented in ethanol-precipitated extracts in comparison to those that were TCA-precipitated. In contrast, only some single proteins were overrepresented. Comparson of 2D gels of precipitated platelet proteins with gels of nonprecipitated platelet proteins showed that only 75% of all proteins were present in a similar concentration. About 12% of all proteins were underrepresented in 2D gels of precipitated proteins, while an additional 12% were overrepresented. This was observed for both precipitation methods. Such discrepancies may be due to irreversible denaturation of some proteins in the precipitation step that cannot be solubilized afterward. Other proteins seem to be enriched by precipitation, probably due to a better extraction from membranes or complexes. These results suggest that every protein extraction method results in a characteristic set of proteins and thereby contributes to the preparation variables in platelet proteomics.

## 3.7 CONCLUDING REMARKS

In an analysis of multiple signals, such as in proteomics, technical variations comprise qualitative as well as quantitative variations. Results from proteomic analyses of different blood samples taken from the same donor vary regarding the number of proteins that are found and their quantities. The variables described above contribute to both kinds of variation. For example, variation in the quality of the separation of platelets from other blood cells (e.g., which part of the supernatant was taken after pelleting RBC and WBC) may result in presence or absence of WBC proteins in the proteome analysis. Variations in the separation of platelets from plasma may lead to different quantities of platelet-associated plasma proteins. The technical variation of the final proteomic analysis is the sum of variations associated with sample preparation and to the proteome analysis itself. The latter variation depends strongly on the proteomic method that is used. We have investigated the biological and technical variation in platelet proteomics using the 2D differential in gel electrophoresis (2DDIGE) [54]. In this study we analyzed the variation in 2D proteome maps of six different blood samples taken from the same donor and run in six duplicates. Variations among the replicates were also analyzed. Figure 3.4 compares these two variations for every spot

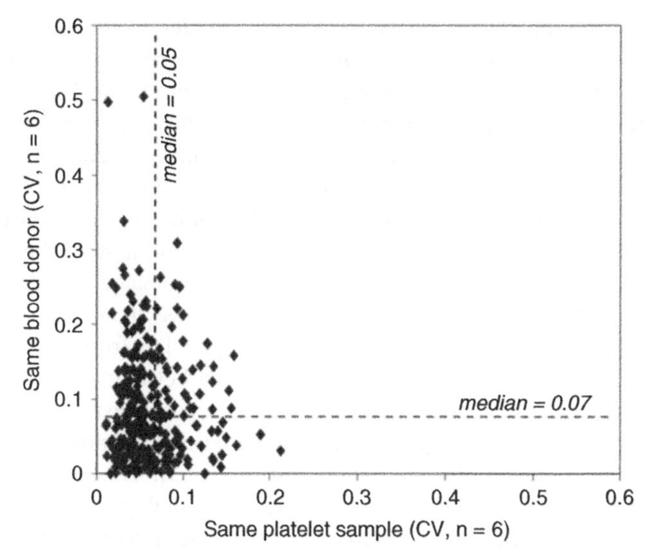


**Figure 3.4** Comparison of the quantitative technical variation. Platelet protein extracts were prepared 6 times from the same donor. Then they were labeled with either Cy3 or Cy5 and separated together with a Cy2-labeled internal standard on three 2D gels (pI range 4–7). The gel was scanned for all three channels, and 484 spots were detected using DeCyder 6.5 software. Each Cy3 or Cy5 spot fluorescent volume was divided by the corresponding Cy2 volume. The $y$ axis indicates the mean CV over six ratios for each spot. The $x$ axis indicates the mean CV over six ratios of the same spots detected on three other gels. On those gels six identical platelet protein extracts were separated and visualized in the same way. The medians over all spots per group are indicated by dashed lines.

on the 2D gel. The coefficient of variation (CV) is not normally distributed. Some individual spots show extraordinarily high variation, while most of them group around the median value (the reason for this asymmetry is discussed in detail in the study by winkler et al. [54]). The median CV of the analysis of the same sample is about 30% lower than that of the analysis of different blood samples from the same donor. These results indicate that the quantitative technical variation due to 2DDIGE contributes more to the total variation than do those due to platelet preparation. This could be achieved only by considering all aspects of platelet preparation indicated in this chapter. From the statistical perspective, any technical variation can be reduced by increasing the number of replicates as long as it is normally distributed. However, systematic variations cannot be eliminated by replicates. When, for example, a study compares platelets from septic patients with those from healthy volunteers, a systematic variation can easily occur at the step of platelet separation from other blood cells. Septic patients have a strong leukocytosis, which may result in a contamination of the PRP with WBC in the samples of this experimental group. Blood samples from healthy volunteers can be more easily divided into PRP and other blood cell fractions. The final proteomic analysis of such samples would probably find a number of disease-associated proteins that in reality derive from WBC and not from platelets. This example shows that the researcher should be aware of the potential variables in platelet preparation described above. A clear definition of the protocol used is essential. Which method is best may depend on the specific study, that is, on the type of patients and the type of analysis that is planned with the prepared sample. In addition, it is important to control for the quality of every single preparation step. The quality of the first preparation step, the separation from other blood cells, can be monitored using an automated cell counter. The contamination with WBC or RBC should contribute less than 0.1% of the total cell number. A quality control for the separation of platelets from plasma proteins is more difficult because no clear line can be drawn between platelet proteome and plasma proteins. There is always a certain degree of plasma proteins in the platelet sample. In our laboratory we calculate the ratio of the amount of albumin and β-actin in all samples of a study (see also Fig. 3.3). This helps us detect a systemic increase in plasma proteins in one experimental group. In such a case it has to be carefully determined whether this increase derives from preparative errors or is of biological relevance.

## ACKNOWLEDGMENTS

We thank Michael Diestinger, Rita Babeluk, Wolfgang Winkler, Christine Brostjan, and Alice Assinger for allowing us to use their data and for helpful discussions.

## REFERENCES

1. Miyazaki Y, Nomura S, Miyake T, Kagawa H, Kitada C, Taniguchi H, Komiyama Y, Fujimura Y, Ikeda Y, Fukuhara S, High shear stress can initiate both platelet aggregation and shedding of procoagulant containing microparticles, *Blood* **88**(9):3456–3464 (1996).
2. García A, Senis YA, Antrobus R, Hughes CE, Dwek RA, Watson SP, Zitzmann N, A global proteomics approach identifies novel phosphorylated signaling proteins in GPVI-activated platelets: Involvement of G6f, a novel platelet Grb2-binding membrane adapter, *Proteomics* **6**(19):5332–5343 (2006).
3. Maguire PB, Wynne KJ, Harney DF, O'Donoghue NM, Stephens G, Fitzgerald DJ, Identification of the phosphotyrosine proteome from thrombin activated platelets, *Proteomics* **2**(6):642–648 (2002).
4. Nicholson NS, Panzer-Knodle SG, Haas NF, Taite BB, Szalony JA, Page JD, Feigen LP, Lansky DM, Salyers AK, Assessment of platelet function assays, *Am. Heart J* **135**(5 Pt. 2 Suppl.):S170–S178 (1998).
5. Everts PA, Knape JT, Weibrich G, Schonberger JP, Hoffmann J, Overdevest EP, Box HA, van Zundert A, Platelet-rich plasma and platelet gel: a review, *J. Extracorp. Technol*. **38**(2):174–187 (2006).

6. Kawamoto EM, Munhoz CD, Glezer I, Bahia VS, Caramelli P, Nitrini R, Gorjao R, Curi R, Scavone C, Marcourakis T, Oxidative state in platelets and erythrocytes in aging and Alzheimer's disease, *Neurobiol. Aging* **26**(6):857–864 (2005).
7. Malone KM, Ellis SP, Currier D, John Mann J, Platelet 5-HT2A receptor subresponsivity and lethality of attempted suicide in depressed in-patients, *Int. J. Neuropsychopharmacol.* **10**(3):335–343 (2007).
8. Kotzailias N, Andonovski T, Dukic A, Serebruany VL, Jilma B, Antiplatelet activity during coadministration of the selective serotonin reuptake inhibitor paroxetine and aspirin in male smokers: A randomized, placebo-controlled, double-blind trial, *J. Clin. Pharmacol.* **46**(4):468–475 (2006).
9. Pignatelli P, Pulcinelli FM, Lenti L, Gazzaniga PP, Violi F, Vitamin E inhibits collagen-induced platelet activation by blunting hydrogen peroxide, *Arterioscler. Thromb. Vasc. Biol.* **19**(10):2542–2547 (1999).
10. Cordova C, Musca A, Violi F, Perrone A, Alessandri C, Influence of ascorbic acid on platelet aggregation *in vitro* and *in vivo*, *Atherosclerosis* **41**(1):15–19 (Jan. 1982).
11. Shen MY, Hsiao G, Liu CL, Fong TH, Lin KH, Chou DS, Sheu JR, Inhibitory mechanisms of resveratrol in platelet activation: Pivotal roles of p38 MAPK and NO/cyclic GMP, *Br. J. Haematol.* **139**(3):475–485 (2007).
12. O'Kennedy N, Crosbie L, Whelan S, Luther V, Horgan G, Broom JI, Webb DJ, Duttaroy AK, Effects of tomato extract on platelet function: A double-blinded crossover study in healthy humans, *Am. J. Clin. Nutr.* **84**(3):561–569 (2006).
13. Heptinstall S, May J, Fox S, Kwik-Uribe C, Zhao L, Cocoa flavanols and platelet and leukocyte function: Recent *in vitro* and *ex vivo* studies in healthy adults, *J. Cardiovasc. Pharmacol.* **47** (Suppl. 2): S197–S205; discussion S206–S209 (2006).
14. Aldemir H, Kilic N, The effect of time of day and exercise on platelet functions and platelet-neutrophil aggregates in healthy male subjects, *Mol. Cell. Biochem.* **280**(1–2):119–124 (2005).
15. Undar L, Akkoc N, Alakavuklar MN, Cehreli C, Undar L, Flow cytometric analysis of circadian changes in platelet activation using anti-GMP-140 monoclonal antibody, *Chronobiol. Int.* **16**(3):335–342 (1999).
16. Lippi G, Franchini M, Montagnana M, Salvagno GL, Poli G, Guidi GC, Quality and reliability of routine coagulation testing: Can we trust that sample? *Blood Coagul. Fibrinol.* **17**(7):513–519 (2006).
17. Toshima H, Sugihara H, Hamano H, Sato M, Yamamoto M, Yamazaki S, Yamada Y, Taki M, Izumi S, Hoshi K, Fusegawa Y, Satoh K, Ozaki Y, Kurihara S, Spontaneous platelet aggregation in normal subject assessed by a laser light scattering method: An attempt at standardization, *Platelets* **19**(4):293–299 (2008).
18. Ramstrom S, Clotting time analysis of citrated blood samples is strongly affected by the tube used for blood sampling, *Blood Coagul. Fibrinol.* **16**(6):447–452 (2005).
19. Filip DJ, Aster RH, Relative hemostatic effectiveness of human platelets stored at 4 degrees and 22 degrees C, *J. Lab. Clin. Med.* **91**(4):618–624 (1978).
20. Schmitz G, Rothe G, Ruf A, Barlage S, Tschope D, Clemetson KJ, Goodall AH, Michelson AD, Nurden AT, Shankey TV, European Working Group on Clinical Cell Analysis: Consensus protocol for the flow cytometric characterisation of platelet function, *Thromb. Haemost.* **79**(5):885–896 (1998).

21. Macey MG, Carty E, Webb L, Chapman ES, Zelmanovic D, Okrongly D, Rampton DS, Newland AC, Use of mean platelet component to measure platelet activation on the ADVIA 120 haematology system, *Cytometry* **38**(5):250–255 (1999).
22. O'Malley T, Ludlam CA, Fox KA, Elton RA, Measurement of platelet volume using a variety of different anticoagulant and antiplatelet mixtures, *Blood Coagul. Fibrinol.* **7**(4):431–436 (1996).
23. Krueger LA, Barnard MR, Frelinger AL 3rd, Furman MI, Michelson AD, Immunophenotypic analysis of platelets, *Curr. Protoc. Cytom.* **6**(6) (Chapter 6, Unit 6):10(2002).
24. Michelson AD, Platelet activation by thrombin can be directly measured in whole blood through the use of the peptide GPRP and flow cytometry: Methods and clinical applications, *Blood Coagul. Fibrinol.* **5**(1):121–131 (1994).
25. Mody M, Lazarus AH, Semple JW, Freedman J, Preanalytical requirements for flow cytometric evaluation of platelet activation: Choice of anticoagulant, *Transfus. Med.* **9**(2):147–154 (1999).
26. Zellner M, Winkler W, Hayden H, Diestinger M, Eliasen M, Gesslbauer B, Miller I, Chang M, Kungl A, Roth E, Oehler R, Quantitative validation of different protein precipitation methods in proteome analysis of blood platelets, *Electrophoresis* **26**(12):2481–2489 (2005).
27. García BA, Smalley DM, Cho H, Shabanowitz J, Ley K, Hunt DF, The platelet microparticle proteome, *J. Proteome Res.* **4**(5):1516–1521 (Sept.–Oct. 2005).
28. Thiele T, Steil L, Gebhard S, Scharf C, Hammer E, Brigulla M, Lubenow N, Clemetson KJ, Volker U, Greinacher A, Profiling of alterations in platelet proteins during storage of platelet concentrates, *Transfusion* **47**(7):1221–1233 (2007).
29. Walkowiak B, Kaminska M, Okroj W, Tanski W, Sobol A, Zbrog Z, Przybyszewska-Doros I, The blood platelet proteome is changed in UREMIC patients, *Platelets* **18**(5):386–388 (2007).
30. Senis YA, Tomlinson MG, García A, Dumon S, Heath VL, Herbert J, Cobbold SP, Spalton JC, Ayman S, Antrobus R, Zitzmann N, Bicknell R, Frampton J, Authi KS, Martin A, Wakelam MJ, Watson SP, A comprehensive proteomics and genomics analysis reveals novel transmembrane proteins in human platelets and mouse megakaryocytes including G6b-B, a novel immunoreceptor tyrosine-based inhibitory motif protein, *Mol. Cell. Proteom.* **6**(3):548–564 (2007).
31. Marcus K, Immler D, Sternberger J, Meyer HE, Identification of platelet proteins separated by two-dimensional gel electrophoresis and analyzed by matrix assisted laser desorption/ionization-time of flight-mass spectrometry and detection of tyrosine-phosphorylated proteins, *Electrophoresis* **21**(13):2622–2636 (2000).
32. Zahedi RP, Lewandrowski U, Wiesner J, Wortelkamp S, Moebius J, Schutz C, Walter U, Gambaryan S, Sickmann A, Phosphoproteome of resting human platelets, *J. Proteome Res.* **7**(2):526–534 (2008).
33. Hoffmeister KM, Felbinger TW, Falet H, Denis CV, Bergmeier W, Mayadas TN, von Andrian UH, Wagner DD, Stossel TP, Hartwig JH, The clearance mechanism of chilled blood platelets, *Cell* **112**(1):87–97 (2003).
34. Rolf N, Knoefler R, Suttorp M, Kluter H, Bugert P, Optimized procedure for platelet RNA profiling from blood samples with limited platelet numbers, *Clin. Chem.* **51**(6):1078–1080 (2005).

35. Gnatenko DV, Zhu W, Bahou WF, Multiplexed genetic profiling of human blood platelets using fluorescent microspheres, *Thromb. Haemost*. **100**(5):929–936 (2008).
36. Calhoun WJ, Reed HE, Moest DR, Stevens CA, Enhanced superoxide production by alveolar macrophages and air-space cells, airway inflammation, and alveolar macrophage density changes after segmental antigen bronchoprovocation in allergic subjects, *Am. Rev. Respir. Dis*. **145**(2 Pt. 1):317–325 (1992).
37. Springer D, Miller J, Spinelli S, Pasa-Tolic L, Purvine S, Daly D, Zangar R, Jin S, Blumberg N, Francis C, Taubman M, Casey A, Wittlin S, Phipps R, Platelet proteome changes associated with diabetes and during platelet storage for transfusion, *J. Proteome Res*. **8**(5):2261–2272 (2009).
38. Korporaal SJ, Relou IA, van Eck M, Strasser V, Bezemer M, Gorter G, van Berkel TJ, Nimpf J, Akkerman JW, Lenting PJ, Binding of low density lipoprotein to platelet apolipoprotein E receptor 2′ results in phosphorylation of p38MAPK, *J. Biol. Chem*. **279**(50):52526–52534 (2004).
39. Zellner M, Babeluk R, Diestinger M, Pirchegger P, Skeledzic S, Oehler R, Fluorescence-based Western blotting for quantitation of protein biomarkers in clinical samples, *Electrophoresis* **29**(17):3621–3627 (2008).
40. White JG, Clawson CC, The surface-connected canalicular system of blood platelets—a fenestrated membrane system, *Am. J. Pathol*. **101**(2):353–364 (Nov. 1980).
41. Flaumenhaft R, Molecular basis of platelet granule secretion, *Arterioscler. Thromb. Vasc. Biol*. **23**(7):1152–1160 (2003).
42. White JG, Escolar G, The blood platelet open canalicular system: A two-way street, *Eur. J. Cell. Biol*. **56**(2):233–242 (1991).
43. Watson SP, Bahou WF, Fitzgerald D, Ouwehand W, Rao AK, Leavitt AD, Mapping the platelet proteome: A report of the ISTH Platelet Physiology Subcommittee, *J. Thromb. Haemost*. **3**(9):2098–2101 (2005).
44. Mustard JF, Perry DW, Ardlie NG, Packham MA, Preparation of suspensions of washed platelets from humans, *Br. J. Haematol*. **22**(2):193–204 (1972).
45. Wong JW, McRedmond JP, Cagney G, Activity profiling of platelets by chemical proteomics, *Proteomics* **9**(1):40–50 (2009).
46. Gevaert K, Ghesquiere B, Staes A, Martens L, Van Damme J, Thomas GR, Vandekerckhove J, Reversible labeling of cysteine-containing peptides allows their specific chromatographic isolation for non-gel proteome studies, *Proteomics* **4**(4):897–908 (2004).
47. Oehler R, Hefel B, Roth E, Determination of cell volume changes by an inulin-urea assay in 96-well plates: a comparison with coulter counter analysis, *Anal. Biochem*. **241**(2):269–271 (1996).
48. Rai AJ, Gelfand CA, Haywood BC, Warunek DJ, Yi J, Schuchard MD, Mehigh RJ, Cockrill SL, Scott GB, Tammen H, Schulz-Knappe P, Speicher DW, Vitzthum F, Haab BB, Siest G, Chan DW, HUPO Plasma Proteome Project specimen collection and handling: Towards the standardization of parameters for plasma proteome samples, *Proteomics* **5**(13):3262–3277 (2005).
49. García A, Prabhakar S, Brock CJ, Pearce AC, Dwek RA, Watson SP, Hebestreit HF, Zitzmann N, Extensive analysis of the human platelet proteome by two-dimensional gel electrophoresis and mass spectrometry, *Proteomics* **4**(3):656–668 (2004).

50. O'Neill EE, Brock CJ, von Kriegsheim AF, Pearce AC, Dwek RA, Watson SP, Hebestreit HF, Towards complete analysis of the platelet proteome, *Proteomics* **2**(3):288–305 (2002).
51. Rendu F, Marche P, Maclouf J, Girard A, Levy-Toledano S, Triphosphoinositide breakdown and dense body release as the earliest events in thrombin-induced activation of human platelets, *Biochem. Biophys. Res. Commun*. **116**(2):513–519 (1983).
52. Walkowiak B, Kralisz U, Michalec L, Majewska E, Koziolkiewicz W, Ligocka A, Cierniewski CS, Comparison of platelet aggregability and P-selectin surface expression on platelets isolated by different methods, *Thromb. Res*. **99**(5):495–502 (2000).
53. Tangen O, Berman HJ, Marfey P, Gel filtration. A new technique for separation of blood platelets from plasma, *Thromb. Diath. Haemorrh*. **25**(2):268–278 (1971).
54. Winkler W, Zellner M, Diestinger M, Babeluk R, Marchetti M, Goll A, Zehetmayer S, Bauer P, Rappold E, Miller I, Roth E, Allmaier G, Oehler R, Biological variation of the platelet proteome in the elderly population and its implication for biomarker research, *Mol. Cell. Proteom*. **7**(1):193–203 (2008).

# PART II

# ANALYSIS OF THE PLATELET PROTEOME: GLOBAL APPROACHES AND SUBPROTEOMES

# 4

# TWO-DIMENSIONAL GEL ELECTROPHORESIS: BASIC PRINCIPLES AND APPLICATION TO PLATELET SIGNALING STUDIES

ÁNGEL GARCÍA

**Abstract**

Since the late 1990s, proteomics has been widely applied to the study of signaling cascades in human platelets. Whole platelet lysates and tyrosine phosphorylated proteins have been studied by a combination of two-dimensional gel electrophoresis (2DGE) and sodium dodecyl sulfate–polyacrylamide gel electrophoresis (SDS-PAGE) for protein separation, and mass spectrometry for protein identification. This combined approach has led to the identification of novel signaling proteins, some of which are good candidates to play a significant role in platelet signaling and activation. The present chapter introduces the reader to the basis of 2DGE and reviews the application of 2DGE-based proteomics to the study of the platelet proteome, focusing on signaling studies.

## 4.1 INTRODUCTION: TWO-DIMENSIONAL GEL ELECTROPHORESIS (2DGE) IN PROTEOMICS RESEARCH

### 4.1.1 Historic Perspective, Advantages, and Disadvantages of 2DGE

Two-dimensional gel electrophoresis (2DGE) is the most widely applied protein separation technique in proteomics. It combines two distinct procedures for

*Platelet Proteomics: Principles, Analysis and Applications*, First Edition.
Edited by Ángel García and Yotis A. Senis.

protein separation: isoelectric focusing (IEF), which separates proteins according to their isoelectric point (pI) in the first dimension; and sodium dodecylsulfate–polyacrylamide gel electrophoresis (SDS-PAGE), which separates proteins according to their molecular mass in the second dimension. This separation method offers great resolution, allowing the separation of thousands of proteins at a time in one large 2D gel. Since it was first described in 1975 [1,2], the 2DGE method has developed over the years, and today the use of immobilized pH gradient (IPG) strip gels for the first dimension and large polyacrylamide gradient gels for the second dimension offers a powerful tool for high-resolution protein separation.

A historical moment for proteomic research took place in the late 1980s, with the development of "soft" ionization methods for mass spectrometry, such as matrix-assisted laser desorption/ionization (MALDI) and electrospray ionization (ESI) [3,4]. Those methods solved the problem of generating ions from large, nonvolatile analytes, such as proteins and peptides, without significant analyte fragmentation, which is the reason why they are called "soft" ionization methods. Together with the sequencing of the human genome in 2001, they were key for the development of modern proteomics [5]. Indeed, the combination of 2DGE for protein separation and mass spectrometry for protein identification formed the core of proteomics in the early days. Despite the promising alternative or complementary technologies [e.g., multidimensional protein identification technology (MudPIT), stable isotope labeling, protein or antibody arrays] that have emerged more recently (see Chapter 2), 2DGE is still the only technique that can be routinely applied for large differential analysis studies involving complex protein mixtures such as whole-cell lysates.

The 2DGE method delivers a map of intact proteins, which reflects changes in protein expression levels and posttranslational modifications (PTMs). This is in contrast with methods based on multidimensional liquid chromatography–tandem mass spectrometry ($LC^n$-MS/MS), where information on the relative molecular mass ($M_r$) and isoelectric point (pI) is lost and where stable isotope labeling is required for quantitative analysis [6]. A schematic illustration of the working range of 2DGE—as displayed in Figure 4.1—can explain why it is one of the most frequently chosen separation methods for the analysis of proteomes. In high-resolution studies, several thousand proteins can be separated, displayed, and stored in one gel. Proteins in the size range of 10–200 kDa and with pI 3–11 can be analyzed, including relatively hydrophobic proteins, since the separation takes place under completely denaturing conditions. However, very hydrophobic proteins as well as many high-molecular-weight or very basic proteins are resolved with difficulty by 2DGE. These limitations are due primarily to solubility problems in the IEF sample buffer [7]. In addition, low-abundance proteins may be masked in the 2D gel by more abundant proteins, which is a consequence of the high dynamic range of the proteome. More recent technical improvements in 2DGE-based proteomics include improved solubilization of membrane proteins, enrichment of low-abundance proteins by sample prefractionation, novel

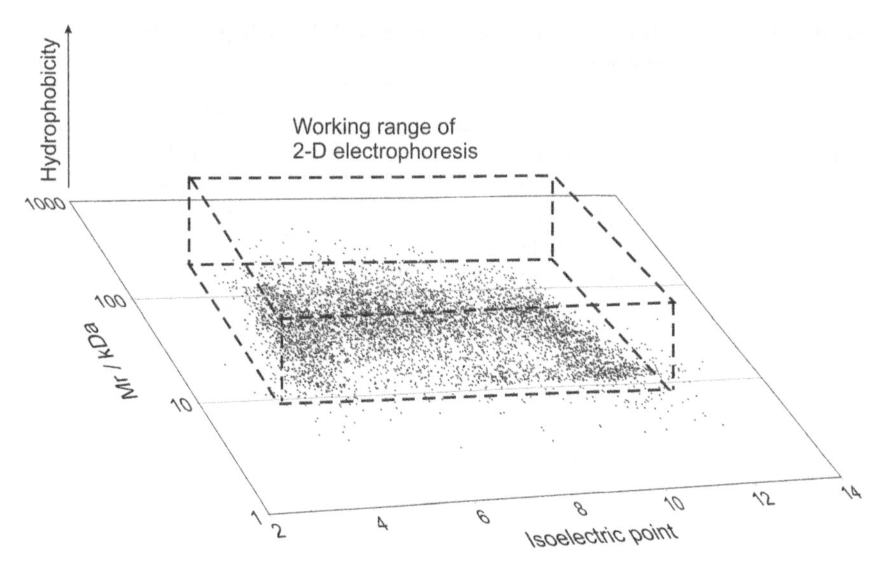


**Figure 4.1** Estimated working range of 2DGE for separating highly complex protein mixtures. (From Westermeier et al. [7]. Copyright Wiley-VCH Verlag GmbH & Co. KGaA. Reproduced with permission.)

highly sensitive water-soluble fluorescent dyes for protein staining, and narrow-pI-range 2D gels ("zoom" gels) for a detailed analysis of the region of interest [7,8]. When the limitations highlighted above are an issue, the use of prefractionation techniques in combination with 1D SDS-PAGE and LC-MS/MS, or $LC^n$-MS/MS, is a valid alternative [9]. Table 4.1 summarizes the main advantages and disadvantages of the protein separation methods most commonly used in proteomics.

### 4.1.2 Sample Preparation for 2DGE

***4.1.2.1 Protein Extraction*** The proteomics workflow based on the use of 2DGE and mass spectrometry starts with the sample preparation. This is a critical step that should be standardized for each type of biological sample to be studied. The initial aim is to obtain a sufficient amount of pure protein for analysis. As an example, a minimum of 500 μg of protein per gel is needed for a micropreparative high-resolution 2DGE analysis, based on 18–24-cm IPG strips for the first dimension.

As in any proteomics study, especially those involving gels, the whole process preceding the MS step should ideally be done in a cleanroom, with positive pressure, to avoid keratin contaminations that would prevent identification of the proteins of interest, especially low-abundance proteins. During this first and critical step proteins are isolated from the biological sample (e.g., platelets). To avoid protein losses, the treatment of the sample must be kept to a minimum; to avoid protein modification, the sample should be kept as cold as possible; and

**TABLE 4.1 Main Advantages and Disadvantages of Three Separation Techniques Used in Proteomics**

| Separation Method | Advantages | Disadvantages |
|---|---|---|
| 2DGE | Very sensitive<br>High resolution; it is possible to detect one protein per spot, and thousands of spots per gel<br>Information about PTMs ($M_r$, pI)<br>Ideal for differential analysis (e.g., signaling studies, healthy vs. diseased); quality and reproducibility improved with DIGE | Complex and labor-intensive<br>Complex sample preparation<br>Reproducibility, especially interlaboratory<br>Problems to analyze very large, very basic, and very hydrophobic proteins |
| SDS-PAGE | Compatible with SDS and other ionic detergents in sample buffer<br>Compatible with most prefractionation techniques<br>Allows analysis of membrane proteins | Presence of several proteins in a gel band<br>No information about PTMs (only $M_r$, no pI) |
| $LC^n$ coupled to MS/MS | Very high sensitivity<br>Minimized sample handling<br>Rapid and readily automated | Lack of pI and $M_r$ estimate of proteins<br>Requires significant computing resources for data analysis and gives information on limited number of proteins |

to avoid losses and modifications, the time should be kept as short as possible. Quite often, protein isolation may involve sub-cellular and/or protein prefractionation, protein enrichment (especially for the analysis of low-abundance proteins), and/or protein precipitation methods [e.g., trichloroacetic acid (TCA)/acetone protein precipitation]. Those procedures should be carried out according to the abovementioned rules. When dealing with clinical samples for biomarker discovery, special care should be taken in collection, storage, and processing of the samples (see Chapter 3).

It is also important to ensure that the protein sample is pure and free of contaminants and disturbing compounds, such as salts, lipids, polysaccharides, and nucleic acids, which can interfere with the isoelectric focusing (IEF) step (first dimension). Polysaccharides (especially those carrying a charge) and nucleic acids can interact with proteins and give rise to streaky 2D patterns. A common method for removal of these molecules is precipitation of proteins with acetone or TCA/acetone. Another recommendation for the removal of nucleic acids is the digestion by a mixture of protease-free RNAses and DNAses. High salt concentrations in the sample (>100 mM) will disturb the IEF and lead to streaky patterns. In extreme cases, virtual cessation of IEF may occur as a result of salt fronts. Salt removal can be achieved by spin dialysis, or precipitation of proteins with TCA/acetone [10]. Therefore, in the case of cells that are washed with PBS prior to lysis, PBS should be carefully removed before adding the 2D sample buffer. Moreover, amphoteric buffers in cell cultures, such as HEPES [*N*-(2-hydroxyethyl)piperazine-*N*′-ethanesulfonic acid], overbuffer the gradient in the areas of their pIs, which results in vertically narrow areas without protein spots. High amounts of lipids may interact with membrane proteins and consume detergents. Delipidation of lipid-rich biological samples can be accomplished by extraction with organic solvents (e.g., cold acetone), which can be done following protein precipitation (e.g., precipitation with 10–20% TCA in acetone followed by washes in cold acetone).

***4.1.2.2 Sample Buffer*** Before the electrophoresis, proteins must be dissolved in a sample buffer appropriated for 2DGE. This must be a buffer where the proteins are completely solubilized, without forming any aggregates or complexes, but maintaining their native charge intact, so that they can be separated in the first dimension by IEF. Therefore, the buffer cannot be as aggressive as the typical SDS-based buffer used for a standard SDS-PAGE, and the samples cannot be boiled prior to the electrophoresis. The basic components of a 2DGE sample buffer, which must be solubilized in milliQ water, are as follows:

- *Detergent.* CHAPS is a zwitterionic detergent preferred to others such as Triton X-100 and Nonidet NP-40 because of its higher purity. It increases the solubility of hydrophobic proteins and works well in combination with NDSB-256 (nondetergent sulfobetaine), another zwitterionic detergent. The combination of both improves the resolution of the analysis leading to the detection of a higher number of spots in the gel. As mentioned above, anionic detergents, such as SDS, cannot be used because they mask the native charge of the proteins. A maximum amount of 0.2% SDS can be added to improve the representation of membrane proteins in the gel; however, that is the critical maximum SDS concentration that can be used.
- *Chaotropic Agents.* These are used to disrupt hydrogen bonds, leading to protein unfolding and denaturation. The typical agent used to achieve this is urea, which can be combined with thiourea, due to the ability of the latter to break hydrophobic interactions. In fact, a combination of 5–7 M urea and 2 M thiourea, in conjunction with appropriate detergents, is currently one of the best solutions for solubilizing hydrophobic proteins for 2DGE analysis [7,11].

- *Reducing Agents.* Reducing agents are necessary for cleavage of intra- and intermolecular disulfide bonds to achieve complete protein unfolding. They are used to prevent different oxidation steps of the proteins. The most commonly used reducing agents are dithiotreitol (DTT) and dithioerythritol (DTE), which are applied in excess (concentrations $\leq$100 mM). Tributylphosphine (TBP) is a neutral reductant proposed by Herbert et al. [12] as an alternative to DTT. Despite disadvantages such as low solubility in water, short half-life ($t_{1/2}$), and toxicity, a combination of a low-dose TBP (2 mM) with a medium dose of DTT (65 mM), is a good choice for the sample buffer [10]. 2-Mercaptoethanol should not be used because of its buffering effect above pH 8 [13]. This would cause horizontal streaks in the area between pH 8 and 9.
- *Carrier Ampholytes.* Carrier ampholytes improve the solubility of proteins during the IEF by substituting ionic buffers. In a mixture, they are charged, migrating to their pI during the IEF, where they become uncharged. The amount and composition of the ampholytes mixture to be added to the sample depend on the pH range of the IPG strips to be used and must be optimized for each sample.
- *Protease and Phosphatase Inhibitors.* These are very important for prevention of protein degradation and preserve phosphorylations, an important PTM for protein activity. Depending on how proteins are isolated, these inhibitors may be added during the sample preparation (e.g., cell lysis), and not to the sample buffer. Typical protease inhibitors include phenylmethylsulfonyl fluoride (PMSF) or the water-soluble form 4-(2-aminoethyl)benzenesulfonylfluoride hydrochloride (AEBSF), aprotinin, benzamidine, leupeptin, and pepstatin A. These compounds are commercially available as cocktails of various concentrations [10]. Typical phosphatase inhibitors include sodium fluoride and sodium orthovanadate.
- *Dyes.* Low amounts of the anionic dye bromophenol blue are used as a control for monitoring the electrophoresis running conditions.
- *Basic Proteins.* For zoom 2D gels focusing on the basic region of the proteome (e.g., pI 6–11), isopropanol should be added to the buffer at a final concentration of 10% (v/v) [10].

### 4.1.3 Stepwise 2DGE

There are several protocols for 2DGE that vary fundamentally with the size and type of the gels and the equipment that is used [14–17]. For instance, GE Healthcare (Uppsala, Sweden) offers an integrated platform that includes reagents for sample preparation, equipment for IEF (Multiphor or IPGPhor), equipment for second dimension (e.g. Ettan Dalt 6), robot cutters, and image analysis software. Differences in equipment, protocols, and reagents are the main factors affecting interlaboratory reproducibility in 2DGE. Regardless of the platform used, 2DGE comprises four main steps, which may take up to 3 days to complete, including

(1) rehydration, (2) IEF (first dimension), (3) equilibration, and (4) SDS-PAGE (second dimension).

***4.1.3.1 Rehydration and Isoelectric Focusing*** *Isoelectric Focusing* (IEF) is the first dimension of protein separation, where proteins are separated according to their charge (pI) in IPG gel strips. Different range of IPG strips can be used for analytical or micropreparative studies. Maximum resolutions can be achieved with 18–24 cm narrow-pH-range IPG strips, producing the so-called zoom gels. For a micropreparative study using those strips, a minimum of 500 μg of protein per strip is recommended. The IEF protocols vary depending on the equipment and the IPG strip size (7, 11, 18, or 24 cm) [14]. The electrophoresis takes place horizontally. Prior to IEF, the IPG dry strips must be rehydrated (usually overnight) to their original thickness of 0.5 mm. This can be done with the sample already dissolved in rehydration buffer (the 2D sample buffer mentioned above). This is called *sample in-gel rehydration*. Alternatively, IPG strip rehydration can be done with the buffer alone, followed by sample application by cup loading. Since a maximum of only 150 μL of sample can be applied by the latter method, a modification has been developed consisting of the application of higher sample volumes by means of a paper bridge [7]. The gel strips are covered with *paraffin oil* during rehydration and IEF to prevent drying of the strips, crystallization of the urea, and oxygen and carbon dioxide uptake.

In general, for micropreparative 2DGE, passive in-gel rehydration is often preferred, although it is not advisable if the sample contains very high-molecular-weight, very alkaline, and/or very hydrophobic proteins, since these are not efficiently taken up by the gel. For sample in-gel rehydration, a solution with the sample dissolved in rehydration buffer is pipetted into the grooves of the reswelling tray. Normally, 450–500 μL of solution is used for 240 mm long × 3 mm wide IPG dry strips. For shorter strips, the volume must be adjusted accordingly. Once the sample is in the groove, the IPG strip is applied (gel side down) without trapping air bubbles. The IPG strip, which must be movable and not stick to the tray, is then covered with *paraffin oil* and rehydrated overnight at approximately 20°C [14]. For cup loading, IPG dry strips are reswollen in sample rehydration buffer without the sample. Samples are subsequently dissolved in sample buffer, applied into disposable plastic or silicone rubber cups, and placed onto the surface of the IPG strip. Optimal results are obtained when the samples are applied near the anode or the cathode. When using basic pH gradients such as IPGs 6–11 or 9–12, anodic application is mandatory.

An alternative to the methods described above is active rehydration prior to IEF, where a small voltage (30–50 V) is applied for 12–16 h to help introduce the sample into the IPG strip. This is done with the gel facing down and is immediately followed by the IEF. This procedure is a good alternative for analytical electrophoresis, but not as good for micropreparative analyses because for large amounts of protein, IEF works better with the IPG strip gel facing up. In addition, not all electrophoresis equipment is suited for an active rehydration procedure.

The amount of protein that can be loaded onto a single IPG strip to achieve optimal resolution, maximum spot numbers, and minimal streaking/background smearing depends on parameters such as the pH gradient used (wide or narrow), separation distance, and protein complexity of the sample. Maximum resolution can be achieved when using zoom narrow-pH-range gels (e.g., 4–5, 5–6, 4–7, 6–11). For micropreparative purposes, between 500 and 1000 μg of protein can be loaded on 18–24 IPG strips for a good resolution, but it is possible to load even more, although generally, with whole-cell lysates, the gel would seem overloaded for amounts above 2 mg of protein. IEF is carried out in specialist equipment at high voltages (normally at constant voltages of 3500–8000 V). For micropreparative studies, IEF takes place with the gel surface of the IPG strip facing up. This has the advantage of employing filter pads at the gel ends, which take up salt ions and proteins with pI values lying outside the pH gradient of the IPG strip. Another benefit is the open surface just covered with paraffin oil; when highly abundant proteins form ridges, there will be no mechanical pressure on them [7].

The electrophoretic conditions of the IEF vary dramatically depending on the equipment, the amount of protein (and its composition), and the type of strip. Therefore, standardization is always required when initiating a new study. Generally, for micropreparative analyses in 18–24-cm strips, and 3500 V at the steady state, the IEF may take over 20 h [14]. Since immobilized pH gradients have very low conductivity, the current is usually limited to 50–70 μA per strip. An initial step of 300 V for 2 h is recommended. Temperature during the IEF should be strictly controlled. Running the strips at 20°C is optimal, because it is above the temperature where crystallization of urea can occur, and below the overheating temperature, which can cause carbamylation of proteins during IEF. Active temperature control is necessary. Simply adjusting the temperature of the room is not sufficient. Following IEF, IPG strips can be stored between −20°C and −80°C.

***4.1.3.2 Equilibration and SDS-PAGE*** *Equilibration* Before the second dimension separation, it is essential that the IPG strips be equilibrated for 10–15 min in a buffer containing SDS to ensure that they acquire the negative charge needed for their separation in the second dimension. A recommended equilibration buffer contains 6 M urea, 1–2% (w/v) DTT, 50 mM Tris-HCl, pH 6.8 (pH 8 for basic gels, such as 6–11), 2% (w/v) SDS, 30% (w/v) glycerol, and bromophenol blue. A good alternative is to use 4 M urea and 2 M thiourea instead of 6 M urea [10]. Thiourea helps to achieve a more efficient transfer of hydrophobic proteins, although it may cause vertical streaks in the 2D pattern. Sometimes, a second equilibration step may be carried out consisting of a 15 min incubation of the strip in the abovementioned buffer containing 4% (w/v) iodoacetamide instead of DTT. The iodoacetamide alkylates sulfhydryl groups and prevents their reoxidation. Nevertheless, proteins of interest can be reduced within the gel and alkylated following electrophoresis, prior to protein digestion and MS analysis.

*Second Dimension: SDS-PAGE* This is normally performed in large vertical systems where up to 6–12 gels can run in parallel. Typically, gel sizes of 20 × 26 cm (length × width) and a gel thickness of 1.0-mm are recommended. The latter favors faster separations, leading to sharper spots. In addition, 1.0-mm gels produce a protein/gel ratio more favorable for tryptic digestions, resulting in higher peptide yield. Moreover, optimal resolution can be achieved with gradient gels. In 2DGE, it is not necessary to use stacking gels for SDS-PAGE, as the protein zones within the IPG strips are already concentrated and the low-polyacrylamide-concentration IEF gel acts as a stacking gel. The running buffer for SDS-PAGE contains Tris, glycine, and SDS [10].

To set up the second dimension, a few milliliters of 0.5% melted agarose (in SDS-PAGE running buffer) are poured on top of the corresponding 2D gel and the IPG strip is immediately immersed into the melted agarose until reaching contact with the top of the gel. A knife or spatula can be used to help achieve good contact between the IPG strip and the top of the 2D gel before the agarose polymerizes. Optimal running conditions are determined empirically and will vary depending on the size of the gel and the electrophoresis equipment used [7].

### 4.1.4 Postelectrophoretic and DIGE Staining

Following electrophoresis, gels are fixed in a solution containing methanol or ethanol and acetic acid, and stained. Protocols for the procedures described above vary depending on the dye that is used. Various dyes are used for proteomics research: pre- or post-electrophoretic, fluorescent or nonfluorescent, more or less sensitive, and more or less expensive [18].

Postelectrophoretic protein staining is done by intercalation of fluorophores into the SDS micelles coating the proteins, or by direct electrostatic interaction with the proteins. Coomassie brilliant blue (CBB) staining methods have found widespread use because they are relatively cheap, easy to use, and compatible with MS. Moreover, because it is a visible staining, a visible scanner is sufficient to capture the images. The downside of CBB is the low sensitivity (range 200–500 ng of protein per spot), which prevents detection of low-abundance proteins. Silver staining methods are far more sensitive than CBB, and have a detection limit as low as 0.1 ng protein per spot. However, their linearity and reproducibility are not as good as CBB stains, making them less suitable for quantitative analysis. In addition, silver staining methods are quite labor-intensive and complex, and some are not compatible with MS [15].

In modern proteomics, fluorescent dyes are the protein stain of choice, providing high-quality images for quantitative differential analysis. A good example is the ruthenium-based dye SYPRO Ruby. Following a fixing step in 10% methanol/7% acetic acid for 1 h, gel staining can be accomplished within a few hours or overnight in a single-step procedure. Following a brief wash for 1 h in 10% methanol/7% acetic acid, the gels are scanned using a charge-coupled device (CCD) camera or a laser-based scanner, such as the Typhoon (GE Healthcare). The detection limit of SYPRO Ruby is approximately 1–2 ng protein per spot, and the linear dynamic range of quantitation is about three orders of magnitude [14]. Further, as is the case with other fluorescent dyes, SYPRO Ruby is

MS-compatible. In addition to fluorescent dyes for total protein staining, PTM-specific fluorescent dyes are also available, for instance, for staining glycosylated or phosphorylated proteins. As an example, the phosphoprotein dye Pro-Q Diamond (Invitrogen–Molecular Probes), was introduced relatively recently and is now widely used [15].

An alternative to postelectrophoresis staining is the covalent derivatization of proteins with fluorophores prior to IEF. This is the basis of a method called *differential in-gel electrophoresis* (DIGE) [19]. With this approach, the proteins of two samples are labeled prior to IEF with spectrally distinct fluorescent cyanine minimal dyes (CyDyes). The samples are mixed together and run on the same 2D gel. In this way, the proteins of different samples migrate under identical conditions and the gel-to-gel variations are eliminated. After electrophoresis the gel is scanned with a fluorescent imager at different wavelengths. Quantification is possible because there is a linear relationship between a labeled protein and the signal measured with the imager. This is because on excitation of the dye by monochromatic light, the dye emits light in proportion to the amount of labeled compound in the sample. Three spectrally distinct dyes are available for DIGE: Cy2, Cy3, and Cy5. They can be excited with blue, red, and green lasers, respectively.

Two different methods of labeling are employed in DIGE: lysine and cysteine tagging. By far the mostly frequently applied method is labeling of the lysine. The pI values of the cyanine dye labeled proteins remain unaffected, because the cyanine dyes compensate for the loss of the positive charge of the lysyl residues. However, the $M_r$ increases by 434–464 Da (depending on the dye molecule) per labeled lysyl residue. In practice, approximately only 3–5% of protein is labeled. Since the bulk of the protein remains unlabeled, the slight increase in $M_r$ sometimes presents a problem for spot excision for MS analysis, particularly with low-$M_r$ proteins. One alternative is to stain the separated proteins additionally with SYPRO Ruby or CBB prior to spot picking.

### 4.1.5 Image Analysis

Following gel staining and scanning of a complete set of samples, the samples are now ready for image analysis. One of the key objectives in proteomics is to identify the differential expression between groups of samples run on a series of 2D gels. This is done by dedicated software able to detect differences due to spots that appear or disappear or have changed in abundance (increased or decreased in size and intensity). These differentially regulated features, highlighting changes in protein abundance and PTMs, conform the group of proteins of interest that will be identified by MS. Image analysis is a key step, as is the bottleneck in 2DGE-based proteomics. It is very expensive and time-consuming and requires detailed evaluation of the results for the researcher to be confident that changes are consistent in order to move to the MS analysis step. A variety of image analysis software packages are available, such as MELANIE (GeneBio), SameSpots (nonlinear dynamics), DeCyder (GE Healthcare), and REDFIN (Ludesi). DeCyder was especially designed to analyze DIGE experiments, although the other software mentioned above can also be used for DIGE as well as for single-protein stain experiments. Costs are a fundamental limitation when conducting image analysis. A good choice of the software package and experience with it before carrying out the analysis are fundamental aspects in achieving positive results.

Following the image analysis, spots of interest are excised from the gels—either manually or by robot cutters—and the proteins in-gel trypsin digested and identified by MS using powerful search engine tools (e.g., Mascot, Sequest) and databases such as Swiss-Prot or NCBI (see Chapter 2).

## 4.2 ANALYSIS OF PLATELET PROTEOME BY 2DGE-BASED PROTEOMICS

### 4.2.1 Basic Aspects and General Proteome Mapping

As mentioned in Chapter 1, since platelets do not have a nucleus, proteomics is a perfect tool with which to approach their biochemistry. Platelets can be obtained in large quantities from relatively small amounts of blood. Thus, 100 mL of blood yield approximately $1-2 \times 10^{10}$ platelets, and from $2 \times 10^{10}$ platelets it is possible to obtain 16–24 mg of protein, depending on the protein extraction method [8,20]. As an example, a high-resolution 2D gel (18 × 20 cm) may need a minimum of 500 μg of protein for a proper analysis. It is important to consider the origin of the platelets. For proteomics experiments involving healthy platelets, these platelets should be obtained from healthy donors not on antiplatelet medication (e.g., aspirin) for at least 10 days prior to the experiment. It is also advisable to isolate the platelets immediately following the blood donation, to avoid changes in their physiology and viability. Procedures and recommendations for platelet sample preparation related to proteomics in platelet research for biomarker discovery can be found in Chapter 3 of this book.

Since the 1990s proteomics has been widely applied to platelet research. The first platelet proteomic studies that combined 2DGE for protein separation and MS for protein identification are quite recent. It was in 2000 when Marcus and collaborators reported the first study of platelet cytosolic proteins by 2DGE and MALDI-TOF-MS. They identified 186 protein features, mostly cytoskeletal, that constituted the first pI 3–10 platelet proteome map [21]. A few years later, two reports from the same research group established in the Glycobiology Institute at Oxford University (UK) constituted the broadest investigation so far on the 2DGE-based human proteome [22,23]. Both reports provided a high-resolution 2DGE proteome map comprising more than 2300 different protein features (Fig. 4.2). Proteins were separated by 2DGE using narrow pH gradients during IEF (4–5, 5–6, 4–7, and 6–11), and 9–16% PAGE gradient gels for the second dimension (18 × 18 cm). Gels were stained with a highly sensitive fluorescent dye, and following image analysis, the corresponding protein spots were excised, in-gel trypsin digested, and analyzed by LC-MS/MS. In order to further conceptualize the functional groups of proteins represented in the 2D proteome map, 50% of the features detected were analyzed by MS. O'Neill and coworkers focused on the analysis of the more acidic fraction of the proteome (pI 4–5) [22], whereas García and collaborators analyzed the pI 5–11 region [23]. Overall, more than 1000 proteins were identified, corresponding to 411 open reading frames (ORFs). The list was rich in proteins involved in signaling (24%) and protein synthesis and degradation (22%). The large number of signaling proteins, which included kinases, G proteins, adapters, and protein substrates, was somehow expected in view of the ability of platelets to undergo powerful and rapid activation following damage to the vasculature. The presence of a large number of proteins involved in translation, transcription, and regulation of the cell cycle was unusual in light of the limited degree of protein synthesis that takes

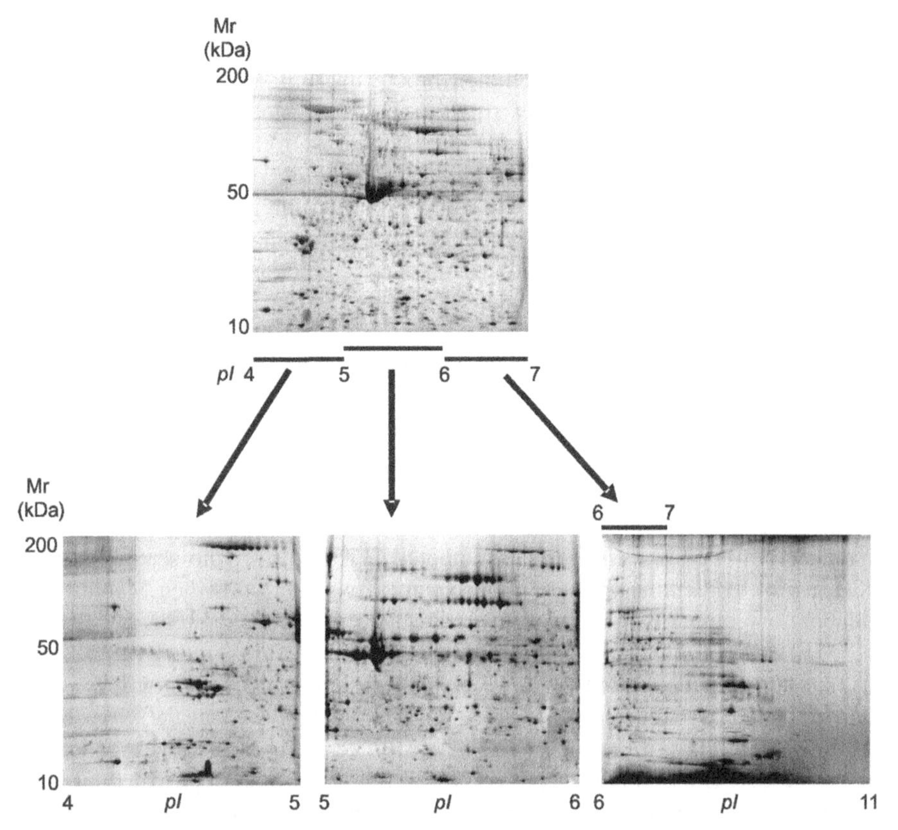


**Figure 4.2** 2DGE proteome map of the human platelet. The narrow-pI-range gels 4–5, 5–6, 4–7, and 6–11 are shown. Protein identifications are available at the following Web-based database: `http://www.bioch.ox.ac.uk/glycob/ogp`. (Modified from García et al. [23].)

place in platelets. It is unclear whether these proteins are functionally relevant or were incorporated into the platelet only during the budding process from the megakaryocyte. Nevertheless, more recent studies suggest that protein synthesis is a relevant issue in platelets [24]. Strikingly, approximately 45% of the proteins identified by these two reports had never been reported in platelets previously, including 15 hypothetical proteins that had not been described in any other cell type, which emphasizes the power of the approach.

It is difficult to estimate the extent of coverage of the platelet proteome that has been achieved by using 2DGE. A typical eukaryotic cell contains 20,000–25,000 genes, and it has been speculated that a third of them are expressed in most cell types. However, many of these genes are expressed at very low levels, which is why the analysis of rare transcripts remains problematic [25]. Since platelets are anucleate, it is even more difficult to estimate the total number of proteins that they contain.

As was mentioned above, very basic and/or hydrophobic proteins escape the 2DGE approach. This why membrane proteins were underrepresented in the 2DGE proteome study carried out at by García and coworkers at Oxford [23]. To overcome this problem, membrane proteins can be isolated by various enrichment methods and then separated by 1D SDS-PAGE, and/or multidimensional LC, and identified by MS (see Chapters 2 and 5). The platelet proteome that escaped to the global 2DGE-based approach was covered in more recent years by direct LC-MS/MS studies (see Chapters 2 and 8) and by subcellular prefractionation approaches combined with LC-MS/MS analysis, as described in Chapters 6 and 7.

### 4.2.2 Differential Proteomic Analysis for Biomarker Discovery

Despite the rapid evolution of proteomics based on MudPIT technology (see Chapter 2), 2DGE is still the method of choice for many platelet proteomics studies because of its versatility for differential analysis studies where various groups of samples are compared (see Fig. 4.3). This finding is reflected not only in this chapter but also in others where 2DGE is used in clinical proteomic studies: in transfusion medicine (Chapter 13) or in cardiovascular proteomics (Chapter 14). Our group at the Universidade de Santiago routinely uses 2DGE-based proteomics to search for platelet biomarkers in acute coronary syndromes (ACSs). In an initial study, the proteome of platelets from patients with non-ST elevation (NSTE) ACS was compared with that from matched controls with chronic ischemic cardiomyopathy. Proteins were separated in large 2D gels, using 24-cm pI 4–7 IPG strips for the first dimension and large 10% SDS-PAGE gel for the second. Following protein staining with SYPRO Ruby, and differential image analysis, 40 differentially regulated features were detected. From those, 40 proteins were identified by MALDI-MS/MS corresponding to 22 different genes. Major groups of proteins identified corresponded to cytoskeletal signaling and proteins either secreted or involved in vesicles or secretory trafficking pathways. The study highlights proteins involved in αIIbβ3 and GPVI signaling as

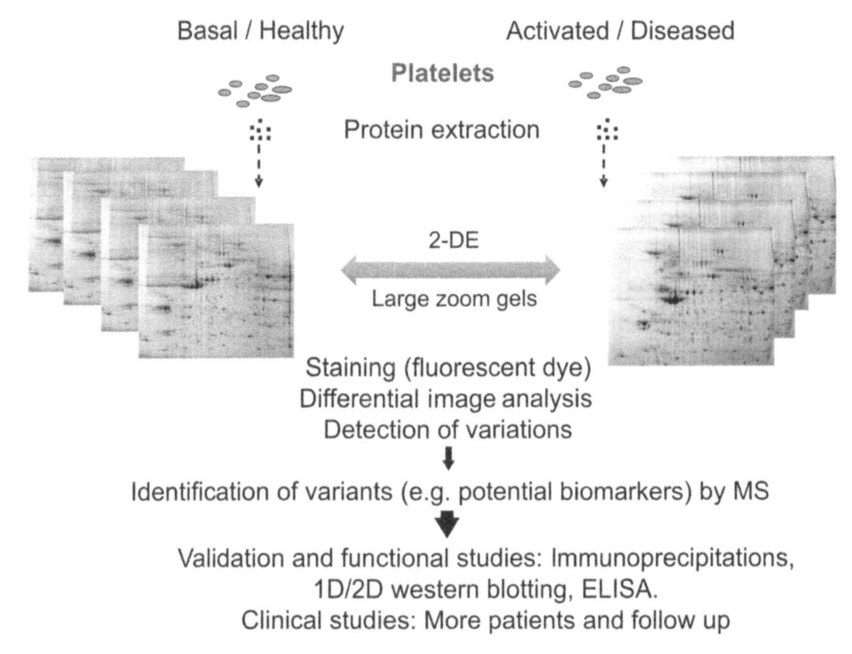


**Figure 4.3** Experimental design for a 2DGE-based differential proteome analysis approach to study platelet signaling pathways (basal vs. activated) or for biomarker discovery studies (healthy vs. diseased). Following the electrophoresis, gels are stained with a fluorescent dye (e.g., SYPRO Ruby). An alternative to this post-electrophoretic staining is the DIGE approach described in Section 4.1.4.

differentially regulated in NSTE-ACS [26]. Another study from a different group also used 2DGE to compare the proteome of platelets from aspirin-resistant and aspirin-sensitive stable coronary ischemic patients [27]. The conclusion is that platelets from both groups of patients differ in the expression levels of proteins associated with mechanisms such as energetic metabolism, cytoskeleton, oxidative stress, and cell survival, which may be associated with their different ability to respond to aspirin. The studies described above highlight the potential of platelet proteomics to identify possible biomarkers in coronary artery disease.

### 4.2.3 Differential Proteomic Analysis for Platelet Signaling

***4.2.3.1 2DGE-Based Studies*** One of the most active fields in platelet research is the study of signaling cascades. As highlighted in Chapter 1, ligand-mediated receptor occupancy/clustering initiates a signaling cascade leading to downstream effects, which in the case of platelets are related mostly to activation, shape change, adhesion, and aggregation. A detailed proteome analysis of signaling cascades in human platelets could potentially lead to a better understanding of the mechanisms that underlie platelet activation and function. One of the main ways in which signaling cascades are regulated is by reversible PTMs, with phosphorylation playing a major role. Indeed, as highlighted in Chapter 2, a variety of methods have been developed for use prior to MS analysis for the enrichment and separation of phosphorylated and glycosylated proteins and peptides. In addition, several MS-based methods have been developed to map phosphorylation sites in signaling cascades [28,29].

Following the general proteome mapping exercises described above, gel-based proteomics was immediately applied to the study of platelet signaling, comparing the proteome of basal and activated platelets (Fig. 4.3) [30]. Most of the studies were based on 2DGE for protein separation. In more recent studies immunoprecipitations were combined with SDS-PAGE to improve the coverage of selective groups of low-abundance proteins and those more difficult to analyze by 2DGE (e.g., very hydrophobic proteins, such as membrane proteins). The main platelet signaling studies by proteomics are summarized in Table 4.2, and explained in more detail below.

Among the main intracellular signaling cascades related to platelet activation/aggregation, the thrombin receptor signaling pathway has been the most widely investigated by proteomics so far. In those studies, 2DGE was the primary method of choice in most cases. Thus, in 1998, Immler et al. identified by LC-MS/MS a group of spots on 2D gels that were differentially regulated in thrombin-activated platelets. Those features were related to different isoforms and phosphorylation states of myosin light chain [31]. Two years later, Gevaert and collaborators studied cytoskeletal preparations of basal and thrombin-stimulated platelets by a combination of 2DGE and MALDI-MS. They reported the identification of 27 proteins, most of which were F-actin binding proteins, that translocate to the cytoskeleton on thrombin stimulation [32]. A later study by Maguire and collaborators used an immunoprecipitation-based approach to analyze the phosphotyrosine proteome of thrombin-activated platelets [33]. Following immunoprecipitations using an antiphosphotyrosine antibody, proteins were separated by 2DGE and identified by MALDI-TOF. Of the 67 proteins detected in the thrombin-activated proteome when compared to resting platelets, 10 were successfully identified, including FAK and Syk.

In 2003, Marcus and collaborators combined different separation and analytical methods to obtain a more comprehensive proteome analysis of thrombin-activated platelets [34]. In one approach, proteins were radiolabeled with $^{32}$P and separated by 2DGE using different pI ranges. Phosphorylated proteins were detected by autoradiography. As part of the same study, Marcus and colleagues also separated the proteins by 2DGE following inmunoprecipitation with the 4G10 antiphosphotyrosine antibody. Differentially regulated proteins were identified by MALDI-TOF and nano-LC-ESI-MS/MS. Overall, the authors identified 55 proteins in the pI 4–7 and 6–11 ranges, most of which corresponded to different isoforms of pleckstrin and cytoskeletal proteins.

One more recent approach that has been used with success to analyze signaling cascades in platelets involves the separation of the whole proteome, without

**TABLE 4.2 Main Contributions to Platelet Signaling Studies by Proteomics[a]**

| Study | Separation Method | MS Approach | Key Achievements |
|---|---|---|---|
| Gevaert et al., [32] | 2DGE (pI 3–10) | MALDI-TOF | 27 proteins identified that translocate to the cytoskeleton on thrombin stimulation |
| Maguire et al., [33] | IP plus 2DGE (pI 3–10) | MALDI-TOF | Immunoprecipitation-based approach to analyze the phosphotyosine proteome from thrombin-activated platelets; 10 proteins identified by a combination of MS and Western blotting |
| Marcus et al., [34] | 2DGE (zoom gels) IP plus 1D SDS-PAGE | MALDI-TOF LC-MS/MS (ion trap) | Differential analysis of phosphorylated proteins in resting and thrombin-stimulated platelets; 77 differentially regulated protein features detected, 55 identified; several Ser- and Thr-phosphorylation sites mapped on five proteins |
| García et al., [35] | 2DGE (zoom gels) | LC-MS/MS (QTOF) | Differential proteome analysis of resting and TRAP-activated platelets; 62 differentially regulated protein features detected, 41 identified; identifications included novel platelet signaling proteins, such as Dok2, and first mapping of phosphorylation sites on RGS18 |
| García et al., [36] | 2DGE (zoom gels) IP plus 1D SDS-PAGE | LC-MS/MS (QTOF) | Differential proteome analysis of resting and CRP-activated platelets (GPVI signaling cascade); overall, 96 differentially regulated proteins were identified, including novel platelet signaling proteins, such as the adapters Dok1, and the type 1 transmembrane protein G6f; G6f phosphorylation on Tyr281 mapped by MS/MS, leading to an interaction with Grb2 |
| Senis et al., [37] | IP plus 1D SDS-PAGE | LC-MS/MS (QTOF) | 27 signaling proteins tyrosine phosphorylated in response to $\alpha IIb\beta 3$ outside-in signaling, including Dok1 and the novel adapter from the same family, Dok3; tyrosine phosphorylation of both proteins was primarily Src-independent downstream of the integrin, leading to an interaction with Grb2 and SHIP1 (potential negative regulation of integrin signaling); first time this is shown in platelets |

*Notation*: 2DGE—two-dimensional gel electrophoresis; SDS-PAGE—sodium dodecylsulfate–polyacrylamide gel electrophoresis; IP—immunoprecipitation; MALDI—matrix-assisted laser desorption ionization—Q—quadrupole; TOF—time of flight; LC-MS/MS—liquid chromatography–tandem mass spectrometry; TRAP—thrombin receptor activating peptide; GPVI—glycoprotein VI; Dok—downstream of tyrosine kinase; RGS18—regulator of G-protein signaling 18.
[a]Methods for protein separation and mass spectrometric analysis are listed in each case together with the key achievements.

any pre-fractionation, by high-resolution 2DGE to detect differentially regulated features that can be analyzed by MS. García and colleagues have used this approach to investigate intracellular signaling cascades in platelets stimulated with thrombin receptor activating peptide (TRAP), which is able to activate the main thrombin receptor PAR-1 [35]. Working under nonaggregating conditions, basal and TRAP activated platelet proteins were extracted by a very sensitive method and were separated by narrow-range pI 4–7 and 6–11 high-resolution 2D gels (18 × 18 cm). Gels were stained with a fluorescent dye prior to differential image analysis. That process led to the identification of 62 differentially regulated protein features. From these features, 41 were identified by LC-MS/MS and were found to derive from 31 different genes, most of which corresponded to signaling proteins. Several of the proteins identified had not previously been reported in platelets, including the adapter downstream of tyrosine kinase 2 (Dok2). Further studies revealed that the change in mobility of Dok2 was due to tyrosine phosphorylation. García and collaborators also provided the first demonstration of phosphorylation of the regulator of G-protein signaling, RGS18, and mapped one of the phosphorylation sites by MS/MS [35]. This report set the stage for future studies and illustrates the potential of 2DGE-based proteomics to study platelet signaling.

Following the success of the abovementioned study, García and colleagues carried out a similar platelet signaling proteomics study, this time by analyzing the proteome of collagen-related peptide (CRP)-activated platelets compared to the proteome of unstimulated platelets [36]. Glycoprotein VI (GPVI) is the major activation inducing receptor for collagen in platelets. CRP is an invaluable GPVI-specific agonist that has been extensively used to elucidate the GPVI signaling cascade. Since tyrosine kinases and phosphatases play a fundamental role in the GPVI signaling cascade, study of the tyrosine phosphoproteome was one of the main objectives of the study. García et al. used two main approaches for protein separation in order to address the issue described above: (1) 2DGE (pI 4–7 and 6–11) of whole-cell lysates, similar to the approach previously used by the same group to study PAR1 signaling cascade; and (2) phosphotyrosine immunoprecipitations (with agarose-conjugated 4G10 monoclonal antibody) followed by 1D PAGE. In both cases, proteins were identified by LC-MS/MS. By using this global approach, 96 proteins were found to undergo PTMs in response to CRP in human platelets, including 11 novel platelet proteins, such as the adapters Dok1, SPIN90, and osteoclast stimulating factor 1 (OSF1). Interestingly, the 2DGE-based approach yielded the detection of 111 differentially regulated protein features, corresponding to 72 different ORFs. Overall there was a shift of many proteins toward a more acidic region of the proteome following CRP stimulation, which was consistent with protein phosphorylation. A high proportion of the proteins identified through the 2DGE differential analysis were signaling (36%) and cytoskeletal (14%) proteins, consistent with the critical role played by these two groups of proteins in mediating platelet function and their regulation by phosphorylation. Some signaling proteins identified were adapters, including Dok1, Dok2, Gads, Grb2, and SKAP-HOM. However, the 2DGE approach failed to identify several proteins well known to participate in the GPVI signaling cascade. This could be a consequence of a number of factors, including a low level of expression of many tyrosine phosphorylated signaling proteins, their collocalization in the gels with other more highly expressed proteins, changes in spot volume below the selected threshold of twofold, or inability of certain proteins to be properly resolved by 2DGE because of solubility problems during IEF.

***4.2.3.2 Phosphotyrosine Immunoprecipitations and 1D SDS-PAGE*** In order to overcome the abovementioned limitations of the 2DGE-based study of the GPVI signaling cascade, García and colleagues decided to use a complementary approach. Central to this modified approach was the introduction of an affinity step prior to electrophoresis [36]. Indeed, as was mentioned above, the GPVI signaling cascade was investigated in detail by using a combination of phosphotyrosine immunoprecipitation and 1D PAGE followed by MS analysis. An advantage of this approach is the availability of reasonably good antiphosphotyrosine antibodies, such as the monoclonal antibody 4G10. Agarose-conjugated 4G10 was used to reduce masking by the heavy chain. In this study, phosphotyrosine proteins immunoprecipitated from basal and CRP-stimulated platelets were separated on 4–12% Bis-Tris gels and stained with the specific phosphoprotein gel stain Pro-Q Diamond (Invitrogen), which detects all sites of protein phosphorylation. Equivalent gels were run in parallel and stained for total protein with a fluorescent dye similar to SYPRO Ruby. Specific bands corresponding to tyrosine-phosphorylated proteins were excised from the gels and the proteins were in-gel trypsin-digested and identified by LC-MS/MS. This approach identified 30 different proteins specifically recruited to a phosphotyrosine signaling complex in response to CRP. Most of the known GPVI signaling proteins were identified, including three novel proteins, namely, β-Pix, the SH3 adapter protein SPIN90, and the type 1 transmembrane protein G6f. Interestingly, G6f was found to be specifically phosphorylated on Tyr281 in response to platelet activation with CRP, providing a docking site for the adapter Grb2 [36]. All the above data were validated by traditional biochemistical approaches (e.g., immunoprecipitations and Western blotting), which provided valuable information on the novel proteins identified and led to more detailed mechanistic studies.

The success of the abovementioned study in dissecting the GPVI signaling pathway set the stage for future similar studies. In 2009, the same group of researchers, based at Birmingham (UK), Oxford (UK), and Santiago de Compostela (Spain), carried out a detailed proteomic analysis of integrin αIIbβ3 outside-in signaling in human platelets [37]. Since the integrin αIIbβ3 signaling pathway is also based on tyrosine phosphorylation events, the chosen experimental approach was also based on phosphotyrosine immunoprecipitations. Washed platelets (1.5 mL at $5 \times 10^8$ $mL^{-1}$) were plated on either a BSA- or fibrinogen-coated surface for 45 min at 37°C. Proteins were immunoprecipitated with an antiphosphotyrosine antibody (4G10, agarose-conjugated). Six identical immunoprecipitations from lysates from the adhesion assays [bovine serum albumin (BSA)-nonadherent and fibrinogen-adhered human platelets] were pooled and resolved on 4–12% NuPAGE Bis-Tris gradient gels (Invitrogen). Following gel staining with a fluorescent dye equivalent to SYPRO Ruby, bands of interest were excised, protein-trypsin-digested, and analyzed by MS. The approach led to the identification of 27 proteins, 17 of which were not previously known to be part of a tyrosine phosphorylation-based signaling complex downstream of αIIbβ3. The group of proteins identified included the novel immunoreceptors G6f and G6b-B, and two members of the Dok family of adapters, Dok1 and Dok3, which underwent increased tyrosine phosphorylation following platelet spreading on fibrinogen. More recently, the Dok1 PTB domain was shown to bind to the cytoplasmic tail of the integrin β3 subunit. However, in contrast to the talin PTB domain, Dok1 negatively regulates integrin activation [38]. Oxley and colleagues demonstrated that the inhibitory effect mediated by Dok1 on αIIbβ3 signaling occurs through phosphorylation of the β3 integrin Tyr747. Phosphorylation of this site favors formation of a Dok1–integrin complex, over a talin–integrin complex [39]. These findings demonstrate how phosphorylation can act as a "molecular switch" that modulates integrin activation, and illustrates the importance of Dok proteins in integrin signaling. García and colleagues therefore focused on Dok proteins for further mechanistic and functional studies. Indeed, they showed that tyrosine phosphorylation of Dok1 and Dok3 was primarily Src kinase-independent downstream of the integrin. Moreover, both proteins inducibly interacted with Grb2 and SHIP1 in fibrinogen-spread platelets [37]. That was the first time that the abovementioned interactions were shown in platelets. In addition, the same study compared the role of Dok proteins downstream of the integrin αIIbβ3 and GPVI (following CRP or collagen stimulation) and showed that both Dok1 and Dok3 are differentially phosphorylated downstream of both receptors. Previous reports on B and T cells led the authors to hypothesize that Dok1 and Dok3 participate in a multi-molecular signaling complex, together with SHIP1 and Grb2, which may negatively regulate αIIbβ3 outside-in signaling.

## 4.3 CONCLUDING REMARKS

Since the 1990s proteomics has been widely applied to platelet signaling research. Despite the recent advances in multidimensional LC for protein separation prior to MS, gel-based separation methods have been, and still are, a preferable choice for differential proteomic analyses. In that context, high-resolution 2DGE has proved to be an invaluable tool for the differential analyses comparing basal and activated platelets. That approach has been applied not only to signaling studies but also to differential analyses comparing healthy and disease platelets. Platelet proteins can be prefractionated, or enriched in a certain subset of proteins, prior to separation and MS analysis. In the case of signaling cascades where tyrosine phosphorylation plays a fundamental role, the use of the 4G10 antibody (agarose-conjugated) for immunoprecipitations, followed by protein separation by 1D PAGE, has proved to be very successful in providing additional information to the 2DGE approach. Although the field is still young, in more recent years proteomics has proved to be an invaluable tool for dissecting the main platelet signaling cascades. This technology led to the identification of novel proteins that were or still are being studied further at a more mechanistic level. The primary objective of those studies is that some of the novel proteins identified may play a key role in platelet activation and be a good pharmacological target for thrombotic disease.

## ACKNOWLEDGMENTS

The author is a Ramón y Cajal Research fellow [Spanish Ministry of Science and Innovation (MICINN)], and the work presented in this chapter was supported by MICINN Grant SAF2007-61773 and the Fundación Mutua Madrileña (Spain).

## REFERENCES

1. O'Farrell PH, High resolution two-dimensional electrophoresis of proteins, *J. Biol. Chem*. **250**(10):4007–4021 (1975).
2. Klose J, Protein mapping by combined isoelectric focusing and electrophoresis of mouse tissues. A novel approach to testing for induced point mutations in mammals, *Humangenetik* **26**(3):231–243 (1975).
3. Karas M, Hillenkamp F, Laser desorption ionization of proteins with molecular mass exceeding 10000 daltons, *Anal. Chem*. **60**(20):2299–2301 (1988).
4. Fenn JB, Mann M, Meng CK, Wong SF, Whitehouse CM, Electrospray ionization for the mass spectrometry of large biomolecules, *Science* **246**(4926):64–71 (1989).
5. Aebersold R, Mann M, Mass spectrometry-based proteomics, *Nature* **422**(6928): 198–207 (2003).
6. Domon B, Aebersold R, Mass spectrometry and protein analysis, *Science* **312**(5771):212–217 (2006).
7. Westermeier R, Naven T, Höpker HR, *Proteomics in Practice: A Guide to Successful Experimental Design*, 2nd ed., Wiley-VCH, Weinheim, 2008.
8. García A, Watson SP, Dwek RA, Zitzmann N, Applying proteomics technology to platelet research, *Mass Spectrom. Rev*. **24**(6):918–930 (2005).
9. Washburn MP, Wolters D, Yates III JR, Large-scale analysis of the yeast proteome by multidimensional protein identification technology, *Nat. Biotechnol*. **19**(3):242–247 (2001).
10. García A, Two-dimensional gel electrophoresis in platelet proteomics research, *Meth. Mol. Med*. **139**:339–353 (2007).
11. Rabilloud T, Solubilization of proteins in 2DE: An outline, *Meth. Mol. Biol*. **519**:19–30 (2009).
12. Herbert BR, Molloy MP, Gooley AA, Walsh BJ, Bryson WG, Williams KL, Improved protein solubility in two-dimensional electrophoresis using tributyl phosphine as reducing agent, *Electrophoresis* **19**(5):845–851 (1998).
13. Righetti PG, Tudor G, Gianazza E, Effect of 2-mercaptoethanol on pH gradients in isoelectric focusing, *J. Biochem. Biophys. Meth*. **6**(3):219–227 (1982).

14. Görg A, Obermaier C, Boguth G, Harder A, Scheibe B, Wildgruber R, Weiss W, The current state of two-dimensional electrophoresis with immobilized pH gradients, *Electrophoresis* **21**(6):1037–1053 (2000).
15. Görg A, Weiss W, Dunn MJ, Current two-dimensional electrophoresis technology for proteomics, *Proteomics* **4**(12):3665–3685 (2004).
16. Weiss W, Görg A, High-resolution two-dimensional electrophoresis, *Meth. Mol. Biol.* **564**:13–32 (2009).
17. Carrette O, Burkhard PR, Sanchez JC, Hochstrasser DF, State-of-the-art two-dimensional gel electrophoresis: A key tool of proteomics research, *Nat. Protoc.* **1**(2):812–823 (2006).
18. Miller I, Crawford J, Gianazza E, Protein stains for proteomic applications: Which, when, why? *Proteomics* **6**(20):5385–5408 (2006).
19. Viswanathan S, Unlü M, Minden JS, Two-dimensional difference gel electrophoresis, *Nat. Protoc.* **1**(3):1351–1358 (2006).
20. García A, Senis Y, Tomlinson MG, Watson SP, Platelet genomics and proteomics, in Michelson AD, ed., *Platelets*, 2nd ed., Elsevier/Academic Press, 2007, pp. 99–116.
21. Marcus K, Moebius J, Meyer HE, Differential analysis of phosphorylated proteins in resting and thrombin-stimulated human platelets, *Anal. Bioanal. Chem.* **376**(7):973–993 (2003).
22. O'Neill EE, Brock CJ, von Kriegsheim AF, Pearce AC, Dwek RA, Watson SP, Hebestreit HF, Towards complete analysis of the platelet proteome, *Proteomics* **2**(3):288–305 (2002).
23. García A, Prabhakar S, Brock CJ, Pearce AC, Dwek RA, Watson SP, Hebestreit HF, Zitzmann N, Extensive analysis of the human platelet proteome by two-dimensional gel electrophoresis and mass spectrometry, *Proteomics* **4**(3):656–668 (2004).
24. Weyrich AS, Schwertz H, Kraiss LW, Zimmerman GA, Protein synthesis by platelets: Historical and new perspectives, *J. Thromb. Haemost.* **7**(11):1759–1766 (2009).
25. Kuznetsov VA, Knott GD, Bonner RF, General statistics of stochastic process of gene expression in eukaryotic cells, *Genetics* **161**(3):1321–1332 (2002).
26. Fernández Parguiña A, Grigorian-Shamajian L, Agra RM, Teijeira-Fernández E, Rosa I, Alonso J, Vinuela-Roldán JE, Seoane A, González-Juanatey JR, García A, Proteins involved in platelet signaling are differentially regulated in acute coronary syndrome: a proteomic study. *PLoS* One **5**(10):e13404 (2010).
27. Mateos-Cáceres PJ, Macaya C, Azcona L, Modrego J, Mahillo E, Bernardo E, Fernandez-Ortiz A, López-Farré AJ, Different expression of proteins in platelets from aspirin-resistant and aspirin-sensitive patients, *Thromb. Haemost.* **103**(1):160–170 (2010).
28. Loughrey Chen S, Huddleston MJ, Shou W, Deshaies RJ, Annan RS, Carr SA, Mass spectrometry-based methods for phosphorylation site mapping of hyperphosphorylated proteins applied to Net1, a regulator of exit from mitosis in yeast, *Mol. Cell. Proteom.* **1**(3):186–196 (2002).
29. Steen H, Kuster B, Fernandez M, Pandey A, Mann M, Tyrosine phosphroylation mapping of the epidermal growth factor receptor signaling pathway, *J. Biol. Chem.* **277**(2):1031–1039 (2002).
30. García A, Proteome analysis of signaling cascades in human platelets, *Blood Cells Mol. Dis.* **36**(2):152–156 (2006).

31. Immler D, Gremm D, Kirsch D, Spengler B, Presek P, Meyer HE, Identification of phosphorylated proteins from thrombin-activated human platelets isolated by two-dimensional gel electrophoresis by electrospray ionization-tandem mass spectrometry (ESI-MS/MS) and liquid chromatography-electrospray ionization-mass spectrometry (LC-ESI-MS), *Electrophoresis* **19**(6):1015–1023 (1998).
32. Gevaert K, Eggermont L, Demol H, Vandekerckhove J, A fast a convenient MALDI-MS based proteomic approach: Identification of components scaffolded by actin cytoskeleton of activated human thrombocytes, *J. Biotechnol*. **78**(3):259–269 (2000).
33. Maguire PB, Wynne KJ, Harney DF, O'Donoghue NM, Stephens G, Fitzgerald DJ, Identification of the phosphotyrosine proteome from thrombin activated platelets, *Proteomics* **2**(6):642–648 (2002).
34. Marcus K, Moebius J, Meyer HE, Differential analysis of phosphorylated proteins in resting and thrombin-stimulated human platelets, *Anal. Bioanal. Chem*. **376**(7):973–993 (2003).
35. García A, Prabhakar S, Hughan S, Anderson TW, Brock CJ, Pearce AC, Dwek RA, Watson SP, Hebestreit HF, Zitzmann N, Differential proteome analysis of TRAP-activated platelets: Involvement of DOK-2 and phosphorylation of RGS proteins, *Blood* **103**:2088–2095 (2004).
36. García A, Senis YA, Antrobus R, Hughes CE, Dwek RA, Watson SP, Zitzmann N, A global proteomics approach identifies novel phosphorylated signaling proteins in GPVI-activated platelets: Involvement of G6f, a novel platelet Grb2-binding membrane adapter, *Proteomics* **6**(19):5332–5343 (2006).
37. Senis YA, Antrobus R, Severin S, Parguiña AF, Rosa I, Zitzmann N, Watson SP, García A, Proteomic analysis of integrin alphaIIbbeta3 outside-in signaling reveals Src-kinase-independent phosphorylation of Dok-1 and Dok-3 leading to SHIP-1 interactions, *J. Thromb. Haemost*. **7**(10):1718–1726 (2009).
38. Wegener KL, Partridge AW, Han J, Pickford AR, Liddington RC, Ginsberg MH, Campbell ID, Structural basis of integrin activation by talin, *Cell* **128**(1):171–182 (2007).
39. Oxley CL, Anthis NJ, Lowe ED, Vakonakis I, Campbell ID, Wegener KL, An integrin phosphorylation switch: The effect of beta3 integrin tail phosphorylation on Dok1 and talin binding, *J. Biol. Chem*. **283**(9):5420–5426 (2008).

# 5

# THE PLATELET MEMBRANE PROTEOME

YOTIS A. SENIS

**Abstract**

The platelet membrane proteome encompasses all proteins embedded in or associated with the various platelet lipid bilayers. Mapping the platelet plasma membrane proteome is of particular interest in terms of both understanding platelet biology and function, and the development of novel antiplatelet therapies. More recent developments in membrane proteomics strategies have greatly enhanced the identification of low-abundance hydrophobic receptors and membrane-associated signaling proteins. In this chapter, theoretical and practical aspects of platelet membrane proteomics are discussed. Future areas of investigation are also highlighted.

## 5.1 INTRODUCTION: CLASSIFICATION OF MEMBRANE PROTEINS

It is estimated that more than half of all proteins interact either directly or indirectly with membranes in cells [1,2]. Membranes are typically composed of 50% proteins by mass, which translates to a lipid/protein ratio of 50 lipid molecules to one protein molecule [1]. However, this ratio varies depending on the cell or organelle from 82% lipids and 18% proteins in the case of myelin, to 25% lipids and 75% proteins in mitochondria [1]. Membrane proteins are classified as either *integral* or *peripheral membrane proteins* (IMPs or PMPs, respectively)

*Platelet Proteomics: Principles, Analysis and Applications*, First Edition.
Edited by Ángel García and Yotis A. Senis.

(Fig. 5.1). IMPs are permanently attached to the membrane and are divided into transmembrane proteins that span the entire membrane, and *monotopic* proteins that are permanently attached to only one side of the membrane, such as glycosylphosphoatidylinositol (GPI)-anchored receptors (Fig. 5.1). Transmembrane IMPs are further divided into *bitopic* and *polytopic* proteins that contain either a single or multiple transmembrane domains (TMDs), respectively [3]. All transmembrane proteins contain a hydrophobic membrane spanning domain

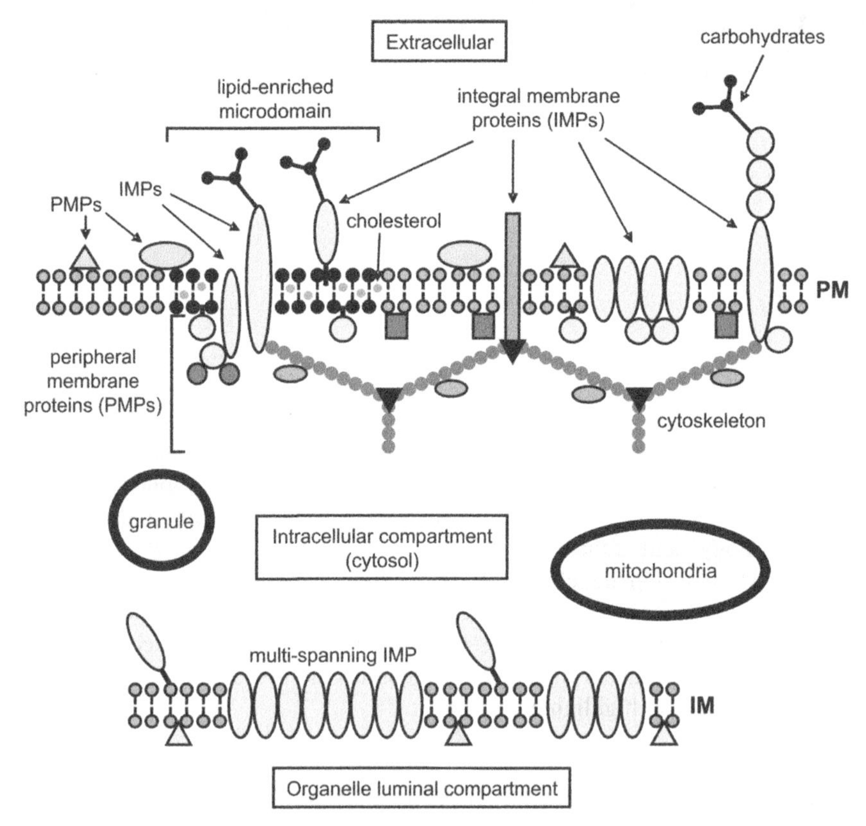


**Figure 5.1** Integral and peripheral membrane proteins. Integral membrane proteins (IMPs) are permanently attached to the membrane. IMPs are divided into transmembrane proteins that span the entire membrane, and integral *monotopic* proteins that are permanently attached to only one side of the membrane. Transmembrane proteins may be single- or multispanning (*bitopic* or *polytopic*, respectively). Peripheral membrane proteins (PMPs) are temporarily associated with either lipid molecules or IMPs through a combination of hydrophobic, electrostatic, and other transient noncovalent interactions. The cytoskeleton interacts directly and indirectly with the cytosolic surface of the plasma membrane (PM). The PM is compartmentalized into functionally distinct microdomains with unique lipid and protein compositions (e.g., cholesterol and activatory receptors). Intracellular membranes (IMs) have distinct lipid and protein compositions compared with the PM. Extracellular membrane proteins are often glycosylated (carbohydrates).

that has either an α-helical or β-barrel structure [3]. α-Helical proteins are found in all cellular membranes, with the exception of the outer membrane of Gram-negative bacteria, whereas β-barrel proteins are expressed only in specific membranes (e.g., mitochondria). According to a bioinformatics study in 2009, it is estimated that 27% of the total human proteome are α-helical transmembrane proteins [2]. In contrast, PMPs are temporarily associated with either lipid molecules or IMPs embedded in the membrane through a combination of hydrophobic, electrostatic, and other transient noncovalent interactions [2]. Many of the interactions of PMPs with the PM take place in response to cell activation, either as a result of a change in the lipid composition of the membrane or through post-translational modification of an IMP. For example, tyrosine phosphorylation of the integral membrane adapter protein LAT leads to the recruitment of the SH2-domain-containing cytosolic proteins phospholipase Cγ2, Grb2, and Gads to the cytosolic surface of the PM [4]. Similarly, increased levels of phosphatidylinositol 3,4,5-trisphosphate (PIP3) in the intracellular surface of the PM leads to recruitment of PH-domain-containing cytosolic proteins [5], while increased expression of phosphatidylserine (PS) on the extracellular surface of the PM recruits components of the coagulation cascade [6]. Since most of these interactions have low affinity, PMPs can be dissociated from membranes with high salt and high pH conditions. (*Note*: In the author's experience, contaminating cytoskeletal PMPs cannot be completely removed.) IMPs and PMPs may be posttranslationally modified with lipid side-chains added (myristoyl, plamitoyl, prenyl, or GPI) that help localize and anchor the protein to the membrane.

Transmembrane PM proteins can be divided into three major functional groups: receptors, transporters, and enzymes [2]. Receptors mediate cellular responses on binding of a ligand, most of which belong to one of four superfamilies: G-protein-coupled receptors (901 proteins), immunoglobulin and related receptors (149 proteins), receptor-type tyrosine kinases (72 proteins), and scavenger and related receptors (63 proteins) [2]. Transporters move substances across membranes through electrochemical gradients or energy from chemical reactions [2]. The three major functional classes are: channels (247 proteins), solute carriers (393 proteins), and active transporters (81 proteins) [2]. Transmembrane enzymes are grouped into six major classes: oxidoreductases (123 proteins), transferases (194 proteins), hydrolases (178 proteins), lyases (17 proteins), isomerases (8 proteins), and ligases (7 proteins) [2]. Transmembrane proteins that do not fit into any of the three major functional classes are classified as miscellaneous and can be further subclassified as ligands (57 proteins), structural/adhesion (187 proteins), proteins of unknown function (181 proteins), and other (272 proteins) [2].

Identifying the complete repertoire of proteins expressed on the surface of resting and activated platelets has important implications for understanding how platelets function under normal physiological conditions and under pathological conditions. However, as illustrated above, establishing the criteria to classify a protein as a membrane protein, or indeed demonstrate its association with the membrane, is subject to debate and experimental conditions. It has been estimated that as many as 70% of all known drug targets are PM [7]. Therefore, mapping the platelet PM proteome will undoubtedly lead to a better understanding of the molecular mechanisms regulating platelet activation and thrombosis, and identification of novel antithrombotic drug targets.

## 5.2 CHALLENGES OF MEMBRANE PROTEOMICS

The two major difficulties in identifying α-helical IMPs are (1) their hydrophobic properties, which make them difficult to solubilize and analyze by MS; and (2) their low abundance [3,8,9]. Large proteins containing multiple hydrophobic TMDs do not solubilize well in the isoelectric focusing (IEF) sample buffer, and those that do solubilize are prone to precipitate at their pI [3]. The limitation in the dynamic range of detection of proteomics is also an issue as many membrane proteins are typically present at low levels. Another complication is high abundance, contaminating PMPs attached to the inner surface of the PM. These factors therefore render it essential to enrich surface proteins prior to resolving by a chromatographic technique that does not involve gel-based IEF.

The most successful platelet membrane proteomics studies to date employed the following general strategy: membrane enrichment, removal of already annotated PMPs, protein/peptide separation by 1DGE or nonclassical 2DGE (defined in Section 5.4), band excision and in-gel trypsinization, and peptide identification by LC-MS/MS (Fig. 5.2). More recently, membrane enrichment followed by either combined fractional diagonal chromatography (COFRADIC) or the commonly used "shotgun" approach, multidimensional protein identification technology (MudPIT), have also been successfully applied to identifying platelet IMPs (see Fig. 5.2 and Chapter 8) [10]. It is now well established that membrane enrichment and protein separation by nonclassical 2DGE are critical steps for successful IMP identification. Also important is removal of associated PMPs and peptide separation by either 1D or 2D liquid chromatography. Optimizing the chromatographic step used to introduce the sample into the mass spectrometer improves IMP identification and must be carefully considered when devising a strategy for identifying IMPs [8]. Membrane enrichment reduces sample complexity by removing abundant PMPs, which enhances the dynamic range of the sample and increases the chances of identifying low-abundance proteins. However, in the case of the PM, the associated proteins may be of interest as important signaling and cytoskeletal proteins interact with the PM in resting and activated conditions.

## 5.3 ENRICHING MEMBRANE PROTEINS

Since membrane enrichment is such a critical step for identifying IMPs, several of the most commonly used techniques are described below. A summary is provided in Table 5.1, along with advantages and disadvantages of each approach.

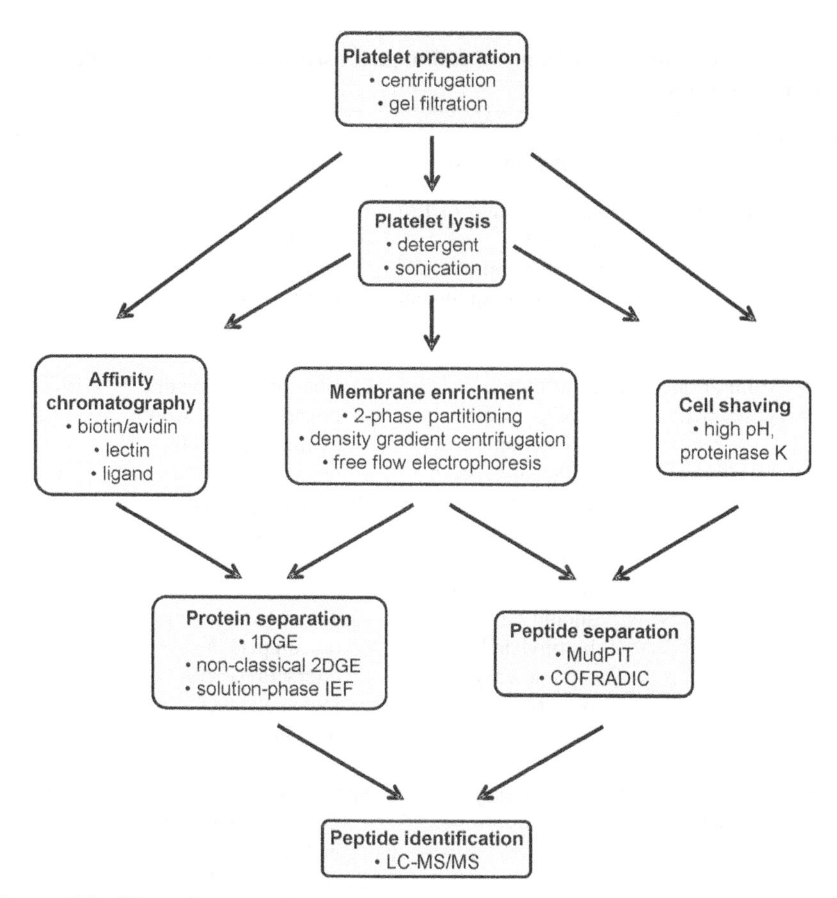


**Figure 5.2** Flow diagram of steps involved in characterizing the platelet membrane proteome. The backbone of the strategy involves (1) platelet preparation, (2) enrichment of membrane proteins, (3) protein/peptide separation, (4) peptide identification by LC-MS/MS [1DGE—one-dimensionalgel electrophoresis; 2DGE—two-dimensionalgel electrophoresis; IEF—isoelectric focusing; MudPIT—multidimensional protein identification technology; COFRADIC—combined fractional diagonal chromatography; LC-MS/MS—liquid chromatography–tandem mass spectrometry].

### 5.3.1 Chemical Precipitation and Density Gradient Centrifugation

Chemical precipitation and/or density gradient centrifugation is one of the most common approaches for enriching biological membranes. Aqueous two-phase purification is such an approach, in which membrane proteins are separated from soluble proteins according to hydrophobicity, by differential centrifugation or density sedimentation through a defined two-polymer system [11]. Poly(ethylene glycol) (PEG) 3350/dextran T-500 is one of the most common two-polymer systems. In two-phase partitioning, vesicles are sorted according to their hydrophobicity/hydrophilicity and net surface charge, which may be partially due to their

**TABLE 5.1 Membrane Enrichment Techniques Used in Platelet Membrane Proteomics Studies**

| Enrichment Technique | Advantages | Disadvantages | Reference(s) |
|---|---|---|---|
| Aqueous 2-phase partitioning | Good separation of PM and IM<br>Does not require specialist equipment | High-salt, high-pH conditions are required to remove PM associated PMPs | 10,46 |
| Biotin–avidin affinity chromatography | High-affinity biotin–avidin interaction allows stringent wash conditions to remove PMPs<br>All surface proteins with primary amine should be biotinylated<br>Cheap and easy to perform | Can be difficult to elute bound proteins from avidin/streptavidin resin<br>Biotinylating reagent may enter cell and biotinylate cyotoslic proteins<br>Abundant proteins may outcompete interactions with low abundance proteins | 19,37 |
| Lectin affinity chomratography | Cheap and easy to perform | Specific subpopulations of receptors will be isolated<br>Competition with secreted glycoproteins<br>Stringent wash conditions may elute some glycosylated IMPs | 19,45 |
| Freeflow electrophoresis | Good separation of PM and IM | High-salt, high-pH conditions are required to remove PM associated PMPs<br>Labor-intensive<br>Requires specialist equipment | 19 |

*Notation*: IM—intracellular membrane; IMP—integral membrane protein; PM—plasma membrane; PMP—peripheral membrane protein.

phospholipid composition. In the PEG/dextran system plasma membranes (PMs) have the highest affinity for the more hydrophobic upper PEG phase compared with vesicles composed of other membranes. Additional wash steps using high salt and high pH conditions can be incorporated to remove associated proteins once the membranes have been isolated. Lewandrowski et al. used a two-phase system consisting of 6.3% PEG 3350 and dextran T-500 to obtain enriched platelet PMs that were subsequently washed with 100 mM sodium carbonate, pH 11.5, to remove associated proteins [10]. Although density gradient ultracentrifugation methods have been used extensively for many years, they do not have the resolution to provide highly purified PM fractions, and therefore can only provide enrichment (Table 5.1).

### 5.3.2 Affinity Chromatography

Affinity chromatography is another approach that is routinely used to isolate membrane proteins. Compounds (e.g., plant lectins) that recognize unique extracellular features of surface proteins are used to pull down a broad spectrum of IMPs from complex cell lysates (Table 5.1). Another commonly used technique for the global identification of IMPs involves biotinylation of surface proteins in intact cells followed by purification with immobilized avidin. Affinity chromatography generally provides more purified IMPs compared with chemical and density gradient precipitation. However, fewer IMPs are usually identified, due to the specificity of the approach, competition for binding sites, and incomplete elution. The biotin–avidin method has been more successful at identifying IMPs than has lectin affinity chromatography, which is better suited for the analysis of glycosylation sites of surface membrane proteins. More refined affinity chromatography approaches can also be employed for receptor hunting by coupling the ligand of interest to a solid-phase matrix and using it to pull down the receptor from a complex cell lysate.

***5.3.2.1 Lectin Affinity Chromatography*** Lectin affinity chromatography takes advantage of the fact that lectins interact specifically and reversibly with sugar residues and the majority of cell surface proteins are glycosylated to varying degrees. Bound proteins can be competed off the lectin column with the sugar residue or an ionic gradient. Since different lectins target distinct classes of proteins, this finding can be utilized to isolate subpopulations of receptors. For example, wheatgerm agglutinin (WGA), binds molecules containing *N*-acetyl-β-D-glucosamine residues, whereas concanavalin A (ConA) binds molecules containing α-D-mannosyl residues, α-D-glucosyl residues, and branched mannoses [12]. One drawback of lectin affinity chromatography is that not all surface proteins are glycosylated or have an appropriate sugar residue to support binding. An additional consideration is that, in a detergent-solubilized cell lysate, the lectin will bind proteins present in intracellular secretory granules, as these also tend to be highly glycosylated. This can result in secretory proteins outcompeting low-abundance IMPs for binding sites.

*5.3.2.2* ***Biotin–Avidin Affinity Chromatography*** An alternative approach to lectin affinity chromatography is biotin–avidin affinity chromatography. This technique involves biotin-labeling surface proteins with a water-soluble, membrane-impermeable reagent, followed by disrupting the cells either with a detergent or by sonication and affinity isolating the biotinylated proteins with immobilized avidin [13]. Sulfosuccinimidyl-2-[biotinamido]ethyl-1,3-dithiopropionate (sulfo-NHS-SS-biotin) is a commercially available biotinylation reagent that is membrane-impermeable and therefore limits contamination of PM with intracellular proteins. The labeling is based on the reaction of primary amines, primarily on lysine residues on the extracellular region of membrane proteins with hydroxysulfosuccinimide esters on the biotinylation reagent. This approach has the advantages that the majority of surface proteins are biotinylated and that the interaction with avidin is of very high affinity ($k_d \sim 10^{-15}$ M), enabling use of stringent washing conditions and therefore removal of many associated proteins. The use of modified forms of biotinylation reagents and avidin make it less likely that intracellular proteins will be labeled and easier to release bound proteins prior to analysis [14]. This approach has been successfully used by a number of groups to identify surface proteins on both prokaryotic and eukaryotic cells [15–19].

*5.3.2.3* ***Ligand Receptor Affinity Chromatography*** Affinity chromatography can also be used to identify specific receptors if the affinity of the interaction is sufficiently high to withstand detergent-mediated solublization of the cell and washing conditions to remove nonspecifically associated proteins. For example, the novel platelet hemi-immunoreceptor tyrosine-based activation motif (hemITAM) receptor C-type lectin-like receptor 2 (CLEC-2) was identified in this way. Search for CLEC-2 was hypothesis-driven as the snake toxin rhodocytin was shown to induce platelet activation and aggregation of platelets lacking GPVI, GP1bα, and α2β1 [20]. CLEC-2 was pulled down from platelet lysates prepared using rhodocytin conjugated to a solid-phase matrix [21]. Subsequent biochemical and functional studies in cell lines confirmed that CLEC-2 is in fact the platelet receptor of rhodocytin, plays an important role in lymphatic vascular development, and has been implicated in hemostasis and thrombosis, although the latter is controversial [21–26].

*5.3.2.4* ***Antibody Affinity Chromatography*** Antibody affinity chromatography has also been used to identify novel platelet surface receptors. Examples include PEAR1, G6b-B, and G6f, all of which are phosphorylated in activated platelets, and were identified by immunoprecipitation from platelet lysates using an antiphosphotyrosine antibody, followed by separation by 1DGE, isolation of bands, and identification by LC-MS/MS [27–29].

### 5.3.3 Freeflow Electrophoresis

Freeflow electrophoresis (FFE) involves separation of the PM from IMs by electrophoretic means. Intact cells are first treated with neuraminidase, which reduces the negative charge of the PM by removal of sialic acid residues from glycosidic sidechains on platelet surface glycoproteins [30]. The mixture of plasma and intracellular membranes is separated from the granules by sorbitol density gradient centrifugation. The PM/IM fraction is subsequently separated by FFE. Although this approach works well, it is labor-intensive and requires specialized equipment. A further limitation of this approach is that membrane-associated cytoskeletal proteins remain attached to the PM and can mask low-abundance receptors, as is the case with aqueous two-phase partitioning (described previously). This can be partially circumvented through the use of high-salt, high-pH washes. FFE significantly outperformed WGA and biotin–avidin affinity chromatography in terms of platelet PM and IMPs identified, despite a higher level of contamination with membrane-associated PMPs [19].

### 5.3.4 Membrane Shaving

The aim of this technique is to circumvent the problematic hydrophobic regions of IMPs and focus on analyzing the hydrophilic extracellular regions of surface-exposed membrane proteins. In theory this technique would provide information about the surface topology of the cell and eliminate the need for solubilization of hydrophobic proteins. Although the concept has been around for a long time, devising an effective method has been difficult, mainly because of cell instability during the protease treatment and subsequent centrifugation to remove cells. Cell lysis results in significant contamination by abundant cytoplasmic proteins. More recently, a method has been reported whereby membranes are induced to form open vesicles by mechanical agitation in high-pH buffer [31]. Subsequent treatment with the nonspecific protease, proteinase K, digests all surface-exposed proteins and soluble proteins. A nonspecific enzyme is used so that cleavage of extracellular regions is nonselective. Digested, soluble peptides are then analyzed by MudPIT in which LC-MS/MS was performed at elevated temperatures [31,32]. However, problems associated with cell or vesicle lysis and large numbers of peptides being generated with the nonspecific protease K have hampered more extensive use of this technique [11]. It has yet to be employed for platelet membrane proteomics.

## 5.4 PROTEIN SEPARATION STRATEGIES

The main purpose of separating proteins prior to MS is to reduce sample complexity and enhance the chances of identifying low-abundance IMP and signaling proteins. Since the classical 2D gel-based protein separation approach (IEF followed by SDS-PAGE) is not amenable to separating proteins containing hydrophobic TMDs, other gel-based strategies are being employed, including 1DGE and nonclassical 2D approaches. Underrepresentation of hydrophobic TMDs in any of the gel-based protein separation techniques described below may be due to (1) the lack of Lys and Arg residues in hydrophobic TMDs and/or structural inaccessibility or (2) the inability to extract long hydrophobic peptides from the gel.

Solution-phase IEF is a gel-free method for separating proteins that has yet to be extensively applied to the separation of IMPs, but holds promise. Anion and cation exchange and reversed-phase liquid chromatography are routinely used for separation of peptides as part of shotgun proteomics strategies of complex protein digests, but can also be applied for protein separation. Brief descriptions of protein separation strategies are given below, along with advantages and disadvantages of each.

### 5.4.1 Classical 2DGE

Isoelectric focusing (IEF) followed by SDS-PAGE has long been the "classical" method for analysis of protein mixtures. Resolved protein spots are excised from gels, subjected to in-gel trypsin digestion, and identified by MS. However, the classical 2DGE approach is not compatible with the identification of IMPs, as hydrophobic IMPs tend to precipitate out of solution when their pI is reached, preventing transfer into the second SDS-PAGE dimension. Strong detergents such as SDS used for solublizing IMPs are incompatible with the low-ionic-strength requirements of IEF. Ionic compounds in solution will also severely alter protein migration. As a result, IEF solubilisation buffers contain nonionic or zwitterionic compounds that are not as effective at solubilizing IMPs. Although alternative detergents can be used to alleviate some of these challenges, classical 2DGE remains problematic for separating IMPs. This is commonly recognized by the proteomics community and a variety of alternative gel- and solution-based techniques are now applied in conjunction with MS for the identification of IMPs.

### 5.4.2 Nonclassical 2DGE

Three nonclassical 2DGE techniques that have proved to be applicable for resolving IMPs are (1) blue native (BN)/SDS-PAGE, (2) 16-benzyldimethyl-*n*-hexadecylammonium chloride (16-BAC)/SDS-PAGE, and (3) SDS/SDS-PAGE [33]. BN/SDS-PAGE was originally developed to determine the mass and oligomeric state of mitochondrial membrane proteins [34]. Protein–protein interactions are often maintained using this method as mild solubilization and running conditions are used. Proteins are initially resolved in the presence of the anionic dye Coomassie brilliant blue G-250, which binds hydrophobic proteins, conferring a net negative charge, allowing electrophoretic mobility and aqueous solubility, and reducing protein aggregation and precipitation. Samples are subsequently separated by SDS-PAGE in the second dimension, which resolves the individual components of each complex. BN/SDS-PAGE has proved to be more successful than traditional 2DGE at resolving IMPs, including multispanning IMPs [3]. However, this technique has yet to be applied to resolving platelet IMPs.

16-BAC/SDS-PAGE separates proteins according to molecular mass in a discontinuous acidic gradient (pH 4–1.5) using the cationic detergent 16-BAC in the first dimension and standard SDS-PAGE in the second dimension [33,35]. 16-BAC is better at solubilizing membrane proteins compared with nonionic/zwitterionic detergents used for IEF, but not as good as SDS. This technique has been used to identify mitochondrial membrane proteins as well as human platelet membrane proteins [35,36].

### 5.4.3 SDS/SDS-PAGE

A third 2DGE approach is SDS/SDS-PAGE, where a low percentage acrylamide gel is used to separate proteins in the first dimension and a high acrylamide gel is used to separate proteins in the second dimension [33]. This variant of 2DGE has not been as widely implemented as BN/SDS-PAGE or 16-BAC/SDS-PAGE; however, it reportedly outperformed 16-BAC/SDS-PAGE in the identification of IMPs [37].

***5.4.3.1 One-Dimensional Gel Electrophoresis (1DGE)*** This method has been demonstrated by several groups to provide adequate protein separation for protein identification by LC-MS/MS. It also outperforms any of the 2DGE approaches that it has been compared with for identification of IMPs, and has the advantage of being well established, easy to use, and highly reproducible. 1DGE also has the capacity to separate proteins with a wide range of molecular masses, pI values, and hydorphobicities, mainly due to the SDS used to solubilize and denature proteins. 1DGE in conjunction with LC-MS/MS is routinely reported to identify ~125–200 proteins, with ~20–55% IMPs [3]. In some cases, the 2DGE-LC-MS/MS platform even rivals shotgun approaches in terms of the number of proteins identified. The highest identification rate using this platform came from Park et al., who identified >1300 proteins (60% of which were integral, anchored, or associated membrane proteins) in the membrane fraction isolated from human brain samples [38].

***5.4.3.2 Solution-Phase IEF*** Solution-phase IEF is a gel-free method of separating proteins based on pI values that has the potential to circumvent the precipitation and transfer problems associated with gel-based IEF [3]. However, it has yet to be extensively applied to the separation of membrane proteins. Protein precipitation should in theory be minimized compared with gel-based IEF. Solution-phase IEF fractionated proteins can be further separated by gel-based techniques, as was done by Kaiser et al. and Tucker et al. in their studies investigating platelet PMPs in resting, CRP- and thrombin-activated platelets, respectively [39,40].

## 5.5 SHOTGUN STRATEGIES FOR IDENTIFYING INTEGRAL MEMBRANE PROTEINS

Given that most protein spots and bands contain multiple proteins, separation of peptides prior to MS analysis is a critical factor for maximizing protein identification by gel-based protein separation methods. Peptides from low-abundance IMPs may be masked if peptides are not resolved, which is why 1DGE is coupled to μLC prior to MS/MS. "Shotgun" proteomics refers to the gel-free separation of complex peptide mixtures followed by MS analysis. A variety of peptide separation techniques are used either alone or in combination, including size exclusion, anion exchange, hydrophobic or hydrophilic interaction chromatography, capillary electrophoresis, capillary IEF, and FFE [9].

MudPIT is the most widely used shotgun approach. It involves peptide separation by strong cation exchange and reversed-phase (RP)-LC followed by MS/MS analysis [9,31,41]. MudPIT has been used extensively to analyze membranes from various sub-cellular compartments, including PMs [10], rat nuclear membrane [42], Golgi apparatus [43], and yeast mitochondria [44]. Thousands of protein identifications are typically reported, representing two- to fivefold more capacity than 1DGE-LC-MS/MS, with IMP enrichment ranging from ~20% to 65% at most, which is comparable to the 1DGE approach [3].

## 5.6 MAPPING THE PLATELET MEMBRANE PROTEOME

In initial platelet proteome studies utilizing the classical 2DGE-LC-MS/MS platform, IMPs and associated signaling PMPs were severally underrepresented (Table 5.2). The main reasons for these shortcomings were (1) lack of membrane enrichment and (2) incompatibility of hydrophobic proteins with classical 2DGE. More specialized techniques involving various membrane enrichment strategies followed by 1DGE-LC-MS/MS greatly enhanced identification of membrane proteins (Table 5.2). More recently, use of membrane enrichment in conjunction with MudPIT and COFRADIC approaches has resulted in a major advance in the identification of platelet IMPs and PMPs in resting human platelets (see Table 5.2 and Chapter 8). With the vast improvements in IMP identification by MS more recently, more focused studies have been undertaken, investigating glycosylation sites of platelet IMPs, and PMP recruitment to the PM in activated platelets. Findings from some of the landmark platelet membrane proteomics studies to date are summarized below.

### 5.6.1 Shortcomings of Early Classical Platelet Proteomics Studies

Until relatively the past few years, there was a dramatic underrepresentation of membrane proteins identified in platelets using the classical 2DGE-LC-MS/MS platform. Only the most abundant platelet PM proteins, namely, the integrin αIIbβ3 and GPIb-IX-V, were being identified. In one of the first global platelet proteomics studies by O'Neill et al., which focused on proteins in the 4–5 pI range, only two IMPs were identified out of 284 protein features from 123 different open reading frames [45]. In a subsequent study from the same group, García et. al. extended their search to proteins in the 5–11 pI range. In this case, out of 760 protein features identified, corresponding to 311 open reading frames, only 3% or 9 transmembrane proteins were identified [46]. Results were not much better even when a non-gel-based protein separation strategy was used. Gevaert et al. identified only 11 proteins that contain at least one TMD of 163 proteins encoded by different open reading frames using the COFRADIC proteomic approach (described in Chapter 8) [47]. The main reasons for these

**TABLE 5.2 Platelet Membrane Proteome Strategies**

| Study | Objective | Membrane Protein Enrichment | Protein/ Peptide Separation | Number of TMD-Containing Proteins Identified |
|---|---|---|---|---|
| O'Neill et al. [42] | Global platelet proteome | None | 2DGE | 2 |
| García et al. [43] | Global platelet proteome | None | 2DGE | 9 |
| Geavert et al. [44] | Global platelet proteome | None | COFRADIC | 11 |
| Moebius et al. [33] | Global platelet membrane proteome | Aqueous 2-phase partitioning | 1DGE, 16BAC/PAGE | 131 |
| Lewandrowski et. al. [45] | Global platelet *N*-glycosylated proteome | ConA affinity chromatography; hydrazine affinity capture | None | 20 |
| Lewandrowski et. al. [46] | Global platelet *N*-glycosylated proteome | Aqueous 2-phase partitioning | SCX | 68 |
| Guerrier et. al. [56] | Global low-abundance platelet proteome | None | Combinatorial ligand library, 2DGE | 0 |
| Senis et al. [19] | Global platelet membrane proteome | WGA chromatography; biotion/avidin chromatography; FFE | 1DGE | 136 |
| Lewandrowski et al. [10] | Global platelet membrane proteome | Aqueous 2-phase partitioning | 1DGE, MudPIT, COFRADIC | 626 |

*Notation*: 1DGE—one-dimensional electrophoresis; 2DGE—two-dimensional electrophoresis; 16BAC—16-benzyldimethyl-*n*-hexadecylammonium chloride; COFRADIC—combined fractional diagonal chromatography; FFE—freeflow electrophoresis; MudPIT—multidimensional protein identification technology; TMD—transmembrane domain; WGA—wheatgerm agglutinin; SCX-strong cation exchange.

shortcomings were the lack of membrane enrichment and the use of classical 2DGE in the first two cases.

### 5.6.2 Modified Approaches to Mapping of the Platelet Membrane Proteome

The first platelet membrane proteomics study, by Moebius et al. in 2005 [36], employed membrane enrichment and 1DGE rather than 2DGE to improve IMP identification. These changes resulted in a sixfold increase in the number of IMPs identified, compared with previous studies [36]. Washed human platelets were prepared and treated with neuraminidase to remove sialyic acid sidechains and reduce the net negative charge on the platelets [36]. Platelets were subsequently mechanically lysed by sonication and membranes enriched by discontinuous sorbital gradient centrifugation. Proteins in the crude membrane pellet were solubilized in carbonate buffer/Triton X-114 and precipitated with TCA/acetone. Samples were resolved by 16-BAC/SDS-PAGE, gels were silver-stained, bands excised, and proteins digested with trypsin. Peptides were separated by nano-HPLC and identified by electrospray ionization (ESI)ion trap, ESI-QTOF, or ESI linear ion trap mass spectrometers. All proteins identified were checked for TMDs using two algorithms, SOUSI and TMHMM. This study was groundbreaking in that 83 PM proteins were identified, several of which had never before been shown to be expressed in platelets, including G6b-A and several hypothetical and unknown proteins [36]. In addition, 48 TMD-containing proteins were identified that are localized in other cellular compartments, including mitochondria, endoplasmic reticulum (ER), and vesicles. This was the first time that low/intermediate-abundance membrane proteins were identified, including GPVI (3000–5000 copies/platelet) and the integrin subunits α2, α5 and α6, which are expressed at 1000–3000 copies each. A shortcoming of this study was that no G-protein-coupled receptors were identified. This is likely due to the low abundance and low solubility of multispanning proteins, and because hydrophobic regions seldom contain any basic amino acids, leading to long tryptic fragments, which cannot be analyzed classically by RP separation.

Twenty platelet TMD-containing proteins (including the novel platelet immunoreceptor G6f) were identified in a subsequent study investigating *N*-glycosylation sites in human platelet glycoproteins (see also Chapter 2) [48]. Lewandrowski et al. [48] used two techniques to enrich platelet *N*-glycosylation sites: (1) lectin (ConA) affinity chromatography (2) and chemical derivatization of glycan residues and trapping of glycosylasted proteins on hydrazide-functionalized resins. In total, 41 different glycoproteins were identified using these techniques that were almost exclusively surface transmembrane or secreted proteins. It is not surprising that so few PM proteins were identified using these techniques, as not all surface glycoproteins are *N*-glycosylated and more abundant proteins will outperform less abundant proteins for binding to either the ConA or hydrazide resins. This study demonstrated that lectin affinity enrichment of membrane proteins enhances the identification IMPs by mass spectrometry. However, a novel strategy devised by the same group termed *enhanced N-glycosylation site analysis using strong cation exchange enrichment* (ENSAS) involving aqueous two-phase membrane partitioning followed by strong cation exchange chromatography outperformed the abovementioned approaches for identification of platelet PM proteins *N*-glycosylation sites, identifying 68 TMD-containing proteins [49].

The next major global platelet membrane proteomics study, by Senis et al. [19], followed a strategy similar to that of Moebius et al., in that membranes were first enriched then resolved by a gel-based technique not involving IEF [19]. It differed in that different membrane enrichment strategies were used and proteins were resolved by 1DGE rather than a nonclassical 2DGE approach. The three membrane enrichment strategies used in the Senis et al. study included (1) wheatgerm agglutinin (WGA) affinity chromatography, (2) biotin–avidin affinity chromatography, and (3) FFE. Tryptic fragments of enriched proteins were identified by nano-HPLC ion trap and QTOF mass spectrometers [19]. A total of 21 peripheral membrane (PM) and 2 intracellular membrane (IM) proteins were identified by two or more peptide hits when WGA was used to enrich membranes. This was comparable to the number of TMD-containing proteins identified by Sickmann's group using ConA to affinity-enrich PM proteins [48]. Although lectin affinity enrichment greatly enhanced the number of IMPs identified compared with classical 2DGE proteomics approaches, there was still a large deficit in the number of known platelet membrane proteins being identified using this approach. Probable reasons for this deficit include (1) competition between highly abundant surface glycoproteins and secreted proteins, (2) the specificity of lectins for carbohydrate sidechains, and (3) low-affinity interactions in some cases.

Because of the deficiencies of the WGA affinity enrichment strategy, Senis et al. also employed biotin–avidin affinity chromatography, which would in theory pull down all biotinylated surface proteins [19]. The membrane-insoluble biotinylating reagent sulfo-NHS-SS-biotin was used to biotinylate surface proteins and thereby limit labeling of intracellular proteins. However, a substantial number (>100 proteins) of intracellular proteins were still obtained, most of which are known PMPs. This was most likely due to the mild washing conditions that would not have removed all PMPs. NeutrAvidin™ beads were used rather than avidin or streptavidin beads to facilitate removal of bound proteins through the reducing agent DTT. Despite outperforming WGA affinity chromatography, biotin–avidin affinity chromatography still did not identify all known platelet surface IMPs. In total, 35 PM, 14 IM, and 5 TMD-containing proteins of unknown localization were identified by two or more peptide hits.

The third membrane enrichment approach employed by Senis et al. was FFE [19]. Advantages of this approach include separation of the entire PM from IMs; therefore, there should be little or no loss of IMPs from the PM fraction, and IMs can also be analyzed separately. Disadvantages include the fact that specialist equipment is required and that PMPs remain attached to the PM, low-abundance IMPs and signaling proteins may be masked. PMP contamination can be reduced by washing the PM fraction with high-salt and high-pH conditions. A total of 35 PM, 30 IM, and 10 TMD-containing proteins of unknown locations were identified in the FFE-generated PM fraction by two or more peptide hits compared with 31 PM, 66 IM, and 20 TMD-containing proteins of unknown location in the FFE-generated IM fraction. Significantly, only 2 of the 44 proteins identified only in the FFE-IM fraction were known PM proteins. The presence of IM proteins in the PM fraction, and vice versa, is therefore most likely due to the presence of proteins in both membrane compartments as well as a degree of cross-contamination. The majority of the IM proteins identified are expressed in the ER.

In total, these three approaches identified 46 PM, 68 IM, and 22 TMD-containing proteins of unknown compartmentalization on the basis of identification of two or more unique peptides by MS/MS. In total, 60% of proteins were identified by more than one enrichment method. A little over a third (17) of the proteins identified by all of the enrichment methods are well known, highly expressed platelet surface proteins. As with the study by Moebius et al., only a small proportion (17%) of the identified PM proteins contained more than one predicted TMD. No seven-transmembrane G-protein-coupled receptors were identified; however, three tetraspanins, CD9, Tspan9, and Tspan33, were identified and several multispanning IM proteins were identified, including calcium-transporting ATPase type 2C, $IP_3$ receptors, and SERCA2A, which are predicted to contain eight, six, and seven TMDs, respectively, suggesting that the lack of identification of G-protein-coupled receptors may be due, in part, to their low abundance. At the time, it was estimated that just under 100 of the identified TMD-containing proteins reported in the study had not previously been described in platelets on the basis of biochemical and functional data, demonstrating the advantage of using the three separate enrichment techniques.

As part of the study by Senis et al., 45 proteins were identified on the basis of a single unique peptide using one or more of the three enrichment techniques employed in this study. Several of these proteins are known to be expressed in platelets, including the $\alpha5$ integrin subunit, CLEC-2, LAMP1, and CD68. However, the estimated false-positive rate for this group of proteins was 5%, which is significantly higher than that for proteins identified by two peptide hits (<0.025%), demonstrating the need for supporting biochemical, cellular, or functional data to confirm their expression in platelets.

Two receptor-like PM proteins identified in this study of particular interest are the PTP CD148 (also referred to as DEP1, PTPRJ, and rPTP$\eta$) and the ITIM-containing receptor G6b-B. Although CD148 had previously been reported in platelets, its functional role in platelets was not known [50,51]. G6b-A and -B isoforms had previously been identified in other platelet proteomics studies, but neither had been confirmed to be expressed in platelets by other means [36,52]. Anti-G6b-B-specific polyclonal antibodies raised against unique peptides in the cytoplasmic tail of G6b-B were used to confirm its expression in platelets by Western blotting [19]. CD148 was hypothesized to fulfill the role of a CD45-like PTP in platelets, and regulate ITAM and integrin receptor signaling; G6b-B was hypothesized to inhibit ITAM receptor signaling. Follow-up studies of both CD148 and G6b-B have demonstrated that CD148 is a critical positive regulator of Src family kinases in platelets, and that G6b-B negatively regulates GPVI and CLEC-2 signaling [53–55]. These two examples highlight the usefulness and importance of proteomics data in initiating novel areas of platelet research.

### 5.6.3 The Gold Standard to Date

The most comprehensive platelet membrane proteome study to date by Lewandowski et al. combined aqueous two-phase partitioning of platelet membranes with protein identification by shotgun and COFRADIC approaches [10].

The shotgun approaches included (1) 1DGE-LC-MS/MS and (2) MudPIT. Two-phase membrane partitioning followed by this three-pronged peptide identification strategy proved to be highly successful. A total of 1282 proteins were identified in the PM fraction, of which 498 proteins were identified using the peptide-centric COFRADIC approach on a single-peptide basis, and 1202 proteins were identified by the combined results of the two shotgun-based approaches. In total, 418 proteins were identified by the two shotgun-based approaches and COFRADIC, demonstrating the complementarity of the identification approaches. All proteins were identified by a minimum of two valid peptide identifications using the two shotgun approaches, which, together with the search parameters used, gave a false-positive rate of $<1\%$, making it a highly reliable and accurate dataset.

Previous studies analyzing whole-platelet lysates by classical 2DGE proteomics approaches identified 63 proteins containing predicted TMDs, whereas the more focused study by Lewandrowski et al. identified 626 proteins containing predicted TMDs, 30 of which contained 7 TMDs, including G-protein-coupled receptors such as PAR1, PAR4, $P2Y_1$, $P2Y_{12}$, and CXCR4. Nearly 100 proteins were predicted to have $\geq 8$ TMDs, including solute carriers and calcium channels. This was a vast improvement from any of the previous platelet membrane proteome studies that did not identify any G-protein-coupled receptors and relatively few multispanning TMD-containing proteins. Presumably, this is due at least partly to the carbonate extraction step used to reduce contamination with the PM-associated cytoskeleton.

COFRADIC enabled identification of proteins with a range of levels of expression, from highly abundant proteins, such as integrin $\alpha IIb\beta 3$ ($>$80,000 copies on the surface of resting platelets), down to intermediate/low-abundance proteins, such as the integrins $\alpha 5$ (1000–2000 copies) and $\alpha 6$ (1000–2000 copies), and the G-protein-coupled receptor PAR1 (1200 copies). Approximately half of the 498 proteins identified by COFRADIC contained at least one predicted TMD, which was substantially better than the 66 predicted TMD-containing proteins out of 385 proteins identified in a previous COFRADIC-based study of whole-platelet lysates [52], demonstrating the advantage of the two-phase partitioning-based membrane purification for identifying membrane proteins. Although only 80 proteins were identified exclusively by COFRADIC, this approach also provides *N*-terminal sequence information, which may not be known for some proteins and can be used for functional experiments or antibody production.

The subcellular distribution of proteins was estimated with GoMiner and Ontologizer algorithms. On the basis of these bioinformatics analyses, of the 1282 proteins identified in the PM fraction, 371 were predicted to be PM components, 142 of which are IMPs. In addition, a range of proteins from other membrane compartments were also identified, including 199 from the ER, 148 from the Golgi apparatus, and 140 from vesicles. A comparison of the summed proteins identified in the ER, Golgi apparatus, and vesicles with those in the PM revealed a major overlap of 116 proteins. Presumably these proteins are present in all of these compartments and may be shuttled between these organelles. However, many of the proteins identified were uncharacterized and were not assigned a subcellular location.

The majority of the proteins identified in the PM compartment were classified as possessing signal transduction activity (156 proteins) or as receptors (104 proteins) [10]. Of the receptors identified, 13 were G-protein-coupled receptors, six of which (AVPR1A, GPR92, $P2Y_1$, $P2Y_{12}$, PAR1, and PTGIR) had only recently been identified in platelets by quantitative PCR [56]. This study was also highly successful in identifying proteins involved in cell adhesion and aggregate formation, two critical functions of platelets. In total, 86 proteins implicated in cell adhesion and 47 proteins present or involved in cell junctions were identified, including all the platelet intregins and key components of their signaling apparatus.

Besides G-protein-coupled receptors, numerous other classes of multispanning PM proteins identified in this study included tetraspanins, which contain four TMDs and have membrane ordering functions in various cell types; and ion transporters and channels, which allow the movement of ions and small molecules across various membranes. The identification of 91 ion transporters, most of which have $\geq$8 TMDs, demonstrates the superiority of this strategy for identifying platelet multispanning IMPs. The reason for this vast improvement is likely multifactorial and reflects the applicability of the entire strategy rather than a specific step for identifying multispanning IMPs.

Another novel and informative aspect of this study was the relative quantification of platelet PM receptors. A number of techniques have been developed for the quantification of the relative and absolute levels of proteins in a sample by MS. These approaches often rely on the use of chemical labeling, however, other label-free techniques have also gained credibility. One of these techniques, termed the *exponentially modified protein abundance index* (emPAI), utilizes spectral counting. Detection of peptides by MS is partially dependent on the concentration of peptides present in the sample. Assuming that abundant peptides are detected more often, an approximation of the relative abundance of proteins can be based on the number of peptide hits identified for each protein. Lewandrowski et al. used a modified version of the emPAI index to generate approximate estimates of platelet protein abundance, which correlated well with published levels [10]. These data can therefore be used to differentiate between high-, medium-, and low-abundance proteins. Interestingly, the novel platelet ITIM receptor G6b-B had a high emPAI score of 122 and ranked among the most abundant receptors on the platelet surface (estimated $\sim$20,000 copies per platelet). However, it should be noted that although a general correlation was found between emPAI and copy numbers of surface receptors, absolute copy numbers of individual proteins may not fit with its position in the emPAI ranking. Another potential problem with this type of semiquantitative analysis is that different copy numbers of receptors are reported in the literature depending on the assay used.

Finally, Lewandrowski et al. used the STRING (Search Tool for the Retrieval of Interacting Genes/Proteins) algorithm for network analysis for the complete list of 1282 proteins. Proteins were clustered into groups; the central ones were adhesion-mediated receptors and kinases or effectors. Other groupings included (1) actin- and cytoskeleton-associated proteins, such as Arp2/3 complex, vinculin, and VASP (vasodilator-stimulated phosphoprotein); (2) vesicle-associated SNAPs (synaptosome-associated proteins) and SNAREs [soluble NSF (*N*-ethylmaleimide-sensitive factor) attachment (protein) receptors]; (3) proteins related to metabolism; (4) glycosylation; and (5) proteins of mitochondrial origin. A total of 858 protein–protein interactions were revealed by STRING. However, a large number of proteins were not associated with other membrane or soluble proteins.

## 5.7 IDENTIFYING PERIPHERAL MEMBRANE PROTEINS IN ACTIVATED PLATELETS

The platelet PM is a dynamic compartment that undergoes major changes following platelet activation. These changes include reorganization of the PM into functionally distinct microdomains (lipid rafts and tetraspanin-enriched microdomains); shedding (GPVI and GPIb-IX-V), internalization (PAR1/4), and up-regulation (P-selectin and integrin αIIbβ3) of receptors; recruitment of signaling proteins to the cytosolic surface of the PM; and membrane flipping. Characterizing these changes has important implications for understanding how platelets respond to their surroundings.

GPVI clustering initiates a series of tyrosine phosphorylation events that leads to the recruitment of numerous signaling proteins to the cytosolic surface of the PM [57]. Signaling proteins can bind either directly to phospholipids in the PM via PH domains, or indirectly via a variety of protein–protein interactions mediated by SH3 and SH2 domains that bind to proline-rich regions and phosphorylated tyrosine residues in IMPs, respectively. Similarly, thrombin-mediated cleavage of PAR1/4 triggers recruitment of signaling proteins to the PM [58]. GPVI- and PAR1/4-mediated platelet activation also induces increased PS exposure on the outer layer of the PM, which acts as a docking site for coagulation factors. Although both the GPVI and PAR1/4 signaling pathways are well characterized, new signaling components are constantly being discovered that fine-tune both pathways.

Challenges that hamper the identification of new signaling proteins recruited to the PM on platelet activation include (1) low relative abundance; and (2) the transient, low-affinity nature of the interactions. The inner surface of the PM is lined with an abundance of cytoskeletal proteins, including actin, tubulin, and filamin that can easily mask the presence of low levels of signaling proteins. In addition, low-affinity interactions can be easily disrupted even with mild solublization conditions. Gibbins and coworkers devised two different approaches for enriching PMPs recruited to the PM of platelets stimulated with the GPVI-specific aganist collagen-related peptide (CRP) and thrombin [39,40]. CRP-stimulated platelets were lysed by repeated freeze–thawing in a physiological buffer and gentle sonication, without the use of detergents and high-salt buffers that would likely result in the loss of PMPs. All platelet membranes were subsequently pelleted using high-speed centrifugation and PMPs eluted from membranes with 100 mM sodium carbonate. A gentle elution step was used to avoid eluting IMPs and tighly associated cytoskeletal proteins that would have increased sample complexity. Phosphatase and kinase inhibitors (sodium vanadate and staurosporine, respectively) were included during the lysis step to prevent changes in phosphorylation that could disrupt the delicate balance of membrane-associated proteins. Sample complexity was further reduced by a 2D system, involving liquid-phase IEF in the first dimension and SDS-PAGE in the second dimension. A total of 105 proteins were identified by FT-ICR in 26 bands that change in intensity in CRP-stimulated platelets, 44% of which were signaling proteins. Other proteins identified included cytoskeletal structural or regulatory proteins (19%), secreted proteins (10%), and proteins involved in metabolism (26%). Only 1% of the proteins identified were transmembrane proteins (P-selectin), demonstrating the relative specificity of the elution step. Surprisingly, the collagen chaperone protein Hsp47 (colligin) was detected in the PMP fraction of CRP-stimulated platelets and was shown to be exposed on the surface of activated platelets [39]. Functional evidence demonstrated that surface-bound Hsp47 facilitates collagen-mediated platelet aggregation. These findings highlight the importance of further characterizing PM-associated PMPs in resting and activated platelets. Although the strategy outlined in this study provides an excellent starting point, a more sensitive approach is necessary as numerous GPVI signaling proteins known to be recruited to the PM (e.g., Syk, PLC$\gamma$2, SLP76) failed to be identified. Perhaps the combinatorial ligand library affinity chromatography approach described by Guerrier et al. for concentrating and identifying low-abundance soluble platelet proteins can be applied to identifying low-abundance PMPs recruited to the PM of activated platelets [59].

A different strategy was used by Tucker and colleagues in a comparative analysis of the membrane proteomes of resting and thrombin-stimulated platelets [40]. Platelet PMs were enriched by biotin–avidin affinity chromatography following freeze–thawing and sonication in 0.1% (v/v) Triton X-100. PMPs were eluted from biotinylated PMs bound to NeutrAvidin™-conjugated beads with 4% (w/v) DTT and 8% (w/v) CHAPs. Sample complexity was further reduced by solution-phase IEF followed by SDS-PAGE. Image analysis of silver-stained gels revealed 65 bands of increased intensity in the thrombin-stimulated surface fraction compared to the resting surface fraction, from which 88 different proteins were identified [40]. Of these proteins, 25% were identified as IMPs, 31% were known signaling proteins, and 8% were cytoskeleton proteins. One of the signaling proteins identified was the adapter HIP-55. Further analysis of HIP55 in platelets demonstrated increased association with Syk and the $\beta$3 integrin subunit on collagen and thrombin stimulation, suggesting that HIP55 may be involved in regulating $\alpha$IIb$\beta$3 signaling. Platelets from HIP55-deficient mice exhibited reduced fibrinogen binding in response to thrombin, which supports a role of HIP55 in regulating integrin signaling [40].

## 5.8 FUTURE DIRECTIONS

Although much progress has been made in mapping the platelet PM proteome, assembling a truly complete map, including all IMPs, PMPs, and post-translational modifications, is far from complete. Future milestones include the following:

1. Identifying all IMPs on the surface of resting human platelets. One limitation of proteomics is not knowing when a proteome is complete. Estimates can be based on comparisons with transcriptomics analysis of human megakaryocyte and platelets.
2. All IMPs should be quantified on the surface of resting human platelets. Although most commonly used quantitative proteomics techniques are designed for nucleated cells, new techniques are being developed that can be applied to anucleated platelets, such as spectral counting, as was demonstrated by Lewandrowski et al. [10].
3. All posttranslational modifications need to be identified.
4. All PMPs associated with the outer and inner surfaces of resting human platelets need to be identified. Intracellular membranes can also be analyzed in a similar manner with time. These comprehensive maps of the platelet PM and IMs can then be used to study changes following platelet activation with different agonists and also for comparison with platelets from patients with platelet-based bleeding disorders. Such maps would also be invaluable for systems biologists to model signaling networks.

Other areas for future study should include (1) characterizing the composition of PM microdomains (described in more detail below); (2) trafficking and compartmentalization of PMPs in activated platelets; (3) interaction proteomics; (4) mapping the mouse platelet membrane proteome, as it is the main animal model for investigating the molecular mechanisms regulating platelet function and thrombosis; and (5) mapping the human and mouse megakaryocyte membrane proteomes.

### 5.8.1 Membrane Microdomain Proteomics

Plasma membranes are not uniform lipid bilayers that contain randomly distributed proteins, as predicted by the original *fluid mosaic model* [60], but rather are compartmentalized into functionally distinct regions that exhibit unique lipid and protein compositions. Two distinct types of microdomains with unique and overlapping features are lipid rafts (also referred to as *lipid-* and *glycolipid-enriched microdomains*) and tetraspanin-enriched microdomains. These two types of microdomains have been implicated in a variety of cellular functions in health and disease. Lipid rafts have a specific protein and lipid composition, and are enriched in cholesterol and glycosphingolipids, causing phase separation from the bulk PM (Fig. 5.1). Tetraspannin-enriched microdomains are similar to lipid rafts in that they are also enriched in cholesterol and partition into low-density fractions. However, tetraspanin-enriched microdomains display a number of critical biochemical properties that distinguish them from conventional lipid rafts, including differences in detergent-mediated solubility, sensitivity to cholesterol depletion, and composition. Tetraspanins are proteolipids, exhibiting palmitoylation on their juxtamembrane cysteines, which has been shown to mediate tetraspanin-tetraspanin interactions, most likely via interactions with cholesterol [61]. Humans contain 32 tetraspanins, 19 of which have been identified in mouse megakaryocytes by mRNA profiling and 13 (CD9/37/63/81/82/151, TSN2/4/9/14/15/18/33) by proteomics [10,62]. In terms of protein composition, both lipid rafts and tetraspanin-enriched microdomains contain several G proteins; however, lipid rafts typically contain GPI-linked proteins and caveolin, whereas tetraspanin-enriched microdomains do not [61].

The main reason for studying lipid rafts and tetraspanin-enriched microdomaisn in platelets is because an increasing body of evidence demonstrates that both types of microdomains are important centers for the integration and interaction of signaling complexes and trafficking [61,63,64]. Identifying the protein and lipid constituents of these microdomains therefore has important implications for understanding how signaling is initiated and regulated in platelets. Several studies have investigated the partitioning of specific receptors and signaling molecules in lipid and nonlipid rafts in resting and activated platelets by Western blotting [65–67]. Some of the main findings to date indicate that GPVI signaling is dependent on lipid rafts [65], LAT is a constitutive component of lipid rafts [65], the integrin $\alpha IIb\beta 3$ does not require lipid rafts to signal [65], and lipid rafts facilitate clustering of PECAM1 and GPVI [66]. There have been over 25 proteomics studies investigating the protein composition of lipid rafts and tetraspanin-enriched microdomains in a variety of cell types; however, none have been done in platelets or megakaryocytes. As in other proteomics studies, the key to success is enrichment of either lipid rafts or tetraspanin-enriched microdomains, with limited contamination from bulk PM and cytosolic proteins [61,68,69]. Conditions used for lysing cells and fractionating membrane components must be carefully considered in order to ensure enrichment of the desired microdomain [61,69]. Further, protein markers for different fractions must be used in each experiment, to ensure purity of microdomains. Another commonly used control to demonstrate localization of proteins to lipid rafts is cholesterol depletion, which disrupts lipid rafts but has no effect on tetraspanin-enriched microdomains [69].

## 5.9 CONCLUDING REMARKS

The two main challenges of platelet membrane proteomics are (1) the hydrophobic properties of IMPs, which makes them difficult to isolate and analyze by MS; (2) the low abundance of some IMPs and PMPs, which can be easily masked by more abundant IMPs and cytosolic and cytoskeletal proteins. These challenges are being overcome through new and improved techniques designed specifically for isolating and detecting hydrophobic IMPs by MS/MS. Strategies are also being developed for quantifying proteins by MS/MS that can be applied to platelets and isolating PMPs from different membrane compartments. As the number of platelet membrane proteins that we identify increases, so will our understanding of the molecular mechanisms regulating platelet activation and thrombosis. New proteomics-based methods will also emerge for diagnosing the molecular basis of platelet-based bleeding disorders, as will antithrombotic drug targets.

## ACKNOWLEDGMENTS

Yotis Senis is a British Heart Foundation Intermediate Basic Science Research Fellow. His research is supported by generous funding from the British Heart Foundation.

## REFERENCES

1. Tan S, Tan HT, Chung MC, Membrane proteins and membrane proteomics, *Proteomics* **8**(19):3924–3932 (2008).
2. Almen MS, Nordstrom KJ, Fredriksson R, Schioth HB, Mapping the human membrane proteome: A majority of the human membrane proteins can be classified according to function and evolutionary origin, *BMC Biol.* **7**:50 (2009).
3. Speers AE, Wu CC, Proteomics of integral membrane proteins—theory and application, *Chem. Rev.* **107**(8):3687–3714 (2007).
4. Asazuma N, Wilde JI, Berlanga O, Leduc M, Leo A, Schweighoffer E, et al., Interaction of linker for activation of T cells with multiple adapter proteins in platelets activated by the glycoprotein VI-selective ligand, convulxin, *J. Biol. Chem.* **275**(43):33427–33434 (2000).
5. Vanhaesebroeck B, Leevers SJ, Ahmadi K, Timms J, Katso R, Driscoll PC, et al., Synthesis and function of 3-phosphorylated inositol lipids, *Annu. Rev. Biochem.* **70**:535–602 (2001).
6. Monroe DM, Hoffman M, Roberts HR, Platelets and thrombin generation, *Arterioscler. Thromb. Vasc. Biol.* **22**(9):1381–1389 (2002).
7. Hopkins AL, Groom CR, The druggable genome, *Nat. Rev. Drug Discov.* **1**(9):727–730 (2002).
8. Lu B, McClatchy DB, Kim JY, Yates JR 3rd, Strategies for shotgun identification of integral membrane proteins by tandem mass spectrometry, *Proteomics* **8**(19):3947–3955 (2008).
9. Lu B, Xu T, Park SK, Yates JR 3rd, Shotgun protein identification and quantification by mass spectrometry, *Meth. Mol. Biol.* **564**:261–288 (2009).
10. Lewandrowski U, Wortelkamp S, Lohrig K, Zahedi RP, Wolters DA, Walter U, et al., Platelet membrane proteomics: A novel repository for functional research, *Blood* **114**(1):e10-9 (2009).
11. Cordwell SJ, Thingholm TE, Technologies for plasma membrane proteomics, *Proteomics* **10**(4):611–627 (2010).

12. Robertson ER, Kennedy JF, Glycoproteins: A consideration of the potential problems and their solutions with respect to purification and characterisation, *Bioseparation* **6**(1):1–15 (1996).
13. Jang JH, Hanash S, Profiling of the cell surface proteome, *Proteomics* **3**(10): 1947–1954 (2003).
14. Gauthier DJ, Gibbs BF, Rabah N, Lazure C, Utilization of a new biotinylation reagent in the development of a nondiscriminatory investigative approach for the study of cell surface proteins, *Proteomics* **4**(12):3783–3790 (2004).
15. Sabarth N, Lamer S, Zimny-Arndt U, Jungblut PR, Meyer TF, Bumann D, Identification of surface proteins of Helicobacter pylori by selective biotinylation, affinity purification, and two-dimensional gel electrophoresis, *J. Biol. Chem.* **277**(31):27896–27902 (2002).
16. Shin BK, Wang H, Yim AM, Le Naour F, Brichory F, Jang JH, et al., Global profiling of the cell surface proteome of cancer cells uncovers an abundance of proteins with chaperone function, *J. Biol. Chem.* **278**(9):7607–7616.
17. Zhao Y, Zhang W, Kho Y, Proteomic analysis of integral plasma membrane proteins, *Anal. Chem.* **76**(7):1817–1823 (2004).
18. Peirce MJ, Wait R, Begum S, Saklatvala J, Cope AP, Expression profiling of lymphocyte plasma membrane proteins, *Mol. Cell. Proteom.* **3**(1):56–65 (2004).
19. Senis YA, Tomlinson MG, García A, Dumon S, Heath VL, Herbert J, et al., A comprehensive proteomics and genomics analysis reveals novel transmembrane proteins in human platelets and mouse megakaryocytes including G6b-B, a novel immunoreceptor tyrosine-based inhibitory motif protein, *Mol. Cell. Proteom.* **6**(3):548–564 (2007).
20. Bergmeier W, Rabie T, Strehl A, Piffath CL, Prostredna M, Wagner DD, et al., GPVI down-regulation in murine platelets through metalloproteinase-dependent shedding, *Thromb. Haemost.* **91**(5):951–959 (2004).
21. Suzuki-Inoue K, Fuller GL, García A, Eble JA, Pohlmann S, Inoue O, et al., A novel Syk-dependent mechanism of platelet activation by the C-type lectin receptor CLEC-2, *Blood* **107**(2):542–549 (2006).
22. Fuller GL, Williams JA, Tomlinson MG, Eble JA, Hanna SL, Pohlmann S, et al., The C-type lectin receptors CLEC-2 and Dectin-1, but not DC-SIGN, signal via a novel YXXL-dependent signaling cascade, *J. Biol. Chem.* **282**(17):12397–12409 (2007).
23. May F, Hagedorn I, Pleines I, Bender M, Vogtle T, Eble J, et al., CLEC-2 is an essential platelet-activating receptor in hemostasis and thrombosis, *Blood* **114**(16):3464–3472 (2009).
24. Bertozzi CC, Schmaier AA, Mericko P, Hess PR, Zou Z, Chen M, et al., Platelets regulate lymphatic vascular development through CLEC-2-SLP-76 signaling, *Blood* **116**:661–670 (2010).
25. Suzuki-Inoue K, Inoue O, Ding G, Nishimura S, Hokamura K, Eto K, Kashiwagi H, Tomiyama Y, Yatomi Y, Umemura K, Shin Y, Hirashima M, Ozaki Y, Essential *in vivo* roles of the C-type lectin receptor CLEC-2: Embryonic/neonatal lethality of clec-2-deficient mice by blood/lymphatic misconnections and impaired thrombus formation of clec-deficient platelets *J. Biol. Chem* **285**:24494–24507 (2010).
26. Hughes CE, Navarro-Náñez I, Finney BA, Mourão-Sá D, Pollitt AY, Wetson SP, CLEC-2 is not required for platelet aggregation at arteriolar shear, *J. Thromb. Haemost.* **8**:2328–2332 (2010).

27. Nanda N, Bao M, Lin H, Clauser K, Komuves L, Quertermous T, et al., Platelet endothelial aggregation receptor 1 (PEAR1), a novel epidermal growth factor repeat-containing transmembrane receptor, participates in platelet contact-induced activation, *J. Biol. Chem.* **280**(26):24680–24689 (2005).

28. García A, Senis YA, Antrobus R, Hughes CE, Dwek RA, Watson SP, et al., A global proteomics approach identifies novel phosphorylated signaling proteins in GPVI-activated platelets: Involvement of G6f, a novel platelet Grb2-binding membrane adapter, *Proteomics* **6**(19):5332–5343 (2006).

29. Senis YA, Antrobus R, Severin S, Parguiña AF, Rosa I, Zitzmann N, et al., Proteomic analysis of integrin alphaIIbbeta3 outside-in signaling reveals Src-kinase-independent phosphorylation of Dok-1 and Dok-3 leading to SHIP-1 interactions, *J. Thromb. Haemost.* **7**(10):1718–1726 (2009).

30. Authi KS, Preparation of highly purified human platelet plasma intracellular membranes using high voltage free flow electrophoresis and methods to study $Ca^{2+}$ regulation, in Watson SP, Authi KS, eds., *Platelets: A Practical Approach*, Oxford Univ. Press, 1996, pp. 91–109.

31. Wu CC, MacCoss MJ, Howell KE, Yates JR 3rd, A method for the comprehensive proteomic analysis of membrane proteins, *Nat. Biotechnol.* **21**(5):532–538 (2003).

32. Speers AE, Blackler AR, Wu CC, Shotgun analysis of integral membrane proteins facilitated by elevated temperature, *Anal. Chem.* **79**(12):4613–4620, (2007).

33. Burre J, Wittig I, Schagger H, Non-classical 2-D electrophoresis, *Meth. Mol. Biol.* **564**:33–57 (2009).

34. Schagger H, Cramer WA, von Jagow G, Analysis of molecular masses and oligomeric states of protein complexes by blue native electrophoresis and isolation of membrane protein complexes by two-dimensional native electrophoresis, *Anal. Biochem.* **217**(2):220–230 (1994).

35. Zahedi RP, Meisinger C, Sickmann A, Two-dimensional benzyldimethyl-n-hexadecylammonium chloride/SDS-PAGE for membrane proteomics, *Proteomics* **5**(14):3581–3588 (2005).

36. Moebius J, Zahedi RP, Lewandrowski U, Berger C, Walter U, Sickmann A, The human platelet membrane proteome reveals several new potential membrane proteins, *Mol. Cell. Proteom.* **4**(11):1754–1761 (2005).

37. Burre J, Beckhaus T, Schagger H, Corvey C, Hofmann S, Karas M, et al., Analysis of the synaptic vesicle proteome using three gel-based protein separation techniques, *Proteomics* **6**(23):6250–6262 (2006).

38. Park YM, Kim JY, Kwon KH, Lee SK, Kim YH, Kim SY, et al., Profiling human brain proteome by multi-dimensional separations coupled with MS, *Proteomics* **6**(18):4978–4986 (2006).

39. Kaiser WJ, Holbrook LM, Tucker KL, Stanley RG, Gibbins JM, A functional proteomic method for the enrichment of peripheral membrane proteins reveals the collagen binding protein Hsp47 is exposed on the surface of activated human platelets, *J. Proteome Res.* **8**(6):2903–2914 (2009).

40. Tucker KL, Kaiser WJ, Bergeron AL, Hu H, Dong JF, Tan TH, et al., Proteomic analysis of resting and thrombin-stimulated platelets reveals the translocation and functional relevance of HIP-55 in platelets, *Proteomics* **9**(18):4340–4354 (2009).

41. Wu CC, Yates JR 3rd, The application of mass spectrometry to membrane proteomics, *Nat. Biotechnol.* **21**(3):262–267 (2003).

42. Schirmer EC, Florens L, Guan T, Yates JR 3rd, Gerace L, Nuclear membrane proteins with potential disease links found by subtractive proteomics, *Science* **301**(5638):1380–1382 (2003).
43. Wu CC, MacCoss MJ, Mardones G, Finnigan C, Mogelsvang S, Yates JR 3rd, et al., Organellar proteomics reveals Golgi arginine dimethylation, *Mol. Biol. Cell* **15**(6):2907–2919 (2004).
44. Reinders J, Sickmann A, Proteomics of yeast mitochondria, *Meth. Mol. Biol.* **372**:543–557 (2007).
45. O'Neill EE, Brock CJ, von Kriegsheim AF, Pearce AC, Dwek RA, Watson SP, et al., Towards complete analysis of the platelet proteome, *Proteomics* **2**(3):288–305 (2002).
46. García A, Prabhakar S, Brock CJ, Pearce AC, Dwek RA, Watson SP, et al., Extensive analysis of the human platelet proteome by two-dimensional gel electrophoresis and mass spectrometry, *Proteomics* **4**(3):656–668 (2004).
47. Gevaert K, Ghesquiere B, Staes A, Martens L, Van Damme J, Thomas GR, et al., Reversible labeling of cysteine-containing peptides allows their specific chromatographic isolation for non-gel proteome studies, *Proteomics* **4**(4):897–908 (2004).
48. Lewandrowski U, Moebius J, Walter U, Sickmann A, Elucidation of N-glycosylation sites on human platelet proteins: A glycoproteomic approach, *Mol. Cell. Proteom.* **5**(2):226–233 (2006).
49. Lewandrowski U, Zahedi RP, Moebius J, Walter U, Sickmann A, Enhanced N-glycosylation site analysis of sialoglycopeptides by strong cation exchange prefractionation applied to platelet plasma membranes, *Mol. Cell. Proteom.* **6**(11):1933–1941 (2007).
50. Borges LG, Seifert RA, Grant FJ, Hart CE, Disteche CM, Edelhoff S, et al., Cloning and characterization of rat density-enhanced phosphatase-1, a protein tyrosine phosphatase expressed by vascular cells, *Circ. Res.* **79**(3):570–580 (1996).
51. de la Fuente-García MA, Nicolas JM, Freed JH, Palou E, Thomas AP, Vilella R, et al., CD148 is a membrane protein tyrosine phosphatase present in all hematopoietic lineages and is involved in signal transduction on lymphocytes, *Blood* **91**(8):2800–2809 (1998).
52. Martens L, Van Damme P, Van Damme J, Staes A, Timmerman E, Ghesquiere B, et al., The human platelet proteome mapped by peptide-centric proteomics: A functional protein profile, *Proteomics* **5**(12):3193–3204 (2005).
53. Newland SA, Macaulay IC, Floto AR, de Vet EC, Ouwehand WH, Watkins NA, et al., The novel inhibitory receptor G6B is expressed on the surface of platelets and attenuates platelet function *in vitro*, *Blood* **109**(11):4806–4809 (2007).
54. Senis YA, Tomlinson MG, Ellison S, Mazharian A, Lim J, Zhao Y, et al., The tyrosine phosphatase CD148 is an essential positive regulator of platelet activation and thrombosis, *Blood* **113**(20):4942–4954 (2009).
55. Mori J, Pearce AC, Spalton JC, Grygielska B, Eble JA, Tomlinson MG, et al., G6b-B inhibits constitutive and agonist-induced signaling by glycoprotein VI and CLEC-2, *J. Biol. Chem.* **283**(51):35419–35427 (2008).
56. Amisten S, Braun OO, Bengtsson A, Erlinge D, Gene expression profiling for the identification of G-protein coupled receptors in human platelets, *Thromb. Res.* **122**(1):47–57 (2008).

57. Nieswandt B, Watson SP, Platelet-collagen interaction: Is GPVI the central receptor? *Blood* **102**(2):449–461 (2003).
58. Coughlin SR, Thrombin signaling and protease-activated receptors, *Nature* **407**(6801):258–264 (2000).
59. Guerrier L, Claverol S, Fortis F, Rinalducci S, Timperio AM, Antonioli P, et al., Exploring the platelet proteome via combinatorial, hexapeptide ligand libraries, *J. Proteome Res.* **6**(11):4290–4303 (2007).
60. Singer SJ, Nicolson GL, The fluid mosaic model of the structure of cell membranes, *Science* **175**(23):720–731 (1972).
61. Le Naour F, Andre M, Boucheix C, Rubinstein E, Membrane microdomains and proteomics: Lessons from tetraspanin microdomains and comparison with lipid rafts, *Proteomics* **6**(24):6447–6454 (2006).
62. Protty MB, Watkins NA, Colombo D, Thomas SG, Heath VL, Herbert JM, et al., Identification of Tspan9 as a novel platelet tetraspanin and the collagen receptor GPVI as a component of tetraspanin microdomains, *Biochem. J.* **417**(1):391–400 (2009).
63. Simons K, Toomre D, Lipid rafts and signal transduction, *Nat. Rev. Mol. Cell. Biol.* **1**(1):31–39 (2000).
64. Helms JB, Zurzolo C, Lipids as targeting signals: Lipid rafts and intracellular trafficking, *Traffic* (Copenhagen) **5**(4):247–254 (2004).
65. Wonerow P, Obergfell A, Wilde JI, Bobe R, Asazuma N, Brdicka T, et al., Differential role of glycolipid-enriched membrane domains in glycoprotein VI- and integrin-mediated phospholipase Cgamma2 regulation in platelets, *Biochem. J.* **364**(Pt. 3):755–765 (2002).
66. Lee FA, van Lier M, Relou IA, Foley L, Akkerman JW, Heijnen HF, et al., Lipid rafts facilitate the interaction of PECAM-1 with the glycoprotein VI-FcR gamma-chain complex in human platelets, *J. Biol. Chem.* **281**(51):39330–39338 (2006).
67. Munday AD, Gaus K, Lopez JA, The platelet glycoprotein Ib-IX-V complex anchors lipid rafts to the membrane skeleton: implications for activation-dependent cytoskeletal translocation of signaling molecules, *J. Thromb. Haemost.* **8**(1):163–172 (2010).
68. Zheng YZ, Foster LJ, Biochemical and proteomic approaches for the study of membrane microdomains, *J. Proteom.* **72**(1):12–22 (2009).
69. Sprenger RR, Horrevoets AJ, The ins and outs of lipid domain proteomics, *Proteomics* **7**(16):2895–2903 (2007).

# 6

# PROTEOMICS OF PLATELET GRANULES, ORGANELLES, AND RELEASATE

JAMES P. MCREDMOND

**Abstract**

Platelets contain subcellular organelles with particular relevance in thrombosis and inflammation. In particular, the contents of platelet dense granules and $\alpha$-granules contribute to thrombosis, coagulation, atherosclerosis, and other pathological consequences of platelet activation. Proteomic analysis of these organelles helps to determine the roles of individual proteins and relationships between them. Isolating compartments also increases the depth to which the platelet proteome may be probed. Appropriately designed proteomic studies provide insights into the mechanisms and consequences of platelet responses.

## 6.1 INTRODUCTION: PLATELET GRANULES

One of the platelet responses to activation is the secretion of granules, releasing their contents into plasma, with consequences for thrombosis, atherosclerosis, and other pathological processes. The combined set of proteins released from platelets, regardless of original source, has been termed the *releasate*. More recent proteomic studies have greatly expanded our knowledge of the protein content of platelet granules, enhancing our understanding of how they are regulated and

*Platelet Proteomics: Principles, Analysis and Applications*, First Edition.
Edited by Ángel García and Yotis A. Senis.

the consequences of the platelet release response. Following a brief discussion of platelet granules (readers are directed to the cited references for more detail), these proteomic studies are reviewed here.

### 6.1.1 α-Granules and Dense Granules

Two main classes of granules, α-granules and dense granules, contain predominantly small molecules and proteins, respectively (Fig. 6.1). Both types of granules are derived from megakaryocyte multivesicular bodies [1], and are packaged into proplatelet extensions of the megakaryocyte following transport along microtubules [2]. Platelets also contain lysosomes, but secretion of their contents is not thought to be physiological [1].

α-Granules contain a wide variety of proteins. Some of these are packaged into contents in the megakaryocte during the platelet production process. Of these, some [such as platelet factor 4 (PF4)] are completely platelet-specific and not found in other cell lineages [1]. More recent work suggests that there are distinct subsets of α-granules with differential sorting of selected proteins into each [3]. Other α-granule contents are taken up from plasma by endocytosis or receptor-mediated uptake and transported into the α-granule. Some of the mechanisms governing sorting of proteins into granules have been elucidated [4]. Thus, the granule contents overlaps with plasma proteome, but also contains additional contents. The α-granule membrane shares many proteins with the platelet plasma membrane, which may reflect fusion of endocytotic bodies with the granule.

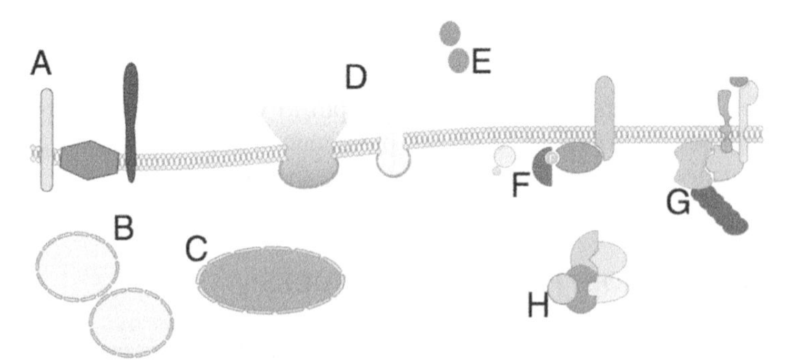


**Figure 6.1** Sources of proteins for platelet organelle proteomic studies. Various platelet compartments or functional protein sets may be usefully isolated for proteomic analysis. In many cases, comparison of proteins recovered from resting and activated platelets is particularly informative. Specific examples of each of these are discussed in the text. Structures: A—plasma membrane proteins (internal membrane proteins may also be isolated); B—α-granules; C—dense granules; D—soluble released proteins, predominantly from α- and dense granules; E—platelet microparticles; F—membrane-associated proteins; G—protein complexes bound to particular membrane receptors; H—intracellular protein complexes.

Platelet dense granules (sometimes called dense core or δ-granules) are so-called because of there relative opacity under electron microscopy. This is due to their non-protein contents, which includes ADP, ATP, serotonin, calcium, and polyphosphate [5]. Several proteins are known to be found on the dense granule membrane, including transporters, GTP-binding proteins, markers such as LAMP1 and LAMP2, and proteins shared with the platelet plasma membrane, such as αIIbβ3 and P-selectin [6].

### 6.1.2 Release Response

Platelet activation by strong agonists results in secretion of both types of granules, with mobilization of calcium from intracellular stores required. In addition to delivering the granule contents to the platelet exterior, this exposes the granule membranes on the surface, increasing the platelet surface area (increasing the number of available copies of receptors such as αIIbβ3), and exposes previously hidden receptors such as P-selectin.

Granule contents contribute to propagation of platelet activation by delivering agonists such as ADP and serotonin (from dense granules) to the plasma. α-granule proteins contribute to processes such as coagulation, inflammation, host defense, wound healing, atherosclerosis, and angiogenesis [4]. Thus the platelet secretome contributes to a range of physiological and pathological processes, spanning timescales from fractions of a second to years.

Secretion may be linked to other platelet activation responses, including shape change and cytoskeletal rearrangement. Effective platelet spreading and shape change may require the additional membrane area derived from fusion of granules with the plasma membrane, while the resting platelet cytoskeleton may act as a barrier to secretion, with reorganization of it facilitating secretion of α-granules and dense granules to different extents [7].

The molecular mechanisms governing granule release are now well understood. As with other secretion events, fusion of the granule and surface membranes is driven by associations between so-called SNARE [soluble NSF (*N*-ethylmaleimide-sensitive factor) attachment protein receptors] proteins on the cytoplasmic face of each membrane. SNAREs associated with particular granules in the platelet include members of the VAMP, syntaxin, and SNAP families [4]. Elucidating the details of how SNARE interactions are regulated by Rab, Munc, and other protein families will provide insights into the control of platelet secretion, including selective secretion of α-granules and dense granules and possibly subsets of each.

### 6.1.3 Storage Pool Disorders

*Storage pool deficiencies* (SPDs) are genetic disorders that are distinct from disorders relating to platelet signalling that may also affect granule secretion. SPDs are rare, and elucidating their mechanisms gives insights into the mechanisms and control of platelet granule release, which may, in turn, lead to therapies to limit platelet positive-feedback responses as well as longer-term consequences of secretion. However, since many platelet secretory responses share mechanisms with secretion in other systems (such as neurotransmitter release and insulin secretion), this may be a challenging therapeutic strategy. Platelet SPDs may affect predominantly α-granules [α-SPD, such as gray platelet syndrome (GPS) and Quebec platelet disorder (QPD)], dense granules [δ-SPD, such as Hermansky–Pudlak syndrome (HPS) and Chediak–Higashi syndrome (CHS)] or both (α,δ-SPD, described in a single patient [8]). The symptoms of SPDs illustrate the importance of platelet secretory granules in normal hemostasis.

Other rare genetic disorders with a platelet involvement may have improperly formed granules as one of their pathological components [such as thrombocytopenia and absent radius (TAR) syndrome, Wiskott–Aldrich syndrome [6], and Griscelli syndrome [9]].

***6.1.3.1 α-SPD*** Platelets from patients with GPS appear to lack α-granules and the stored proteins normally found therein. The platelets appear gray under a May–Grunwald–Giemsa stain, with a tendency for prolonged bleeding and abnormal responses to thrombin and collagen. Leakage of proteins normally packaged into granules may lead to megakaryocyte and bone marrow abnormalities [10], and aberrant expression of P-selectin on megakaryocytes may lead to unusual interactions with neutrophils. Proteins normally endocytosed from plasma may be found in small platelet granules [1].

Decreased platelet α-granule contents are one manifestation of Quebec platelet disorder (QPD), an inherited disease with delayed bleeding and increased fibrinolysis. QPD is caused by a duplication of the urokinase gene, leading to increased u-PA-triggered plasmin activity in, and degradation of the contents of, α-granules in megakaryocytes [11].

***6.1.3.2 δ-SPD*** Hermansky–Pudlak and Chediak–Higashi syndromes are autosomal recessive diseases, characterized by albinism affecting the skin, eyes, and hair, due to a defect in melanocytes related to the platelet lesion. This consists of an impaired secretion-dependent platelet response, due to abnormalities in genes involved in dense granule formation. Genes implicated in the various forms of HPS include ADTB3A, involved in vesicular trafficking, and HPS proteins 1 and 3–7, components of mulitprotein BLOCs (biogenesis of lysosome-related organelle complexes). While HPS is a manageable disease, patients with CHS suffer from infections and fatal lymphoproliferative diseases. CHS is caused by defects in LYST, involved in lipid-related protein trafficking [9].

Type 2 Griscelli syndrome is characterized by hypopigmentation, platelet dense granule deficiency, and T-lymphocyte granule defects, and has been linked to mutations in Rab27a [12].

A novel δ-SPD with extant dense granules containing reduced adenosine nucleotide, but not serotonin levels, has been ascribed to mutations in the nucleotide transporter gene MRP4 [13]. A rare and less characterized δ-SPD is the "empty sack" syndrome, in which granules are apparent but both serotonin and adenosine nucleotide contents are reduced [14].

***6.1.3.3 Animal Models of Storage Pool Deficiencies*** An array of mutant mice (gunmetal, ashen, pale ear, light ear, pearl, mocha, pallid) exhibit some of the characteristics of SPDs [1], with reduced contents in α- and dense granules and partial albinism [12]. The gunmetal mouse defect has been traced to a Rab geranylgeranyl transferase gene, which might be expected to affect vesicle trafficking, with Rab27a particularly abundant in platelets and showing abnormal localization (failure to associate with membranes) in gunmetal mice [15]. A mutation in Rab27a itself causes the related ashen mouse model, characterized by a dense granule deficiency [10] and used as a model of type 2 Griscelli syndrome. However, more recent studies suggest—in mouse models at least—that the defect in packaging serotonin and ADP into platelet dense granules is due to mutations in the gene Slc35d3, encoding a nucleotide sugar transporter, while the granule defects in other cells are due to Rab27a mutations [16].

## 6.2 PROTEOMIC STUDIES OF GRANULES: THE ORGANELLE APPROACH

Platelets are certainly less complex than other human cell types—most notably, they lack a nucleus—and this will be reflected in the platelet transcriptome [17] and proteome. However, this does not mean that the platelet proteome cannot challenge the resolving power of even the most modern proteomic approaches. For instance, platelets contain proteins that span a vast range of abundance levels, from rare signaling molecules (of interest as potential drug targets) to abundant cytoskeletal proteins and αIIbβ3—one of the most densely expressed receptors in the body [18]. The platelet proteome may also contain more distinct platelet-specific proteins than previously thought. There are atypical splice variants of proteins expressed in platelets [19], perhaps due to the unusual way in which transcription is used by these cells to control protein expression [20].

A well-established approach to reducing the complexity of a particular system prior to proteomic analysis is to isolate particular cellular fractions, compartments, or organelles of interest and characterize these in detail [21]. Several benefits accrue from this approach, which has expanded our understanding of the contents and function of platelet granules and other organelles:

1. The number of distinct proteins and peptides in the mixture analyzed is reduced, increasing the proportion that may be identified, since very complex mixtures are not exhaustively characterized by current approaches [22]. This increases the likelihood of finding proteins novel to the organelle or fraction of interest.
2. This protein/peptide reduction is due partly to the second advantage of the organelle approach, the removal of highly abundant proteins. The most abundant proteins and peptides take up chromatographic space and MS analysis time that could be used to separate and identify peptides and proteins of more interest to the experimenter. Separating fractions of interest

from those that contain highly abundant proteins therefore allows the identification and quantification of more proteins of lower abundance. Of course, this may not be possible if, as for the platelet membrane or cytoskeleton, the fraction of interest is the location of highly abundant proteins.

3. The strategy of separation of fractions or organelles intrinsically yields information on the subcellular location of the proteins identified therein, with consequences for their ontological classification, possible interaction partners, and so on. When combined with preparations that alter the cell state so as to change the protein composition of a particular organelle or compartment, this yields even more benefits. For instance, the proteins enriched in the membrane of activated platelets, distinct from those present in resting platelet membranes, are of particular interest [23].

Even if they are not physically connected, some fractions have functional relationships, and it makes sense to analyze them collectively. For instance, the set of proteins secreted by activated platelets (the releasate) is derived from multiple discrete organelles. Released proteins with different platelet origins have been found in human atherosclerotic plaques [24,25].

In this chapter, studies that use these principles will be discussed to clarify the role of particular subproteomes of the platelet, namely, the platelet releasate and granules.

### 6.2.1 Platelet-Specific Considerations

The first step in any platelet proteomics study is the platelet isolation process. Unless suitable precautions are taken at this stage, contaminants from plasma proteins or other blood cells may be introduced. However, care must also be taken throughout this procedure to ensure that platelet activation is limited, since activation introduced by experimental manipulation will invalidate many later findings. Indeed, studies of all organelles will be affected by activation of the platelet, since changes in protein location, organization, and affinity for other proteins is widespread during activation. In particular, the protein composition of granules and the releasate will clearly change during platelet activation, since granule release is a major repsonse of platelets to activation. Many of the studies described below explicitly compare fractions from resting and activated platelets. One complication in verifying the purity of samples and ascribing a particular source to individual proteins is that many platelet α-granule proteins are also found in plasma, and may be taken up into the platelet from these circulating proteins [1]. Organelles also often copurify with cytoskeletal proteins [21] that may cloud proteomic analysis.

## 6.3 INDIVIDUAL STUDIES

Proteomic studies of note that have described biochemical organelles of the human platelet will be discussed in the following sections, focusing on the preparative and experimental strategy used, and the usefulness of the resultant findings. For more details on the experimental techniques used, readers are directed to the original papers cited below and appropriate references cited therein.

### 6.3.1 Platelet Releasate

The total platelet *releasate* or *secretome* consists of all the proteins and other biomolecules released by an activated platelet (Fig. 6.1). This is a functional rather than a morphological characterization, since releasate proteins may be derived from α-granules, dense granules, microparticles/microvesicles, and exosomes. The releasate may thus be expected to include the secreted contents of all the secretory organelles of the platelet. Typically, the releasate is prepared from washed platelets stimulated with a strong agonist such as thrombin receptor activating peptide (TRAP) or thrombin. However, other platelet stimuli may result in a selective release of only some organelle contents. Prior to any proteomic analysis, the identity of many secreted proteins was known, and many of these were plasma-derived [1]. However, proteomics has greatly expanded our understanding of the functional importance of this area, and there are now clear indications that proteins of the platelet releasate are found in human atherosclerotic plaque, where they may contribute to its pathogenesis [24,25].

***6.3.1.1 MudPIT Analysis of the Platelet Releasate*** The first explicitly proteomic approach to characterizing the repertoire of proteins released by platelets was that of Coppinger and coworkers in 2004 [24]. Platelets were activated with thrombin to induce a complete platelet release response, and the releasate separated by centrifugation and was cleared of microparticles by ultracentrifugation. Western blotting for released thrombospondin and the integrin $\alpha_{IIb}$ showed that this approach yielded a releasate free of contaminating membrane components, and there was minimal release from control platelets. An initial approach described in their paper [24], of 2D gel separation followed by protein identification by MALDI-MS, identified only a small number of relatively abundant proteins in releasate. A subsequent MudPIT (multidimensional protein identification technology) approach, using succesive salt elutions from a two-phase LC column loaded with releasate, was more successful. MS/MS results were analyzed using Sequest, with 81 proteins identified in at least two of three replicate experiments. No transmembrane or signaling proteins were found, underlining the successful separation of released proteins by this approach. Some cytoskeletal proteins were found in the releasate, and despite skepticism at the time, some of these have since been confirmed [26].

Of the 81 releasate proteins identified, 37% were previously known to be secreted or shed from platelets, and a further 35% were then known to be released from a variety of other cell types, such as immune cells and liver (many liver-produced plasma proteins may be taken up by platelets). The remaining proteins were newly described as secretory proteins in any system, and few were previously found in platelets. A corresponding message for 68% of the 81 releasate proteins was found in platelet RNA [27], arguing against uptake from blood plasma as an exclusive source for these. Strikingly, proteins corresponding to 18 of the 50 most abundant platelet messages were found in the platelet releasate, suggesting that synthesis of released platelet proteins is an important function of the megakaryocyte/platelet lineage.

A role for some newly described released proteins in human disease was sought by performing immunohistochemical staining for them in human atherosclerotic plaques derived from patients undergoing endarterectomy or revascularization. As well as the previously described platelet factor 4 and CD41, three novel proteins were found in lesions (but not normal tissue) from diseased vessels: secretogranin III, which is homologous to inflammatory proteins; cyclophilin A, which is an autocrine stimulant of vascular smooth muscle; and calumenin, which may promote vascular calcification. Together, the presence of these released platelet proteins in lesions suggests multiple functional pathways through which platelet releasate may progress to atherosclerosis.

***6.3.1.2 Effect of Aspirin on the Platelet Releasate*** The platelet releasate was revisited by this group in a study investigating the effect of antiplatelet agents on the profile of released proteins evoked by several platelet-activating agents [28]. Since aspirin is known to reduce thrombosis when used chronically [29], and released proteins are found in atherosclerotic plaques [24,25], the effect of aspirin on releasate is of interest. Platelets, in the absence or presence of aspirin, were stimulated with submaximal concentrations of ADP, collagen, or TRAP for 3 min. In another experiment, the effect of aspirin on the collagen releasate at 0.5, 3, and 20 min was determined. The releasate was isolated as previously [24]. Proteins were separated using a standard GeLC approach [30,31], and identified by Sequest, searching against an IPI database. Protein abundances were determined semiquantitatively using spectral counting and measurement of ion currents.

Comparison of stained gels of releasate obtained under various conditions showed a distinct releasate profile for each agonist, with reduction in the amount of protein released in each case that was attributed to pretreatment with aspirin. This effect was confirmed using semiquantitative abundance measurements (spectral counts and extracted ion currents) from representative proteins and was seen in each donor examined. Notably, the same proteins were broadly identified as released from control and aspirin-treated activated platelets, but the amount of each protein released was reduced by aspirin pretreatment. Normalized, log-transformed spectral counts were used for clustering analysis of released proteins, which revealed the differential secretion effect of different agonists. There is a set of more abundant proteins whose secretion is provoked by all three agonists, while lower-abundance proteins are released only in response to TRAP. In relation to the timecourse study, the gross effect of aspirin on collagen-induced release was most apparent at 0.5 and 3 min, with little inhibition of release by aspirin at 20 min. Additional platelet-activating stimuli from outside-in signaling are invoked by the authors to explain this failure of aspirin to inhibit secretion in the face of prolonged activation. The authors suggest that the inhibitory effect of aspirin, together with the timecourse study, imply a second stage of secretion on mild platelet stimulation, with this second wave dependent on additional signals, which may be provided by thromboxane $A_2$, or outside-in signaling. Also, the ability of aspirin to reduce the amount, but not greatly affect the profile of proteins released, suggests different populations of α-granules in platelets with different regulatory mechanisms.

***6.3.1.3 Quantitative Analysis of Platelet Releasate*** A separate approach to identifying differences in proteomic samples is the use of 2D difference gel electrophoresis (DIGE) analysis, and this has been applied to the thrombin-evoked platelet releasate [32]. A standard platelet-rich plasma preparation was used, which was cleaned by gentle centrifugation, and washed platelets generated in the presence of $PGE_1$ and EDTA. Flow cytometry of P-selectin was used to verify that there was minimal activation of the control sample. Platelets were activated with 0.5 U/mL thrombin for 3 min without stirring, and releasate isolated by centrifugation. In a standard DIGE approach [33], seven donors' samples were pooled in various combinations, with control and activated releasates cross-labeled with Cy3 and Cy5, and a mixed pooled sample labeled with Cy2, followed by 2D gel separation of proteins. The separately labeled pool of all samples allows matching of separate gels and controls for nonbiological variations. Automated image acquisition and preliminary processing were performed prior to MALDI-MS analysis of digested, excised spots. Proteins were identified by peptide mass fingerprinting, searching against NCBI and Swiss-Prot databases.

Interindividual differences were reported in the amount of protein recovered from resting and activated samples. This may reflect trapping of released proteins in fibrin clots, genuine variability between individuals, or incomplete platelet activation under the conditions used (3 min of activation by thrombin without stirring or use of fibrin polymerisation inhibitors). Different degrees of activation may result in differential secretion [28]. Approximately 1100 out of 2500 spots on each gel passed initial filtering. Creation of a master gel resulted in 824 spots, with 36 found to be differentially expressed, 21 of which increased with activation. Some known released proteins [such as platelet factor 4 (PF4) and von Willebrand factor (vWF)] are outside the molecular weight range of the gels used. This study confirmed previous findings of released proteins such as α-tubulin, transthyretin, and apolipoprotein $A_1$, as well as vinculin and WD-repeat protein, seen previously [24], and protein disulfide isomerase (PDI [34,35]). Also confirmed was the finding of nuclear proteins released from platelets [34], including the novel finding of lamin A release. This was confirmed by Western blot to be present at molecular weights suggesting activation of the apoptosis pathway. Proteins found to be decreased with activation include plasma proteins such as immunoglobulin, transferrin, fibrinogen and, apolipoprotein A—these may be artifacts of the normalisation of total protein prior to the labeling of samples for gel analysis. The authors note that this approach is not suited to the cataloguing of large numbers of proteins. While there is a low overall identification of proteins (since only differential spots are analyzed), this approach focuses on these differential proteins. Since thrombin has the potential to activate additional signal transduction pathways compared to other agonists (e.g., by signaling through GPIb and PAR4), thrombin activation may also result in greater secretion from the platelet. Whether this may account for the secretion of lamin A in this study is unclear, although it should be noted that the concentration of thrombin used is unlikely to cause significant cleavage and activation of PAR4 [36]. Questions remain over the success of maintaining platelet quiescence in control samples in this study, or conversely, incomplete activation and elicitation of a secretion response and recovery of secreted proteins. The presence of proteins in the resting releasate is ascribed by the authors to a background rate of turnover of platelet secretory granules.

***6.3.1.4 High-Mass-Accuracy Analysis of the Platelet Secretome*** It is to be expected that more recent studies will improve on the findings of earlier ones, due to improvements in mass spectrometry instruments, databases, and methodologies, as well as simply due to increased numbers of experiments focused on particular biological targets. A more recent study of the TRAP-induced platelet secretome has increased the number of proteins that may be confidently ascribed to this platelet subproteome [26]. The authors suggest that the platelet releasate may be a useful source of markers for the presence of cancer cells or of proteins that may influence cancer cell function [3,37]. A standard washed platelet preparation was used, with platelets from three separate donors stimulated with 5 μM TRAP for 3–5 min. Following washing and concentration through a 10-kDa cutoff filter, proteins were separated and digested by a standard GeLC approach [30,31]. Mass spectrometry was with an LTQ-FT instrument, resulting in measurement of peptide fragments at a higher mass accuracy than in any previous comparable study. Peptide spectra were searched with Sequest against an IPI database. Filtering and manual curation were applied to the search results.

A total of 716 proteins were identified, with a set of 225 found in all three replicates deemed the *core platelet secretome*. This set overlaps considerably with earlier studies of the total platelet secretome [24,28] as well as studies of platelet secretory compartments, including microparticles [34] and α-granules [38]. With so many proteins identified, many functional classes are apparent, several of which are noteworthy, in the opinion of this author. Many proteins with a cytoskeletal function/location were noted, including actin, myosin, tubulins, actin capping proteins, actinins, talin, URP2 (kindlin 3), and filamin. While the presence of these proteins in such proteomics experiments is often ascribed to cell lysis, the consistency and abundance with which they appear suggests a genuine physiological presence in platelet releasate (when evoked by thrombin or TRAP), most likely in microparticles. Regulatory proteins involved in RNA translation were also seen in this study (EEF1G, EIF4A1). Given their unusual location in platelets [39], their presence in releasate is perhaps not so surprising. The degree to which factors involved in initiation, propagation, execution, and control of the coagulation system are represented in the platelet releasate is striking. Some of the relevant proteins seen in this study include kininogen; factors V, XIII, and II (prothrombin); fibrinogen and vWF; and protein S, antithrombin, plasminogen, and plasminogen activator inhibitor (PAI) 1. A few other individual proteins are of interest: protein disulfide isomerases (PDIs) are speculated to play a role in initiating coagulation [40], the aminopeptidase NPEPPS has been shown to be present in platelets as an atypical splice variant at the protein and RNA level [19], and the intracellular chloride channel previously identified by DIGE analysis of thrombin-invoked releasate [32] was confirmed. This study represents an impressive progression over previous work. The lack of complete overlap with earlier studies suggests that additional, well-powered investigations of the platelet secretome are required before we can consider this subproteome to be fully characterized [22].

### 6.3.2 α-Granules

While α-granules are the major contributors to the protein component of the platelet releasate, only one study has specifically studied their proteome in detail. Maynard and coworkers [38] used ultrasonication to lyse platelets, followed by sucrose gradient centrifugation to separate subcellular fractions of the lysed cells. One fraction was highly enriched in α-granules (as determined by electron microscopy and Western blotting for proteins characteristic of various platelet fractions), although there were also traces of other platelet components, and evidence for proteins associated with other compartments. Thus, there was substantial enrichment, but not complete separation of α-granules from other platelet components, as is true to varying degrees for all organelle isolation procedures. Following enrichment, α-granule proteins were identified using a standard GelC-MS/MS approach [30,31]. MS results were searched against a Uniprot database using Mascot, and proteins were deemed to be present if two distinct peptides were identified in each of two out of three replicate experiments. In total, 65 proteins known to be associated with mitochondria were manually removed from the analysis, since mitochondrial contamination is a possibility in sucrose gradient separation of lysed cells.

Of the 219 proteins identified, 36 were previously associated with α-granules [41]. Of the identified proteins, 50 were among the 81 previously found in platelet releasate [24]; and of 44 proteins deemed by the authors to be novel findings for α-granules, 26 were also found in the membrane fraction. Few of these were found in platelet releasate, illustrating the complementarity of these approaches. Membrane-bound proteins may play a role in regulating granule release, or provide additional functionality when expressed on the surface of activated platelets. The presence of several of these 44 proteins was confirmed in the α-granules of intact platelets by electron microscopy. These include Scamp2, likely involved in regulating secretion, found on the α-granule and plasma membranes; laminin α5 (LAMA5), which may be secreted as a temporary extracellular matrix protein at sites of wounding, found intraluminally in platelet α-granules; endothelial cell-selective adhesion molecule (ESAM), a junctional adhesion molecule, which may mediate tight platelet–platelet and platelet–endothelial cell interactions [42];

and amyloid protein–like protein 2 (APLP2), which joins other Alzheimer's disease–related proteins (A4, presenilin) previously found in α-granules, located on granule membrane and intraluminal regions. The authors suggest that platelets may be useful in the diagnosis and monitoring of Alzheimer's disease.

As the most protein-rich secretory organelle in the platelet, the α-granule is of great interest as a source of biologically active molecules that may play a role in atherosclerosis and thrombosis. Targeted studies in megakaryocytes or mature platelets may reveal more about granule formation, how granule secretion is regulated, as well as differences between circulating proteins and their corresponding versions in granules.

### 6.3.3 Dense Granules

The specific soluble protein content of the platelet dense granule has now been characterized [25]. To overcome the low protein content of this organelle, large amounts of expired donated pooled platelets were used. Following cell removal and pelleting, platelets were gently lysed and cleared, and then platelet contents applied to a density column. After centrifugation, the enrichment of dense granules in the expected fraction was confirmed by blotting for specific markers and electron microscopy to visualize granules. Twin proteomic approaches were taken, with 2D gels and MALDI-MS identifying 20 proteins, and LC-MS/MS identifying 35, with a total of 40 proteins seen.

All proteins identified in this study had previously been reported in platelets, and most had been identified as being released from platelets [25]. Since releasate contains proteins from a variety of intracellular sources, this organelle-focused study complements such functional studies to identify the likely source of components of the platelet releasate. The proteins identified belong to various functional categories. Cytoskeletal proteins may be present, due to the mechanics of granule formation and secretion, or as part of the process by which polyphosphate is concentrated in dense granules. Many proteins with a prominent function in platelets (e.g., PF4, β-thromboglobulin, fibrinogen) are shared with α-granules, suggesting some redundancy in the platelet release response. Some glycolytic proteins, also identified in the platelet releasate [24], have also been reported in lysosomel-related organelles from other cells, such as melanocytes, and so are unlikely to be due to simple contamination. One surprising protein found in dense granules, 14-3-3ς was shown to be present in human atheroscelrotic plaque, but not normal vessel, a result that mirrors earlier findings of total platelet releasate [24]. The 14-3-3ς protein is known to regulate GPIb function, and this protein family has a broader role as adapter proteins, often modifying the location or function of proteins on phosphorylation [43]. However, the authors note several other studies showing extracellular roles for 14-3-3 proteins released from cells, including keratinocytes, osteoblasts, algae, and plant root tips [25]. Thus there may well be an unheralded role for this abundant platelet protein beyond the confines of the platelet plasma membrane. This study illustrates that even where no large lists of proteins, or novel proteins are identified, well-designed and well-executed proteomic work may still produce novel findings that challenge our current understanding of platelet function.

### 6.3.4 Microparticles

Microparticles are small cell-derived vesicles of phospholipid bilayer found in blood plasma. Microparticles are now seen as functional cellular derivatives rather than products of cell damage or apoptosis. Normally, 90% of plasma microparticles are platelet-derived [44]. These act as a major source of negatively charged phospholipid for the propagation of the coagulation cascade [45], while monocyte microparticles provide tissue factor for the initiation of coagulation [46]. Since they may form a component of the proteins released from platelets, they are briefly reviewed here, while proteomic analysis of platelet microparticles is discussed in detail in Chapter 7.

***6.3.4.1 Characterization of Platelet Microparticles*** The first proteomic characterization of platelet microparticles was carried out by García and colleagues [34]. Following pelleting of activated platelets, microparticles were separated from soluble released proteins by ultracentrifugation (giving a fraction complementary to releasate studies [24,26]). In this study, 578 proteins were identified by a standard GeLC-MS/MS approach with 380 proteins not previously seen in platelets. There was considerable overlap with a contemporaneous proteomic study of total plasma microparticles [47]. Among the expected proteins found were platelet glycoproteins, coagulation factors, and cytokines. Novel proteins included several then-uncharacterized proteins that have since been ascribed an endoplasmic reticulum location (PRO1855, KIAA0152, and KIAA0851).

***6.3.4.2 Comparison of Platelet and Plasma Microparticles*** The same group revisited the analysis of platelet microparticles in a later work [48], comparing platelet and plasma microparticles; 75% of the platelet microparticle proteins had previously been identified [34]. The abundance of proteins in the platelet and plasma microparticles were compared using spectral counting and an isotope-coded affinity tag (ICAT) approach [49]. Since most plasma microparticles are platelet-derived, a small number of additional proteins were expected in the plasma microparticle set. Indeed, spectral counting identified 21 proteins as abundant in plasma but essentially absent from platelet microparticles. The ICAT approach largely confirmed these results.

The differentially expressed proteins fall into several categories that may be related to their mechanism of generation, including complement functionality, apoptosis-related proteins, and iron transport. Of note, the authors speculate that one such protein, tenascin C, may be released in microparticles from cells associated with healing of small vascular injuries. Platelets are also recruited early to such sites, and it would be of interest if differential analysis of microparticles of platelets versus other specific cell types (rather than the entire plasma microparticle pool) could reveal proteins involved in different stages of the wound healing process, with possible implications for the development of atherosclerosis. If there is an important functional role for such proteins in wound healing, it is unclear why there would be multiple platelet mechanisms (microparticles, granules) for their release at sites of injury.

### 6.3.5 Platelet Plasma Membrane

The platelet plasma membrane is of particular interest, since it is the interface between the platelet and its environment, where pro- and antiaggregatory signals are received and, to some extent, processed. Integral or membrane-associated proteins include receptors for adhesion, aggregation, and extracellular signals, as well as transduction proteins involved in signal mediation, integration, and amplification. Platelet membrane proteomic studies are discussed in detail in Chapter 5.

To date (as of 2010), there are no proteomic studies focusing on organelle-specific membranes, although it is clear that there are distinctions between their protein contents, with P-selectin-limited granule membranes, and its presence on the platelet surface used as a marker of secretion. Characterization of further differences in protein content, between the platelet surface membrane and granule or other cellular compartment membranes, such as the platelet surface canicular system, would be of interest but may be technically challenging.

## 6.4 ORGANELLE SEPARATION

All the studies described above include some cross-contamination of purified organelles by proteins from other locations. Technical improvements will continue to be made in the approaches applied to platelet fractions used in proteomic studies. A magnetic immunoaffinity approach has been described for the purification of platelet organelles [50] with considerably improved enrichment of plasma membrane, α-granules and dense granules when compared to density gradient centrifugation. Lysed platelets are incubated with primary antibodies specific for a membrane marker of the required organelle. After washing, magnetic beads coupled to a secondary antibody are used to separate the target organelle from the remainder of the cell. Specific markers for each fraction show the improved specificity compared to other approaches. The authors note that some crossover between plasma membrane and intracellular compartments is inevitable. There may be difficulties of scale in obtaining sufficient sample by this approach for in-depth proteomic analysis, but it does demonstrate that improvements in preparation, together with advances in protein and peptide separation, and mass spectrometry, will yield better platelet organelle proteomic studies in the future.

## 6.5 CLINICAL UTILITY OF ORGANELLE PROTEOMICS

Platelet proteomic studies are still primarily a research tool, but as the techniques involved become more widely used, standard protocols developed, and consensus reached on what constitutes normal and abnormal findings, these techniques may become useful in clinical diagnosis of storage pool diseases. For instance, quantitative proteomics has been used to help diagnose Quebec platelet disorder, through measurement of decreased amounts of platelet $\alpha$-granule proteins in affected individuals [51]. Since this study compared abundance of common platelet proteins, the use of relatively straightforward abundance measurements (based on spectral counts and scores) rather than absolute quantification of labelled samples is probably valid. More broadly, quantitative analysis of platelet releasate may identify a continuum of response levels, with some individuals being more prone to show secretion of platelet granule contents. Such differences might be important in the development of disease, and used to target at-risk individuals for specific therapies.

## 6.6 CONCLUDING REMARKS

Without doubt, there will be improvements in the future in terms of experimental procedures to isolate organelles, as well as advances in protein separation, mass spectrometry, and bioinformatics. Nevertheless, even current technologies may be usefully employed in the study of platelet organelle proteomes. For example, existing proteomic mass spectral data may contain additional unrecognised identifications that may be revealed by searching novel, computationally derived databases of human proteins [19].

Certainly existing technologies have incomplete penetration of the platelet organelle proteome. Targeted studies can reveal the presence and functional role of individual proteins more clearly. For instance, the peroxisome proliferator activated receptor (PPAR) $\gamma$ and retinoid X receptor (RXR) are found in platelets and released in microparticles [52], but are not described in the quite comprehensive proteomic study of that compartment reviewed above [34]. These proteins are transcription factors, but their release from platelets may affect cells of the immune system, similarly to the paradigm described for Bcl3 [53].

It is now clear that processing of mRNA is handled differently in platelets compared to other cells, with novel splice variants demonstrated [19,54,55]. This implies that even organelles well described in other cells (such as mitochondria) may have platelet-specific components, and that focussed platelet studies are warranted. The differences found in particular organelles or fractions between resting and activated platelets are also substantially uncharacterized. For instance, studies contrasting resting and activated membrane skeletons would be of great utility for identifying proteins involved in regulating shape change and signal transduction cascades. Other areas that may be usefully explored are protein complexes that might be considered functional organelles (such as the spliceosome [20] or integrin-binding proteins [56]) that are not yet characterized proteomically.

Since the repertoire of platelet functions may continue to surprise us, we should not be slow to undertake additional, well-designed proteomic studies to better understand the platelet's components, nor should we be too skeptical of the sometimes unexpected proteins that we identify therein.

## ACKNOWLEDGMENT

A critical reading of this work by Dr. Gerard Cagney is much appreciated.

## REFERENCES

1. King SM, Reed GL, Development of platelet secretory granules, *Semin. Cell Dev. Biol*. **13**:293–302 (2002).
2. Richardson JL, Shivdasani RA, Boers C, Hartwig JH, Italiano JE, Mechanisms of organelle transport and capture along proplatelets during platelet production, *Blood* **106**:4066–4075 (2005).
3. Italiano J, Richardson JL, Patel-Hett S, Battinelli E, Zaslavsky A, Short S, Ryeom S, Folkman J, Klement GL, Angiogenesis is regulated by a novel mechanism: Pro- and anti-angiogenic proteins are organized into separate platelet α-granules and differentialy released, *Blood* **111**:1227–1233 (2008).
4. Blair P, Flaumenhaft R, Platelet alpha-granules: Basic biology and clinical correlates, *Blood Rev*. **23**:177–189 (2009).
5. Ruiz FA, Lea CR, Oldfield E, Docampo R, Human platelet dense granules contain polyphosphate and are similar to acidocalcisomes of bacteria and unicellular eukaryotes, *J. Biol. Chem*. **279**:44250–44257 (2004).
6. McNicol A, Israels SJ, Platelet dense granules: Structure, function and implications for haemostasis, *Thromb. Res*. **95**:1–18 (1999).
7. Flaumenhaft R, Molecular basis of platelet granule secretion, *Arterioscler. Thromb. Vasc. Biol*. **23**:1152–1160 (2003).
8. Weiss HJ, Witte LD, Kaplan KL, Lages BA, Chernoff A, Nossel HL, Goodman DS, Baumgartner HR, Heterogeneity in storage pool deficiency: Studies on granule-bound substances in 18 patients including variants deficient in alpha-granules, platelet factor 4, beta-thromboglobulin, and platelet-derived growth factor, *Blood* **54**:1296–1319 (1979).
9. Nurden AT, Qualitative disorders of platelets and megakaryocytes, *J. Thromb. Haemost*. **3**:1773–1782 (2005).
10. Nurden AT, Nurden P, The gray platelet syndrome: Clinical spectrum of the disease, *Blood Rev*. **21**:21–36 (2007).
11. Paterson AD, Rommens JM, Bharaj B, Blavignac J, Wong I, Diamandis M, Waye JS, Rivard GE, Hayward CP, Persons with Quebec platelet disorder have a tandem duplication of PLAU, the urokinase plasminogen activator gene, *Blood* **115**:1264–1266 (2010).
12. Novak EK, Gautam R, Reddington M, Collinson LM, Copeland NG, Jenkins NA, McGarry MP, Swank RT, The regulation of platelet-dense granules by Rab27a in the ashen mouse, a model of Hermansky-Pudlak and Griscelli syndromes, is granule-specific and dependent on genetic background, *Blood* **100**:128–135 (2002).
13. Jedlitschky G, Cattaneo M, Lubenow LE, Rosskopf D, Lecchi A, Artoni A, Motta G, Niessen J, Kroemer HK, Greinacher A, Role of MRP4 (ABCC4) in platelet adenine nucleotide-storage: Evidence from patients with delta-storage pool deficiencies, *Am. J. Pathol*. **176**:1097–1103 (2010).

14. McNicol A, Israels SJ, Robertson C, Gerrard JM, The empty sack syndrome: A platelet storage pool deficiency associated with empty dense granules, *Br. J. Haematol.* **86**:574–582 (1994).
15. Zhang Q, Zhen L, Li W, Novak EK, Collinson LM, Jang EK, Haslam RJ, Elliott RW, Swank RT, Cell-specific abnormal prenylation of Rab proteins in platelets and melanocytes of the gunmetal mouse, *Br. J. Haematol.* **117**:414–423 (2002).
16. Chintala S, Tan J, Gautam R, Rusiniak ME, Guo X, Li W, Gahl WA, Huizing M, Spritz RA, Hutton S, Novak EK, Swank RT, The Slc35d3 gene, encoding an orphan nucleotide sugar transporter, regulates platelet-dense granules, *Blood* **109**:1533–1540 (2007).
17. Dittrich M, Birschmann I, Pfrang J, Herterich S, Smolenski AP, Walter U, Dandekar T, Analysis of SAGE data in human platelets: Features of the transcriptome in an anucleate cell, *Thromb. Haemost.* **95**:643–651 (2006).
18. Born G, Patrono C, Antiplatelet drugs, *Br. J. Pharmacol.* **147**:S241–S251 (2006).
19. Power KA, McRedmond JP, de Stefani A, Gallagher WM, O'Gaora P, High-throughput proteomics detection of novel splice isoforms in human platelets, *PLoS* ONE **4**: e5001.
20. Denis MM, Tolley ND, Bunting M, Schwertz H, Jiang H, Lindemann S, Yost CC, Rubner FJ, Albertine KH, Swoboda KJ, Fratto CM, Tolley E, Kraiss LW, McIntyre TM, Zimmerman GA, Weyrich AS, Escaping the nuclear confines: Signal-dependent pre-mRNA splicing in anucleate platelets, *Cell* **122**:379–391 (2005).
21. Yates JR, Gilchrist A, Howell K, Bergeron JJ, Proteomics of organelles and large cellular structures, *Nat. Rev. Mol. Cell. Biol.* **6**:702–714 (2005).
22. Liu H, Sadygov RG, Yates JR, A model for random sampling and estimation of relative protein abundance in shotgun proteomics, *Anal. Chem.* **76**:4193–4201 (2004).
23. Kaiser WM, Holbrook LM, Tucker KL, Stanley RG, Gibbins JM, A functional proteomic method for the enrichment of peripheral membrane proteins reveals the collagen binding protein Hsp47 is exposed on the surface of activated human platelets, *J. Proteome Res.* **8**:2903–2914 (2009).
24. Coppinger JA, Cagney G, Toomey S, Kislinger T, Belton O, McRedmond JP, Cahill DJ, Emili A, Fitzgerald DJ, Maguire PB, Characterization of the proteins released from activated platelets leads to localization of novel platelet proteins in human atherosclerotic lesions, *Blood* **103**:2096–2104 (2004).
25. Hernandez-Ruiz L, Valverde F, Ruiz FA, Organellar proteomics of human platelet dense granules reveals that 14-3-3ζ is a granule protein related to atherosclerosis, *J. Proteome Res.* **6**:4449–4457 (2007).
26. Piersma S, Broxterman H, Kapci M, de Haas R, Hoekman K, Verheul H, Jiménez C, Proteomics of the TRAP-induced platelet releasate, *J. Proteom.* **72**:91–109 (2009).
27. McRedmond JP, Park SD, Reilly DF, Coppinger JA, Maguire PB, Shields DC, Fitzgerald DJ, Integration of proteomics and genomics in platelets: A profile of platelet proteins and platelet-specific genes, *Mol. Cell. Proteom.* **3**:133–144 (2004).
28. Coppinger JA, O'Connor R, Wynne K, Flanagan M, Sullivan M, Maguire PB, Fitzgerald DJ, Cagney G, Moderation of the platelet releasate response by aspirin, *Blood* **109**:4786–4792 (2007).
29. Patrono C, García Rodríguez LA, Landolfi R, Baigent C, Low-dose aspirin for the prevention of atherothrombosis, *N. Engl. J. Med.* **353**:2373–2383 (2005).

30. Schirle M, Heurtier MA, Kuster B, Profiling core proteomes of human cell lines by one-dimensional PAGE and liquid chromatography-tandem mass spectrometry, *Mol. Cell. Proteom.* **2**:1297–1305 (2003).

31. Shevchenko A, Jensen ON, Podtelejnikov AV, Sagliocco F, Wilm M, Vorm O, Mortensen P, Shevchenko A, Boucherie H, Mann M, Linking genome and proteome by mass spectrometry: Large-scale identification of yeast proteins from two dimensional gels, *Proc. Natl. Acad. Sci. USA* **93**:14440–14445 (1996).

32. Della Corte A, Maugeri N, Pampuch A, Cerletti C, de Gaetano G, Rotilio D, Application of 2-dimensional difference gel electrophoresis (2D-DIGE) to the study of thrombin-activated human platelet secretome, *Platelets* **19**:43–50 (2008).

33. Timms JF, Cramer R, Difference gel electrophoresis, *Proteomics* **8**:4886–4897 (2008).

34. García BA, Smalley DM, Cho H, Shabanowitz J, Ley K, Hunt DF, The platelet microparticle proteome, *J. Proteome Res.* **4**:1516–1521 (2005).

35. Chen K, Lin Y, Detwiler TC, Protein disulfide isomerase activity is released by activated platelets, *Blood* **79**:2226–2228 (1992).

36. Kahn ML, Nakanishi-Matsui M, Shapiro MJ, Ishihara H, Coughlin SR, Protease-activated receptors 1 and 4 mediate activation of human platelets by thrombin, *J. Clin. Invest.* **103**:879–887 (1999).

37. Kopp H, Hooper A, Broekman MJ, Avecilla S, Petit I, Luo M, Milde T, Ramos C, Zhang F, Kopp T, Bornstein P, Jin D, Marcus A, Rafii S, Thrombospondins deployed by thrombopoietic cells determine angiogenic switch and extent of revascularization, *J. Clin. Invest.* **116**:3277–3291 (2006).

38. Maynard DM, Heijnen HF, Horne MK, White JG, Gahl WA, Proteomic analysis of platelet α-granules using mass spectrometry, *J. Thromb. Haemost.* **5**:1945–1955 (2007).

39. Lindemann S, Tolley ND, Eyre JR, Kraiss LW, Mahoney TM, Weyrich AS, Integrins regulate the intracellular distribution of eukaryotic initiation factor 4E in platelets. A checkpoint for translational control, *J. Biol. Chem.* **276**:33947–33951 (2001).

40. Reinhardt C, Von Brühl M, Manukyan D, Grahl L, Lorenz M, Altmann B, Dlugai S, Hess S, Konrad I, Orschiedt L, Mackman N, Ruddock L, Massberg S, Engelmann B, Protein disulfide isomerase acts as an injury response signal that enhances fibrin generation *via* tissue factor activation, *J. Clin. Invest.* **118**:1110–1122 (2008).

41. Rendu F, Brohard-Bohn B, The platelet release reaction: granules' constituents, secretion and functions, *Platelets* **12**:261–273 (2001).

42. Brass LF, Zhu L, Stalker TJ, Minding the gaps to promote thrombus growth and stability, *J. Clin. Invest.* **115**:3385–3392 (2005).

43. Mackintosh C, Dynamic interactions between 14-3-3 proteins and phosphoproteins regulate diverse cellular processes, *Biochem. J.* **381**:329–342 (2004).

44. Horstman LL, Ahn YS, Platelet microparticles: A wide-angle perspective, *Crit. Rev. Oncol. Hematol.* **30**:111–142 (1999).

45. Wolf P, The nature and significance of platelet products in human plasma, *Br. J. Haematol.* **13**:269–288 (1967).

46. Falati S, Liu Q, Gross P, Merrill-Skoloff G, Chou J, Vandendries E, Celi A, Croce K, Furie BC, Furie B, Accumulation of tissue factor into developing thrombi *in vivo* is dependent upon microparticle P-selectin glycoprotein ligand 1 and platelet P-selectin, *J. Exp. Med.* **197**:1585–1598 (2003).

47. Jin M, Drwal G, Bourgeois T, Saltz J, Wu HM, Distinct proteome features of plasma microparticles, *Proteomics* **5**:1940–1952 (2005).
48. Smalley DM, Root KE, Cho H, Ross MM, Ley K, Proteomic discovery of 21 proteins expressed in human plasma-derived but not platelet-derived microparticles, *Thromb. Haemost*. **97**:67–80 (2007).
49. Gygi SP, Rist B, Gerber SA, Turecek F, Gelb MH, Aebersold R, Quantitative analysis of complex protein mixtures using isotope-coded affinity tags, *Nat. Biotechnol*. **17**:994–999 (1999).
50. Niessen J, Jedlitschky G, Grube M, Bien S, Strobel U, Ritter CA, Greinacher A, Kroemer HK, Subfractionation and purification of intracellular granule-structures of human platelets: An improved method based on magnetic sorting, *J. Immunol. Meth*. **328**:89–96 (2007).
51. Maurer-Spurej E, Kahr W, Carter C, Pittendreigh C, The value of proteomics for the diagnosis of a platelet-related bleeding disorder, *Platelets* **19**:342–351 (2008).
52. Ray D, Spinelli S, Pollock S, Murant T, O'Brien JJ, Blumberg N, Francis C, Taubman M, Phipps R, Peroxisome proliferator-activated receptor gamma and retinoid X receptor transcription factors are released from activated human platelets and shed in microparticles, *Thromb. Haemost*. **99**:86–95 (2008).
53. Weyrich AS, Dixon DA, Pabla R, Elstad MR, McIntyre TM, Prescott SM, Zimmerman GA, Signal-dependent translation of a regulatory protein, Bcl-3, in activated human platelets, *Proc. Natl. Acad. Sci. USA* **95**:5556–5561 (1998).
54. Censarek P, Steger G, Paolini C, Hohlfeld T, Grosser T, Zimmermann N, Fleckenstein D, Schrör K, Weber AA, Alternative splicing of platelet cyclooxygenase-2 mRNA in patients after coronary artery bypass grafting, *Thromb. Haemost*. **98**:1309–1315 (2007).
55. Newland SA, Macaulay IC, Floto AR, de Vet EC, Ouwehand WH, Watkins NA, Lyons PA, Campbell DR, The novel inhibitory receptor G6B is expressed on the surface of platelets and attenuates platelet function *in vitro*, *Blood* **109**:4806–4809 (2007).
56. Daxecker H, Raab M, Bernard E, Devocelle M, Treumann A, Moran N, A peptide affinity column for the identification of integrin alpha(IIb)-binding proteins, *Anal. Biochem*. **374**:203–212 (2008).

# 7

# THE PLATELET MICROPARTICLE PROTEOME

DAVID M. SMALLEY

**Abstract**

Over the last several years there have been great advances in technologies and methodologies that have allowed us to examine proteomes in great detail. This chapter discusses the approach and results for platelet and plasma microparticles. While these particles were originally considered to be simply cell debris, more recent studies demonstrate they have important roles in normal and disease processes. By examining the proteome of platelet derived microparticles, we may gain a better understanding of a wide variety of physiological and pathological processes.

## 7.1 INTRODUCTION

Microparticles (MPs) are small subcellular membrane-enclosed vesicles released by all cell types examined to date (as of 2010), especially when the cells are activated or under stress. Microparticles include ectosomes, exosomes, and apoptotic bodies. Ectosomes are generated from ectocytosis (or blebbing) of the plasma membrane and vary in size. For platelets, all microvesicles ranging from 100 nm to 1.0 μm are generally considered to be ectsomes [1]. Platelet-derived ectosomes were initially identified in plasma when investigators were trying to isolate a blood clotting factor [2]. This factor could be removed from plasma

*Platelet Proteomics: Principles, Analysis and Applications*, First Edition.
Edited by Ángel García and Yotis A. Senis.

by high-speed centrifugation, and its activity turned out to be due to negatively charged phospholipids, predominantly phosphatidylserine, on the surface of these microvesicles. This negatively charged surface binds calcium ions that help localize and orient several other important clotting factors, enhancing their activity. Among these are factor Xa and prothrombin, leading to the generation of thrombin. Exosomes are another class of microparticles released by platelets [1]. They are generated from the fusion of intracellular multivesicular endosomes with the cell surface, which are then released. From most cell types exosomes are uniform in size, but in platelets they range from approximately 40 and 100 nm [1]. These are similar to the size of multivesicular bodies and α-granules located within the platelets suggesting their possible origin. Apoptotic body-derived microparticles are released by most cells. In platelets, the distinction between apoptosis and ectosome formation is unclear, and these processes may be the same. Both processes involve many of the same pathways required for plasma membrane blebbing, such as increased intracellular calcium and activation of caspase 3. Apoptosis also generally involves alterations in nuclear fragmentation, chromatin condensation, and chromosomal DNA fragmentation, none of which can occur in platelets. While release of ectosomes from nucleated cells can occur without apoptosis, it is unclear whether platelets can release ectosomes and still survive. More recent evidence indicates these are distinct processes [3]. Initially microparticles (both ectosomes and exosomes) were considered to be cellular debris with little functional importance. Ectosomes were actually referred to as "platelet dust," and exosomes were discovered as a byproduct of downregulation of the transferring receptor on erythrocytes [4]. There has now been significant work on microparticles generated from a variety of cell types, and it is clear that they participate in a variety of cellular processes, including blood coagulation, intercellular communication, and complement activation [5,6]. Platelet-derived microparticles, however, have not received much attention, even though they are the most predominant microparticle in the plasma [7]. To obtain a better understanding of some of the possible roles of this type of microparticle, we examined the proteome of these microvesicles, and compared it with the proteome of microvesicles isolated from the plasma (plasma microparticles). This chapter describes our results and their possible biological significance, and compares the findings to those other approaches used to examine the function and importance of these microvesicles.

## 7.2 ANALYSIS OF MICROPARTICLES GENERATED FROM PLATELETS FOLLOWING ACTIVATION WITH ADP

Platelets release a variety of types of microvesicles following activation. These originate from the blebbing of the cell membrane and release of organelles, such as α-granules and dense granules. While the term *microparticle* has traditionally been used to refer to only those particles originating from blebbing of the cell surface, in most cases their cellular origin was not examined and assumed to be based on their size and the presence of surface markers. The term *ectosomes* more clearly describes these vesicles [8], and the term *microparticles* (or microvesicles) will be used in this chapter to refer to all membrane-enclosed particles released from cells. To examine the proteome of microvesicles generated from human platelets, platelets were isolated and activated with ADP. The microvesicles generated were isolated and analyzed as depicted in Figure 7.1 and as described in detail previously [9]. Human blood was collected in $\frac{1}{10}$th volume of acid–citrate–dextrose (ACD; 85 mM trisodium citrate, 83 mM dextrose, 21 mM citric acid) and platelets were isolated by differential centrifugation. They were washed extensively to remove unbound plasma proteins, and activated with 10 μM ADP for 10 min at 37°C. Microvesicles were separated from platelets by differential centrifugation (710*g* for 15 min) and were then isolated

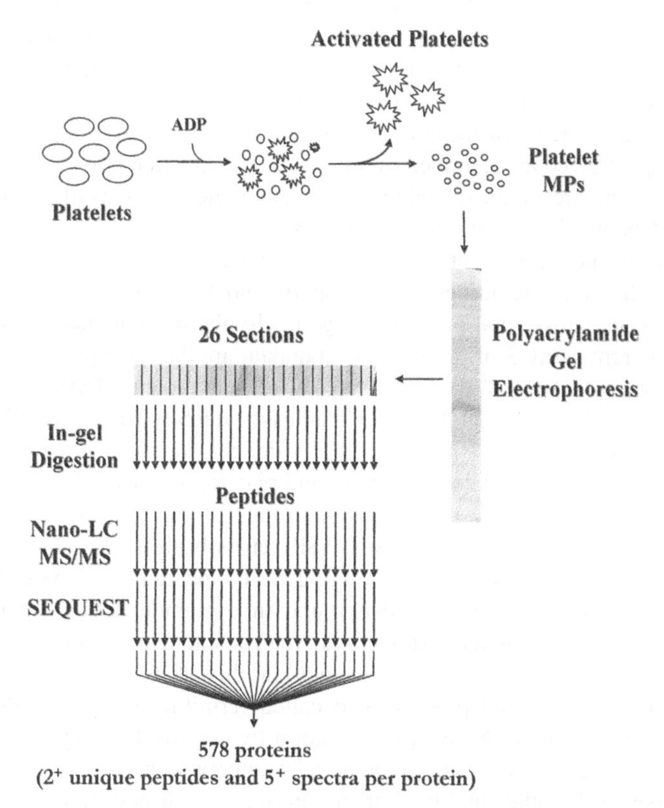


**Figure 7.1** In-depth analysis of platelet microparticles. Isolated platelets were activated with ADP for 10 min to generate microparticles (platelet microparticles). The microparticles were isolated by differential centrifugation and partially separated by PAGE. The gel lane was cut into 26 sections, and the proteins in each section were digested to generate peptides. The peptides were extracted and analyzed by nano-LC-MS/MS. A total of 578 proteins were identified with a minimum of two peptides and a total of five spectra per protein.

by ultracentrifugation (150,000*g* for 90 min). The microparticles were solubilized in Laemmli buffer and applied to a polyacrylamide gel. Following staining, the gel was cut into 26 sections. The proteins in each section were digested with trypsin, and the peptides were extracted. The peptides from each gel fraction were applied to a nano-LC C18 precolumn [360 μm outer diameter (o.d.) × 75 μm inner diameter (i.d.)]. Following rinsing, the precolumn was connected to a C18 analytical column (360 μm o.d. × 50 μm i.d.) constructed with an integrated electrospray emitter tip. Peptides were eluted by increasing the acetonitrile content directly into a Finnigan LTQ ion trap mass spectrometer (Thermo Electron, San Jose, CA) at a flow rate of 60 nL/min. The mass spectrometer was operated in the data-dependent mode where an initial mass spectrometry (MS) scan recorded the mass-to-charge ($m/z$) ratios of parent ions over the mass range 300–2000 Da. The 10 most abundant ions were selected for subsequent collision-activated dissociation (CAD). All MS/MS data were searched against a human database using the Sequest algorithm with highly selective criteria. This included identification of at least two unique peptides from a given protein using Sequest, and visual inspection of at least one of these peptides to verify sequence. We found a total of 578 distinct proteins in these microparticles. The 50 most abundant proteins based on spectral count (number of mass spectra identified for peptides from a given protein) are given in Table 7.1. A complete list is available from the American Chemical Society (`http://pubs.acs.org`) [9].

Of these 50 most abundant proteins, cytoskeletal or cytoskeletal-related ones account for the bulk, including 7 of the top 10, and 16 of the 50. This is not surprising because cytoskeletal proteins are generally the most abundant proteins in the cell. The remaining 3 of the 10 most abundant are thrombospondin 1 (TSP1), von Willebrand factor (vWF), and UNC112-related protein. TSP1 is stored in the α-granules of human platelets, and is released quickly from the cell on activation [10]. Its function has not been totally delineated, but it is believed to be involved with increasing platelet–platelet and platelet–extracellular matrix interactions. It may function by reinforcing interplatelet interactions through direct fibrinogen–TSP1–fibrinogen and TSP1-TSP1 cross-bridges [11]. In addition, it may also play a role in modulating angiogenesis and tumor growth [12]. The von Willebrand Factor (vWF) is also found in α-granules and is involved in platelet–platelet and platelet–extracellular matrix interactions and is also produced by endothelial cells, and deficiency in this protein leads to bleeding disorders. UNC112-related protein, also called fermitin family homolog 3, or kindlin 3, is essential for β-integrin activation by directly binding to regions of its tail, and therefore is required for platelet aggregation [13]. A surprisingly large number of the other highly abundant proteins are generally not considered to be of platelet origin. These include albumin, coagulation factor XIII, and the fibrinogen chains α, β, and γ, all of which are highly abundant plasma proteins, and which are probably taken up into platelets via endocytosis and stored in the α-granules. Both of the subunits of integrin αIIbβ3, the most abundant proteins on the surface of platelets [14], are also near the top of the list, suggesting that many of these microvesicles are ectosomes.

**TABLE 7.1 The 50 Most Abundant Proteins in Platelet MPs Based on Spectral Count**

| Cell Surface Proteins | Gene | GI | Swiss Pro | Function | Molc Class | Spectral Count (Rank) | Number of Different Peptides |
|---|---|---|---|---|---|---|---|
| Talin 1 | TLN1 | 14916725 | Q9Y490 | Participates in assembly of actin filaments, regulates integrin activation | Cytoskeletal protein | 3102 (1) | 92 |
| γ-Filamin | FLNC | 4557597 | Q14315 | Plays a central role in muscle cells; may be involved in reorganizing actin cytoskeleton in other cells | Cytoskeletal protein | 2383 (2) | 87 |
| β-Actin | ACTB | 14250401 | P60709 | Major component of cytoskeleton | Cytoskeletal associated protein | 1826 (3) | 4 |
| Thrombospondin 1 | THBS1 | 135717 | P07996 | An adhesive glycoprotein that mediates cell–cell and cell–matrix interaction | Extracellular matrix protein | 1522 (4) | 44 |
| Vinculin | VCL | 24657579 | P18206 | Anchors F-actin to membrane | Cell cycle control protein | 1461 (5) | 43 |
| $\alpha_1$-Actinin | ACTN1 | 28193204 | P12814 | F-Actin crosslinking protein that is thought to anchor actin to a variety of intracellular structures | Cytoskeletal associated protein | 1382 (6) | 25 |
| Myosin 9 | MYH9 | 29436380 | Q86XU5 | Involved in muscle contraction and cell mobility | Cytoskeletal protein | 1111 (7) | 41 |
| Von Willebrand factor | VWF | 37947 | P04275 | Participates in platelet–vessel wall interactions by binding GP1B and coagulation factor VIII | Coagulation factor | 854 (8) | 82 |
| UNC112-related protein 2 | URP2 | 21752646 | Q86UX7 | May be involved in cell adhesion | Unknown | 745 (9) | 45 |

*(continued)*

**TABLE 7.1** *(Continued)*

| Cell Surface Proteins | Gene | GI | Swiss Pro | Function | Molc Class | Spectral Count (Rank) | Number of Different Peptides |
|---|---|---|---|---|---|---|---|
| β-Tubulin 1, class VI | TUBB1 | 13562114 | P07437 | Major constituent of microtubules; binds two moles of GTP, one on the β chain and one on the α-chain | Cytoskeletal protein | 614 (10) | 24 |
| β-Glycoprotein IIIa | ITGB3 | 2443452 | P05106 | Part of receptors for many extracellular proteins incuding prothrombin, thrombospondin, vitronectin, and vWF | Cell surface receptor | 592 (11) | 34 |
| α-Integrin IIb | ITGA2B | 35520 | P08514 | Heterodimer with $\beta_3$-integrin, receptor for fibronectin, fibrinogen, plasminogen, prothrombin, thrombospondin, etc. | Cell surface receptor | 590 (12) | 20 |
| Coagulation factor XIII | F13A1 | 15987821 | P00488 | Factor XIII is activated by thrombin and calcium ion to crosslink fibrin chains, thus stabilizing the fibrin clot | Coagulation factor | 560 (13) | 23 |
| Tubulin, β-polypeptide paralog | MGC8685 | 2661079 | Q9BVA1 | Major constituent of microtubules | Structural protein | 544 (14) | 19 |
| Fibrinogen, α-chain | FGA | 11761629 | P02671 | Blood clotting factor | Coagulation factor | 534 (15) | 33 |
| β-Fibrinogen | FGB | 11761631 | P02675 | Blood clotting factor | Coagulation factor | 492 (16) | 32 |
| Pyruvate kinase 3, isoform 1 | PKM2 | 31416989 | P14618 | Catalyzes the production of phosphoenolpyruvate from pyruvate and ATP | Enzyme: phosphotransferase | 460 (17) | 26 |

| | | | | | | | |
|---|---|---|---|---|---|---|---|
| Gelsolin isoform 1 | GSN | 4504165 | P06396 | Calcium-regulated, actin-modulating protein that binds the barbed ends of actin monomers or filaments | Cytoskeletal protein | 392 (18) | 25 |
| γ-Fibrinogen | FGG | 31874240 | P02679 | Blood clotting factor | Coagulation factor | 369 (19) | 19 |
| WD repeat-containing protein 1, isoform 1 | WDR1 | 9257257 | O75083 | WD repeats are approximately 30- to 40-amino acid domains containing several conserved residues that bind actin | Unclassified | 283 (20) | 24 |
| Albumin | ALB | 7959791 | P02768 | Most abundent plasma protein, binds fatty acids and hormones; regulates the colloidal osmotic pressure | Transport/cargo protein | 277 (21) | 10 |
| Glyceraldehyde 3 phosphate dehydrogenase | GAPD | 35053 | P04406 | Catalyzes first step of second phase of glycolysis by phosphorylating D-glyceraldehyde | Enzyme: dehydrogenase | 246 (22) | 19 |
| 14-3-3ζ protein | YWHAZ | 21735625 | P63104 | Activates tyrosine and tryptophan hydroxylases and strongly activates protein kinase C | Adapter molecule | 239 (23) | 4 |
| Heatshock 70 protein 8 | HSPA8 | 13938297 | P11142 | Belongs to heatshock protein 70 family | Heatshock protein | 229 (24) | 15 |
| Profilin 1 | PFN1 | 30584265 | P07737 | Ubiquitous actin monomer binding protein | Cytoskeletal protein | 210 (25) | 20 |
| RAP1B | RAP1B | 31874240 | P61224 | Small G protein that regulates integrin activation | GTPase | 205 (26) | 11 |

(*continued*)

**TABLE 7.1** ***(Continued)***

| Cell Surface Proteins | Gene | GI | Swiss Pro | Function | Molc Class | Spectral Count (Rank) | Number of Different Peptides |
|---|---|---|---|---|---|---|---|
| $\alpha_1$-Actin skeletal muscle protein | ACTA1 | 30908859 | Q7Z7J6 | Involved in muscle contraction | Cytoskeletal associated protein | 200 (27) | 6 |
| Ras suppressor protein 1, isoform 1 | RSU1 | 6912638 | Q15404 | Potentially plays a role in the Ras signal transduction pathway; capable of suppressing v-Ras transformation *in vitro* | Unclassified | 190 (28) | 12 |
| Adenylylcyclase-associated protein | CAP1 | 5453595 | Q01518 | May have a regulatory bifunctional role | Unclassified | 185 (29) | 28 |
| Tropomyosin 3 | TPM3 | 21751375 | P06753 | Binds to actin filaments in muscle and nonmuscle cells; plays a role in contraction | Cytoskeletal associated protein | 181 (30) | 14 |
| Nuclear chloride ion channel protein | CLIC1 | 2073569 | O00299 | Exhibits both nuclear and plasma membrane chloride ion channel activity | Intracellular ligand gated channel | 173 (31) | 12 |
| Transgelin 2 | TAGLN2 | 4507357 | Q9H4P0/P37802 | Unknown; homolog of the protein transgelin, which is one of the earliest markers of differentiated smooth muscle | Unclassified | 172 (32) | 15 |
| Multimerin 1 | MMRN1 | 45269141 | Q13201 | Carrier protein for platelet (but not plasma) factor V/Va, which may play a role in the storage and stabilization of factor V | Transport/cargo protein | 171 (33) | 19 |

| | | | | | | | |
|---|---|---|---|---|---|---|---|
| PDZ and LIM domain 1 (elfin) | PDLIM1 | 13994151 | O00151 | Cytoskeletal protein that may act as an adapter that brings other proteins to cytoskeleton | Cytoskeletal protein | 171 (34) | 18 |
| Serum deprivation response protein | SDPR | 4759082 | O95810 | Major component of caveolae and a substrate of protein kinase C | Serine protease | 171 (35) | 15 |
| Heatshock protein gp96 | TRA1 | 15010550 | P14625 | Molecular chaperone that functions in processing and transport of secreted proteins | Heatshock protein | 166 (36) | 16 |
| α-Enolase | Hs.433455 | 2661039 | Q71V37 | Enzyme of glycolysis, converting 2-phospho-D-glycerate to phosphoenolpyruvate | Enzyme: hydratase | 165 (37) | 16 |
| Fructose bisphosphate aldolase A | ALDOA | 38197498 | P04075 | Catalyzes conversion of fructose 1,6-bisphosphate to glycerone phosphate and D-glyceraldehyde 3-phosphate in glycolysis | Enzyme: aldolase | 163 (38) | 16 |
| Hemoglobin δ-chain | HBD | 4504351 | P02042 | Oxygen transport | Transport/cargo protein | 162 (39) | 14 |
| Glutathione transferase | GSTP1 | 4504183 | P09211 | Conjugation of reduced glutathione to a wide number of exogenous and endogenous hydrophobic electrophiles | Enzyme: glutathione transferase | 151 (40) | 11 |
| Heatshock 70-kDa protein 5 | HSPA5 | 16507237 | P11021 | Chaperone; glucose-regulated protein | Chaperone | 148 (41) | 23 |
| Peptidylprolyl isomerase A | PPIA | 6679439 | P62937 | Accelerates folding of proteins by catalyzing *cis–trans* isomerization of proline imidic peptide bonds | Chaperone | 146 (42) | 8 |

*(continued)*

**TABLE 7.1** *(Continued)*

| Cell Surface Proteins | Gene | GI | Swiss Pro | Function | Molc Class | Spectral Count (Rank) | Number of Different Peptides |
|---|---|---|---|---|---|---|---|
| Hemoglobin β-chain | HBB | 4504349 | P02023 | Oxygen transport | Transport/cargo protein | 142 (43) | 10 |
| Calpain-1, large subunit | CAPN1 | 14250593 | P07384 | Proteases that catalyze limited proteolysis of substrates involved in cytoskeletal remodeling and signal tranduction | Cysteine protease | 140 (44) | 25 |
| Hemoglobin $\alpha_1$-globin chain | HBA1 | 13195586 | Q9BX83 | Oxygen transporter activity | Transport/cargo protein | 132 (45) | 13 |
| Heatshock 90-kDa protein, α | HSPCA | 32486 | P07900 | Molecular chaperone with possible ATPase activity | Chaperone | 128 (46) | 11 |
| β-Tropomyosin | TPM4 | 10441386 | P02561 | Binds to actin filaments and plays a central role in muscle contraction | Cytoskeletal associated protein | 123 (47) | 8 |
| Cofilin 1 | CFL1 | 5031635 | P23528 | Reversibly controls actin polymerization and depolymerization in a pH-sensitive manner | Cytoskeletal protein | 122 (48) | 8 |
| Pleckstrin | PLEK | 4505879 | P08567 | Major protein kinase C substrate of platelets; its exact function is not known | Unknown | 122 (49) | 16 |
| β-Tubulin 4Q | TUBB4Q | 12643363 | Q99867 | Major constituent of microtubules | Structural protein | 118 (50) | 7 |

Two excellent studies have examined the proteome of the platelet membrane, but neither distinguished the plasma membrane from internal membranes [15,16]. Additionally, it is known that the protein composition of certain granules such as α-granules mirrors that of the plasma membrane [17]. Maynard et al. performed an in-depth proteomic analysis of α-granules and identified 284 proteins, 44 of which were previously not known to be present [18]. It is interesting that 8 of the 10 most abundant proteins in α-granules were also some of the most abundant proteins that we identified in the platelet microvesicles. It is unclear whether this is more an indication (1) of the similarity of the plasma membrane and the membranes of microvesicles or (2) that the major source of the microparticles isolated following activation with ADP was α-granules. That α-granules may originate from the plasma membrane [19] suggests the former. Another potential source of the microvesicles is the dense bodies released from platelets. While there are only a few of these granules per platelet, they are known to be enriched in a number of small molecules [20]. The soluble proteins from these granules were examined by Hernandez-Ruiz et al. using proteomic techniques [21]. They identified 40 proteins, most of which are in our 50 most abundant list. These include many of the cytoskeletal proteins, TSP1, albumin, and the fibrinogen α- and β-chains. In fact, it included 16 of the 25 most abundant proteins that we identified in the microvesicles, and there appears to be a correlation between the number of peptides that they discovered and our spectral count. There are a few exceptions. They did not detect vWF, even by Western blotting, or the proteins from integrin αIIbβ3. However, they focused on soluble proteins only. From all these studies, it appears that the proteome from isolated platelet microvesicles is remarkably similar to proteomes of both α-granules and dense granules. It seems very likely that release of both types of granules and blebbing of the plasma membrane are all important sources of platelet-derived microparticles.

## 7.3 PLATELET MICROPARTICLES IN THE PLASMA

The presence, cellular origin, and amount of microvesicles in the plasma have been determined largely using flow cytometry or other immunoassay-based detection methods [22]. When ectosomes are released from the cellular surface, they have negatively charged phosphatidylserine on the exterior surface. Using conventional flow cytometry methods, it is difficult to distinguish unlabeled microvesicles from noise. Therefore, the ectosomes are generally fluorescently labeled with annexin V, which binds tightly to the phosphatidylserine. This allows detection of at least the larger ectosomes. The cellular sources of these particles are assumed to be based on the presence of cell-specific surface markers. For example, microparticles with CD41 (glycoprotein IIb) expression are believed to be generated by platelets. Using this method, microparticles from platelets, erythrocytes, endothelial cells, neutrophils, lymphocytes and even smooth muscle cells have been detected in the plasma [23]. In healthy individuals, a majority of these microparticles were believed to originate from platelets. However, there are a number of problems associated with these methods: (1) many microvesicles may be too small to be detected by conventional flow cytometry [20], (2) some microparticles express phosphatidylserine, and therefore are not examined—this may be particularly true for those generated under pathological conditions [24], and (3) the presence (or absence) of a surface marker on a microvesicle is not necessarily indicative of cellular origin. Microvesicles are able to exchange cell surface markers with cells [25,26], and the cell surface markers may be highly dependent on the method of activation. Therefore lack or presence of a specific marker may lead to erroneous conclusions on cellular origin. For example, $CD41^+$-microvesicles are some of the most abundant and well-characterized microparticles in the blood, and have been assumed to be generated from activated platelets. However, a more recent report suggests that many actually originate from megakaryocytes, at least in healthy individuals [27]. While much of our understanding of microvesicles in the plasma originates from flow cytometry and other immunoassay detection methods, the application of other technologies to explore plasma microparticles may be extremely beneficial. Therefore, we examined the proteome of plasma MPs using advanced proteomic techniques, and compared it to the proteome from platelet microparticles.

### 7.3.1 Isolation of Microparticles

Plasma and platelet microparticles were isolated as depicted in Figure 7.2. Blood from three healthy individuals was collected by venipuncture into $\frac{1}{10}$th volume of ACD. Platelet-rich plasma was obtained by centrifugation at 110$g$ for 15 min. Platelets were pelleted by centrifugation at 710$g$ for 15 min and the supernatant, platelet-poor plasma was retained for isolation of plasma microparticles. The platelet pellet was washed 3 times, resuspended in 10 mL of Tyrode's buffer, and centrifuged one additional time at 110$g$ to remove remaining red blood cells and dead cells. ADP (10 μM final concentration) was added to the platelet suspension for 10 min to generate platelet microparticles. Platelets were removed by centrifugation (710$g$ for 15 min), and platelet-derived microparticles were pelleted by centrifugation at 150,000$g$ for 90 min at 10°C. Plasma-derived microparticles were isolated by gel filtration chromatography followed by ultracentrifugation of the platelet as described [28]. Platelet-poor plasma was centrifuged twice at 710$g$ and 25°C for 15 min to remove residual cells and cell debris. This plasma was then applied to a Sephacryl S-500 HR (GE Healthcare) gel filtration column. The microparticle-containing fractions were concentrated by ultracentrifugation at 150,000$g$ for 90 min at 10°C. These methods generated platelet and plasma microparticles with sufficient purity and absence of plasma contamination for subsequent analysis. We compared the proteomes of platelet and plasma microparticles using two different mass spectrometry approaches: a spectral count and a labeled approach.

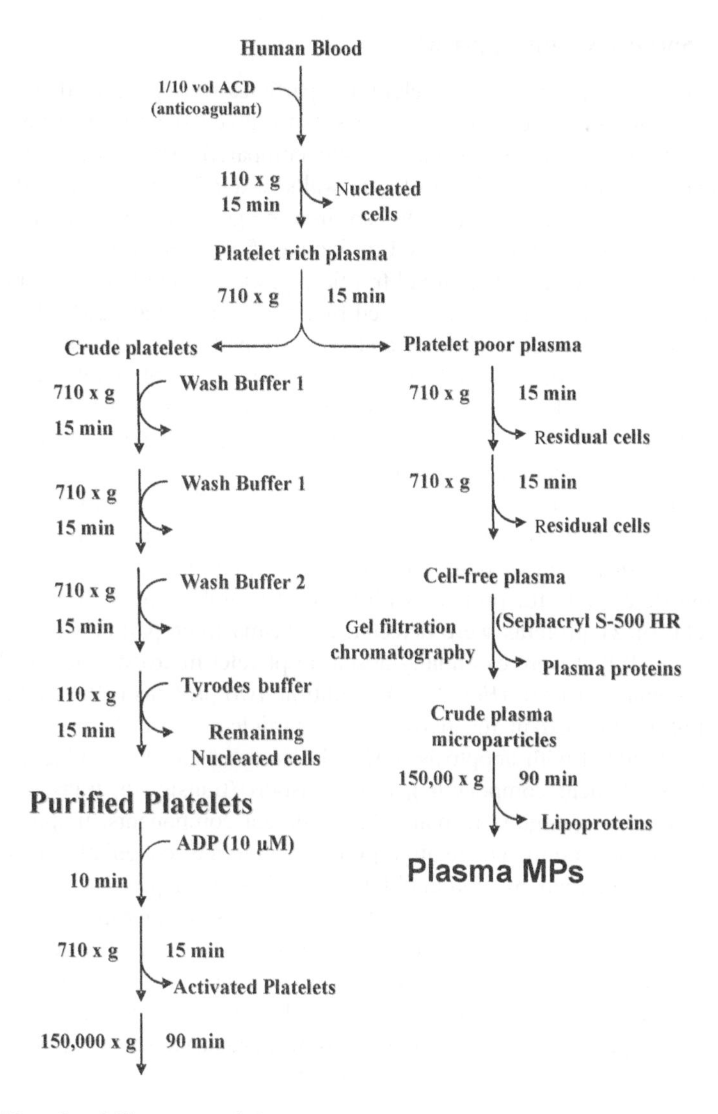


**Figure 7.2** Comparison of platelet microparticles and plasma microparticles. Platelet-rich plasma was generated from human blood. The platelets were isolated by differential centrifugation, washed extensively, and activated with ADP. The microparticles produced (platelet microparticles) were isolated. The platelet-poor plasma was used as the source of plasma microparticles. Following removal of residual cells, the bulk of the plasma proteins were removed by gel filtration chromatography. A majority of the lipoproteins were removed, and the plasma microparticles were concentrated using ultracentrifugation.

### 7.3.2 Spectral Count Approach

In this approach, plasma and platelet microparticles were processed and analyzed separately, and then the number of times that a given spectra identified peptides from a protein in the two samples were compared. The plasma and platelet microparticles were applied to different wells of a 7.5% acrylamide SDS-PAGE and electrophoresed approximately 1 cm into the gel. The gel section containing the proteins was excised and placed in fixative (50% methanol, 12% acetic acid, 0.05% formalin) for 2 h. The in-gel tryptic digestion of the lanes and the peptide extraction were performed as described by Shevchenko et al. [29]. The extracted peptide solutions were lyophilized and reconstituted to 20 μL with 0.1% acetic acid for mass spectrometry analysis. A total of three pairs of platelet- and plasma-derived microparticle peptides were generated, and each of these samples was analyzed by LC-MS/MS twice as described above with the following modification. The gel lane was not sectioned but was used as one sample. The total number of spectra identifying peptides from a particular protein were calculated. Manual verification of peptide sequence was performed on MS/MS spectra only for proteins suspected of being differentially expressed. Differential expression was considered only for proteins with a total spectral count of 10 or more.

A total of 21 proteins were detected in plasma microparticles (total spectral count of $\geq 10$) that were essentially absent in platelet microparticles (with a total spectral count of 1 or 0) (Fig. 7.3). In addition, two proteins (vWF and albumin) were highly enriched in the plasma microparticles. These 21 proteins include several associated with apoptosis (CD5-like antigen, galectin 3 binding protein, several complement components), iron transport (transferrin, transferrin receptor, haptoglobin), immune response (complement components, immunoglobulin J, and κ-chains), and the coagulation process (protein S, coagulation factor VIII). Several were verified by immunoblotting [28]. Eleven proteins were elevated in the platelet microparticles compared to the plasma microparticles. Because most plasma microvesicles were thought to originate from platelets, these differences were initially believed to result simply from increased abundance of certain proteins, particularly vWF, in the plasma MPs. However, the notion that most plasma microparticles actually originate from megakaryocytes [27] suggests that differences may be due to the cells of origin.

### 7.3.3 Labeled Approach

In this approach, plasma and platelet microparticles were isolated and labeled with ICAT (isotope-coded affinity tag) reagent (Applied Biosystems). This reagent labels cystein residues and has a cleavable biotin tag. Once the tagged proteins are separated from the untagged peptides, the biotin part of the molecule is cleaved, leaving an isotopically labeled tag. This tag has either all carbon as the [$^{12}$C] isotope or nine of the carbons replaced with [$^{13}$C], resulting in an additional mass of 227 or 236 atomic mass units. The proteins in these two microparticle preparations were solubilized in 1% SDS, differentially labeled with either the heavy or light tag, and mixed. The peptide mixture was applied to the gel,

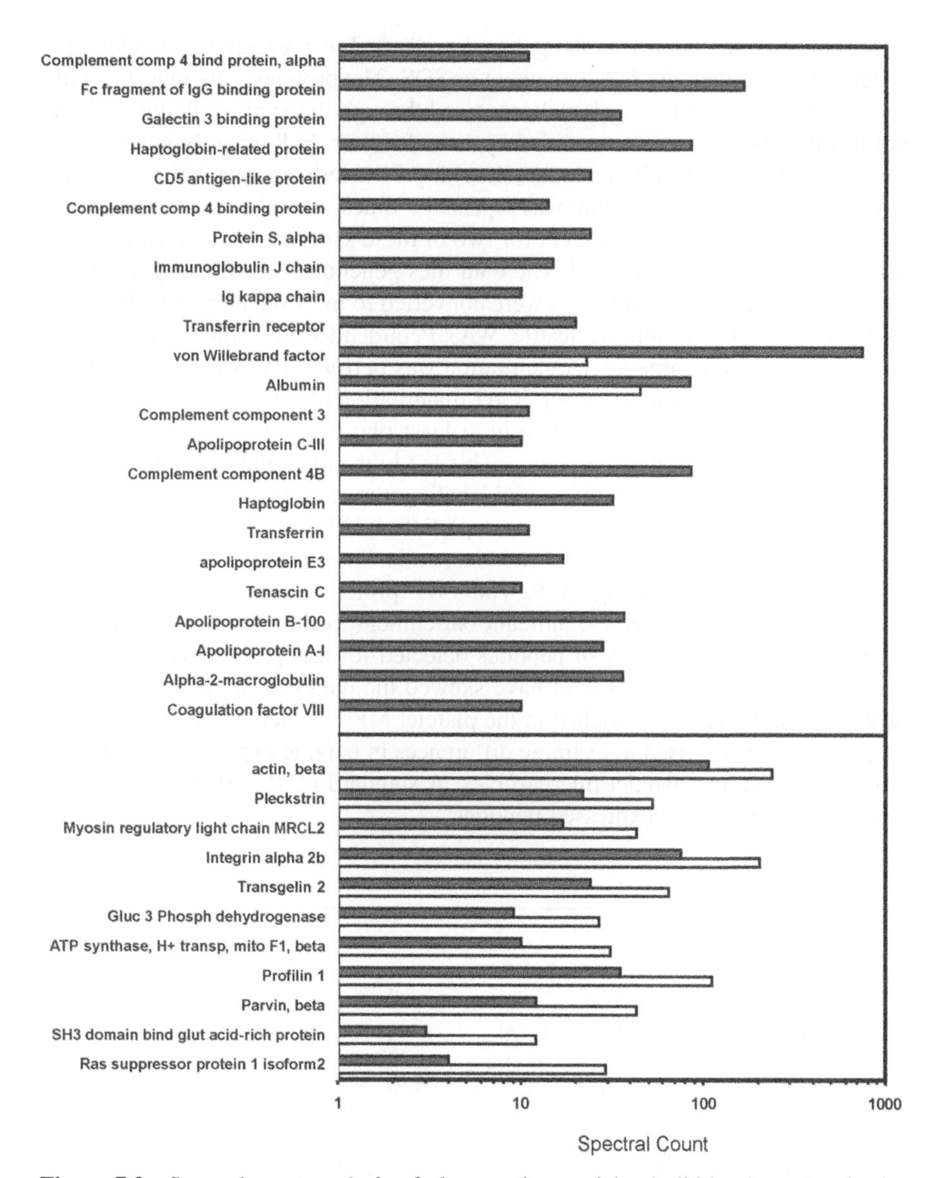


**Figure 7.3** Spectral count analysis of plasma microparticles (solid bars) versus platelet microparticles (open bars). Proteins shown in top part of the scale are enriched in the plasma microparticles versus the platelet microparticles; proteins shown in bottom part of the scale are enriched in the platelet microparticles versus the plasma microparticles. Note that log scale diminishes apparent differences between samples, for example, albumin (spectral count of 85 vs. 45) and vWF (745 vs. 23).

electrophoresed, and cut from the gel as described above, except that the loading buffer did not contain the reducing agent or SDS. The proteins were digested with trypsin, extracted from the gel, and processed through the avidin column, and the biotin was removed as recommended by manufacturer. Following lyophilization, the samples were reconstituted to 20 μL with 0.1% acetic acid for MS analysis as described above. This procedure was repeated 3 times with plasma microparticles labeled with the light ICAT reagent for two of these samples and labeled with the heavy ICAT reagent in the third. The data files generated (`.RAW`) using XCalibur Software (Thermo Electron Corp.) were converted to mzXML files using ReAdW (Institute for Systems Biology, Seattle, WA). Peptide identification was performed using Sequest, and comparative quantitation was performed using MSight (Swiss Institute of Bioinformatics) [28,30]. ICAT quantitation results were reported when good peak quantitation was possible in at least two of the three ICAT runs and labeled alternatively in these two runs. This led to the quantitation of 94 peptides (Fig. 7.4) [28]. Utilizing the differential labeling, the authors calculated, the ratios of plasma microparticle to platelet microparticle ICAT-peptide abundances, performed a $\log_2$ transformation of these ratios, and adjusted it to generate an overall $\log_2$ score of 0.00, excluding vWF. The vWF protein was excluded because it was evident that there was a significant enrichment in the plasma micrparticles. Because of the large number of peptides detected for this protein and the extent of enrichment, including it would have skewed the data and made it appear that most other proteins were enriched in the platelet MPs. To determine significance, the paired $t$-test was used to examine differences in peptide expression intensities between plasma and platelet micrparticles. A standard of $P < 0.05$ was used to determine differentially expressed peptides.

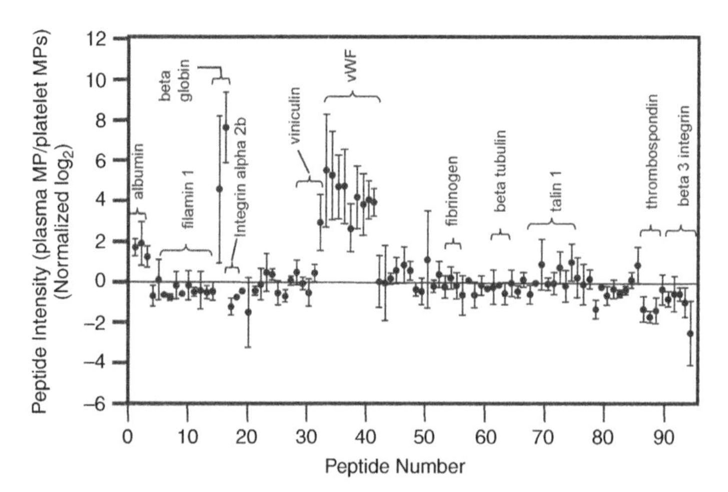


**Figure 7.4** Relative ion intensities of ICAT-labeled peptides. Peptides from platelet microparticles and plasma microparticles were differentially labeled using ICAT reagent. Differences in ion intensity for 96 peptides observed in at least two of three ICAT analyses are shown.

Of the 94 peptides quantified, a majority appeared to be unchanged. Differences were seen in all the peptides from albumin (three peptides), vWF (nine peptides), β-globin (two peptides), and one of the two peptides from viniculin, which appeared to be overexpressed in plasma microparticles versus platelet microparticles. Petides from albumin were elevated between two- and fourfold, which roughly agrees with the spectral count data. All peptides from vWF were elevated, most between 14- and 45-fold, which parallels the 32-fold increase observed in the spectral count approach. A Western blot performed on these two proteins indicated that both the spectral count and labeled approaches actually underestimate the large difference in the vWF composition in platelet versus plasma microparticles [28]. From a theoretical perspective this would be expected. In spectral counting one tries to analyze a peptide only once per LC run, and for ion intensity comparisons using ICAT, one uses the background noise ion intensities if no peaks are obvious. Two peptides from β-globin were elevated over twofold by the labeled method, but increased by only ~50% with the spectral count method and were not statistically significant. The reason for this discrepancy is unclear but may result from the more quantitiative nature of the labeled approach. The only protein not consistent with the two approaches was one of the two peptides of viniculin. This may be due to overlapping peaks, cleaved proteins, or incorrect protein identification. The spectral count analysis and the peptides identified suggests incorrect protein identification.

### 7.3.4 Proteins Enriched in the Plasma Microparticles versus the Platelet Microparticles

A total of 21 proteins were found predominantly in the plasma microparticles, and not in the platelet microparticles in healthy individuals [28]. These proteins can generally be classified into the groups described below.

***7.3.4.1 Complement and Apoptosis-Related Proteins*** Four of the proteins in plasma microparticles (complement 3, complement 4B, complement 4 binding proteins α and β) are from the complement pathway. One mechanism by which nucleated cells prevent complement-induced cell death is shedding the complement components off the membrane [8]. The complement components detected in our study could represent the normal process of the shedding of microparticles to avoid excessive apopotois. C4BP (C4 binding protein) is a potent circulating complement inhibitor that inhibits both the classical and alternative pathways, and with a role in localizing complement regulatory activity to the surface of apoptotic cells, which undergo ectocytosis to release microparticles [35]. Protein S, while known more commonly for its role in hemostasis, has been shown to stimulate the phagocytosis of apoptotic cells [36]. CD5-like antigen and galectin 3 binding protein also are associated with inhibition of apoptosis [37–40]. The presence of all these proteins in plasma microparticles suggests that some are derived from apoptotic cells or that microparticles are shed to prevent apoptosis. These two mechanisms of production cannot be distinguished in our study.

*7.3.4.2* ***Transport/Hemoglobin*** Five proteins enriched in plasma microparticles are related to iron transport and hemoglobin clearance: transferrin, transferrin receptor, haptoglobin, haptoglobin-related protein, and hemoglobin. The transferrin receptor is downregulated from the surface of reticulocytes by internalization, followed by its release in exosomes [5,31]. Transferrin may simply bind this receptor. Haptoglobin is an abundant protein that binds free hemoglobin. It is possible that haptoglobin removes hemoglobin by association with plasma microparticles. Hemoglobin was detected by our analysis in plasma and platelet microparticles, suggesting that haptoglobin–hemoglobin complexes may associate with microparticles.

*7.3.4.3* ***Immunoglobulins*** The third category of plasma proteins overexpressed in plasma microparticles are immunoglobulins, represented by immunoglobulin (Ig) J chain and κ-chain. IgJ. These may reflect the subclinical activation of the classical complement pathway at a low level through IgM autoantibodies, leading to the formation of microparticles from unknown cells (not platelets) that remove IgJ and associated complement components from the affected cell membrane.

*7.3.4.4* ***Lipoproteins*** Unlike the other proteins that appear to be enriched in the plasma microparticles, these most likely represent contamination of plasma microparticles with the larger plasma lipoproteins. Chylomicrons (apoA-I and C-III), chylomicron remnants (apoE), VLDL (apoA-I, B100, and C-III) and IDL (apoB100 and E) fractions appear in overlapping fractions with microparticles. Since these lipoproteins, especially chylomicrons, are similar in size to microvesicles, they will be enriched in the gel filtration fractionation. While the ultracentrifugation will remove a large percentage of them, residual contamination would still likely be present.

*7.3.4.5* ***Blood Coagulation Factors*** Protein S, factor VIII, and vWF were enriched in plasma microparticles. While the best-known role of protein S is as an anticoagulant, it also is associated with apoptosis as described above, and this likely explains its presence in plasma microparticles. Factor VIII is procoagulant, and while it has been reported that it binds directly to protein S [46], the reason for its presence in plasma microparticles is unclear. The vWF protein is a platelet and endothelial cell product that binds to P-selectin [47], GPIb, [48] and GPIIb/IIIa [49]. Since we previously detected vWF in platelet microparticles [20], its presence in both plasma and platelet microparticles in this study was no surprise. However, its dramatic enrichment in plasma microparticles was unexpected and suggests that endothelial vWF associates with plasma microparticles or even promotes their formation. A study has demonstrated that microparticles are indeed formed from platelets bound to immobilized vWF under high shear stress, *in vitro* [50]. Therefore, the enrichment of this protein probably represents vWF from endothelial cells. The vWF protein on the surface of activated endothelial cells or as part of the extracellular matrix acts an anchor for pulling microparticles from the platelet surface. Then these microparticles may be cleaved from the surface, releasing the microparticles attached to the vWF.

***7.3.4.6 Other Proteins*** The Fc fragment of IgG binding protein (FCGBP) is a protein with unknown function produced by intestinal and colonic goblet cells [53,54]. Because of to its cellular source and its ability to bind IgG, it has been hypothesized to assist in preventing antigen invasion into the mucosa [53,55]. Kobayashi et al. [56] found that FCGBP could inhibit a complement-mediated reaction and suggested that it may play an important role in immunological defense of mucosal surfaces [55,57]. While this follows our theory of ectosome production discussed previously, there is little evidence to support the connection between FCGBP and the prevention of cell death in vascular tissue. This is particularly true considering the location of FCGBP production. However, the detection of this protein in plasma microparticles is intriguing, especially when considering its prevalence (spectral count of 167, including 54 unique peptides). The remaining protein, tenascin C, is an extracellular matrix protein associated with provisional and tumor matrices but not found in most mature tissues [58]. It is possible that tenascin C is secreted at sites of small injuries and wound healing, and cells with rapid turnover such as leukocytes and wound fibroblasts may shed tenascin C associated with microparticles. Alternatively, tenascin C may form aggregates that migrate in the microparticle fraction. Its presence was unexpected in the plasma microparticles fraction, and further studies are needed to examine its implications.

## 7.4 SIGNIFICANCE OF PLATELET (AND/OR MEGAKARYOCYTE) MICROPARTICLES

As in all proteomic research, obtaining data is only the beginning. In most cases, using these data as a tool to better understand the biology and normal and pathophysiology of biological systems must be the primary goal. An elevated level of platelet-derived microparticles has been reported for a wide variety of diseases (Table 7.2). Several of these will be discussed in more detail, including their possible contribution to the pathophysiological process.

### 7.4.1 Platelet Microparticles in Cardiovascular Disease

Since platelet activation plays a pivotal role in most cardiovascular diseases and causes the release of platelet microparticles, it is not surprising that platelet numbers are elevated. The microparticles themselves are involved in thrombosis, providing a negatively charged surface for clot formation. However, there is evidence to suggest that their participation may even be more extensive. For example, platelet microparticles can act as carriers of platelet activating factor [32] and RANTES [33,34], and actually deposit these inflammatory factors onto the vessel wall. We did observe RANTES from platelet microparticles in our previous study [9], but platelet activating factor is not a protein and therefore was not detected using our methodology. However, other inflammatory mediators, such as platelet factor 4, CXC7, and $\beta_1$-TGF were found. While a wide variety of inflammatory mediators are involved with the atherosclerotic process, the presence of these factors on plasma microparticles may have significant implications in treatment of this disease. If it were possible to inactivate these microparticle-bound

**TABLE 7.2 Elevations in Platelet MPs in Disease and Other Physiological Changes**

| Event | Reference(s) |
|---|---|
| *Cardiovascular and Blood-Related* | |
| Atherothrombotic events | 38 |
| Acute coronary syndromes | 24, 39 |
| Hypertension | 40 |
| Cisplatin-induced stroke | 41 |
| Severe aortic valve stenosis | 42 |
| Carotid atherosclerosis | 43 |
| Peripheral arterial disease | 44 |
| Stable coronary artery disease | 45 |
| Unstable angina | 46 |
| Acute vasculitis | 47 |
| Percutaneous transluminal coronary angioplasty | 48 |
| Nonvalvular atrial fibrillation | 49 |
| Stent-induced vascular inflammation | 50 |
| Venous thromboembolism | 51 |
| *Metabolism-Related* | |
| Hyperlipidemic following stroke | 52 |
| Postprandial hypertriglyceridemia | 43 |
| Newly menopausal women | 53 |
| Obesity | 54 |
| Aging | 44 |
| Diabetes (type 1) | 55 |
| Diabetes (type 2) | 56 |
| Metabolic syndrome | 57 |
| Premature coronary calcification in newly menopausal women | 58 |
| *Other* | |
| Breast cancer | 59 |
| Colorectal cancer (tissue-factor-positive) | 60 |
| Women with recurrent spontaneous abortion | 61 |
| Atopic dermatitis | 62 |
| Chronic hepatitis C | 63 |
| Minimally symptomatic obstructive sleep apnea | 64 |
| Osteonecrosis | 65 |
| Preterm newborns | 66 |
| Total knee arthroplasty | 67 |
| Thalassemia | 68 |
| Ulcerative colitis | 69 |
| Rheumatoid arthritis | 70 |
| Sepsis | 71 |
| Systemic lupus erythematosus | 72 |
| Systemic sclerosis | 73 |
| Alcoholic liver disease | 74 |
| Autoimmune thrombocytopenias | 75 |
| Normal pregnancy | 76 |
| Preeclampsia | 76 |
| Multiple sclerosis | 77 |

inflammatory mediators without inhibiting the soluble ones, it might be possible to block this pathway without systemic inhibition of the inflammatory process.

### 7.4.2 Platelet Microparticles in Cancer

Since the mid-1800s, cancer-associated thrombosis has been well recognized. In fact, cancer-associated thrombosis is the second leading cause of mortality in affected patients, and therefore platelet activation and increased platelet microparticles in the plasma are expected [35]. More recent evidence indicates that platelet microparticles may play a role in tumor development and invasiveness. For example, platelet microparticles have been shown to promote invasiveness of prostate cancer cells by upregulating the production of matrix metalloproteinase 2 (MMP2) [36]. Others have shown that they are also able to induce angiogenesis [37]. Careful examination of the platelet microparticle proteome may highlight proteins associated with this process that should be further examined.

## 7.5 CONCLUDING REMARKS

Since 2000, great advances in technologies and methodologies have allowed us to examine proteomes in great detail. This chapter discussed the approach and results for platelet and plasma microparticles. While these particles were originally considered to be simply cell debris, more recent studies demonstrate they have important roles in normal and disease processes. By examining the proteome of platelet derived microparticles, we may gain a better understanding of a wide variety of physiological and pathological processes.

## ACKNOWLEDGMENTS

The assistance of Dr. Roger Phipps in technical reading of this manuscript is greatly appreciated.

## REFERENCES

1. Heijnen HF, Schiel AE, Fijnheer R, Geuze HJ, Sixma JJ., Activated platelets release two types of membrane vesicles: Microvesicles by surface shedding and exosomes derived from exocytosis of multivesicular bodies and alpha-granules, *Blood* **94**(11):3791–3799 (1999).
2. Wolf P, The nature and significance of platelet products in human plasma, *Br. J. Haematol*. **13**(3):269–288 (1967).
3. Leytin V, Allen DJ, Mutlu A, Gyulkhandanyan AV, Mykhaylov S, Freedman J, Mitochondrial control of platelet apoptosis: Effect of cyclosporin A, an inhibitor of the mitochondrial permeability transition pore, *Lab. Invest*. **89**(4):374–384 (2009).
4. Pan BT, Johnstone RM, Fate of the transferrin receptor during maturation of sheep reticulocytes in vitro: Selective externalization of the receptor, *Cell* **33**(3):967–978 (1983).
5. van Niel G, Porto-Carreiro I, Simoes S, Raposo G, Exosomes: A common pathway for a specialized function, *J. Biochem*. **140**(1):13–21 (2006).

6. Piccin A, Murphy WG, Smith OP, Circulating microparticles: Pathophysiology and clinical implications, *Blood Rev*. **21**(3):157–171 (2007).
7. Horstman LL, Ahn YS, Platelet microparticles: A wide-angle perspective, *Crit. Rev. Oncol. Hematol*. **30**(2):111–142 (1999).
8. Pilzer D, Gasser O, Moskovich O, Schifferli JA, Fishelson Z, Emission of membrane vesicles: Roles in complement resistance, immunity and cancer, *Springer Semin Immunopathol* **27**(3):375–387 (2005).
9. García BA, Smalley DM, Cho H, Shabanowitz J, Ley K, Hunt DF, The platelet microparticle proteome, *J. Proteome Res*. **4**(5):1516–1521 (2005).
10. Baenziger NL, Brodie GN, Majerus PW, Isolation and properties of a thrombin-sensitive protein of human platelets, *J. Biol. Chem*. **247**(9):2723–2731 (1972).
11. Bonnefoy A, Hantgan R, Legrand C, Frojmovic MM, A model of platelet aggregation involving multiple interactions of thrombospondin-1, fibrinogen, and GPIIbIIIa receptor, *J. Biol. Chem*. **276**(8):5605–5612 (2001).
12. Hyder SM, Liang Y, Wu J, Estrogen regulation of thrombospondin-1 in human breast cancer cells, *Int. J. Cancer* **125**(5):1045–1053.
13. Moser M, Nieswandt B, Ussar S, Pozgajova M, Fassler R, Kindlin-3 is essential for integrin activation and platelet aggregation, *Nat. Med*. **14**(3):325–330 (2008).
14. Wagner CL, Mascelli MA, Neblock DS, Weisman HF, Coller BS, Jordan RE, Analysis of GPIIb/IIIa receptor number by quantification of 7E3 binding to human platelets, *Blood* **88**(3):907–914 (1996).
15. Moebius J, Zahedi RP, Lewandrowski U, Berger C, Walter U, Sickmann A, The human platelet membrane proteome reveals several new potential membrane proteins, *Mol. Cell. Proteom*. **4**(11):1754–1761 (2005).
16. Senis YA, Tomlinson MG, García A, Dumon S, Heath VL, Herbert J, Cobbold SP, Spalton JC, Ayman S, Antrobus R, Zitzmann N, Bicknell R, Frampton J, Authi KS, Martin A, Wakelam MJ, Watson SP, A comprehensive proteomics and genomics analysis reveals novel transmembrane proteins in human platelets and mouse megakaryocytes including G6b-B, a novel immunoreceptor tyrosine-based inhibitory motif protein, *Mol. Cell. Proteom*. **6**(3):548–564 (2007).
17. Berger G, Masse JM, Cramer EM, Alpha-granule membrane mirrors the platelet plasma membrane and contains the glycoproteins Ib, IX, and V, *Blood* **87**(4):1385–1395 (1996).
18. Maynard DM, Heijnen HF, Horne MK, White JG, Gahl WA, Proteomic analysis of platelet alpha-granules using mass spectrometry, *J. Thromb. Haemost*. **5**(9):1945–1955 (2007).
19. King SM, Reed GL, Development of platelet secretory granules, *Semin. Cell. Dev. Biol*. **13**(4):293–302 (2002).
20. White JG, The dense bodies of human platelets: inherent electron opacity of the serotonin storage particles, *Blood* **33**(4):598–606 (1969).
21. Hernandez-Ruiz L, Valverde F, Jimenez-Nunez MD, Ocana E, Saez-Benito A, Rodriguez-Martorell J, Bohorquez JC, Serrano A, Ruiz FA, Organellar proteomics of human platelet dense granules reveals that 14-3-3zeta is a granule protein related to atherosclerosis, *J. Proteome Res*. **6**(11):4449–4457 (2007).
22. Jy W, Horstman LL, Jimenez JJ, Ahn YS, Biro E, Nieuwland R, Sturk A, Dignat-George F, Sabatier F, Camoin-Jau L, Sampol J, Hugel B, Zobairi F, Freyssinet

JM, Nomura S, Shet AS, Key NS, Hebbel RP, Measuring circulating cell-derived microparticles, *J. Thromb. Haemost*. **2**(10):1842–1851 (2004).

23. Martinez MC, Tesse A, Zobairi F, Andriantsitohaina R, Shed membrane microparticles from circulating and vascular cells in regulating vascular function, *Am. J. Physiol. Heart Circ. Physiol*. **288**(3):H1004–H1009 (2005).
24. Mallat Z, Benamer H, Hugel B, Benessiano J, Steg PG, Freyssinet JM, Tedgui A, Elevated levels of shed membrane microparticles with procoagulant potential in the peripheral circulating blood of patients with acute coronary syndromes, *Circulation* **101**(8):841–843 (2000).
25. Fritzsching B, Schwer B, Kartenbeck J, Pedal A, Horejsi V, Ott M, Release and intercellular transfer of cell surface CD81 via microparticles, *J. Immunol*. **169**(10):5531–5537 (2002).
26. Mack M, Kleinschmidt A, Bruhl H, Klier C, Nelson PJ, Cihak J, Plachy J, Stangassinger M, Erfle V, Schlondorff D, Transfer of the chemokine receptor CCR5 between cells by membrane-derived microparticles: a mechanism for cellular human immunodeficiency virus 1 infection, *Nat. Med*. **6**(7):769–775 (2000).
27. Flaumenhaft R, Dilks JR, Richardson J, Alden E, Patel-Hett SR, Battinelli E, Klement GL, Sola-Visner M, Italiano JE Jr., Megakaryocyte-derived microparticles: Direct visualization and distinction from platelet-derived microparticles, *Blood* **113**(5):1112–1121 (2009).
28. Smalley DM, Root KE, Cho H, Ross MM, Ley K, Proteomic discovery of 21 proteins expressed in human plasma-derived but not platelet-derived microparticles, *Thromb. Haemost*. **97**(1):67–80 (2007).
29. Shevchenko A, Wilm M, Vorm O, Mann M, Mass spectrometric sequencing of proteins silver-stained polyacrylamide gels, *Anal. Chem*. **68**(5):850–858 (1996).
30. Palagi PM, Walther D, Quadroni M, Catherinet S, Burgess J, Zimmermann-Ivol CG, Sanchez JC, Binz PA, Hochstrasser DF, Appel RD, MSight: An image analysis software for liquid chromatography-mass spectrometry, *Proteomics* **5**(9):2381–2384 (2005).
31. Johnstone RM, Adam M, Hammond JR, Orr L, Turbide C, Vesicle formation during reticulocyte maturation. Association of plasma membrane activities with released vesicles (exosomes), *J. Biol. Chem*. **262**(19):9412–9420 (1987).
32. Iwamoto S, Kawasaki T, Kambayashi J, Ariyoshi H, Monden M, Platelet microparticles: A carrier of platelet-activating factor? *Biochem. Biophys. Res. Commun*. **218**(3):940–944 (1996).
33. Schober A, Manka D, von Hundelshausen P, Huo Y, Hanrath P, Sarembock IJ, Ley K, Weber C, Deposition of platelet RANTES triggering monocyte recruitment requires P-selectin and is involved in neointima formation after arterial injury, *Circulation* **106**(12):1523–1529 (2002).
34. Mause SF, von Hundelshausen P, Zernecke A, Koenen RR, Weber C, Platelet microparticles: A transcellular delivery system for RANTES promoting monocyte recruitment on endothelium, *Arterioscler. Thromb. Vasc. Biol*. **25**(7):1512–1518 (2005).
35. Amin C, Mackman N, Key NS, Microparticles and cancer, *Pathophysiol. Haemost. Thromb*. **36**(3–4):177–183 (2008).

36. Dashevsky O, Varon D, Brill A, Platelet-derived microparticles promote invasiveness of prostate cancer cells via upregulation of MMP-2 production, *Int. J. Cancer* **124**(8):1773–1777 (2009).
37. Kim HK, Song KS, Chung JH, Lee KR, Lee SN. Platelet microparticles induce angiogenesis in vitro, *Br. J. Haematol.* **124**(3):376–384 (2004).
38. Namba M, Tanaka A, Shimada K, Ozeki Y, Uehata S, Sakamoto T, Nishida Y, Nomura S, Yoshikawa J, Circulating platelet-derived microparticles are associated with atherothrombotic events: A marker for vulnerable blood, *Arterioscler. Thromb. Vasc. Biol.* **27**(1):255–256 (2007).
39. Bernal-Mizrachi L, Jy W, Jimenez JJ, Pastor J, Mauro LM, Horstman LL, de Marchena E, Ahn YS, High levels of circulating endothelial microparticles in patients with acute coronary syndromes, *Am. Heart. J.* **145**(6):962–970 (2003).
40. Preston RA, Jy W, Jimenez JJ, Mauro LM, Horstman LL, Valle M, Aime G, Ahn YS, Effects of severe hypertension on endothelial and platelet microparticles, *Hypertension* **41**(2):211–217 (2003).
41. Periard D, Boulanger CM, Eyer S, Amabile N, Pugin P, Gerschheimer C, Hayoz D, Are circulating endothelial-derived and platelet-derived microparticles a pathogenic factor in the cisplatin-induced stroke? *Stroke* **38**(5):1636–1638 (2007).
42. Diehl P, Nagy F, Sossong V, Helbing T, Beyersdorf F, Olschewski M, Bode C, Moser M, Increased levels of circulating microparticles in patients with severe aortic valve stenosis, *Thromb. Haemost.* **99**(4):711–719 (2008).
43. Michelsen AE, Noto AT, Brodin E, Mathiesen EB, Brosstad F, Hansen JB, Elevated levels of platelet microparticles in carotid atherosclerosis and during the postprandial state, *Thromb. Res.* **123**(6):881–886 (2009).
44. van der Zee PM, Biro E, Ko Y, de Winter RJ, Hack CE, Sturk A, Nieuwland R, P-selectin- and CD63-exposing platelet microparticles reflect platelet activation in peripheral arterial disease and myocardial infarction, *Clin. Chem.* **52**(4):657–664 (2006).
45. Tan KT, Tayebjee MH, Macfadyen RJ, Lip GY, Blann AD, Elevated platelet microparticles in stable coronary artery disease are unrelated to disease severity or to indices of inflammation, *Platelets* **16**(6):368–371 (2005).
46. Singh N, Gemmell CH, Daly PA, Yeo EL, Elevated platelet-derived microparticle levels during unstable angina, *Can. J. Cardiol.* **11**(11):1015–1021 (1995).
47. Daniel L, Fakhouri F, Joly D, Mouthon L, Nusbaum P, Grunfeld JP, Schifferli J, Guillevin L, Lesavre P, Halbwachs-Mecarelli L, Increase of circulating neutrophil and platelet microparticles during acute vasculitis and hemodialysis, *Kidney Int.* **69**(8):1416–1423 (2006).
48. Craft JA, Masci PP, Roberts MS, Brighton TA, Garrahy P, Cox S, Marsh NA, Increased platelet-derived microparticles in the coronary circulation of percutaneous transluminal coronary angioplasty patients, *Blood Coagul. Fibrinol.* **15**(6):475–482 (2004).
49. Choudhury A, Chung I, Blann AD, Lip GY, Elevated platelet microparticle levels in nonvalvular atrial fibrillation: relationship to p-selectin and antithrombotic therapy, *Chest* **131**(3):809–815 (2007).
50. Inoue T, Komoda H, Kotooka N, Morooka T, Fujimatsu D, Hikichi Y, Soma R, Uchida T, Node K, Increased circulating platelet-derived microparticles are associated with stent-induced vascular inflammation, *Atherosclerosis* **196**(1):469–476 (2008).

51. Chirinos JA, Heresi GA, Velasquez H, Jy W, Jimenez JJ, Ahn E, Horstman LL, Soriano AO, Zambrano JP, Ahn YS, Elevation of endothelial microparticles, platelets, and leukocyte activation in patients with venous thromboembolism, *J. Am. Coll. Cardiol.* **45**(9):1467–1471 (2005).
52. Pawelczyk M, Baj Z, Chmielewski H, Kaczorowska B, Klimek A, The influence of hyperlipidemia on platelet activity markers in patients after ischemic stroke, *Cerebrovasc. Dis.* **27**(2):131–137 (2009).
53. Jayachandran M, Litwiller RD, Owen WG, Miller VM, Circulating microparticles and endogenous estrogen in newly menopausal women, *Climacteric* **12**(2):177–184 (2009).
54. Murakami T, Horigome H, Tanaka K, Nakata Y, Ohkawara K, Katayama Y, Matsui A, Impact of weight reduction on production of platelet-derived microparticles and fibrinolytic parameters in obesity, *Thromb. Res.* **119**(1):45–53 (2007).
55. Sabatier F, Darmon P, Hugel B, Combes V, Sanmarco M, Velut JG, Arnoux D, Charpiot P, Freyssinet JM, Oliver C, Sampol J, Dignat-George F, Type 1 and type 2 diabetic patients display different patterns of cellular microparticles, *Diabetes* **51**(9):2840–2845 (2002).
56. Kobayashi K, Yagasaki M, Harada N, Chichibu K, Hibi T, Yoshida T, Brown WR, Morikawa M, Detection of Fcgamma binding protein antigen in human sera and its relation with autoimmune diseases, *Immunol Lett.* **79**(3):229–235 (2001).
57. Kim YS, Ho SB, Intestinal goblet cells and mucins in health and disease: recent insights and progress, *Curr Gastroenterol Rep.* **12**(5):319–330 (2010).
58. Koga H, Sugiyama S, Kugiyama K, Fukushima H, Watanabe K, Sakamoto T, Yoshimura M, Jinnouchi H, Ogawa H, Elevated levels of remnant lipoproteins are associated with plasma platelet microparticles in patients with type-2 diabetes mellitus without obstructive coronary artery disease, *Eur. Heart J.* **27**(7):817–823 (2006).
59. Ueba T, Haze T, Sugiyama M, Higuchi M, Asayama H, Karitani Y, Nishikawa T, Yamashita K, Nagami S, Nakayama T, Kanatani K, Nomura S, Level, distribution and correlates of platelet-derived microparticles in healthy individuals with special reference to the metabolic syndrome, *Thromb. Haemost.* **100**(2):280–285 (2008).
60. Jayachandran M, Litwiller RD, Owen WG, Heit JA, Behrenbeck T, Mulvagh SL, Araoz PA, Budoff MJ, Harman SM, Miller VM, Characterization of blood borne microparticles as markers of premature coronary calcification in newly menopausal women, *Am. J. Physiol. Heart Circ. Physiol.* **295**(3):H931–H938 (2008).
61. Toth B, Liebhardt S, Steinig K, Ditsch N, Rank A, Bauerfeind I, Spannagl M, Friese K, Reininger AJ, Platelet-derived microparticles and coagulation activation in breast cancer patients, *Thromb. Haemost.* **100**(4):663–669 (2008).
62. Hron G, Kollars M, Weber H, Sagaster V, Quehenberger P, Eichinger S, Kyrle PA, Weltermann A, Tissue factor-positive microparticles: Cellular origin and association with coagulation activation in patients with colorectal cancer, *Thromb. Haemost.* **97**(1):119–123 (2007).
63. Kaptan K, Beyan C, Ifran A, Pekel A, Platelet-derived microparticle levels in women with recurrent spontaneous abortion, *Int. J. Gynaecol. Obstet.* **102**(3):271–274 (2008).
64. Tamagawa-Mineoka R, Katoh N, Ueda E, Masuda K, Kishimoto S, Platelet-derived microparticles and soluble P-selectin as platelet activation markers in patients with atopic dermatitis, *Clin. Immunol.* **131**(3):495–500 (2009).

65. Fusegawa H, Shiraishi K, Ogasawara F, Shimizu M, Haruki Y, Miyachi H, Matsuzaki S, Ando Y, Platelet activation in patients with chronic hepatitis C, *Tokai J. Exp. Clin. Med.* **27**(4):101–106 (2002).
66. Ayers L, Ferry B, Craig S, Nicoll D, Stradling JR, Kohler M, Circulating cell-derived microparticles in patients with minimally symptomatic obstructive sleep apnoea, *Eur. Respir. J.* **33**(3):574–580 (2009).
67. Kang P, Shen B, Yang J, Pei F, Circulating platelet-derived microparticles and endothelium-derived microparticles may be a potential cause of microthrombosis in patients with osteonecrosis of the femoral head, *Thromb. Res.* **123**(2):367–373 (2008).
68. Wasiluk A, Mantur M, Szczepanski M, Matowicka-Karna J, Kemona H, Warda, J, Platelet-derived microparticles and platelet count in preterm newborns, *Fetal Diagn Ther.* **23**(2):149–152 (2008).
69. Kageyama K, Nakajima Y, Shibasaki M, Hashimoto S, Mizobe T, Increased platelet, leukocyte, and endothelial cell activity are associated with increased coagulability in patients after total knee arthroplasty, *J. Thromb. Haemost.* **5**(4):738–745 (2007).
70. Pattanapanyasat K, Gonwong S, Chaichompoo P, Noulsri E, Lerdwana S, Sukapirom K, Siritanaratkul N, Fucharoen S, Activated platelet-derived microparticles in thalassaemia, *Br. J. Haematol.* **136**(3):462–471 (2007).
71. Pamuk GE, Vural O, Turgut B, Demir M, Umit H, Tezel A, Increased circulating platelet-neutrophil, platelet-monocyte complexes, and platelet activation in patients with ulcerative colitis: a comparative study, *Am. J. Hematol.* **81**(10):753–759 (2006).
72. Knijff-Dutmer EA, Koerts J, Nieuwland R, Kalsbeek-Batenburg EM, van de Laar MA, Elevated levels of platelet microparticles are associated with disease activity in rheumatoid arthritis, *Arthritis Rheum.* **46**(6):1498–1503 (2002).
73. Mostefai HA, Meziani F, Mastronardi ML, Agouni A, Heymes C, Sargentini C, Asfar P, Martinez MC, Andriantsitohaina R, Circulating microparticles from patients with septic shock exert protective role in vascular function, *Am. J. Respir. Crit. Care Med.* **178**(11):1148–1155 (2008).
74. Pereira J, Alfaro G, Goycoolea M, Quiroga T, Ocqueteau M, Massardo L, Perez C, Saez C, Panes O, Matus V, Mezzano D, Circulating platelet-derived microparticles in systemic lupus erythematosus. Association with increased thrombin generation and procoagulant state, *Thromb. Haemost.* **95**(1):94–99 (2006).
75. Guiducci S, Distler JH, Jungel A, Huscher D, Huber LC, Michel BA, Gay RE, Pisetsky DS, Gay S, Matucci-Cerinic M, Distler O, The relationship between plasma microparticles and disease manifestations in patients with systemic sclerosis, *Arthritis Rheum.* **58**(9):2845–2853 (2008).
76. Ogasawara F, Fusegawa H, Haruki Y, Shiraishi K, Watanabe N, Matsuzaki S, Platelet activation in patients with alcoholic liver disease, *Tokai J. Exp. Clin. Med.* **30**(1):41–48 (2005).
77. Jy W, Horstman LL, Arce M, Ahn YS, Clinical significance of platelet microparticles in autoimmune thrombocytopenias, *J. Lab. Clin. Med.* **119**(4):334–345 (1992).
78. Redman CW, Sargent IL, Circulating microparticles in normal pregnancy and pre-eclampsia, *Placenta* **29**(7 Suppl. A):S73–S77 (2008).
79. Sheremata WA, Jy W, Horstman LL, Ahn YS, Alexander JS, Minagar A, Evidence of platelet activation in multiple sclerosis, *J. Neuroinflam.* **5**(27):27 (2008).

# 8

# *N*-TERMINAL COMBINED FRACTIONAL DIAGONAL CHROMATOGRAPHIC (COFRADIC) ANALYSIS OF THE HUMAN PLATELET PROTEOME

Francis Impens, Kenny Helsens, Niklaas Colaert, Lennart Martens, Joël Vandekerckhove, and Kris Gevaert

**Abstract**

In this chapter we focus on the technical improvments that were made to the *N*-terminal combined fractional diagonal chromatography (COFRADIC) protocol over the past five years. We discuss their general impact on *N*-terminal peptide coverage as well as on proteome coverage of human blood platelets by comparing the results of a more recent *N*-terminal proteome analysis with a previously reported dataset. Furthermore, we show that the specificity profile of the main proteolytic activity present in a typical preparation of human platelets is very similar to that of human calpain-1.

## 8.1 INTRODUCTION

In an ideal world proteomics would identify and quantify all protein forms expressed by a genome under given conditions. Although some reports have

*Platelet Proteomics: Principles, Analysis and Applications*, First Edition.
Edited by Ángel García and Yotis A. Senis.

demonstrated that at least this might well be feasible for lower eukaryotes such as baker's yeast [1], in practice, characterized proteome maps of samples from higher eukaryotes, are far from complete. A major problem is that especially in these higher eukaryotes, alternative splicing of immature messenger RNA and a wealth of often reversible protein modifications yield many different protein variants. This complexity means that their actual numbers can be estimated to be somewhere only on the order of hundreds of thousands. Of course, the question as to whether each different protein form originating from a single gene in a genome holds a unique biological function remains open. In addition to the protein diversity introduced by splicing and modification, proteins are expressed in hugely different copy numbers (e.g., estimated to span $\geq$10 orders of magnitude in serum samples [2]), imposing extra demands on the performance of proteome analysis techniques.

Before the advent of mass spectrometry techniques for protein identification and quantification, proteome analysis was performed by separating proteomes by two-dimensional (polyacrylamide) gel electrophoresis (2DGE), originally introduced in 1975 [3]. 2DGE separates proteins in two dimensions: first by their isoelectric point (isoelectric focusing), and subsequently by their molecular weight (SDS-PAGE). One major problem with the original 2DGE technique was cathodic drift [4], which leads to instable pH gradients (especially at the alkaline side) and thus irreproducible 2D gel patterns. This problem was solved by the introduction of immobilized pH gradients in polyacrylamide gels [5], and this, in turn, allowed 2DGE to become a standard proteomics technique in many labs. Difference in-gel electrophoresis (DIGE) was introduced [6], in which two different proteomes are labeled with different fluorophores, allowing them to be separated on the same 2D gel while retaining the ability to visualize the spot pattern for each proteome separately afterward. This methodology further increased the reproducibility of 2DGE, especially when combined with a third fluorophore that enables coanalysis of an internal standard (typically a mixture of equal amounts of the two proteomes under investigation) for spot intensity normalization purposes.

Nonetheless, 2DGE comes with several disadvantages, and it is fair to state that 2D gels reveal primarily well-soluble proteins and proteins with high copy numbers. Work carried out in the lab of Thierry Rabilloud nicely illustrated that integral membrane proteins with several transmembrane helices or small domains are not only difficult to fully extract from their hydrophobic environment but also tend to become insoluble close to their isoelectric points and thus precipitate during IEF [7]. In 2000, the lab of Ruedi Aebersold performed a comprehensive MS analysis of the yeast proteome separated by 2DGE and showed that in proteome maps of non-fractionated proteomes, proteins needed to be present in at least 1000 copies per cell before they could be identified [8]. Mass spectrometry has clearly evolved since 2000 and became much more sensitive for detecting low-copy-number proteins; however, the Aebersold study clearly pointed to a need to distinct protein structures or cell organelles prior to separation if 2D gels are to be used for comprehensive proteome analysis.

To compensate for these shortcomings of 2DGE, so-called gel-free and MS-driven proteomics techniques were introduced around the late 1990s. Such techniques nowadays rely on (1) the high sensitivity and sampling rate of modern mass spectrometers, (2) the fact that complete genome sequences and their derived *in silico* proteome sequences are made publicly available for an increasing number of organisms, and (3) several sophisticated peptide and protein identification algorithms that can search these proteome databases using mass spectrometry data (reviewed in Ref. 9). Pioneering gel-free proteomics work includes the MudPIT (multidimensional protein identification technology) approach devised by the Yates lab [10], which couples different peptide separation techniques, and the isotope-coded affinity tags (ICATs) introduced by the Aebersold lab [11]. MudPIT originally combined strong cation exchange (chromatography) (SCX) with reversed-phase (RP) HPLC peptide separations to fractionate a proteome digest as extensively as possible prior to MS/MS analysis. The overall idea is that increased peptide separation allows more peptides to be observed and subsequently fragmented by mass spectrometers, thereby increasing the overall proteome coverage. Although MudPIT clearly enabled identification of low-copy-number proteins and hydrophobic proteins [10], an important drawback is the general lack of reproducibility caused by the fact that too many different peptides are ionized per timeframe, resulting in a random sampling of peptide ions for MS/MS analysis by the overwhelmed mass spectrometers [12]. ICAT tries to compensate for this random sampling effect by isolating a subset of so-called representative peptides from whole-proteome digests and using these peptides only for LC-MS/MS analysis. Representative peptides are here best defined as peptides containing sufficiently rare amino acids such that the total number of analytes drops significantly on selection while allowing a broad coverage of the entire proteome. This latter property requires that the selected peptides be distributed evenly throughout the proteome, with as many proteins as possible yielding at least one such peptide on digestion. ICAT molecules use a iodoacetamide derivative to target cysteine residues in peptides and tag these with a biotin group, allowing their affinity-based isolation on (strept)avidin beads [11].

Following introduction of ICAT and MudPIT, the field of proteomics witnessed the dawn of many other approaches for gel-free proteome analysis, many of which are based on a reduction in sample complexity by isolating representative peptide sets (reviewed in Ref. 13). Further information on quantitative MS strategies applied to platelet research can be found in Chapter 2.

In 2002 our group introduced a gel-free proteomics technology based on the separation principle of diagonal electrophoresis and diagonal chromatography [14,15]. Our technology is called *combined fractional diagonal chromatography* (COFRADIC) [16], and since its inception different COFRADIC variants have been introduced to study many different aspects of proteome samples.

The common COFRADIC concept is described as follows. An isolated proteome is digested by a protease (typically trypsin), and a predefined subset of the resulting peptide mixture is isolated by consecutive reversed-phase (RP) chromatographic peptide separation steps, and this representative peptide subset is then analyzed by LC-MS/MS. More specifically, following proteome digestion, the peptide mixture is separated initially by RP-HPLC into distinct peptide fractions; this constitutes the primary COFRADIC peptide separation step. Then, each primary fraction (or a combination of sufficiently broadly separated primary fractions) is dried and treated with chemicals or enzymes that are known to specifically target a given amino acid residue or functional group; this step comprises the COFRADIC sorting reaction. This procedure alters the structure of a predefined class of peptides in such a way that they acquire different chromatographic properties and are thus differently retained on RP columns as compared to their elution profile during the primary RP-HPLC separation. Separating such altered primary fractions a second time on the same RP column using the same conditions employed during the primary run will now segregate the altered peptides from the unaltered peptides by their shift in retention time, thus allowing isolation of either one or both classes of peptides (altered and unaltered) that can then be subjected to LC-MS/MS analysis. This complexity reduction of the peptide mixture by isolating a relevant and representative peptide subset, combined with the general advantages of gel-free proteomics approaches, underlies the power of the COFRADIC technique.

Using the original COFRADIC protocol, we isolated methionyl peptides following controlled oxidation with hydrogen peroxide to their sulfoxide counterparts. In this way, methionyl peptides become more hydrophilic and are isolated during a series of secondary COFRADIC separations [16]. An important advantage that COFRADIC holds over, for instance, the ICAT technology, which also isolates representative peptides, is that COFRADIC is intrinsically much more versatile. Indeed, changing the COFRADIC sorting reaction will target other peptide classes and will thus allow isolation of different subsets of peptides. Hence, following publication of the COFRADIC protocol to isolate methionyl peptides, COFRADIC variants were developed that select *N*-terminal peptides [17] and cysteinyl peptides [18], among others. In 2005, we reported on the combined use of these three COFRADIC protocols for a comprehensive study of the proteome of human platelets, which resulted in the largest set of identified platelet proteins to date [19]. More recently, we began to investigate protein processing (and not protein catabolism by proteasomes) as a posttranslational modification (PTM) involved in signal propagation during blood platelet activation (Impens and colleagues, unpublished data). Protein processing is an interesting protein modification since it is generally irreversible and mostly inactivates the function of the target substrate. However, it might also lead to a gain of function of the substrate or (one of) its fragments. In 2005 we showed that since protein processing creates novel protein *N* termini, differential stable isotope labeling combined with *N*-terminal COFRADIC could readily be used to study *in vivo* protein processing [20].

In platelets, calpain-1 especially has been reported to cleave several substrates during platelet activation (reviewed in Ref. 21). Calpain-1 is a calcium-activated cysteine protease ubiquitously expressed in the cytoplasm of most human cell types. Because of its high expression level, platelets continue to serve as the primary source for the purification of calpain-1 since active, full-length recombinant forms of this protease remain unavailable. As part of a larger study to better understand the general role of proteolysis during platelet activation, we set up a pilot experiment to generate an *in vitro* catalog of calpain-1 substrates in human blood platelets. A freeze–thaw blood platelet lysate was split into two aliquots. Purified human calpain-1 was added to one aliquot, while we attempted to inhibit endogenous calpain activation in the second aliquot by adding EDTA. Both samples were incubated for a short time (20 min) at 37°C and afterward differently isotopically labeled *in vitro* (postmetabolically) by butyration of primary amino groups [using *N*-hydroxysuccinimide esters of $^{12}C_4$ (light) or $^{13}C_4$ (heavy) butyric acid]. The labeled aliquots were then mixed together and analyzed by *N*-terminal COFRADIC. Although not fully successful for reasons discussed below, the experiment was set up using the state-of-the-art *N*-terminal COFRADIC protocol and was therefore highly informative for characterizing the *N*-terminal proteome of platelet preparation, which will be discussed further in this chapter. We here focus on the more recent technical improvements that were made to the *N*-terminal COFRADIC protocol [22] and discuss their general impact on *N*-terminal peptide and proteome coverage and compare the results of the new analysis with those of the *N*-terminal COFRADIC platelet proteome screen published in 2005, hereafter referred to as the "old analysis." Furthermore, we show that the specificity profile of the main proteolytic activity present in a typical preparation of human platelets is highly similar to that of human calpain-1.

## 8.2 THE OLD VERSUS THE NEW *N*-TERMINAL COFRADIC PROTOCOL

Compared to the original version of the *N*-terminal COFRADIC protocol published in 2003 [17], new steps have been included to improve overall protocol performance in terms of both sensitivity and specificity. The scheme shown in Figure 8.1 compares the main steps in the *N*-terminal COFRADIC protocols used for the old and new analyses of the human platelet proteome. The major improvements since the publication of the first human platelet proteome mapped by COFRADIC approaches [19] are discussed below. Note that database searches for all MS/MS spectra from both the old and the new analysis were repeated using identical database and search parameter settings.

### 8.2.1 Stable Isotope Tagging of Protein α-Amino Groups Points to *in vivo* Protein Processing

Before trypsin digestion of an isolated proteome, all cysteine bridges are reduced and free thiol groups are blocked by alkylation, typically using iodoacetamide. This step is absolutely necessary since the actual COFRADIC sorting reagent for isolating *N*-terminal peptides, 2,4,6-trinitrobenzenesulfonic acid (TNBS), readily reacts with free thiol groups, a side reaction that would result in unwanted removal of cysteine-containing *N*-terminal peptides from the mixture of analytes.

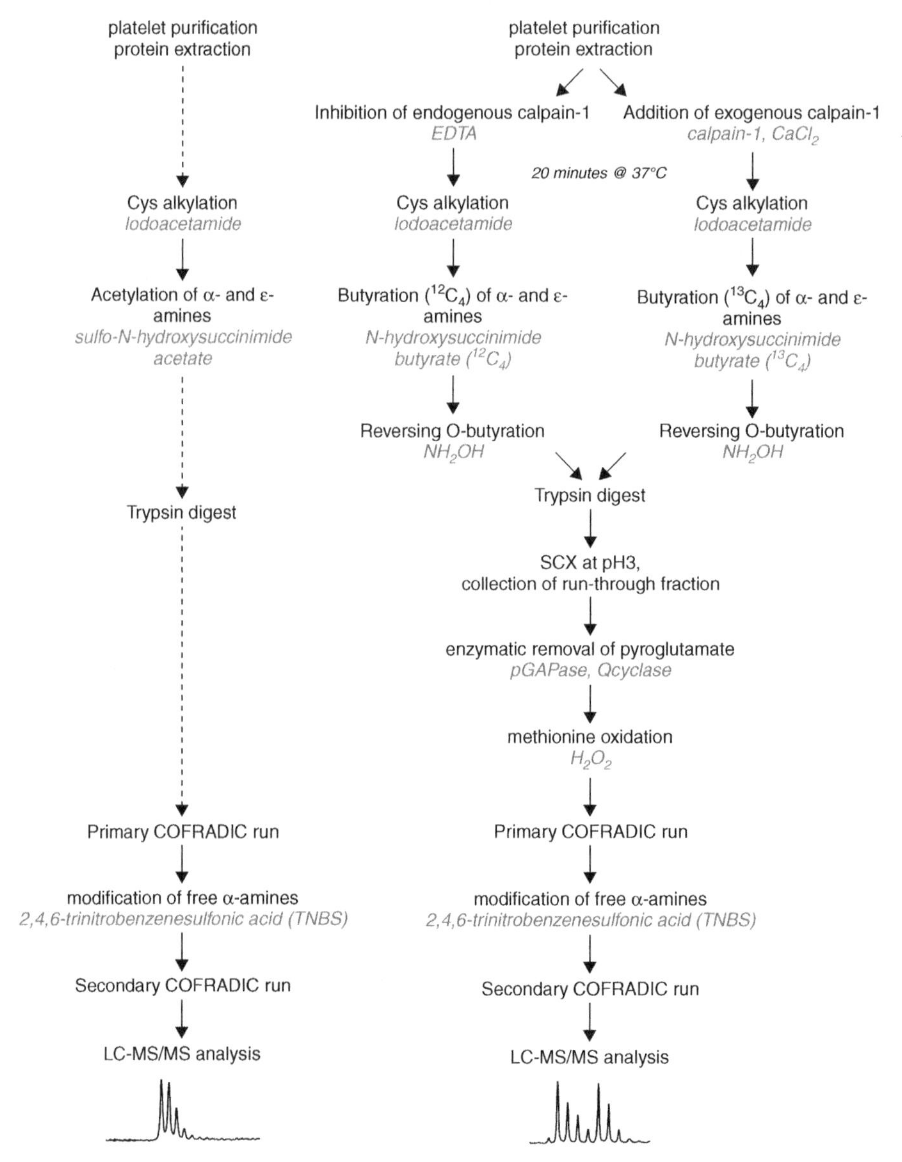


**Figure 8.1** The main steps in the original and novel COFRADIC procedure for sorting *N*-terminal peptides. The workflow of the original *N*-terminal COFRADIC procedure that was used to study the platelet *N*-terminal proteome in 2005 [19] is shown on the left. The additional steps introduced in the novel protocol that was used to reanalyze the platelet proteome (shown on the right) are indicated.

A second *in vitro* protein modification step that is also needed is blocking of primary amino groups—both protein α-amino groups (protein *N* termini) as well as ε-amino groups of lysine sidechains. This was originally achieved through an acetylation reaction with an *N*-hydroxysuccinimide (NHS) ester of acetate. If this modification step would be excluded, *N*-terminal peptides from proteins with an *in vivo* free *N* terminus, as well as lysine-containing *N*-terminal peptides would react with TNBS and would therefore be removed from the final mixture of *N*-terminal peptides. In the new protocol, the acetylation reaction with (sulfo)*N*-hydroxysuccinimide acetate is replaced by reaction with NHS esters of acids carrying deuterium or carbon-13 isotopes that are synthesized according to the protocol published in 2000 [23].

As a result, all primary amino groups are mass-tagged (e.g., using trideuteroacetate) and, following COFRADIC isolation of *N*-terminal peptides, LC-MS/MS analyses, and database searching, such mass tags not only distinguish *in vivo* free from blocked (mainly by acetylation) protein *N* termini [24] but also allow more direct identification of protease substrates [25–28]. It needs to be mentioned here that except for some bioactive peptides, protein α-*N*-acetylation is an abundant co-translational modification on nascent polypeptide chains fueled by the transfer of an acetyl group from acetylcoenzyme A by *N*-acetyltransferase complexes, and is therefore not a posttranslational protein modification. This implies that novel protein *N* termini of protein fragments formed by protease substrate processing are not expected to become acetylated *in vivo*, but rather carry primary *N*-terminal amino groups. Thus, tagging such *N* termini with stable isotopes not only distinguishes *in vivo* blocked from free protein *N* termini (see references cited above) but also allows differential proteome analysis (relative quantification), provided that ample mass spacing between the two peptide variants (isotopomers) exists.

Previous COFRADIC analyses relied on trideuteroacetate and normal acetate for mass labeling for differential proteomics. However, the mass spacing introduced by these compounds is only 3 Da, resulting in overlapping isotopic patterns for peptides with masses higher than ~1000 Da. For even heavier peptides (<2500 Da), this problem becomes so overwhelming that it effectively precludes quantification [24]. For this reason we have synthesized NHS esters of $^{12}C_4$-butyrate and its $^{13}C_4$ counterpart. These NHS esters were found to be as reactive as NHS-acetate esters but introduce an extra mass tag of 4 Da per modified primary amino group and, comparable to the use of oxygen-18 labeling [29], this mass spacing allows relative quantification in a straightforward manner even when low-resolution mass spectrometry is used. It is, however, necessary to mention that clearly not all peptides isolated by the *N*-terminal COFRADIC approach are open to quantification. Indeed, the majority of these *N*-terminal peptides are already *in vivo* blocked by acetylation and cannot be modified by differential butyration. However, as explained above, the peptides of key interest for our studies (i.e., protease substrates) contain primary amino groups and will thus be mass-tagged.

The performance gain obtained by using mass tagged NHS butyrate esters compared to acetylation is illustrated in Table 8.1. In the old COFRADIC screen

**TABLE 8.1 Number of Spectra, Peptides, and Proteins Identified by the Old and New *N*-Terminal COFRADIC Protocols Used to Analyze the Platelet Proteome**

| Activity | Old Protocol (ESI-QTOF) | New Protocol (Calpain Case Study) (ESI Ion Trap) |
|---|---|---|
| Spectra generated | 12,313 | 37,510 |
| Spectra identified | 1,913 | 2,441 |
| Unique peptides identified | 502 | 1,092 |
| Unique proteins identified | 298 | 561 |
| Acetylated peptides starting at position 1 or 2 | 784 (41.0%) | 653 (26.8%) |
| Acetylated peptides starting at position >2 | 393 (20.5%) | 52 (2.1%) |
| Butyrated peptides starting at position 1 or 2 | 0 | 140 (5.7%) |
| Butyrated peptides starting at position >2 | 0 | 1430 (58.6%) |
| Pyroglutamate peptides | 197 (10.3%) | 5 (0.2%) |
| Pyrocarbamidomethyl Cys peptides | 45 (2.4%) | 29 (1.2%) |
| Peptides with a free α-amine | 494 (25.8%) | 132 (5.4%) |
| His-containing peptides | 509 (26.6%) | 57 (2.3%) |
| Missed cleaved peptides (internal Arg) | 124 (6.5%) | 25 (1%) |

about 60% of the identified peptides were acetylated at their *N* termini. Remarkably, one-third of these peptides have start positions >2 in their corresponding protein sequence. Since the acetylation reaction used in the original screen cannot be used to determine whether an acetylation event points to *in vivo* or *in vitro* modification, it is not clear whether such peptides are the result of an alternative start position of protein translation or if they point to proteolytic processing. In contrast, this distinction can readily be made using the new COFRADIC protocol. Here, about 93% of the identified peptides were modified by acetylation or butyration at their *N* termini (see also discussion below). However, it is clear that about two-thirds of these peptides are most likely derived from proteolytic processing (butyrated peptides with start position >2) whereas only a small fraction (2%) potentially points to alternative start positions of the translation machinery (*in vivo* acetylated peptides with start position >2). The high number of peptides pointing to proteolytic processing identified in the more recent analysis suggests a massive activation of protease activity in our control platelet preparation as discussed below. Nevertheless, removal of signal peptides (e.g., from mitochondrial or endoplasmatic reticulum (ER)] proteins is also responsible for a significant fraction of these peptides. As an example, data from the mature *N* terminus of the protein disulfide isomerase A3 (Swiss-Prot accession number P30101), an ER protein, is shown in Figure 8.2. Figure 8.2(a) displays the mass spectrum of the acetylated form of the peptide as identified in the old COFRADIC analysis (upper panel), whereas the lower panel shows the MS spectrum of the butyrated peptide identified using the improved protocol. In the latter case the peptide is present in two forms with a 4 Da mass difference between the two isotopic envelopes as the result of differential butyration.

To further annotate substrates of the exogenous added calpain-1, the ratios of the ion envelope intensities of the differently butyrated peptides from all $^{13}C_4$-butyrated peptides with start position >2 were derived from the MS spectra. Note that the platelet proteome targeted by exogenous calpain 1 was labeled by $^{13}C_4$-butyration. For 58 of these peptides the corresponding light isotopomer was not found, and these were therefore annotated as genuine neo-*N*-terminal peptides formed by the action of exogenously added calpain-1 (Impens and colleagues, unpublished data). However, for the majority of these peptides, light peptides were also observed in the spectra and a peptide isotopomer ratio could be calculated using the peptide envelope intensities. Figure 8.2(b) shows the distribution of the $\log_2$ values of these ratios. Although the distribution is clearly tailored to the left side, due to the action of exogenous on top of endogenous calpain-1 in the heavy sample (discussed below), the median of the $\log_2$ distribution was found to be −0.16, which indicates that butyration occurred to the same extent on both the control and the calpain-1 challenged platelet proteome. Taken together, such a type of postmetabolic protein labeling with stable isotopes is actually preferred and necessary for *N*-terminal COFRADIC platelet proteome samples since platelet proteins cannot be labeled metabolically during cell culture [i.e., by stable isotope labeling by amino acids in cell culture (SILAC) [30]]. In fact, whenever cells can be metabolically labeled, we prefer a SILAC setup using isotopic variants of arginine (all *N*-terminal COFRADIC peptides end on arginine when using trypsin due to blocking of the lysine sidechain) as proteome samples can be brought together relatively early in the whole COFRADIC process, thereby reducing experimentally induced differences.

A well-known side reaction of the use of NHS esters for modifying primary amines is *O*-esterification of the hydroxyl groups of tyrosine, threonine, and serine. More specifically, several reports suggested that this modification leads to stably modified hydroxyl groups in aqueous conditions when the affected amino acid is positioned two amino acids away from a histidine [31]. Hydroxylamine is often used to hydrolyze these esters and thus recycle the original amino acids [32]. This "cleanup" step is also necessary since traces of *O*-esterified peptides will increase the complexity of the mixture of analytes and will interfere with both peptide identification and peptide (protein) quantification. More recently, a simple boiling step was reported to specifically and quantitatively hydrolyze acetyl–ester bonds, leaving acetyl–amide bonds intact [33]. We, indeed, confirmed that this boiling step efficiently reconverted *O*-esterified amino acids when peptides were boiled. However, when working with proteins, we suggest using hydroxylamine since boiled proteins may precipitate, which might interfere with further protein handling.

### 8.2.2 Enrichment of *N*-Terminal Peptides by Strong Cation Exchange (SCX) Chromatography at Low pH

Not a single chemical modification reaction is absolutely quantitative, and not all targeted groups will therefore be modified. When using the original *N*-terminal

COFRADIC protocol [17], we noticed that on average ~10% of all identified peptides were "internal peptides" carrying a primary α-amino group that had failed to react with TNBS. Such peptides are not "proteome-informative" as they report only that their precursor protein was expressed in the analyzed proteome but cannot be used to point to important modifications that are of interest for us

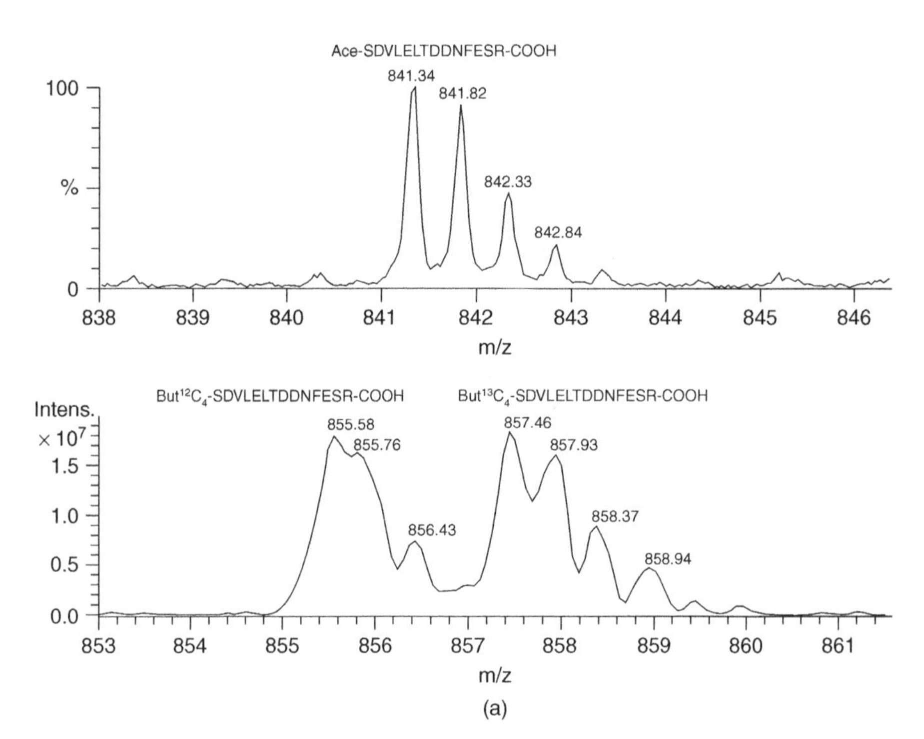


**Figure 8.2** Illustration of peptide mass spectra from the original and the new *N*-terminal COFRADIC analysis. (a) Representative mass spectra from the mature *N* terminus of the ER protein protein disulfide–isomerase A3 (UniProtKB/Swiss-Prot accession number P30101) are shown. The upper panel displays the high-resolution ESI-QTOF MS spectrum recorded during the old analysis. The peptide is acetylated at its *N* terminus, and its start position indicates that the signal peptide was removed according to the UniProtKB/Swiss-Prot annotation. In the lower panel, the low-resolution ESI–ion trap MS spectrum from the same peptide is depicted as recorded during the new analysis. Here the acetyl group is replaced by a butyrate group which means that the modification occurred *in vitro*. Since two isotopic forms of the butyration reagent were used, two peptide envelopes with a mass difference of 4 *m*/*z* can be distinguished in the spectrum. (b) Differential butyration enables relative quantification of *N*-terminal peptides. The light and heavy peptide envelope intensities of butyrated peptides identified with start position >2 were manually looked up in the spectra. The distribution of the $\log_2$ values of the light-over-heavy ratios is shown. The tailing to the left side is the result of higher levels of active calpain-1 in the heavy labeled sample compared to the light labeled sample. The median of the distribution is found to be −0.16, indicating that butyration occurred to the same extent by both forms of butyration reagent.

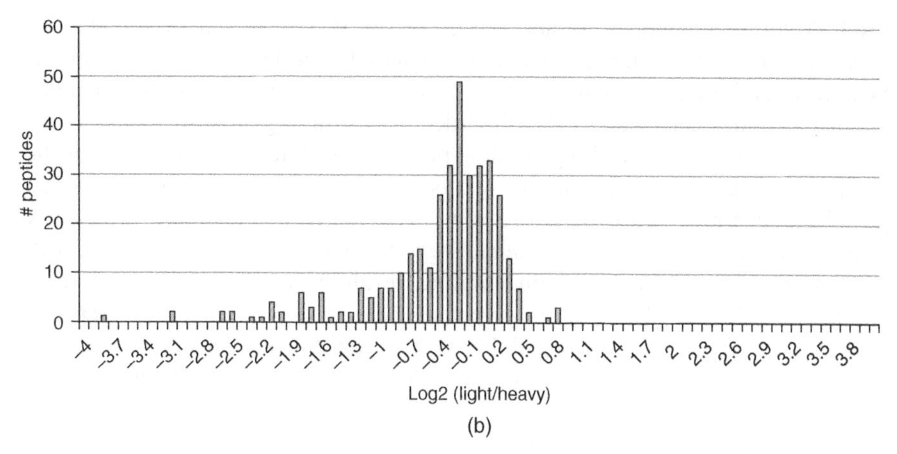


**Figure 8.2** (*Continued*)

such as protein processing and protein *N*-terminal acetylation. Since we wanted to reduce the number of such unwanted TNBS-unmodified peptides, we reckoned that the actual TNBS modification could be made more efficient if the number of peptides with primary α-amino groups would be reduced, or, in other words, if we would be able to enrich *N*-terminal peptides prior to the actual COFRADIC peptide sorting procedure. Previous reports had suggested using SCX at low pH to enrich for *in vivo* blocked *N*-terminal peptides as well as *C*-terminal peptides from relatively simple digests of individual proteins [34,35]. This enrichment is based on the fact that at low pH (pH ~ 3), these peptides carry one less positive charge compared to internal peptides. Indeed, internal tryptic peptides will generally hold two groups that can be positively charged (a free α-amino group and a *C*-terminal arginine or lysine residue), whereas many *N*-terminal peptides will carry a blocked α-amino group by *in vivo* acetylation. Furthermore, few *C*-terminal tryptic peptides will end on arginine or lysine residues. More recently the concept of using SCX to enrich for *N*- and *C*-terminal peptides was applied to whole-proteome digests leading to new insights into the biology and modification of protein termini [36]. However, although *N*-terminal peptides were clearly enriched in this work by the group of Albert Heck, as well as in a later paper from the same group [37], there remained a considerable background of other peptides which suppressed ionization and resulted in the reduced detection of *N*-terminal peptides.

As a result of the above, we decided to use SCX as a preenrichment step for *N*-terminal peptides that are then further enriched by the *N*-terminal COFRADIC procedure [22,38]. By implementing the first steps of the COFRADIC workflow, in particular the blocking of primary amino groups, prior to the SCX separation, we ensure that all *N*-terminal peptides contain blocked α-*N* termini and end on an arginine residue after trypsin digestion. At pH 3, the net charge of such peptides is zero, whereas all internal peptides carry at least one extra positive charge. As a result, *N*-terminal peptides are not retained by the SCX resin and are found in the flow-through, whereas the vast majority of internal peptides will interact with the SCX resin and are thus separated from *N*-terminal peptides. Since the separated internal peptides are now easily discarded, the resulting mixture contains *N*-terminal peptides along with *C*-terminal peptides, peptides starting with a cyclic amino acid (e.g., pyroglutamate, see text below), and a minor number of internal peptides that, for as yet unknown reasons, were not (fully) retained by the SCX resin. Following this SCX *N*-terminal peptide preenrichment step, the TNBS reaction was found to be more quantitative, resulting in a drop of "interfering" internal peptides to $<2\%$ [22]. In our calpain case study, the fraction of peptides with a free α-amine was slightly higher ($\sim5\%$), but this still provides a fivefold increase in selection efficiency over the old *N*-terminal COFRADIC analysis (Table 8.1).

Pre-enriching *N*-terminal peptides by SCX at low pH, however, also comes at a price. All *N*-terminal peptides carrying an additional positively charged residue (i.e., a histidine or an internal arginine as a result of a missed tryptic cleavage; e.g., Arg-Pro bonds) will also be retained by the SCX resin and are thus depleted from the final mixture of *N*-terminal peptides. Although one expects histidine to be a low-abundance amino acid overall—it is the third least abundant amino acid (2.27% of all amino acids; see `http://www.expasy.ch/sprot/relnotes/relstat.html`)—about 22% of all *N*-terminal peptides detectable by MS are predicted to carry at least one histidine in the human proteome. This implies that these peptides will not be detected if SCX is used as a preenrichment step [22]. The fraction of *N*-terminal peptides predicted to contain an internal arginine resulting from a missed cleavage event is much lower, at about 6% [22]. In summary, relying on SCX as a method to preenrich for *N*-terminal peptides results in a predicted removal of about 28% of all theoretical *N*-terminal peptides prior to the actual COFRADIC isolation procedure. In order to achieve comprehensive proteome coverage using *N*-terminal peptides, one might therefore consider combining the old and new *N*-terminal COFRADIC protocols.

### 8.2.3 Depletion of Noninformative Pyroglutamate-Starting Peptides

A frequently overlooked artifact in MS analysis is the spontaneous conversion of *N*-terminal glutamine residues (and to a much lesser extent glutamic acid) into pyroglutamic acid (pyrrolidone carboxylic acid). The rate by which this reaction occurs depends on incubation time, temperature, and buffer conditions and is generally uncontrollable [39]. For the *N*-terminal COFRADIC sorting strategy, the problem is that such pyroglutamate-starting peptides do not react with TNBS and are thus co-sorted with the real *N*-terminal peptides. Depending on the proteome digestion conditions, the final fraction of these noninformative pyroglutamate peptides (non-informative since they seldom point to protein processing events) can be higher than 20%. Furthermore, such peptides not only fail to react with TNBS but are also coenriched by SCX at low pH since their secondary amino group leads to one less positive charge compared to internal tryptic peptides carrying a primary amino group.

To reduce the number of pyroglutamate peptides we recently included an enzymatic step to cleave pyroglutamate from a peptide *N* terminus. The resulting peptide typically starts with a primary amino group and is either retained on the SCX resin (if removal of pyroglutamic acid is performed prior to the SCX step) or modified by TNBS (if pyroglutamic acid is removed following SCX and prior to the first COFRADIC isolation). Pyroglutamylaminopeptidases (pGAPases) are known to efficiently cleave the pyroglutamic acid bond [40] and were used in the past to open up pyroglutamate protein *N* termini for sequencing by the Edman degradation reaction [41,42]. We used a commercial kit containing two enzymes: a glutamine cyclotransferase (Qcyclase), which ensures cyclization of *N*-terminal glutamine residues; and a pGAPase to cleave off pyroglutamate residues (the TAGZyme™ kit from Qiagen, Hilden, Germany). Combining these two enzymes was found to ensure continuous cyclization of *N*-terminal glutamines followed by direct removal by pGAPase. As a result, the number of contaminating pyroglumate tryptic peptides in the final lists of identified peptides dropped dramatically to about 1% or less [22] (only 0.2% shown in Table 8.1). In fact, the majority of remaining pyroglutamate peptides carry a proline as the second amino acid, and this residue is known to inhibit processing by pGAPases [43].

Finally, one can choose to perform this reaction either prior to the SCX step or between this SCX step and the first COFRADIC separation step. However, in order for SCX to proceed efficiently, samples cannot be dissolved in buffers containing even moderate concentrations of salts since these interfere with retention of internal peptides on the SCX resin. Therefore, removal of pyroglutamate is typically performed prior to the first COFRADIC run.

### 8.2.4 Oxidation of Methionine Residues

Similar to the formation of pyroglutamic acid, oxidation of several amino acids can occur spontaneously during sample handling. These artifactual protein modifications might interfere with protein identification and protein quantification since they increase the complexity of the analyte mixture. In this respect, strategies to either uniformly oxidize susceptible amino acids or to reduce them have been recently suggested [44]. A major oxidation-affected residue is methionine, which typically oxidizes to its sulfoxide form. Should this occur between the primary and secondary COFRADIC peptide separation steps, *N*-terminal peptides containing methionine(s) will shift out of their primary collection interval and might thus be missed (indeed, peptides containing methioninesulfoxide are more hydrophilic than their counterparts containing nonoxidized methionine [16]). It is important to note here that methionine is often found at protein *N* termini, with about 30% of all protein *N* termini identified in human proteomes starting with a methionine residue [24]. These *N* termini include substrates from the *N*-acetyltransferase B and C enzyme complexes, known to acetylate certain initiator methionines of proteins.

One way to deal with the outcome of spontaneous methionine oxidation is to enlarge the collection intervals of *N*-terminal peptides during the secondary COFRADIC runs. However, this also increases the number of peptide fractions that need to be analyzed by LC-MS/MS and does not compensate at all for possible quantification problems. Therefore, we decided to introduce a controlled methionine oxidation step with hydrogen peroxide (conditions as described in Ref. 16) prior to the first COFRADIC separation step to convert all methionine residues to methioninesulfoxide residues, thereby rendering them insensitive to further spontaneous oxidative damage.

## 8.3 THE *N*-TERMINAL PLATELET PROTEOME: OLD VERSUS NEW ANALYSIS

As discussed above, the introduction of all these novel steps boosts both the sensitivity and specificity of the *N*-terminal COFRADIC procedure so that nowadays about 95% of all identified MS/MS spectra point to real *N*-terminal peptides (see Ref. 22 and Table 8.1). Applied to the human platelet proteome, the number of 561 unique proteins identified in our more recent screen almost doubles the number from the previous screen (Table 8.1) and is not too far away from the 641 proteins that we reported in 2005 as the combined result of methionine, cysteine, and *N*-terminal COFRADIC analyses [19]. However, a direct comparison of the two *N*-terminal platelet proteome datasets generated in our lab is not straightforward: (1) different protocols were used to generate the datasets, which have their own advantages (e.g., increased number of *N* termini by combining SCX to *N*-terminal COFRADIC) as well their disadvantages (e.g., loss of histidine-containing *N*-terminal peptides on using SCX); and (2) different mass spectrometers were used to analyze the sorted peptides by LC-MS/MS. The first sample was analyzed using an ESI-QTOF instrument with a long duty cycle, thereby generating a relatively low number of high-quality MS/MS spectra. For the more recent analysis a much faster ESI ion trap instrument was used that generated a higher number of fragmentation spectra of somewhat lower quality. This is reflected in the fraction of spectra that were finally identified: 6.5% of ion trap MS/MS spectra versus 15.5% of QTOF MS/MS spectra (see Table 8.1). Thus, since a direct and in-depth comparison of both datasets would be error-prone, we decided to analyze some specific aspects of the datasets as described in the following section.

The following numbers show that the higher number of identified proteins in our more recent study is due primarily to the generation of neo-*N* termini by substrate proteolysis by calpain-1 (see discussion below); we identified 208 proteins by at least one peptide that was butyrated at its *N* terminus and that had a start position >100. These peptides are assumed to be the result of proteolytic cleavage different from the removal of signal peptides, which is generally assumed to reside within roughly the first 50 amino acids of proteins. In contrast, we only identified 112 proteins picked up by their annotated $N$ termini (peptides starting at position 1 or 2). Most interestingly, only 13 proteins were found by both types of peptides, indicating that 195 proteins were identified only following processing.

We further looked up the interactions of the identified proteins as known from protein–protein interaction (PPI) studies to illustrate the higher proteome coverage achieved in our more recent platelet proteome study. PPIs cover a whole variety of actions, including enzyme-mediated modifications such as protein phosphorylation and protein processing. Each of these interactions might therefore alter the function of the targeted protein(s) in case of a binary interaction, or the behavior of a protein module in case of a complex interaction. Blood platelets in particular are well-studied subjects of PPI screens because of their role in the blood clotting cascade. On blood vessel damage, tissue collagen and a collagen-specific glycoprotein Ia/IIa receptor expressed on the platelet cell surface interact, thereby initializing the vital function of platelet activation. Since the 1990s, different methods have been described to study PPIs in a high-throughput fashion. These include the well-known yeast two-hybrid (Y2H) assay [45] or the tandem affinity protein (TAP) purification protocols [46,47]. As high-throughput experiments typically accumulate a large amount of data, it became increasingly important to store (standardized) PPI data into public databases for easy retrieval and analysis. Different laboratories have undertaken this responsibility, and the most prominent public PPI repositories include the Human Protein Reference Database (HPRD) [48], the Molecular Interaction database (MINT) [49], the Biomolecular Interaction Network Database (BIND) [50], and IntAct [51]. As reviewed by Pandey et al., all these databases complement each other in content as well as in end-user functions. Still, the authors show that HPRD, by containing PPI data for about half of the human ORFs, is the most comprehensive public PPI resource [52].

We used Cytoscape [53] to visualize the known interactions stored in HPRD between all proteins from the old and the new dataset. In Figure 8.3(a) two selected subnetworks are shown. The green triangle nodes indicate that most of the proteins were identified only in the recent screen, for instance, in the case of Grb2, an important adaptor protein, and most of its interactors [Fig. 8.3(a)]. For reasons discussed above, it is not surprising that 10 of the 17 Grb2 interacting proteins that were identified only in the new analysis were picked up exclusively with a butyrated, neo-$N$-terminal peptide, indicating that these Grb2 interactors are most probably affected by protease cleavage (by calpain-1; see text below). Additionally, the red-colored nodes show that most of the proteins identified by the old protocol are identified by the new protocol as well. This is illustrated by the network surrounding β-actin as the major component of the platelet cytoskeleton [Fig. 8.3(a)]. From the protein identities in the latter network one would expect that in general these are more abundant proteins compared to the Grb2 subnetwork. An average spectral count per node of respectively 10.2

and 4.8 indicates that this is indeed the case, again illustrating the more in-depth proteome coverage obtained in our recent study.

## 8.4 CALPAIN-1 IS THE MAIN PROTEOLYTIC ACTIVITY IN THE HUMAN PLATELET LYSATE

As mentioned in the introduction, identifying *in vitro* calpain-1 substrates in human platelets was the main objective of our calpain case study. As shown schematically in Figure 8.1, EDTA was added to block calpain activity in one platelet lysate, whereas exogenous purified calpain-1 was added to the other lysate. As mentioned above, we identified 58 heavy butyrated peptides with start position >2 that were present as singletons and are therefore generated by cleavage of exogenous calpain-1. However, in total we identified 939 butyrated peptides with start position >100 for which both light and heavy forms were

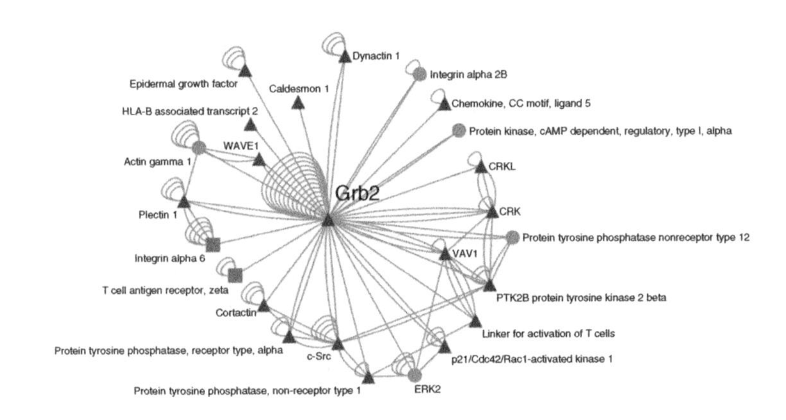


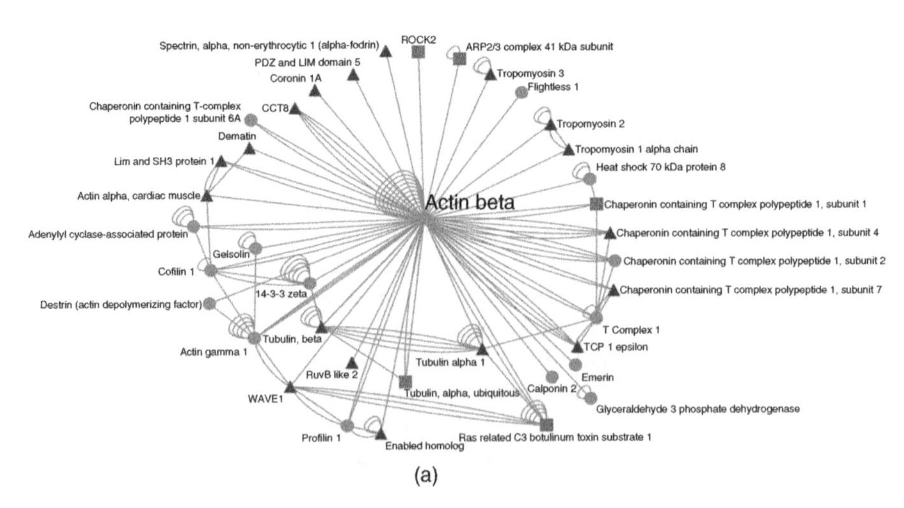


(a)

**Figure 8.3**

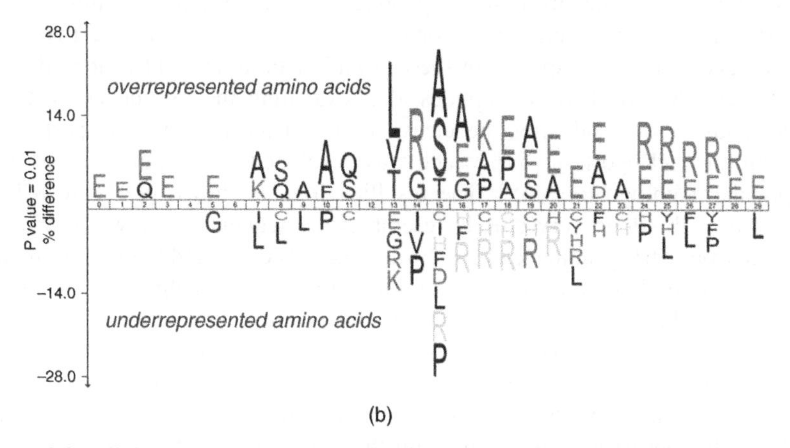


(b)

**Figure 8.3** Higher proteome coverage by the new *N*-terminal COFRADIC protocol. (a) Known interactions between all proteins identified in the old and new *N*-terminal COFRADIC analysis were obtained from the HPRD database and visualized with Cytoscape. The interactors of both Grb2 and β-actin are shown. Triangular nodes and square nodes indicate proteins that were identified in the new or the old analysis, respectively, while circular nodes indicate proteins that were found in both analyses. Grb2 and most of its interactors were uniquely found in the new analysis. Several of these proteins were identified only by the presence of an internal neo-*N*-terminal peptide generated by calpain-1 cleavage (see text). (b) Sequence logo generated by our *in-house*-developed iceLogo tool. Butyrated peptides with start position >100 were used as a positive set and extended with the protein sequence 15 amino acids before and after their start position. The amino acid composition of the human UniProtKB/Swiss-Prot database was used as a baseline set to correct for natural amino acid composition. A random distribution was calculated using a Monte Carlo approach for every amino acid, based on the prevalence of this amino acid in the UniProtKB/Swiss-Prot database and the sample size of the experimental dataset. In this way, a statistical significance can be calculated for every amino acid in the positive set on every position. An amino acid is visualized on the logo when it is significantly ($P$ value $= 0.01$) over- or underrepresented in the positive set compared to UniProtKB/Swiss-Prot. The height of an amino acid at a given position reflects the percentage difference between its occurrence in the experimental set and the UniProtKB/Swiss-Prot composition. The amino acids on the upper part of the logo are the overrepresented amino acids in the calpain-1 substrates, and amino acids on the lower part are underrepresented. Light-gray-colored amino acids indicate that these are completely absent on that position in the experimental dataset. This profile closely resembles the specificity profile that we previously obtained for human calpain-1, with cleavage occurring between positions 14 and 15 in the sequence logo. Our data identify position 13 (P2 position) and 15 (P1′ position) as the most important positions in determining the cleavage site specificity of the protease. Note that from position 15 onward, histidine residues are almost absent because *N*-terminal peptides containing these residues are retained on the SCX column.

present. These peptides are believed to be generated by an endoproteolytic activity rather than resulting from signal peptide removal. Apparently, this activity was present in both experimental conditions given the presence of two clear isotopomers in the mass spectra of these peptides. In an attempt to identify this activity, we aligned all these peptides after extending their sequences with 15 amino acids before and after the cleavage site. This analysis was performed by using our iceLogo tool [54], which was developed as an improvement over the popular Weblogo tool [55]. The resulting logo is shown in Figure 8.3(b). Interestingly, this logo very closely resembles the sequence logo we previously obtained from 561 genuine calpain-1 sites identified in A549 lung carcinoma cells [25]. Note also the previously reported preference of calpain-1 for leucine two positions before the actual cleavage site [position 13 in Fig. 8.3(b)] [56]. The high degree of similarity of this sequence logo with the one of human calpain-1 strongly indicates that this enzyme is responsible for the majority of endoproteolytic activity observed in the platelet lysate on incubation for 20 min at 37°C. This observation is in line with the important role for this protease in platelet function [21]. The fact that this activity was also present in the control setup indicates that the concentration of EDTA (1 mM) used to inhibit endogenous calpain activity in the lysate was too low, although we successfully used this concentration to inhibit calpains in A549 cell lysates. This also implies that the 58 cleavage sites that were uniquely found when exogenous purified calpain was added are most probably kinetically nonfavored substrates of the protease.

## 8.5 CONCLUDING REMARKS

In conclusion, our COFRADIC data on the *N*-terminal proteome of human blood platelets suggest that platelet studies [57] should be undertaken with care since unwanted activation of calpains might strongly interfere with such experiments. Nevertheless, such activation in our hands resulted in a deeper proteome coverage on analysis of the platelet proteome by *N*-terminal COFRADIC. Therefore, we used these data in this chapter to discuss the improvements that were made to the *N*-terminal protocol since the publication of our original work on platelets in 2005 [19].

## ACKNOWLEDGMENTS

FI and KH are Postdoctoral Research Fellows of the Fund for Scientific Research—Flanders (Belgium). The authors further acknowledge support by research grants from the Fund for Scientific Research—Flanders (Belgium) (Projects G.0077.06 and G.0042.07), the Concerted Research Actions (Project BOF07/GOA/012) from Ghent University, and the Inter University Attraction Poles (IUAP06). LM would like to thank Rolf Apweiler and Henning Hermjakob for their support.

## REFERENCES


1. de Godoy LM, Olsen JV, Cox J, Nielsen ML, Hubner NC, Frohlich F, Walther TC, Mann M, Comprehensive mass-spectrometry-based proteome quantification of haploid versus diploid yeast. *Nature* **455**:1251–1254 (2008).
2. Anderson NL, Anderson NG, The human plasma proteome: history, character, and diagnostic prospects, *Mol. Cell. Proteom.* **1**:845–867 (2002).
3. O'Farrell PH, High resolution two-dimensional electrophoresis of proteins, *J. Biol. Chem.* **250**:4007–4021 (1975).
4. O'Farrell PZ, Goodman HM, O'Farrell PH, High resolution two-dimensional electrophoresis of basic as well as acidic proteins, *Cell* **12**:1133–1141 (1977).
5. Bjellqvist B, Ek K, Righetti PG, Gianazza E, Gorg A, Westermeier R, Postel W, Isoelectric focusing in immobilized pH gradients: Principle, methodology and some applications, *J. Biochem. Biophys. Meth.* **6**:317–339 (1982).
6. Unlu M, Morgan ME, Minden JS, Difference gel electrophoresis: A single gel method for detecting changes in protein extracts, *Electrophoresis* **18**:2071–2077 (1997).
7. Santoni V, Molloy M, Rabilloud T, Membrane proteins and proteomics: Un amour impossible? *Electrophoresis* **21**:1054–1070 (2000).
8. Gygi SP, Corthals GL, Zhang Y, Rochon Y, Aebersold R, Evaluation of two-dimensional gel electrophoresis-based proteome analysis technology, *Proc. Natl. Acad. Sci. USA* **97**:9390–9395 (2000).
9. Kapp E, Schutz F. Overview of tandem mass spectrometry (MS/MS) database search algorithms, *Curr. Protoc. Protein Sci.* **25**(25.2) (Chapter 25, Unit 25.2) (2007).
10. Washburn MP, Wolters D, Yates JR 3rd, Large-scale analysis of the yeast proteome by multidimensional protein identification technology, *Nat. Biotechnol.* **19**:242–247 (2001).
11. Gygi SP, Rist B, Gerber SA, Turecek F, Gelb MH, Aebersold R, Quantitative analysis of complex protein mixtures using isotope-coded affinity tags, *Nat. Biotechnol.* **17**:994–999 (1999).
12. Liu H, Sadygov RG, Yates JR 3rd, A model for random sampling and estimation of relative protein abundance in shotgun proteomics, *Anal. Chem.* **76**:4193–4201 (2004).
13. Gevaert K, Van Damme P, Ghesquière B, Impens F, Martens L, Helsens K, Vandekerckhove J, A la carte proteomics with an emphasis on gel-free techniques, *Proteomics* **7**:2698–2718 (2007).
14. Brown JR, Hartley BS, Location of disulphide bridges by diagonal paper electrophoresis. The disulphide bridges of bovine chymotrypsinogen A, *Biochem. J.* **101**:214–228 (1966).
15. Cruickshank WH, Malchy BL, Kaplan H, Diagonal chromatography for the selective purification of tyrosyl peptides, *Can. J. Biochem.* **52**:1013–1017 (1974).
16. Gevaert K, Van Damme J, Goethals M, Thomas GR, Hoorelbeke B, Demol H, Martens L, Puype M, Staes A, Vandekerckhove J, Chromatographic isolation of methionine-containing peptides for gel-free proteome analysis: Identification of more than 800 Escherichia coli proteins, *Mol. Cell. Proteom.* **1**:896–903 (2002).
17. Gevaert K, Goethals M, Martens L, Van Damme J, Staes A, Thomas GR, Vandekerckhove J, Exploring proteomes and analyzing protein processing by mass spectrometric identification of sorted *N*-terminal peptides, *Nat. Biotechnol.* **21**:566–569 (2003).

18. Gevaert K, Ghesquiere B, Staes A, Martens L, Van Damme J, Thomas GR, Vandekerckhove J, Reversible labeling of cysteine-containing peptides allows their specific chromatographic isolation for non-gel proteome studies, *Proteomics* **4**:897–908 (2004).
19. Martens L, Van Damme P, Van Damme J, Staes A, Timmerman E, Ghesquière B, Thomas GR, Vandekerckhove J, Gevaert K, The human platelet proteome mapped by peptide-centric proteomics: A functional protein profile, *Proteomics* **5**:3193–3204 (2005).
20. Van Damme P, Martens L, Van Damme J, Hugelier K, Staes A, Vandekerckhove J, Gevaert K, Caspase-specific and nonspecific in vivo protein processing during Fas-induced apoptosis, *Nat. Meth*. **2**:771–777 (2005).
21. Kuchay SM, Chishti AH, Calpain-mediated regulation of platelet signaling pathways, *Curr. Opin. Hematol*. **14**:249–254 (2007).
22. Staes A, Van Damme P, Helsens K, Demol H, Vandekerckhove J, Gevaert K, Improved recovery of proteome-informative, protein *N*-terminal peptides by combined fractional diagonal chromatography (COFRADIC), *Proteomics* **8**:1362–1370 (2008).
23. Ji J, Chakraborty A, Geng M, Zhang X, Amini A, Bina M, Regnier F, Strategy for qualitative and quantitative analysis in proteomics based on signature peptides, *J. Chromatogr. B Biomed. Sci. Appl*. **745**:197–210 (2000).
24. Arnesen T, Van Damme P, Polevoda B, Helsens K, Evjenth R, Colaert N, Varhaug JE, Vandekerckhove J, Lillehaug JR, Sherman F, Gevaert K. Proteomics analyses reveal the evolutionary conservation and divergence of *N*-terminal acetyltransferases from yeast and humans, *Proc Natl Acad Sci USA* **106**:8157–8162 (2009).
25. Impens F, Van Damme P, Demol H, Van Damme J, Vandekerckhove J, Gevaert K, Mechanistic insight into taxol-induced cell death, *Oncogene* **27**:4580–4591 (2008).
26. Kaiserman D, Buckle AM, Van Damme P, Irving JA, Law RH, Matthews AY, Bashtannyk-Puhalovich T, Langendorf C, Thompson P, Vandekerckhove J, Gevaert K, Whisstock JC, Bird PI, Structure of granzyme C reveals an unusual mechanism of protease autoinhibition, *Proc. Natl. Acad. Sci. USA* **106**:5587–5592 (2009).
27. Lamkanfi M, Kanneganti TD, Van Damme P, Vanden Berghe T, Vanoverberghe I, Vandekerckhove J, Vandenabeele P, Gevaert K, Nunez G, Targeted peptidecentric proteomics reveals caspase-7 as a substrate of the caspase-1 inflammasomes, *Mol. Cell. Proteom*. **7**:2350–2363 (2008).
28. Van Damme P, Maurer-Stroh S, Plasman K, Van Durme J, Colaert N, Timmerman E, De Bock PJ, Goethals M, Rousseau F, Schymkowitz J, Vandekerckhove J, Gevaert K, Analysis of protein processing by *N*-terminal proteomics reveals novel species-specific substrate determinants of granzyme B orthologs, *Mol. Cell. Proteom*. **8**:258–272 (2009).
29. Staes A, Demol H, Van Damme J, Martens L, Vandekerckhove J, Gevaert K, Global differential non-gel proteomics by quantitative and stable labeling of tryptic peptides with oxygen-18, *J. Proteome Res*. **3**:786–791 (2004).
30. Ong SE, Blagoev B, Kratchmarova I, Kristensen DB, Steen H, Pandey A, Mann M, Stable isotope labeling by amino acids in cell culture, SILAC, as a simple and accurate approach to expression proteomics, *Mol. Cell. Proteom*. **1**:376–386 (2002).

31. Miller BT, Kurosky A, Elevated intrinsic reactivity of seryl hydroxyl groups within the linear peptide triads His-Xaa-Ser or Ser-Xaa-His, *Biochem. Biophys. Res. Commun.* **196**:461–467 (1993).

32. Miller BT, Rogers ME, Smith JS, Kurosky A, Identification and characterization of O-biotinylated hydroxy amino acid residues in peptides, *Anal. Biochem.* **219**:240–248 (1994).

33. Abello N, Kerstjens HA, Postma DS, Bischoff R, Selective acylation of primary amines in peptides and proteins, *J. Proteome Res.* **6**:4770–4776 (2007).

34. Kawasaki H, Imajoh S, Suzuki K, Separation of peptides on the basis of the difference in positive charge: Simultaneous isolation of *C*-terminal and blocked *N*-terminal peptides from tryptic digests, *J. Biochem.* (Tokyo) **102**:393–400 (1987).

35. Crimmins DL, Gorka J, Thoma RS, Schwartz BD, Peptide characterization with a sulfoethyl aspartamide column, *J. Chromatogr.* **443**:63–71 (1988).

36. Dormeyer W, Mohammed S, Breukelen B, Krijgsveld J, Heck AJ, Targeted analysis of protein termini, *J. Proteome Res.* **6**:4634–4645 (2007).

37. Taouatas N, Altelaar AF, Drugan MM, Helbig AO, Mohammed S, Heck AJ, SCX-based fractionation of Lys-N generated peptides facilitates the targeted analysis of post-translational modifications, *Mol. Cell. Proteom.* **8**:190–200 (2009).

38. Aivaliotis M, Gevaert K, Falb M, Tebbe A, Konstantinidis K, Bisle B, Klein C, Martens L, Staes A, Timmerman E, Damme JV, Siedler F, Pfeiffer F, Vandekerckhove J, Oesterhelt D, Large-scale identification of *N*-terminal peptides in the halophilic Archaea halobacterium salinarum and Natronomonas pharaonis, *J. Proteome Res.* **6**:2195–2204 (2007).

39. Chung JS, Webster SG, Does the *N*-terminal pyroglutamate residue have any physiological significance for crab hyperglycemic neuropeptides? *Eur. J. Biochem.* **240**:358–364 (1996).

40. Abraham GN, Podell DN, Pyroglutamic acid. Non-metabolic formation, function in proteins and peptides, and characteristics of the enzymes effecting its removal, *Mol. Cell. Biochem.* **38**(Spec. No.):181–190 (1981).

41. Lamb JE, Goldstein IJ, A structural comparison of the A and B subunits of Griffonia simplicifolia I isolectins, *Arch. Biochem. Biophys.* **229**:15–26 (1984).

42. Mozdzanowski J, Bongers J, Anumula K, High-yield deblocking of amino termini of recombinant immunoglobulins with pyroglutamate aminopeptidase, *Anal. Biochem.* **260**:183–187 (1998).

43. Dando PM, Fortunato M, Strand GB, Smith TS, Barrett AJ, Pyroglutamyl-peptidase I: Cloning, sequencing, and characterisation of the recombinant human enzyme, *Protein Exp. Purif.* **28**:111–119 (2003).

44. Froelich JM, Reid GE, The origin and control of *ex vivo* oxidative peptide modifications prior to mass spectrometry analysis, *Proteomics* **8**:1334–1345 (2008).

45. Fields S, Song O, A novel genetic system to detect protein-protein interactions, *Nature* **340**:245–246 (1989).

46. Rigaut G, Shevchenko A, Rutz B, Wilm M, Mann M, Seraphin B, A generic protein purification method for protein complex characterization and proteome exploration, *Nat. Biotechnol.* **17**:1030–1032 (1999).

47. Puig O, Caspary F, Rigaut G, Rutz B, Bouveret E, Bragado-Nilsson E, Wilm M, Seraphin B, The tandem affinity purification (TAP) method: A general procedure of protein complex purification, *Methods* **24**:218–229 (2001).

48. Peri S, Navarro JD, Amanchy R, Kristiansen TZ, Jonnalagadda CK, Surendranath V, Niranjan V, Muthusamy B, Gandhi TK, Gronborg M, Ibarrola N, Deshpande N, Shanker K, Shivashankar HN, Rashmi BP, Ramya MA, Zhao Z, Chandrika KN, Padma N, Harsha HC, Yatish AJ, Kavitha MP, Menezes M, Choudhury DR, Suresh S, Ghosh N, Saravana R, Chandran S, Krishna S, Joy M, Anand SK, Madavan V, Joseph A, Wong GW, Schiemann WP, Constantinescu SN, Huang L, Khosravi-Far R, Steen H, Tewari M, Ghaffari S, Blobe GC, Dang CV, García JG, Pevsner J, Jensen ON, Roepstorff P, Deshpande KS, Chinnaiyan AM, Hamosh A, Chakravarti A, Pandey A, Development of human protein reference database as an initial platform for approaching systems biology in humans, *Genome Res*. **13**:2363–2371 (2003).

49. Zanzoni A, Montecchi-Palazzi L, Quondam M, Ausiello G, Helmer-Citterich M, Cesareni G, MINT: A Molecular INTeraction database, *FEBS Lett*. **513**:135–140 (2002).

50. Alfarano C, Andrade CE, Anthony K, Bahroos N, Bajec M, Bantoft K, Betel D, Bobechko B, Boutilier K, Burgess E, Buzadzija K, Cavero R, D'Abreo C, Donaldson I, Dorairajoo D, Dumontier MJ, Dumontier MR, Earles V, Farrall R, Feldman H, Garderman E, Gong Y, Gonzaga R, Grytsan V, Gryz E, Gu V, Haldorsen E, Halupa A, Haw R, Hrvojic A, Hurrell L, Isserlin R, Jack F, Juma F, Khan A, Kon T, Konopinsky S, Le V, Lee E, Ling S, Magidin M, Moniakis J, Montojo J, Moore S, Muskat B, Ng I, Paraiso JP, Parker B, Pintilie G, Pirone R, Salama JJ, Sgro S, Shan T, Shu Y, Siew J, Skinner D, Snyder K, Stasiuk R, Strumpf D, Tuekam B, Tao S, Wang Z, White M, Willis R, Wolting C, Wong S, Wrong A, Xin C, Yao R, Yates B, Zhang S, Zheng K, Pawson T, Ouellette BF, Hogue CW, The Biomolecular Interaction Network Database and related tools 2005 update, *Nucleic Acids Res*. **33**:D418–D424 (2005).

51. Hermjakob H, Montecchi-Palazzi L, Lewington C, Mudali S, Kerrien S, Orchard S, Vingron M, Roechert B, Roepstorff P, Valencia A, Margalit H, Armstrong J, Bairoch A, Cesareni G, Sherman D, Apweiler R, IntAct: An open source molecular interaction database, *Nucleic Acids Res*. **32**:D452–D455 (2004).

52. Mathivanan S, Periaswamy B, Gandhi TK, Kandasamy K, Suresh S, Mohmood R, Ramachandra YL, Pandey A, An evaluation of human protein-protein interaction data in the public domain, *BMC Bioinform*. **7**(Suppl. 5): S19.

53. Shannon P, Markiel A, Ozier O, Baliga NS, Wang JT, Ramage D, Amin N, Schwikowski B, Ideker T, Cytoscape: A software environment for integrated models of biomolecular interaction networks, *Genome Res*. **13**:2498–2504 (2003).

54. Colaert N, Helsens K, Martens L, Vandekerckhove J, Gevaert K, Improved visualization of protein consensus sequences by iceLogo, *Nat Meth*. **6**:786–787 (2009).

55. Crooks GE, Hon G, Chandonia JM, Brenner SE, WebLogo: A sequence logo generator, *Genome Res*. **14**:1188–1190 (2004).

56. Cuerrier D, Moldoveanu T, Davies PL, Determination of peptide substrate specificity for mu-calpain by a peptide library-based approach: The importance of primed side interactions, *J. Biol. Chem*. **280**:40632–40641 (2005).

57. Lopez-Otin C, Overall CM, Protease degradomics: A new challenge for proteomics, *Nat. Rev. Mol. Cell. Biol*. **3**:509–519 (2002).

# PART III

## INTEGRATED "OMICS" AND APPLICATION TO DISEASE

# 9

# SERIAL ANALYSIS OF GENE EXPRESSION (SAGE) FOR STUDYING THE PLATELET AND MEGAKARYOCYTE TRANSCRIPTOME

MICHAEL G. TOMLINSON

**Abstract**

Serial analysis of gene expression (SAGE) is a method for large-scale transcriptome analysis. The first aim of this chapter is to explain the SAGE methodology and its improvements since its discovery in 1995. The second is to describe the rather meager four papers that have used SAGE to analyze the transcriptomes of platelets and megakaryocytes. Finally, the question will be addressed as to whether SAGE has now been outdated by twenty-first-century technology, namely, by the juggernaut that is next-generation sequencing.

## 9.1 INTRODUCTION

*Serial analysis of gene expression* (SAGE) is a method for large-scale transcriptome analysis. SAGE was so ingeniously conceived by Velculescu and friends [1], and yet most readers of this book will not be able to explain precisely how SAGE works. With this in mind, the objective of this chapter is threefold: (1) to explain

*Platelet Proteomics: Principles, Analysis and Applications*, First Edition.
Edited by Ángel García and Yotis A. Senis.

the SAGE methodology and its improvements in the years since its discovery (in 1995), (2) to describe the rather meager four papers that have used SAGE to analyze the transcriptomes of platelets and megakaryocytes (Table 9.1), and (3) finally, to determine whether SAGE has now been outdated by twenty-first-century technology, namely, by the juggernaut that is next-generation sequencing.

The full complement of all expressed genes in a given cell type is referred to as the *transcriptome*. Characterization of the transcriptome can be regarded as a fundamental first step toward the holy grail of understanding the full complexity of cellular function. Since 1989, the field of genomics has evolved to address this issue using several high-throughput gene expression profiling technologies that measure levels of messenger RNA (mRNA) [2,3]. However, given the remarkable advances in proteomic technologies, outlined in other chapters of this book, and given the fact that it is the proteins that ultimately drive cellular function, why now measure mRNA at all? The answer lies in the fact that genomics is still more sensitive and more high-throughput than proteomics. Moreover, mRNA levels provide key knowledge as to the activity of genes and the resulting state of the cell, and are generally related to the levels of proteins.

The first and simplest method in the genomic era was *expressed sequence tag* (EST) sequencing [4,5]. This relied on the reverse transcription of mRNA into complementary DNA (cDNA), the cloning of cDNAs into suitable vectors, and subsequent sequencing of typically 300–500 base pairs using vector primers. Such sequence enabled the discovery of new genes and facilitated the human genome sequencing project by identifying coding regions in the genome [4,5]. However, the high cost of sequencing large numbers of ESTs made this method unsuitable for full characterization of cellular transcriptomes.

The development of DNA microarrays helped to overcome this problem [6,7]. Microarray technology relies on the property of single-stranded DNA to hybridize to complementary sequences. Defined target DNA, in the form of synthesized oligonucleotides [20–75 base pairs (bp) long] [6] or longer cDNA clones (500–2000 bp) [7], are typically printed onto glass slides, and these DNA microarrays are probed with a fluorescence-labeled pool of cDNA from the desired cell or tissue type. The fluorescent signal is then measured to gain a measure of relative mRNA abundance. However, DNA microarrays are not fully quantitative, since the signal is partially dependent on hybridization efficiencies between the printed target and fluorescence-labelled cDNA, and weakly expressed genes can fall below the sensitivity of detection. Moreover, DNA microarrays cannot be used to identify new genes, since they require a priori knowledge of each sequence that is printed onto the microarray. Indeed, DNA microarrays are more ideally suited to comparing expression profiles between different transcriptomes, rather than quantitatively defining a given transcriptome.

The SAGE method overcame the limitations of EST sequencing and DNA microarrays for transcriptome characterisation [1]. SAGE is essentially a shortcut method of performing EST sequencing that relies on two principles: (1) a short fragment of DNA (the SAGE tag) can be isolated from a defined location

**TABLE 9.1 Published SAGE Analyses of Platelet and Megakaryocyte Transcriptomes**

| Study | SAGE Method | mRNA Source | Number of Genes Identified | Conclusions |
|---|---|---|---|---|
| Kim et al. [10] | MicroSAGE | Human meg ($CD41^+$) and nonmeg ($CD41^-$) fractions from cord blood $CD34^+$ cells expanded *ex vivo* with TPO for 10 days | 20,580 tags from meg, 18,329 tags from nonmeg, representing a total of 8976 genes | These data will be useful for the identification of genes involved in the process of megakaryocytopoiesis |
| Gnatenko et al. [9] | LongSAGE | Human platelets collected by apheresis, gel-filtered, filtered through a 5-μm filter, depleted of $CD45^+$ cells | 2033 total tags, including 233 nonmitochondrial tags representing 126 genes | (1) 89% of tags represented mitochondrial transcripts; (2) SAGE data correlated well with DNA microarray performed in the same study |
| Dittrich et al. [8] | SAGE | Human platelets, collected by apheresis (estimated contamination of one leukocyte per one million platelets) | 25,000 total tags, including 12,609 nonmitochondrial tags representing 2300 different transcripts | (1) 49% of tags represented mitochondrial transcripts; (2) nonmitochondrial transcripts were enriched for "catalytic" and "signal transducer" genes; (3) the most abundant transcripts had relatively long 3′ UTRs and were predicted to be relatively stable |
| Senis et al. [11] | LongSAGE | Mouse megakaryocytes grown *in vitro* from bone marrow progenitors cultured in stem cell factor and thrombopoietin | 53,046 total tags, representing a total of 8000 genes | (1) SAGE appeared more sensitive than proteomics in identifying all of the main classes of platelet/meg transmembrane proteins; (2) On comparison of the SAGE library to other nonmeg libraries, 17 of the 25 most meg-specific genes encoded for transmembrane proteins |

on a mRNA transcript and the sequence used to identify the full transcript, and (2) multiple SAGE tags can be ligated together such that a single sequencing reaction can identify 30 or more tags. Importantly, SAGE is quantitative, and therefore different SAGE libraries, unlike most microarray data, can be directly compared [2,3].

This chapter will introduce SAGE technology, from the initial method through to more recent improvements, and will describe the findings from four publications that have used SAGE to characterize the transcriptome of platelets and their megakaryocyte precursors [8–11]. Finally, the impact of very recent advances in next-generation DNA sequencing on SAGE and platelet/megakaryocyte transcriptome analysis will be discussed.

## 9.2 A SAGE DISCOVERY

In their 1995 paper in the journal *Science* [1], Velculescu and colleagues introduced SAGE as "a technique that allows a rapid, detailed analysis of thousands of transcripts." As outlined above, the first principle of SAGE was that a short sequence from a defined position on a mRNA transcript could be used to accurately identify the parent transcript. Indeed, the authors calculated that even a sequence as short as 9 bp carried enough information to distinguish 262, 144 ($4^9$) different transcripts, assuming random nucleotide usage over this sequence. This number is greater than their estimates of $\leq$140,000 distinct mRNA sequences in the human transcriptome [1]. The second principle of SAGE was that such short sequences, termed *SAGE tags*, could be ligated together to form concatamers to allow the efficient identification of multiple SAGE tags from a single DNA sequencing run. The authors drew the analogy with computers, which serially communicate by transmitting continuous strings of data [1].

Velculescu et al. demonstrated the efficacy of SAGE by the identification of 1000 9-bp tags from pancreas, revealing an expression profile characteristic of pancreatic function together with novel tags representing new transcripts [1]. This was done, as shown in Figure 9.1, by first isolating mRNA and synthesizing double-stranded cDNA using an oligo(dT) primer that included a biotin tag. An "anchoring enzyme," in the form of a restriction endonuclease with a 4-bp recognition sequence (*Nla*III), was then used to cleave the DNA. In theory, such an enzyme should cut every 256 ($4^4$) bp and should therefore cut the vast majority of cDNAs. The most 3′ region of cDNA, containing the polyA tail and biotinylated oligo(dT) primer, was subsequently captured using streptavidin beads. The sample was then divided in half and one of two linker sequences ligated to each sequence via the anchoring enzyme restriction site. Perhaps the most ingenious aspect of this approach was the inclusion of a type IIS restriction site in each linker. Such restriction endonucleases, unlike most others, do not cut their recognition site, but instead cleave at a defined position up to 26 bp away from their recognition site. Therefore the use of this so-called tagging enzyme (*Bsm*FI) results in the release of the linker with a short region of the cDNA,

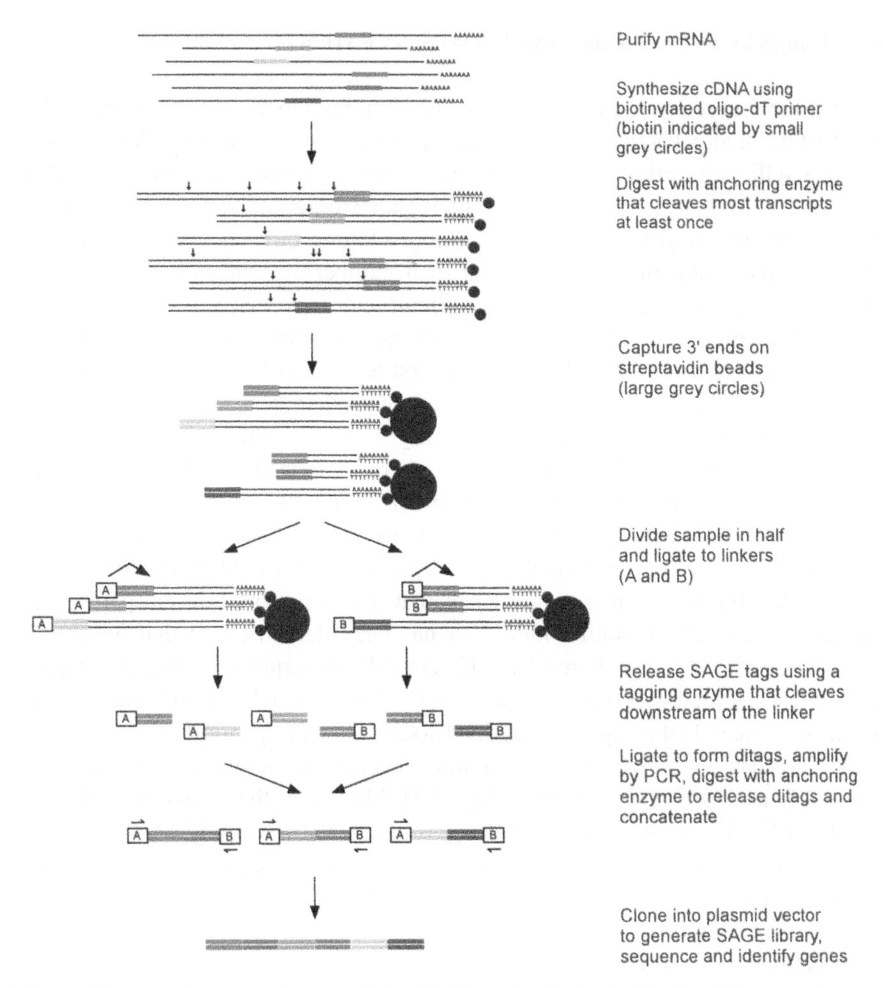


**Figure 9.1** A schematic diagram of the SAGE method. A specific SAGE tag is released from each cDNA and ligated with another SAGE tag to form a ditag. The ditags are PCR-amplified and ligated together to form concatamers. These are sequenced to allow identification of the genes from which the SAGE tags were derived.

which in this study was composed of the anchoring enzyme site and an additional 9 bp. The two pools of released tags of 13 bp were then ligated together, forming ditags, and polymerase chain reaction (PCR) primers specific for each linker region were used to amplify the ditags. These ditags were then released from the linker regions by digestion with the original anchoring enzyme and then ligated together to form concatamers. Finally these concatamers were cloned, to generate a so-called SAGE library, and sequenced (Fig. 9.1). While this method initially appears quite complex, the authors correctly point out that any laboratory with the capacity to perform PCR, and with access to a DNA sequencing facility, has the capacity to carry out SAGE [1].

## 9.3 LongSAGE AND SuperSAGE IMPROVEMENTS

Despite the theoretical capacity of molecular biology laboratories to successfully use SAGE, many encountered technical problems in the years following the initial SAGE publication. These were due to lack of sufficient purity of mRNA and linkers, loss of activity during shipping or storage of the most commonly used anchoring enzyme (*Nla*III), and a poor cloning efficiency in terms of the ratio of clones containing inserts to the total number of clones [2]. This lead to the popular opinion that SAGE was a difficult technique that was rather risky to attempt. However, while this perception still exists to some extent, these problems were solved largely by the release of the first SAGE kit by Invitrogen, namely, the I-SAGE kit. In the author's experience, any competent molecular biologist can use such kits to generate a SAGE library within 2 weeks.

Despite this technical advance, successfully generated SAGE libraries still suffered from several unanticipated problems. First, SAGE tags, such as the commonly used 14-bp tags, showed specificity lower than had been originally predicted. This was because such predictions assumed a random distribution of bases within mRNAs, but it is now clear that this is not true, due in part to the presence of conserved motifs. Indeed, it has been demonstrated that over 30% of 14-bp SAGE tags are shared by different mRNA sequences, thus decreasing the reliability of tag-to-gene annotation [2,3]. A second, related problem was the inability to map 14-bp tags to a defined location in the genome. For example, such a tag has an average of 27 genomic matches, rendering such SAGE tags essentially useless for genome mapping [2,3]. Moreover, the relatively high sensitivity of SAGE identified a relatively high number of novel mRNAs. However, it was very difficult to study these potentially exciting gene products any further, because a knowledge of only 14 bp was rarely sufficient to allow the parent mRNA to be identified experimentally [2,3].

To address these issues, two improved SAGE protocols were developed, named LongSAGE [12] and SuperSAGE [13], which generated longer SAGE tags of 21 and 26 bp, respectively. The LongSAGE method was published in 2002 in the journal *Nature Biotechnology* [12] by the same group that had reported the original SAGE technology in 1995 (Fig. 9.2). LongSAGE uses an alternative type IIS tagging restriction enzyme, *Mme*I, which cuts 17 bp downstream of its 4-bp recognition site, thus yielding a 21-bp SAGE tag. The authors calculated that approximately 99.8% of such tags would be expected to occur only once in the human genome [12]. However, experimentally this number was lower; of over 8000 distinct tags generated from a LongSAGE library of about 28,000 tags from a colorectal cancer cell line, only two-thirds mapped to a unique location in the genome. The remaining third mapped to gene duplications, common domains, or repeat sequences within different genes [12]. Nevertheless, LongSAGE was clearly a dramatic improvement on the original SAGE method and was effective in identifying novel genes and exons. In addition, LongSAGE would, by definition, more reliably identify known genes. Indeed, in the author's experience of the Invitrogen I-SAGE Long kit, mapping of a LongSAGE tag to more than one gene is quite rare.

SuperSAGE, published by Matsumura et al. in 2003 in the journal *Proceedings of the National Academy of Sciences* (*USA*) (Fig. 9.2) [13], represented a further improvement on SAGE and LongSAGE, through the generation of a 26-bp tag using the tagging enzyme *Eco*P151. This allows almost perfect gene to tag annotation, which has allowed the authors to develop two novel applications for SuperSAGE: analysis of the interaction transcriptome and SuperSAGE arrays [14].

The term *interaction transcriptome* refers to the transcriptomes of two or more species that are interacting. An example is the infection of rice leaves by the fungus *Magnaporthe grisea*. Since SuperSAGE tags are long enough to essentially map to just one gene among sequences from all species within the GenBank DNA database, Matsumura et al. successfully used SuperSAGE to simultaneously analyze the rice and *M. grisea* transcriptomes in this model system [13]. It is currently impossible to perform such interaction transcriptomics in any other way, since the mRNA from the two species cannot be separately isolated.

A second utility of SuperSAGE was in the development of SuperSAGE arrays, which combine the power of SAGE to identify novel genes with the advantage of DNA microarrays in simultaneously analyzing multiple mRNA samples. To generate the array, SuperSAGE is first performed. The resulting tags are subsequently used to design the microarray by directly synthesising the 26-bp tags on glass slides. Matsumura and colleagues proved the utility of this approach by generating a 1000-tag microarray from SuperSAGE tags derived from rice leaves and cultured cells [15]. Of those tags found to be significantly differentially expressed by SuperSAGE, 88% were similarly significant on the SuperSAGE array. This technique is of particular use for the study of nonmodel organisms, for which the availability of transcriptome data is currently insufficient to allow microarray design [14]. In addition, SuperSAGE arrays could be useful for the study of cell types for which the transcriptomes are not as well defined as in other cells, such as the platelet or megakaryocyte.

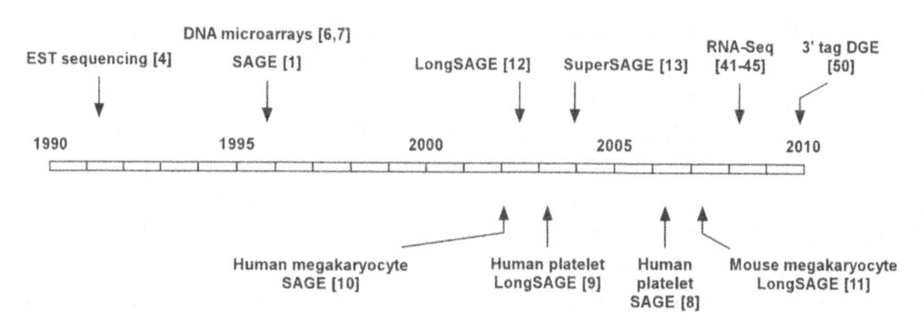


**Figure 9.2** A brief history of transcriptomics. Major advances in transcriptome analysis are presented above the timeline, and the four platelet/megakaryocyte SAGE publications are shown below.

## 9.4 PLATELET SAGE

Since platelets have no nucleus, it was widely assumed that platelets had negligible mRNA levels. This view changed in 1988 when Newman et al. used reverse transcriptase (RT)-PCR to demonstrate the existence of a number of platelet-specific mRNA species [16]. However, a more recent study, using RT-PCR following laser-assisted microdissection to obtain a pure population of platelets, has estimated that a platelet contains only 0.002 fg (Femtogram) of mRNA, as compared to the 12,500-fold more mRNA found in a typical nucleated cell [17]. This highlights the technical difficulty in studying the platelet transcriptome, since even an extremely minor leukocyte contamination of a platelet preparation, bordering on the undetectable, can result in substantial contamination with leukocyte mRNA. These issues are likely to be responsible for the paucity of platelet SAGE analyses versus the relatively large number in other haematopoietic cell types [18].

The first of the two published platelet SAGE libraries (Table 9.1) was generated by Gnatenko et al. [9], who have more recently published a thorough review on their methodology [19]. With the contamination issue in mind, the authors carefully obtained a highly purified population of platelets by first passing gel-filtered platelets through a 5-$\mu$m filter and then depleting cells expressing the leukocyte marker CD45 (not expressed on platelets) using magnetic beads [9]. This method appeared successful since CD45 mRNA was not present in the resulting platelet sample, as detected by RT-PCR. The mRNA was used to make a LongSAGE library from which a total of 2033 tags were sequenced [9]. This is a relatively small number of tags, since most SAGE libraries contain at least 10 times this number, but the decision to sequence no further tags was probably influenced by the fact that 1800 tags (89% of the library) were from mitochondrial genes [9]. The mitochondrial genome encodes 13 genes and two ribosomal subunits, and in the absence of a nuclear genome, actively transcribed mitochondrial genes clearly dominate the platelet transcriptome in terms of mRNA abundance.

Of the remaining 233 nonmitochondrial tags sequenced by Gnatenko et al., 13 were identified by at least 2 tags [9]. The most highly expressed mRNA was $\beta_2$-microglobulin (26 tags), a component of MHC class 1 that is expressed on most cell types, followed by thymosin-$\beta_4$ (21 tags), CXCL4 (16 tags combined from two variants), CXCL7 (7 tags), and GPIb$\beta$ (5 tags) [9]. These latter four are each bone fide platelet proteins. Thymosin-$\beta_4$ is a 5-kDa protein, highly expressed in platelets, that sequesters monomeric (G-)actin, thus providing a pool for the polymerization of filamentous (F-)actin when required, for example, during cytoskeletal reorganization on platelet activation [20]. CXCL4 (platelet factor 4) and CXCL7 (proplatelet basic protein) reside in $\alpha$-granules as the most highly expressed platelet chemokines [21]. CXCL4 induces endothelial cell activation and recruits monocytes, and appears to promote the progression of atherosclerosis, while CXCL7 plays a role in inflammation and potentially vascular repair, by recruiting neutrophils and endothelial progenitor cells, respectively [21]. GPIbβ, a major platelet surface glycoprotein, is a component of the GPIb-IX-V receptor complex that is essential for platelet rolling on von Willebrand factor (vWF), at sites of vessel injury, and which is defective in Bernard–Soulier patients [22]. In addition to these known, highly expressed platelet proteins, the protein kinase C substrate, neurogranin (3 tags), was identified as a novel intracellular platelet protein, a fact confirmed by Western blotting and intracellular flow cytometry [9]. The role of neurogranin in platelets remains an unexplored mystery, in part because it is still largely regarded as a neuron-specific protein, where it appears to play a key role in memory by enhancing the strength of synapses between neurons [23]. A final conclusion from this landmark study was that these nonmitochondrial SAGE data were consistent with a DNA microarray study that was presented in the same publication [9]. Thus SAGE is clearly a useful method for platelet transcriptome analysis, with the caveat that a library of 300,000 SAGE tags would be required to yield a library of nonmitochondrial tags of a typical size of 30,000 tags.

In 2006 a second platelet SAGE library was published by Dittrich et al. [8]. This library contained 25,000 tags and thus was more than 10-fold larger than that previously published. However, this newer study used SAGE rather than LongSAGE; as such, tags are more likely to be incorrectly assigned. Of the 25,000 tags, 51% corresponded to nonmitochondrial tags [8], which is substantially more than the 11% obtained in the earlier study [9]. The reason for this discrepancy is unclear. Contamination with leukocyte mRNA is a possibility, and certainly the authors did not utilize 5-μm filters or CD45 depletion, as used in the previous study to largely exclude such contamination. However, this possibility was discounted by the authors, since they used flow cytometry to demonstrate contamination of less than one leukocyte per one million platelets, and so estimated that leukocyte mRNA should represent only 0.1% of the platelet mRNA [8].

The 12,609 nonmitochondrial tags yielded 2300 distinct transcripts [8]. Highly expressed platelet transcripts were consistent with the previously published platelet LongSAGE library [9] and with a more recent platelet microarray [24]. Indeed, their list of the top 50 most highly expressed genes contained, in common with the previously reported LongSAGE library, $\beta_2$-microglobulin (288 tags), thymosin $\beta_4$ (273 tags), neurogranin (261 tags), GPIbβ (76 tags), CXCL4 (58 tags), and CXCL7 (29 tags) [8]. However, the two most highly expressed genes, Ankrd17 (395 tags) and ELL3 (344 tags), were not represented by 2 or more tags in the Gnatenko et al. study [9]. Possible reasons for this discrepancy are the potential for these to be incorrectly assigned, due to the shorter length of the SAGE tags identified in this study, or because there was insufficient sequence information in the public databases for these to be identified in the Gnatanko et al. study [9], which was published 3 years before that of Dittrich et al. [8]. The latter possibility is likely for Ankrd17 and ELL3, since the majority of sequence information for these genes were deposited after submission of Gnatenko et al. [9]. Ankrd17, that is, ankyrin repeat domain 17, is a ubiquitously expressed intracellular protein that contains ankyrin repeats. Interestingly, deletion of this gene in mice results in embryonic lethality due to hemorrhage, suggesting a possible role for this protein in platelet function [25]. ELL3, that is, elongation factor RNA polymerase II-like 3, is an RNA polymerase transcription factor that was thought to be testis-specific [26]. The role of ELL3 in platelets has not been studied, although a role in megakaryocyte progenitors is more likely, given its transcription factor function.

Detailed analyses of the mRNA species identified by the Dittrich et al. SAGE library revealed a number of interesting features [8]:

1. Platelet mRNAs were significantly enriched for functions in *catalytic activity* and *signal transducer activity*, but were underrepresented in the *structural molecule activity* category, when compared to typical nucleated cells.
2. The most abundant mRNAs had significantly longer 3′ untranslated regions with more stable predicted secondary structures than those in nucleated cells, which is perhaps not surprising given the lack of ongoing transcription in the anucleate platelet, resulting in persistence of only the most stable mRNA species.
3. Regulatory elements were identified in the untranslated regions of certain highly expressed mRNAs, suggesting that these mRNAs might be translated into protein in response to certain signals [8].

Consistent with observation 3 (above) more recent studies have now shown that platelets translate mRNA into protein, as reviewed by Weyrich et al. [27]. Indeed, platelets have the machinery to make protein, in the form of rough endoplasmic reticulum studded with ribosomes. The functionality of this machinery was demonstrated by metabolic labeling of platelets with radiolabeled carbohydrates or amino acids [28]. Weyrich and colleagues have since demonstrated that platelet activation signals induce the translation of certain proteins from mRNA [29] and that platelets retain the ability to splice pre-mRNA into mRNA for interleukin-$1_{\beta}$ [30]. The general importance of this phenomenon is not clear, since platelet activation is typically regarded as a terminal event. However, at least for B-cell lymphoma-3 (Bcl-3), a member of the Iκβα family of transcriptional regulators, a functional effect has been described. In particular, mice deficient in Bcl3 showed partially impaired platelet clot retraction [31], a process that can take several hours and that is important for thrombus remodeling and ultimately wound repair.

## 9.5 MEGAKARYOCYTE SAGE

Two groups have used SAGE to tackle the megakaryocyte transcriptome (Table 9.1) as a somewhat easier alternative to that of the platelet [10,11]. Megakaryocytes are highly transcriptionally active and would be predicted to possess a set of mRNA species similar to that of their platelet offspring. Moreover, more recent advances in the *in vitro* culture of megakaryocytes from cord blood, fetal liver, or bone marrow precursors have solved the problem of megakaryocytes being a relatively rare cell in the adult bone marrow. Indeed, sufficient numbers of megakaryocytes can now be produced to provide enough mRNA for SAGE analysis.

In 2002, the Kim laboratory was the first to probe the megakaryocyte transcriptome using SAGE [10]. Pooled human umbilical cord blood from four donors was used as a source of mononuclear CD34-positive progenitor cells, which were isolated using magnetic beads coated with a CD34 antibody. After 10 days' culture in the presence of thrombopoietin, megakaryocytes were separated from nonmegakaryocytes using magnetic beads coated with an antibody to integrin αIIb. Kim et al. then used a modification of the original SAGE protocol, termed *microSAGE*, which allows construction of a SAGE library from a relatively small amount of total RNA, in this case 10 μg [10]. The key to microSAGE is the minimisation of loss of material that occurs during the frequent transfer of material from tube to tube in the original SAGE protocol. This is achieved by generating the SAGE tags from RNA in a single PCR tube, which is coated in streptavidin to capture the biotinylated oligo-dT primer used in the initial cDNA synthesis step. In addition, extra PCR cycles are performed on the ditags, which is unlikely to generate bias because all ditags are the same length [32].

Kim et al. thus produced a human megakaryocyte SAGE library of 20,580 tags, and a complementary αIIb-negative nonmegakaryocyte library of 18,329 tags [10]. The lists of the top 50 most abundant tags in each library were dominated by the ribosomal proteins that regulate translation. Of perhaps more biological interest were lists of genes that were expressed at either relatively high or relatively low levels in the megakaryocyte versus nonmegakaryocyte libraries. However, and rather surprisingly, these lists did not contain the bone fide megakaryocyte/platelet markers αIIbβ3 (used to purify the megakaryocytes), GPIb-IX-V, or GPVI [10]. This suggests that either the purified αIIb-positive cells were not representative of truly mature megakaryocytes, or that the αIIb-negative cells expressed other megakaryocyte markers. It would be interesting to address this issue by examining the entire dataset, but unfortunately this has not been deposited in a public database.

We published the second megakaryocyte SAGE library in 2007 as part of a combined proteomic/genomic analysis of the platelet/megakaryocyte lineage [11]. Mouse megakaryocytes were grown *in vitro* from bone marrow progenitors, generated by depleting bone marrow cells with magnetic beads coated with antibodies against non-megakaryocyte markers. The progenitors were cultured for 7 days in stem cell factor and thrombopoietin. A highly pure population of megakaryocytes was isolated by two passages through a bovine serum albumin gradient, which yielded over 95% of cells with a ploidy of 64 or 128*n* [33]. In contrast to the SAGE library generated by Kim et al., we used the LongSAGE method using the Invitrogen I-SAGE Long Kit. This method, as already discussed, provides a more accurate gene to tag identification than did the original SAGE [12].

A total of 53,046 tags were sequenced, corresponding to approximately 8300 genes [11]. The high purity of the megakaryocytes was verified by the absence of markers of other hematopoietic cell types in the SAGE library. Moreover, major (αIIb and β3 integrin subunits), intermediate (GPIb-IX-V and CD9), and relatively weakly expressed (GPVI and $P2Y_1$) platelet membrane proteins were identified, with the number of SAGE tags correlating quite well with their known protein levels on the platelet surface. Interestingly, 81% of the proteins identified in the accompanying proteomic analysis of the human platelet surface, were also identified in the mouse megakaryocyte SAGE library, suggesting a high degree of similarity in proteins expressed by each cell type and across the different species. However, the greater sensitivity of SAGE was underscored by the fact that this genomic method identified over 8 times more membrane proteins than the proteomic study [11].

A key feature of SAGE is its quantitative nature. As such, our mouse megakaryocyte SAGE library could be compared with any of the hundreds of SAGE libraries in the NCBI public database. Indeed, we compared our library with 30 other mouse SAGE libraries from a variety of different cell types [11]. Strikingly, megakaryocyte-specific genes demonstrated two main characteristics: (1) there was a tendency for these to be genes that are known to be important for megakaryocyte and/or platelet function. Examples include the αIIb integrin subunit, the C-type lectin-like receptor CLEC-2, the α6 integrin subunit, the GPIb-IX-V complex, the $P2X_1$ ATP receptor, P-selectin, and Mpl, the thrombopoietin receptor; and (2) the relatively megakaryocytic genes were enriched for transmembrane proteins. We concluded that such a SAGE study, in combination with bioinformatic comparisons with other SAGE libraries, is a particularly useful approach for the identification of novel platelet/megakaryocyte membrane proteins [11]. This idea is supported by our more recent reanalysis of the megakaryoctye library (Table 9.2). We compared this library to a range of cell types more extensive than performed previously, by using SAGE libraries from 20 different mouse tissues. The list of relatively megakaryocyte-specific genes included many of the platelet-specific markers identified in the previous analysis [11]. However, new entries included the following: the major platelet chemokines CXCL4 and CXCL7 [21]; the cytoplasmic protein kindlin-3, which is essential for integrin activation and platelet aggregation [34]; the cytoplasmic protein tescalcin, which is essential for megakaryocyte development through regulation of expression of Ets family transcription factors [35]; GPR56, an orphan G-protein-coupled receptor with a role in brain development but unstudied in platelets [36]; three serine protease inhibitors (serpins); the poorly understood CTLA2 cysteine protease inhibitors; and the leukocyte immunoglobulin-like receptor B4 that contains an immunoreceptor tyrosine-based inhibitory motif. Many of these relatively megakaryocyte-specific genes are clearly important in megakaryocyte/platelet biology. We predict that many of the other less well-studied such genes will be similarly important.

One caveat of these genomic analyses of megakaryocytes is the extent to which these cultured cells mimic those mature megakaryocytes that have developed in the very different environment provided by the bone marrow. Indeed, this may be addressed only when future technologies have been developed that allow transcriptome analysis from a single mature megakaryocyte isolated directly from the bone marrow.

**TABLE 9.2 Identification of Relatively Megakaryocyte-Specific Genes[a]**

| Protein Name | Meg Tags per Million | Nonmeg Tags per Million |
|---|---|---|
| α2b integrin | 2330 | 15 |
| β3 integrin | 583 | 0 |
| α6 integrin | 414 | 32 |
| GPIbα | 395 | 0 |
| GPIbβ | 470 | 3 |
| GPV | 169 | 1 |
| GPIX | 169 | 2 |
| GPVI | 113 | 1 |
| FcRγ | 432 | 27 |
| CLEC-2 | 1015 | 17 |
| PAR3 | 865 | 14 |
| PAR4 | 113 | 1 |
| Mpl | 244 | 0 |
| TLT-1 | 226 | 1 |
| CXCL4 chemokine (platelet factor 4) | 9810 | 18 |
| CXCL7 chemokine (proplatelet basic protein) | 1146 | 2 |
| Kindlin-3 | 733 | 14 |
| Tescalcin | 489 | 5 |
| GPR56 | 1109 | 57 |
| Serpin A3G | 526 | 20 |
| Serpin B2 | 1203 | 14 |
| Serpin B10 | 282 | 3 |
| Cytotoxic T-lymphocyte-associated protein 2α | 3740 | 105 |
| Cytotoxic T-lymphocyte-associated protein 2β | 3439 | 103 |
| LILRB4 | 827 | 13 |

[a]The mouse megakaryocyte (meg) LongSAGE library of 53,046 tags [11] was compared to 18 other mouse LongSAGE libraries from the following sources: spleen (3 libraries), thymus (3 libraries), embryonic stem cell, whole embryo, whole body, brain, heart ventricle, kidney, large intestine, lung, ovary, pancreas, skeletal muscle, and testis. These libraries ranged from 51,136 to 155,533 tags. A panel of 25 of the most megakaryocyte-specific genes are listed, alongside the number of megakaryocyte and nonmegakaryocyte genes, each adjusted to the number of tags per one million total tags.

## 9.6 NEXT-GENERATION SEQUENCING

Each of the four platelet/megakaryocyte SAGE libraries, described in the previous two sections, was sequenced using the chain terminator method developed by Frederick Sanger [37]. This method has dominated the world of DNA sequencing since the 1980s, and was used to sequence the genome of humans and many other species. The most commonly used current Sanger method, termed *dye terminator sequencing*, utilizes a template DNA, a primer, DNA polymerase, deoxynucleotides, and fluorescent terminator dideoxynucleotides, which lack the hydroxyl group required to form a 3′-phosphodiester bond with the next nucleotide in the chain. DNA polymerization thus terminates on addition of a fluorescent dideoxynucleotide. Each of the four dideoxynucleotides carries a fluorescent group of distinct excitation and emission wavelength. The products of the polymerization reaction can thus be separated by size, using capillary electrophoresis, and the fluorescent dye can be detected in the form of fluorescent peak trace chromatograms, yielding the DNA sequence of the template [38].

The Sanger method can be considered as *first-generation sequencing*. Despite its success, the method is very expensive if the aim is to generate large amounts of sequencing data sufficient to cover an entire eukaryotic genome. This is because of the high costs of running capillary electrophoresis to separate fluorescent products, the relatively low number of samples that can be run at the same time, and the laborious nature of sample preparation. Therefore, in more recent years, a huge effort has been invested in *next-generation sequencing* methods, which produce vast amounts of sequence data in a relatively cheap manner. Indeed, several companies have used their technology to prove that a human genome can be sequenced in a week or less. The costs of such efforts are still in the order of several thousand dollars, but the idea of sequencing a human genome for $1000 in less than a day is now a realistic prospect [39,40].

The breakthroughs that have allowed next-generation sequencing have come in three main areas, namely, improved template preparation, sequencing/imaging, and data analysis. Templates are typically isolated by randomly breaking large pieces of DNA into smaller fragments, by a process such as high-pressure nitrogen treatment (nebulization). These fragments are then immobilized onto a solid surface, on which millions of sequences can be analysed simultaneously. These templates can be either single-molecule or clonally amplified *in situ* by a PCR-based method. The former method provides more quantitative data, while the latter is more common because of the relative ease of sequencing from a relatively large amount of template. The novel sequencing and imaging methods are central to next-generation sequencing. The precise methods employed by different manufacturers have been nicely reviewed by Metzker [40] and are described in the next paragraph. Finally, improved data analysis has been essential for the handling of millions of sequences from a single next-generation sequencing run. Data analysis might become a limiting factor for future next-generation sequencing methods. Indeed, companies such as Pacific Biosciences are developing novel technologies that are intended to sequence the equivalent of a human genome in four minutes [39].

The first commercial next-generation sequencer was the 454 Genome-Sequencer FLX instrument, produced by Roche in 2005. The sequencing method is based on the detection of released pyrophosphate [41]. The sequencing templates are prepared by fragmenting the DNA and ligating adapters, containing priming sites, to each end. These adapters allow the templates to be captured on beads that contain complementary adapters, and this is done under conditions that favor the capture of one template strand per bead. The template is amplified by a process known as *emulsion PCR*, in which each bead is contained in a water droplet and the PCR reagents immersed in oil. Following PCR amplification, the beads are placed in a picotiter plate at one bead per well. The wells also contain DNA polymerase, ATP sulfurylase, luciferase, luciferin, and apyrase. On addition of a single unlabeled nucleotide, if the nucleotide is incorporated into the template by DNA polymerase, pyrophosphate is released. This is converted into ATP by the ATP sulfurylase, which allows the ATP-driven oxidization of luciferin by luciferase, which results in the production of light. The light is detected by glass fibers which connect each well to a sensitive charge-coupled device (CCD) camera. The apyrase degrades unincorporated nucleotides and ATP, allowing the cycle to be repeated with the next unlabeled nucleotide. The end results are read lengths of approximately 330 bp from between one and two million templates [40].

The most widely used next-generation sequencer is the Illumina/Solexa Genome Analyser, commercialized in 2006, which sequences from up to 200 million clonally amplified sequences on a glass slide using the process of cyclic reversible termination [42]. As the name suggests, the trick is the use of fluorescent terminator dideoxynucleotides that have reversible blocking groups. The sequencing templates are first prepared by fragmenting the DNA and ligating adapters to each end. The DNA is then denatured and the resulting single-stranded DNA attached to a solid support. This is coated with adapters and complementary adapters, allowing each single-stranded DNA template to form a bridge structure through hybridization to the complementary adapter. Several solid-phase PCR cycles are then performed, in which the adapters on the surface serve as primers. This generates clusters of approximately 1000 copies of template DNA. The sequencing process takes the form of a series of cycles, each of which identifies a single nucleotide in the sequence. In each cycle, the incorporation of the fluorescent dideoxynucleotides is detected by highly sensitive total internal reflection fluorescence microscopy. The fluorescent group is then cleaved off and the 3′-hydroxyl group restored, so allowing an additional fluorescent dideoxynucleotide to become incorporated in the following cycle. The read length of this method is 75 or 100 bp [42].

In addition to the Roche/454 and the Illumina/Solexa technologies, three other platforms are now commercially available: (1) the Applied Biosystems ABI SOLiD system, which uses chemistry based on DNA ligation; (2) the Helicos single-molecule sequencing device, named HeliScope; and (3) the Polonator G.007 instrument. Readers are directed to an excellent recent review for a description of the technologies behind these systems [40].

## 9.7 RNA-Seq

The potential applications for next-generation sequencing include the sequencing of human genomes ("personal genomics"), genomewide analysis of chromatin structure and epigenetic marks (ChIP-Seq), and transcriptome analysis (RNA-Seq). The latter is of particular relevance to this chapter because the capacity to obtain millions of RNA sequences could potentially render SAGE obsolete.

RNA-Seq appears likely to revolutionize transcriptome analysis in the next few years [43,44]. A typical RNA-Seq experiment is shown in Figure 9.3. RNA is first extracted and then converted into a library of cDNA fragments. This is usually done by fragmenting the RNA and then generating double-stranded cDNA, but it can also be done by generating double-stranded cDNA first and subsequently fragmenting that. Adapters are then attached to each end of the cDNA followed by next-generation sequencing, typically involving one of the three technologies that currently dominate the market, manufactured by Illumina, 454 Life Science and Applied Biosystems [40].

The year 2008 marked the beginning of the RNA-Seq revolution (Fig. 9.2), and a number of high-profile publications highlighted its utility for transcriptome analysis [45–49]. Two groups used Illumina sequencing to generate approximately 30 million and 122 million sequences of 35 and 40 bp, respectively, from

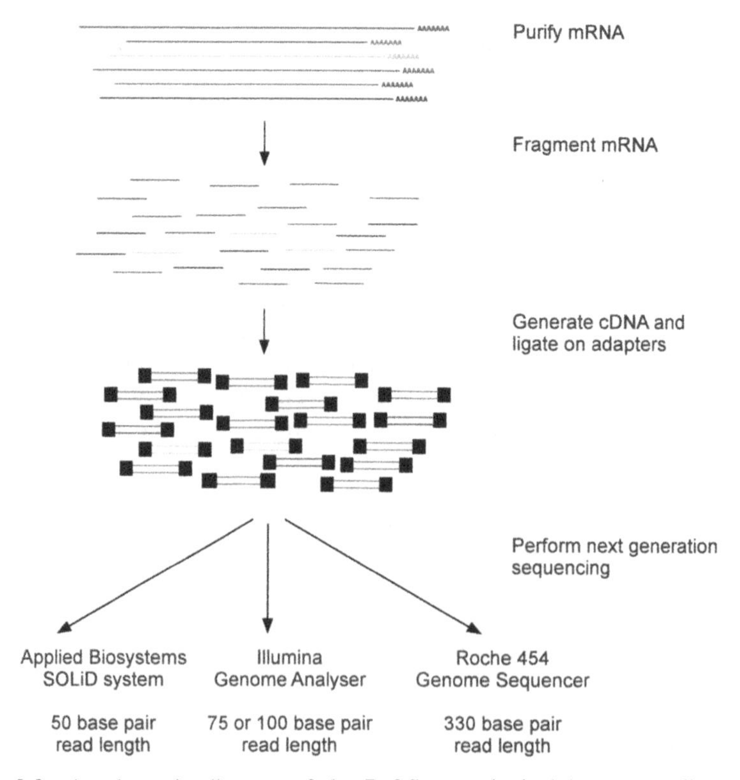


**Figure 9.3** A schematic diagram of the Ref-Seq method. Adapters are ligated onto cDNA fragments that were generated from fragmented RNA. Sequence is obtained using one of three main technologies for next-generation sequencing.

mRNA of the yeast species *Saccharomyces cerevisiae* and *S. pombe*, which suggested that 75% and 90% of their nonrepetitive genomes are transcribed [47,49]. This revealed much greater complexity than previously thought and also allowed an extensive mapping of exon boundaries in the yeast genome. A similar Illumina RNA-Seq approach was used for mouse brain, liver, and skeletal muscle, in which 10–30 million sequences of 25 bp were obtained [46]. These data suggested new and revised gene models, novel transcripts, and new potential precursors of microRNAs, which are rapidly emerging as key posttranscriptional regulators that generally silence genes through binding to complementary sequences in their mRNA. Finally, similar conclusions were drawn from Illumina RNA-Seq analyses of human embryonic kidney and B-cell lines, which each generated approximately 8 million sequences of 27 bp [48].

Compared to RNA-Seq, SAGE suffers two major disadvantages:

1. Because only one SAGE tag is generally sequenced from a defined position on the mRNA strand, SAGE rarely provides information on alternative transcripts. This is a problem, given that as many as 95% of all human genes are now thought to undergo alternative splicing, most often in the form of exon skipping [44,50], and that identifying such events and determining the functional consequences is now seen as a major challenge.
2. SAGE has relied on the relatively expensive Sanger sequencing method. The utilization of next-generation sequencing is certainly a possibility for sequencing SAGE libraries, but the logic in doing this is questionable. Indeed, this approach would not solve the inherent problem of alternative transcript identification. Moreover, most next-generation sequencing technologies produce relatively short lengths of sequence, such that relatively few SAGE tags would be sequenced per template. Since the number of RNA sequences that can be generated by next-generation sequencing is essentially unlimited, why bother to go to the trouble of SAGE library generation to slightly increase this number?

## 9.8 COMBINING SAGE WITH NEXT-GENERATION SEQUENCING

It is perhaps a little premature to regard SAGE as an outdated technology, particularly given the advent of 3′-tag digital gene expression (DGE) [51]. This method uses next-generation sequencing to sequence individual LongSAGE tags, thus avoiding the rather laborious and costly process of concatamerization that is unnecessary for next-generation sequencing. The authors of this study used the sequencing method of Illumina, who provide a 3′-tag DGE service for which only the RNA is provided by the investigator. The efficacy of this method was demonstrated following analysis of two well-characterized commercially available RNA reference libraries [51]. The authors concluded that 3′-tag DGE is highly reproducible between sequencing runs and is relatively cost-effective for fully sequencing a transcriptome. Indeed, 5 million sequences derived from

3′-tag DGE, which is less than a single sequencing lane, are estimated to achieve 90% coverage of the transcriptome, whereas for RNA-Seq the number is 40 million [51]. Of course, 3′-tag DGE data still suffer from the fact that SAGE tags do not normally provide useful information about alternatively spliced transcripts. Nevertheless, 3′-tag DGE may represent the future for cheap and rapid transcriptome analysis, an area currently dominated by DNA microarrays.

## 9.9 CONCLUDING REMARKS

Serial analysis of gene expression has played a pivotal role in the exciting era of genomics but has perhaps been underutilized for the analysis of megakaryocyte and platelet transcriptomes. This is partly due to the general difficulties in performing any genomic study on these cells, in particular the past difficulties in obtaining sufficient megakaryocytes and the relatively small amounts of mRNA in the anucleate platelet. The exciting advent of next-generation sequencing, in the form of RNA-Seq and 3′-tag DGE—the latter of which uses next generation sequencing to sequence individual LongSAGE tags—is now likely to surpass original SAGE and even DNA microarrays for transcriptome analysis. The future application of RNA-Seq and 3′ tag DGE to platelets and megakaryocytes will undoubtedly reveal new insights into this cell lineage through the identification of new alternative transcripts and novel expressed genes.

## ACKNOWLEDGMENTS

The author thanks Victoria Heath for critically reading this chapter and is also grateful to John Herbert and Roy Bicknell for their help with bioinformatic analyses of our mouse megakaryocyte SAGE library.

## REFERENCES

1. Velculescu VE, Zhang L, Vogelstein B, Kinzler KW, Serial analysis of gene expression, *Science* **270**:484–487 (1995).
2. Anisimov SV, Serial analysis of gene expression (SAGE): 13 years of application in research, *Curr. Pharm. Biotechnol*. **9**:338–350 (2008).
3. Wang SM, Long-short-long games in mRNA identification: The length matters, *Curr. Pharm. Biotechnol*. **9**:362–367 (2008).
4. Adams MD, Dubnick M, Kerlavage AR, Moreno R, Kelley JM, Utterback TR, Nagle JW, Fields C, Venter JC, Sequence identification of 2,375 human brain genes, *Nature* **355**:632–634 (1992).
5. Adams MD, Kelley JM, Gocayne JD, Dubnick M, Polymeropoulos MH, Xiao H, Merril CR, Wu A, Olde B, Moreno RF, et al., Complementary DNA sequencing: Expressed sequence tags and human genome project, *Science* **252**:1651–1656 (1991).

6. Lockhart DJ, Dong H, Byrne MC, Follettie MT, Gallo MV, Chee MS, Mittmann M, Wang C, Kobayashi M, Horton H, Brown EL, Expression monitoring by hybridization to high-density oligonucleotide arrays, *Nat. Biotechnol*. **14**:1675–1680 (1996).
7. Schena M, Shalon D, Davis RW, Brown PO, Quantitative monitoring of gene expression patterns with a complementary DNA microarray, *Science* **270**:467–470 (1995).
8. Dittrich M, Birschmann I, Pfrang J, Herterich S, Smolenski A, Walter U, Dandekar T, Analysis of SAGE data in human platelets: Features of the transcriptome in an anucleate cell, *Thromb. Haemost*. **95**:643–651 (2006).
9. Gnatenko DV, Dunn JJ, McCorkle SR, Weissmann D, Perrotta PL, Bahou WF, Transcript profiling of human platelets using microarray and serial analysis of gene expression, *Blood* **101**:2285–2293 (2003).
10. Kim JA, Jung YJ, Seoh JY, Woo SY, Seo JS, Kim HL, Gene expression profile of megakaryocytes from human cord blood CD34(+) cells ex vivo expanded by thrombopoietin, *Stem Cells* **20**:402–416 (2002).
11. Senis YA, Tomlinson MG, García A, Dumon S, Heath VL, Herbert J, Cobbold SP, Spalton JC, Ayman S, Antrobus R, Zitzmann N, Bicknell R, Frampton J, Authi KS, Martin A, Wakelam MJ, Watson SP, A comprehensive proteomics and genomics analysis reveals novel transmembrane proteins in human platelets and mouse megakaryocytes including G6b-B, a novel immunoreceptor tyrosine-based inhibitory motif protein, *Mol. Cell. Proteom*. **6**:548–564 (2007).
12. Saha S, Sparks AB, Rago C, Akmaev V, Wang CJ, Vogelstein B, Kinzler KW, Velculescu VE, Using the transcriptome to annotate the genome, *Nat. Biotechnol*. **20**:508–512 (2002).
13. Matsumura H, Reich S, Ito A, Saitoh H, Kamoun S, Winter P, Kahl G, Reuter M, Kruger DH, Terauchi R, Gene expression analysis of plant host-pathogen interactions by SuperSAGE, *Proc. Natl. Acad. Sci. USA* **100**:15718–15723 (2003).
14. Matsumura H, Kruger DH, Kahl G, Terauchi R, SuperSAGE: A modern platform for genome-wide quantitative transcript profiling, *Curr. Pharm. Biotechnol*. **9**:368–374 (2008).
15. Matsumura H, Bin Nasir KH, Yoshida K, Ito A, Kahl G, Kruger DH, Terauchi R, SuperSAGE array: The direct use of 26-base-pair transcript tags in oligonucleotide arrays, *Nat. Meth*. **3**:469–474 (2006).
16. Newman PJ, Gorski J, White GC, 2nd, Gidwitz S, Cretney CJ, Aster RH, Enzymatic amplification of platelet-specific messenger RNA using the polymerase chain reaction, *J. Clin. Invest*. **82**:739–743 (1988).
17. Fink L, Holschermann H, Kwapiszewska G, Muyal JP, Lengemann B, Bohle RM, Santoso S, Characterization of platelet-specific mRNA by real-time PCR after laser-assisted microdissection, *Thromb. Haemost*. **90**:749–756 (2003).
18. Hashimoto S, Matsushima K, SAGE application in hematological research, *Curr. Pharm. Biotechnol*. **9**:383–391 (2008).
19. Gnatenko DV, Dunn JJ, Schwedes J, Bahou WF, Transcript profiling of human platelets using microarray and serial analysis of gene expression (SAGE), *Meth. Mol. Biol*. **496**:245–272 (2009).
20. Mannherz HG, Hannappel E, The beta-thymosins: Intracellular and extracellular activities of a versatile actin binding protein family, *Cell Motil. Cytoskel*. **66**:839–851 (2009).

21. Gleissner CA, von Hundelshausen P, Ley K, Platelet chemokines in vascular disease, *Arterioscler. Thromb. Vasc. Biol*. **28**:1920–1927 (2008).

22. Bergmeier W, Chauhan AK, Wagner DD, Glycoprotein Ibalpha and von Willebrand factor in primary platelet adhesion and thrombus formation: Lessons from mutant mice, *Thromb. Haemost*. **99**:264–270 (2008).

23. Zhong L, Cherry T, Bies CE, Florence MA, Gerges NZ, Neurogranin enhances synaptic strength through its interaction with calmodulin, *EMBO J*. **28**:3027–3039 (2009).

24. McRedmond JP, Park SD, Reilly DF, Coppinger JA, Maguire PB, Shields DC, Fitzgerald DJ, Integration of proteomics and genomics in platelets: A profile of platelet proteins and platelet-specific genes, *Mol. Cell. Proteom*. **3**:133–144 (2004).

25. Hou SC, Chan LW, Chou YC, Su CY, Chen X, Shih YL, Tsai PC, Shen CK, Yan YT, Ankrd17, an ubiquitously expressed ankyrin factor, is essential for the vascular integrity during embryogenesis, *FEBS Lett*. **583**:2765–2771 (2009).

26. Miller T, Williams K, Johnstone RW, Shilatifard A, Identification, cloning, expression, and biochemical characterization of the testis-specific RNA polymerase II elongation factor ELL3, *J. Biol. Chem*. **275**:32052–32056 (2000).

27. Weyrich AS, Schwertz H, Kraiss LW, Zimmerman GA, Protein synthesis by platelets: Historical and new perspectives, *J. Thromb. Haemost*. **7**:241–246 (2009).

28. Kieffer N, Guichard J, Farcet JP, Vainchenker W, Breton-Gorius J, Biosynthesis of major platelet proteins in human blood platelets, *Eur. J. Biochem*. **164**:189–195 (1987).

29. Weyrich AS, Dixon DA, Pabla R, Elstad MR, McIntyre TM, Prescott SM, Zimmerman GA, Signal-dependent translation of a regulatory protein, Bcl-3, in activated human platelets, *Proc. Natl. Acad. Sci. USA* **95**:5556–5561 (1998).

30. Lindemann S, Tolley ND, Dixon DA, McIntyre TM, Prescott SM, Zimmerman GA, Weyrich AS, Activated platelets mediate inflammatory signaling by regulated interleukin 1beta synthesis, *J. Cell. Biol*. **154**:485–490 (2001).

31. Weyrich AS, Denis MM, Schwertz H, Tolley ND, Foulks J, Spencer E, Kraiss LW, Albertine KH, McIntyre TM, Zimmerman GA, mTOR-dependent synthesis of Bcl-3 controls the retraction of fibrin clots by activated human platelets, *Blood* **109**:1975–1983 (2007).

32. Datson NA, Scaling down SAGE: From miniSAGE to microSAGE, *Curr. Pharm. Biotechnol*. **9**:351–361 (2008).

33. Dumon S, Heath VL, Tomlinson MG, Gottgens B, Frampton J, Differentiation of murine committed megakaryocytic progenitors isolated by a novel strategy reveals the complexity of GATA and Ets factor involvement in megakaryocytopoiesis and an unexpected potential role for GATA-6, *Exp. Hematol*. **34**:654–663 (2006).

34. Moser M, Nieswandt B, Ussar S, Pozgajova M, Fassler R, Kindlin-3 is essential for integrin activation and platelet aggregation, *Nat. Med*. **14**:325–330 (2008).

35. Levay K, Slepak VZ, Tescalcin is an essential factor in megakaryocytic differentiation associated with Ets family gene expression, *J. Clin. Invest*. **117**:2672–2683 (2007).

36. Koirala S, Jin Z, Piao X, Corfas G, GPR56-regulated granule cell adhesion is essential for rostral cerebellar development, *J. Neurosci*. **29**:7439–7449 (2009).

37. Sanger F, Nicklen S, Coulson AR, DNA sequencing with chain-terminating inhibitors, *Proc. Natl. Acad. Sci. USA* **74**:5463–5467 (1977).

38. Smith LM, Sanders JZ, Kaiser RJ, Hughes P, Dodd C, Connell CR, Heiner C, Kent SB, Hood LE, Fluorescence detection in automated DNA sequence analysis, *Nature* **321**:674–679 (1986).

39. Ansorge WJ, Next-generation DNA sequencing techniques, *Nat. Biotechnol.* **25**:195–203 (2009).

40. Metzker ML, Sequencing technologies—the next generation, *Nat. Rev. Genet.* **11**:31–46 (2010).

41. Ronaghi M, Uhlen M, Nyren P, A sequencing method based on real-time pyrophosphate, *Science* **281**:363–365 (1998).

42. Bentley DR, Balasubramanian S, Swerdlow HP, Smith GP, Milton J, Brown CG, Hall KP, Evers DJ, Barnes CL, Bignell HR, Boutell JM, Bryant J, Carter RJ, Keira Cheetham R, Cox AJ, Ellis DJ, Flatbush MR, Gormley NA, Humphray SJ, Irving LJ, Karbelashvili MS, Kirk SM, Li H, Liu X, et al., Accurate whole human genome sequencing using reversible terminator chemistry, *Nature* **456**:53–59 (2008).

43. Marguerat S, Bahler J, RNA-seq: From technology to biology, *Cell. Mol. Life Sci.* **67**:569–579 (2010).

44. Wang Z, Gerstein M, Snyder M, RNA-Seq: A revolutionary tool for transcriptomics, *Nat. Rev. Genet.* **10**:57–63 (2009).

45. Lister R, O'Malley RC, Tonti-Filippini J, Gregory BD, Berry CC, Millar AH, and Ecker JR, Highly integrated single-base resolution maps of the epigenome in Arabidopsis, *Cell* **133**:523–536 (2008).

46. Mortazavi A, Williams BA, McCue K, Schaeffer L, Wold B, Mapping and quantifying mammalian transcriptomes by RNA-Seq, *Nat. Meth.* **5**:621–628 (2008).

47. Nagalakshmi U, Wang Z, Waern K, Shou C, Raha D, Gerstein M, Snyder M, The transcriptional landscape of the yeast genome defined by RNA sequencing, *Science* **320**:1344–1349 (2008).

48. Sultan M, Schulz MH, Richard H, Magen A, Klingenhoff A, Scherf M, Seifert M, Borodina T, Soldatov A, Parkhomchuk D, Schmidt D, O'Keeffe S, Haas S, Vingron M, Lehrach H, Yaspo ML, A global view of gene activity and alternative splicing by deep sequencing of the human transcriptome, *Science* **321**:956–960 (2008).

49. Wilhelm BT, Marguerat S, Watt S, Schubert F, Wood V, Goodhead I, Penkett CJ, Rogers J, Bahler J, Dynamic repertoire of a eukaryotic transcriptome surveyed at single-nucleotide resolution, *Nature* **453**:1239–1243 (2008).

50. Pan Q, Shai O, Lee LJ, Frey BJ, Blencowe BJ, Deep surveying of alternative splicing complexity in the human transcriptome by high-throughput sequencing, *Nat. Genet.* **40**:1413–1415 (2008).

51. Asmann YW, Klee EW, Thompson EA, Perez EA, Middha S, Oberg AL, Therneau TM, Smith DI, Poland GA, Wieben ED, Kocher JP, 3′ tag digital gene expression profiling of human brain and universal reference RNA using Illumina genome analyzer, *BMC Genom.* **10**:531 (2009).

# 10

# THE APPLICATION OF MICROARRAY ANALYSIS AND ITS INTEGRATION WITH PROTEOMICS FOR STUDY OF PLATELET-ASSOCIATED DISORDERS

DMITRI V. GNATENKO AND WADIE F. BAHOU

**Abstract**

Lacking ongoing genomic DNA transcription, platelets represent a unique model system for studying interactions between transcriptome and proteome. This chapter focuses on application of comparative transcriptomic studies for predicting novel proteins and pathways in platelets and for identification of molecular signatures of platelet-related disorders. In the future, integration of transcript and protein profiling approaches will permit the identification of molecular mechanisms of platelet function at a new level.

## 10.1 INTRODUCTION

Platelets offer a unique cellular system to study the relationship between genes and proteins. Indeed, platelets luck nuclear genomic DNA, but retain substantial pool of megakaryocyte-derived mRNAs. These mRNA transcripts in total constitute the platelet *transcriptome* [1–5]. Historically, the platelet mRNA content was considered invariant. More recent discovery of signal-dependent pre-mRNA splicing in platelets [6] reconfigured this traditional paradigm. Splicing allows

*Platelet Proteomics: Principles, Analysis and Applications*, First Edition.
Edited by Ángel García and Yotis A. Senis.

platelets to alter the pool of translatable messages from the same invariable pool of transcripts in response to cellular activation/stimulation, resulting in an altered pool of proteins synthesized in platelets.

The entire pool of platelet proteins constitutes the platelet *proteome*. Simplistic understanding of the relation between genes, their transcripts, and the corresponding proteins may lead to the belief that proteome is completely reflected in transcriptome. However, complex multilevel regulation of both suggests that this may be true for only a limited set of proteins. Advances in gene and protein profiling technologies, coupled with the completion of the human genome project, permit integration of transcript and protein profiling approaches to dissect molecular mechanisms of platelet function in normal state and disease. In this chapter, we provide an overview of human platelet transcriptomic studies to dissect molecular mechanisms of platelet-associated diseases and their integration with proteomic analysis of human platelets.

## 10.2 PLATELET mRNA CONTENT

Generated as cytoplasmic buds from precursor bone marrow megakaryocytes, platelets are anucleate cells that retain small amounts of megakaryocyte-derived mRNAs [7,8]. On average, platelets contain as little as $2 \times 10^{-3}$ fg mRNA/cell (approximately 3–4 logs less RNA than a typical nucleated cell [9]), although younger platelets contain relatively larger amounts of mRNA [10]. Platelets contain rough endoplasmic reticulum and polyribosomes, and retain the ability for protein biosynthesis from cytoplasmic mRNA [11,12]. Quiescent platelets generally display minimal translational activity, although newly formed platelets synthesize various α-granule and membrane glycoproteins, including GPIb and GPIIb/IIIa. Furthermore, stimulation of quiescent platelets by agonists such as α-thrombin increases protein synthesis of various platelet proteins [13]. Messages encoding many platelet proteins are found in the platelet transcriptome, including those for Fc receptors [14], chemokines [15], and coagulation factors [16]. Signal-dependent translation of a regulatory protein, Bcl3, is reported in activated platelets [17].

## 10.3 TRANSCRIPT PROFILING TECHNIQUES

Development of global transcript profiling technologies such as microarray and serial analysis of gene expression (SAGE) in conjunction with completion of the human genome project led to understanding of the complex processes of gene interactions within a living cell [18,19], and contributed to many fields of biological research, including platelet studies.

Microarray analysis adapts artificially constructed grids of known DNA samples such that each element of the grid probes for a specific RNA sequence; these are then used to capture and quantify RNA transcripts [20]. Many different microarray protocols have been developed, but all of them represent a *closed transcript profiling system*—they detect only those transcripts corresponding to specific probes imprinted onto the chip. Transcripts without corresponding probes will be missed. Different microarray systems are commercially available, including "whole genome" microarrays from Affymetrix, Applied Biosystems, and MGW-Biotech AG. In addition, it is now possible for an individual researcher to purchase a good-quality "customized" microarray chip, without the need to design, manufacture, and characterize it personally (Agilent Technologies and other manufacturers).

Unlike microarray, SAGE represents an *open transcript profiling system*, which can theoretically detect any transcript. Chapter 9 provides comprehensive overview of SAGE technology. Briefly, "classical" SAGE [21] relies on the observation that short (<10 bp) sequences (tags) within 3′-mRNAs can stringently discriminate among the ~30,000 genes in the human genome. The sequence of each tag, along with its positional location, uniquely identifies the gene from which it is derived, and differentially expressed genes can be identified in a quantitative manner since the frequency of tag detection reflects the steady-state mRNA level of the cellular transcriptome [21,22] (www.sagenet.org). Several modifications to the original protocols have been devised for (1) generation of longer tags as a means of providing more definitive tag-to-gene identification, (2) efficient identification of low-abundance transcripts using subtractive SAGE techniques, and (3) amplification techniques to circumvent small mRNA starting material [23–25].

Accurate transcript profiling requires quantitative measurement of transcript abundance. Since most microarray protocols are *semiquantitative*, microarray results need validation using other techniques. The most common and well-established technique is quantitative reverse transcription fluorescence-based real-time PCR (Q-PCR). Q-PCR allows monitoring the progress of the PCR as it occurs (i.e., in real time). Data are therefore collected throughout the PCR process, rather than at the end of the PCR. This completely revolutionizes the way one approaches PCR-based quantitation of DNA and RNA. In real-time PCR, reactions are characterized by the point in time during cycling when amplification of a target is first detected rather than the amount of target accumulated after a fixed number of cycles. The higher the starting copy number of the nucleic acid target, the sooner a significant increase in fluorescence is observed. In contrast, an *endpoint assay* measures the amount of accumulated PCR product at the end of the PCR cycle. Several platforms of Q-PCR are commercially available (Tagman, CyberGreen), all of which allow accurate transcript profiling.

More recently, fluorescent microsphere-based assay for transcript profiling has been developed [26]. Unlike traditional microarrays, this technology is built on a bead-array format carried out in solution. It allows *multiplexing*—simultaneous detection of up to 30 individual transcripts in one well—and is based on signal amplification, not transcript amplification, as is Q-PCR. This branched DNA (bDNA) gene detection system is a "sandwich" nucleic hybridization assay that quantifies mRNA directly from cellular lysates by amplifying the reporter signal rather than target transcripts [27,28]. A schema of transcript profiling using

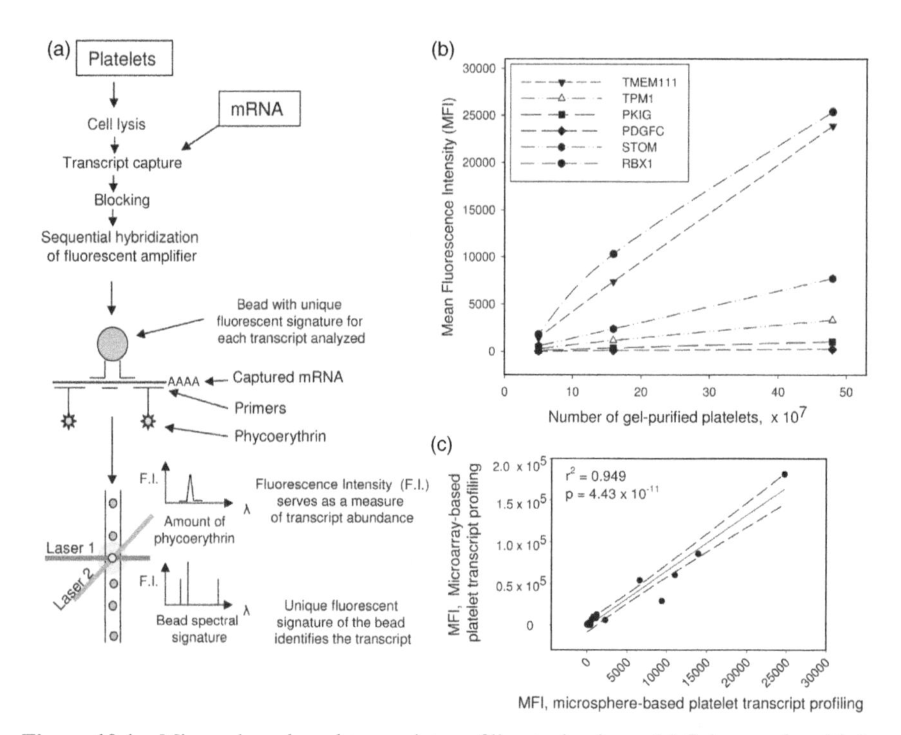


**Figure 10.1** Microsphere-based transcript profiling technology. (a) Schema of multiplex transcript profiling using microspheres. This technology is designed to analyze transcript profiling of intact cells, thus skipping all RNA-manipulation steps. It can also be adapted to analyze mRNA. Oligonucleotides specific to an individual platelet mRNA are coupled to unique microspheres, thereby allowing capture of gene-specific mRNAs on beads with known fluorescent ID. Sequential hybridization steps provide for the binding and amplification of the streptavidin-conjugated fluorescent signal (*R*-phycoerythrin). Two laser beams allow simultaneous detection of individual bead ID and quantification of corresponding mRNA abundances. (b) Microsphere-based profiling of six transcripts in human platelets. Different amounts of gel-filtered platelets ($5 \times 10^7$, $16 \times 10^7$, $48 \times 10^7$) were directly lysed *in vitro*, followed by platelet multiplexing. Gene names are shown in inset. (c) Regression analysis of transcript profiles of 17 platelet-expressed transcripts obtained by microarray or microsphere-based assays demonstrate excellent correlation; 95% confidence intervals are shown.

fluorescent microspheres is shown in Figure 10.1(a). Luminex xMAP$^{TM}$ technology assigns unique fluorescent signatures to individual microspheres. Platelets are lysed in a 96-well plate, and platelet mRNAs are captured on corresponding beads using transcript-specific oligonucleotides. Unbound material is washed, and captured transcripts are labeled with streptavidin-conjugated *R*-phycoerythrin. Transcript profiling is achieved using BioPlex plate reader (Bio-Rad, Hercules, CA) by simultaneous identification of individual bead ID (laser 1) and measurement of the amount of corresponding transcript by fluorescence of phycoerythrin (laser 2). Relative transcript abundance for each individual gene is determined using the raw fluorescent signal intensities obtained from 100 microspheres, and are reported as the mean fluorescent intensity (MFI).

In our laboratory, this technology has been adapted for platelet transcript profiling. Gnatenko et al. selected 17 platelet-expressed transcripts for multiplex profiling [29]. Transcripts were chosen to represent gene abundances at different ranges, as initially delineated using microarray transcript profiling from five healthy control subjects [30]. All 17 transcripts were detected in as little as $5 \times 10^7$ purified platelets (a platelet mass corresponding to ~100 μL of whole blood), including low abundant transcripts [Fig. 10.1(b)]. Furthermore, relative transcript abundance of 17 platelet-expressed transcripts measured by microarray (see Section 10.5.1) and microsphere-based assays demonstrated excellent concordance [Fig. 10.1(c)]. Taken together, these results demonstrate that microsphere-based assay represents novel promising approach to platelet transcript profiling.

## 10.4 OVERVIEW OF PLATELET TRANSCRIPTOME

Initial successful characterization of platelet-derived mRNA transcripts was achieved using construction of platelet-specific cDNA libraries [31] and single-gene polymerase chain reaction (PCR) [32] technology. To date (as of 2010), a limited number of published microarray studies using platelet-derived mRNAs have been described, with generally concordant agreements on transcript quantitation and gene expression patterns [1–5]. Furthermore, it became clear that this approach provides an efficient means to identify novel genes and proteins functionally expressed in human platelets [1]. Not surprisingly, platelets retain fewer transcripts than those found in nucleated cells, ranging from ~1600 to 3000 mRNAs [1–3,5]. The small number of platelet-expressed transcripts reflects the lack of ongoing transcription in the anucleate platelet.

Feasibility of accurate platelet transcript profiling from a single donor using as little as 50 ng of total platelet RNA, or ~40 mL of a whole blood, has been demonstrated [33]. The "switching mechanism at the 5′ end of RNA templates" (SMART) amplification technique was used for mRNA amplification and the reliability, and precision of the amplification was validated by quantitative PCR (Q-PCR) and by parallel hybridization of amplified and nonamplified RNA samples. In this study, reproducible gene profiling gave similar results for 9815 of 9850 represented genes, providing an initial proof of principle that this approach could be applied for platelet-associated human diagnostic studies.

A combination of two complementary transcript profiling techniques—microarray and SAGE—allowed validation of the most abundant platelet transcripts [1]. Initial studies of limited platelet SAGE library (2033 tags) have demonstrated that 89% of platelet RNA tags are mitochondrial (mt) transcripts. This is presumably related to persistent mt-transcription in the absence of nucleus-derived transcripts. Microarray alone could not detect mitochondrial transcripts since specific probes for human mitochondrion were not present on the microarray chip. Analysis of a more comprehensive platelet SAGE library (25,000 tags) revealed that ~50% (12,609 tags) of platelet SAGE tags are nucleus-derived, whereas the remaining 50% are of mitochondrial origin [4,34]. Still, the overrepresentation of mitochondrial transcripts in platelets is considerably greater than that of its closest cell type, skeletal muscle, in which mitochondrial SAGE tags constitute 20–25% [35].

Taken together, these results demonstrate the feasibility of platelet transcript profiling in (1) identification of differentially expressed genes; (2) characterization of novel platelet-expressed genes; and (3) characterization of the "molecular signature" of a disease, which can be used for diagnostics and prognostics or for identification of potential targets for drug development.

## 10.5 PLATELET TRANSCRIPTOME IN NORMAL STATE AND IN DISEASE

### 10.5.1 Genomewide Transcript Profiling: Analyzing Molecular Signatures of Platelet-Associated Diseases

More recent progress in technology have made it feasible to adapt transcriptomic analysis to study platelet transcriptome, identify differences between normal and diseased platelets, and exploit these differences to develop modern diagnostic tests [34,36–38]. Thus, microarray analysis of fibroblasts has been used to study molecular mechanisms involved in the gray platelet syndrome [39]. Leukocytes of patients with polycythemia rubra vera were analyzed by microarray to demonstrate overexpression of transcription factor NF-E2 and evaluate transcriptomewide effects of V617F Jak2 mutation [40,41]. These studies, however, focused not on platelets but on other cell types to analyze platelet-associated disorders using transcriptome analysis.

Direct analysis of platelet transcriptome has been applied to identify transcriptomic differences between normal platelets and platelets of patients with essential thrombocythemia (ET) [30]. ET represents a distinct subtype of myeloproliferative disorder characterized by increased proliferation of megakaryocytes and resultant elevated numbers of circulating platelets [42]. Initially, platelets from 6 ET patients and 5 volunteers were purified and analyzed using HU133A microarray chip, containing probe sets for 22,283 transcripts. Bioinformatic analysis demonstrated distinctly different molecular signatures of normal and ET platelets. Grouped together, ET platelets demonstrated higher numbers of expressed transcripts compared to normal controls, but considerably less than the transcript numbers generally found in nucleated cells [1]. Of the genes classified as marginal or present in a minimum of four microarrays, ET patient samples expressed an average of 3562 transcripts compared to 1668 for normal controls (compared with ~10,500 transcripts identified in all three leukocyte microarrays). More stringent analyses (i.e., marginal or present in all arrays within a single group) extended these differences, with 1840 transcripts expressed in ET platelets versus 1086 transcripts expressed in platelets from healthy controls

($p < 0.03$). An unsupervised, hierarchical clustering analysis demonstrated that all ET platelet samples were grouped together on the basis of similarities of gene expression, with only one normal misclassified as ET [Fig. 10.2(a)]. All platelet samples, ET and normal, were grouped separately from leukocytes. Comparison of leukocyte and platelet transcript profiles allowed delineation of *platelet-restricted genes*—identification of genes whose expression was restricted to platelets ($N = 126$). This subset of genes was generated by excluding genes that were expressed in leukocytes from the list of genes expressed in platelets.

A total of 170 genes were identified by one-way analysis of variance (ANOVA) as differentially expressed between normal and ET platelets, the majority of which (141) were upregulated in ET platelets. Only 29 genes were downregulated in ET platelets. Among those, a single platelet-restricted gene, *HSD17B3* [encoding the type 3 $17_{\beta}$-hydroxysteroid dehydrogenase (17βHSD3)] was expressed in all normal platelet arrays, and uniquely underexpressed in ET compared to normal platelets. This gene belongs to a large family of steroid dehydrogenases; it encodes an enzyme that catalyzes conversion of 4-androstenedione to testosterone. Traditionally this enzyme is regarded as testis-specific; molecular defects of the *HSD17B3* gene are causally implicated in male pseudohermaphroditism [43].

To date, genes encoding 14 types of 17βHSD enzymes have been described [44]. Examination of the microarray data demonstrated limitation of *HSD17B* transcript expression in platelets to three isoforms: *HSD17B3, HSD17B11*, and *HSD17B12*. Expression of *HSD17B11* was low-level and not statistically different between ET and normal platelets. However, there was a striking difference in the patterns of *HSD17B3* and *HSD17B12* expression between ET and normal platelets. *HSD17B3* transcript expression was absent in all six ET patients, whereas elevated transcript levels of *HSD17B12* was observed in the same patient subgroup. In contrast, expression of *HSD17B3* in normal platelets was accompanied by negligible to low-level *HSD17B12* expression.

This was the first evidence that *HSD17B3* and *HSD17B12* are expressed in human platelets and may be involved in essential thrombocythemia. Validation of these findings was completed on two levels: (1) Q-PCR was used to confirm transcript abundance, and (2) functional activity studies were carried out to confirm steroid interconversion. Q-PCR assay was applied to the original ET cohort and an expanded cohort of normal controls, specifically collected to exclude potential gender bias in *HSD17B* gene expression. These results confirmed those found by microarray, demonstrating ~4.5–fold greater *HSD17B3* transcript levels in normal platelets (compared to ET, $p \leq 0.001$) and concomitant ~27–fold greater *HSD17B12* transcript expression in ET platelets (compared to normal, $p \leq 0.03$). Q-PCR analysis of a larger cohort of 20 ET patients (6 original ET patients and 14 newly studied individuals) was entirely concordant for all individuals studied, demonstrating that *17BHSD12 : 17BHSD3* transcript ratios reliably predicted the ET phenotype in all patients studied to date ($p < 0.0001$). Functional 17βHSD3 activity studies demonstrated that platelets retain the capacity to convert testosterone to 4-androstenedione. Furthermore, the high-level expression

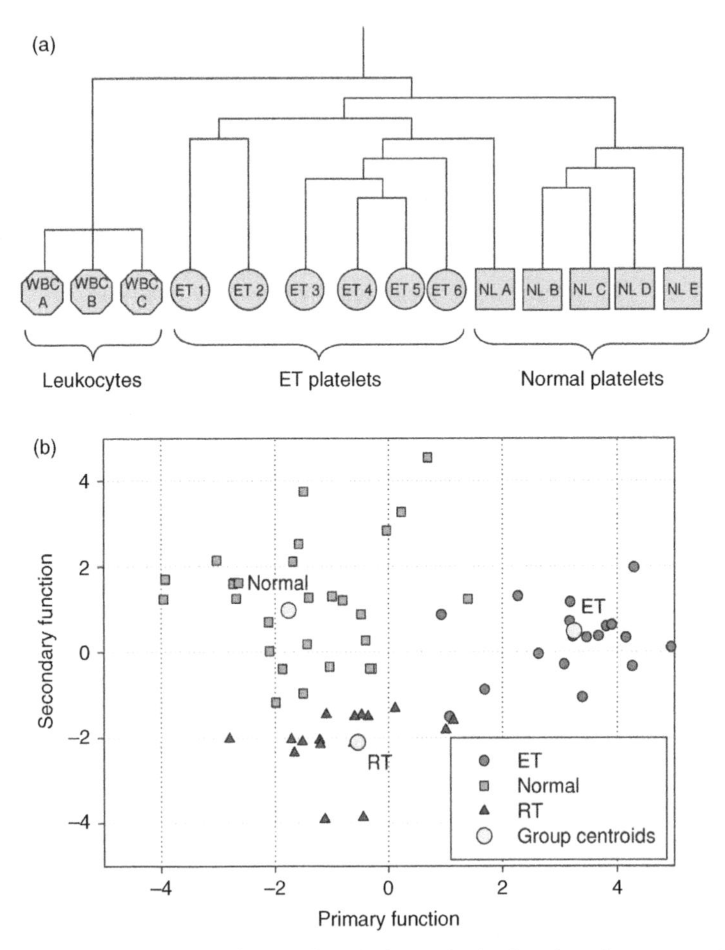


**Figure 10.2** Classification of normal and diseased platelets based on transcript profiling. (a) Gene expression profiles from 14 individual platelet samples (11 apheresis donors [5 normal, NL A–E; 6 patients with essential thrombocythemia (ET)], or 3 normal leukocyte (WBC A–C) donors) are displayed. The relationships among the experimental samples are displayed as dendrograms, in which the pattern and length of the branches depict sample cohort relatedness among the experimental groups. (b) Discriminant analysis of the ET, RT, and normal subjects. Function 1 is the primary canonical discriminant function; function 2 is the secondary canonical discriminant function. Centroids of each group are shown in yellow.

of *HSD17B12* transcript in ET platelets was unassociated with overall enzymatic capabilities in androgen biosynthesis.

Taken together, these data provide the first evidence that genomewide platelet transcript profiling can be adapted to study molecular signatures of platelet-associated diseases. Comparison of normal and ET transcript profiles allowed identification of two transcripts (*17BHSD3* and *17BHSD12)* that had not been described in platelets before. However, the role of steroid metabolism enzymes in platelet function remains to be addressed. Thus, human platelets express both androgen and estrogen β receptors, and platelet function is known to be modulated by gender differences [45], the menstrual cycle [46], and exogenous testosterone [47]. Similarly, hormonal replacement in postmenopausal females, or oral contraceptive use in menstruating females, is known to predispose women to thrombotic diseases [48,49], although the mechanism(s) of this effect remain unclear. The evidence that platelets retain 17βHSD3 activity, express distinct subtypes of 17βHSD enzymes, and demonstrate altered *HSD17B* expression patterns in a disorder known to be association with thrombohemorrhagic risk, provides novel insights into the interplay between sex hormones, platelet function, and vascular diseases.

### 10.5.2 Platelet-Restricted Transcript Analysis: Classification of Thrombocytosis

Commercially available microarrays represent a reliable platform for transcript profiling, because they are designed to provide accurate transcript measurement and include internal controls and standards for reliable data processing and normalization. However, because of their high cost, they cannot be used for large-scale molecular profiling. One alternative approaches includes generation of a custom microarray for a specific scientific task.

On the basis of genomewide platelet transcript profiling data, we have designed, manufactured, and characterized a custom oligonucleotide array that is specifically focused on the analysis of platelet-expressed transcripts. This array includes probes for the following cohorts of genes: (1) a group of platelet-restricted genes with no expression in leukocytes ($N = 126$), (2) a preliminary group of discriminatory genes distinguishing between normal and ET phenotypes ($N = 71$), (3) a list of genes with platelet expression $>$ leukocyte expression by 10-fold ($N = 285$), and (4) a list of genes with leukocyte expression $>$ platelet expression by 10-fold ($N = 43$) (leukocyte contamination control). Several *Arabidopsis* probe elements have also been included to serve as normalization and quantification measures of inter- and intraslide variability [50]. After removal of duplicates, the final list contains 432 genes [along with positive [ribosomal RNAs, housekeeping genes, etc.) and negative (i.e., *Arabidopsis*) controls]. To optimize data analysis, 70-mer oligonucleotide probe sets are spotted on a glass slide in quadruplicate.

Initial genomewide platelet transcript profiling focused on the analysis of two groups of platelet samples—from normal individuals and patients with ET [1,30]. A custom platelet-specific microarray assay was applied to study three groups of platelet samples: normal, ET, and reactive thrombocytosis (RT). The overall objective of this analysis was to study differences in the platelet transcriptome and identify biomarkers that can discriminate among the three groups, thus demonstrating the feasibility of adapting transcript profiling to classification of different classes of thrombocytosis.

Initially, platelet RNA of 63 individuals was analyzed by hybridization to platelet-specific microarray chip—normal controls ($N = 28$), patients with ET ($N = 18$), and patients with RT ($N = 17$). The total platelet RNA was isolated, amplified, labeled with Cy5, and hybridized to the chip. The universal human RNA was used as a reference (amplified, labeled with Cy3, and hybridized to the same chip in parallel). Initial two-way nonparametric ANOVA identified a 131-member gene list of differentially expressed transcripts among the three subgroups ($p < 0.05$), with no gender differences in healthy controls. Stepwise linear discriminant analysis identified a subset of transcripts that segregated the three phenotypic cohorts (ET vs. RT vs. normal) [Fig. 10.2(b)]. The utility of the initial biomarker subset in predicting class was confirmed using a leave-one-out cross-validation analysis, in which each case is classified by the profiles derived from all cases excluding that case [51]. Further studies of expanded cohorts of ET and RT patients and normal controls confirmed initial observations and identified a subset of 11 biomarker genes that discriminated among the three cohorts with 86.3% accuracy, with 93.6% accuracy in two-way class prediction (ET vs. RT) [52]. These data provide the first evidence that gene expression profiles can be used to classify platelet phenotypes using routine phlebotomy, with clinical implications for patients with thrombocytosis. Accurate classification of thrombocytosis is of paramount importance, given the frequency of thrombohemorrhagic complications known to occur in ET (to the exclusion of RT).

## 10.6 FROM TRANSCRIPTOME TO PROTEOME

Although the measurement of transcribed mRNA has proved to be very powerful in the discovery of molecular markers and the elucidation of functional mechanisms, alone it is not sufficient for the characterization of biological systems as a whole [53]. Integration of transcriptomic and proteomic approaches allows analysis of platelet function at the new level. Despite significant progress, information on correlation between mRNA and protein abundance remains controversial. In yeast, comparison of mRNA abundances measured by SAGE with the corresponding protein abundances demonstrated that mRNA expression level is a poor predictor of corresponding protein abundance [54]. Another study of yeast proteome reported good correlation between protein abundance, mRNA abundance, and codon bias [55], demonstrating that for each molecule of well-translated mRNA, there were about 4000 molecules of protein. In human liver, however, the correlation coefficient between mRNA abundance and protein abundance was 0.48; of the 50 most abundant liver mRNAs, 29 encoded secreted proteins, whereas none of the 50 most abundant proteins appeared to be secreted products [56]. More recent progress in characterizing translation factors and their protein–protein and RNA–protein interactions demonstrated a significant effect of posttranscriptional control on protein expression [57].

The more recent availability of platform technologies for high-throughput proteome analysis has led to the emergence of cell-related model system integrating messenger RNA profiling with protein expression data for a nucleated cell. The Pearson correlation coefficients for these data range from 0.46 to 0.76 [58]. In these integrated studies, serial analyses of gene expression and DNA microarrays have been used to quantify the transcriptome, while proteome analysis has been based on two-dimensional gel electrophoresis, isotope-coded affinity tags, and multidimensional protein identification technology. To date, limited integration of genetic and proteomic technologies have been developed, despite more recent evidence from mathematical modeling studies that have demonstrated the need to delineate both mRNA and protein expression levels for optimal definition of intracellular networks [59,60].

Initial attempts to correlate mRNA and protein profiles for platelets have been described [3,61]. These studies demonstrated that up to 69% of secreted and cytosole proteins were detectable at the mRNA level, and similar concordance was obtained using two published datasets, suggesting relatively good correlation between proteome and transcriptome data in terms of protein and transcript detection and identification [3]. Presence of some of the messages that were not previously reported in platelets was confirmed at the protein level by the identification of the corresponding proteins. The authors concluded that despite the absence of gene transcription, the platelet proteome is mirrored in the transcriptome and that transcriptional analysis predicts the presence of novel proteins in the platelet. However, as mentioned above, the correlation between mRNA transcript and protein abundance for platelets is low.

## 10.7 INTEGRATING TRANSCRIPTOME AND PROTEOME OF HUMAN PLATELETS

In ongoing studies in our laboratory, we characterized the relationship between platelet transcripts and the platelet proteins through complementary proteomic and genomic techniques. A normal platelet transcriptome database was generated from five mRNA profiles of highly purified normal human platelets, analyzed by Affymetrix microarray [1,30]. Platelets were collected from volunteer donors ($N = 5$) by apheresis to obtain sufficient RNA for hybridization to the Affymetrix U133A gene chip, and expression data were analyzed using GeneSpring 7.0 software. A transcript was considered "platelet-expressed" if it was "present" or "marginal" in four of five platelet samples. Using these strict criteria, 1640 mRNAs were expressed at significant levels by normal platelets [1]. Relative transcript abundance was established by rank-ordering the unique set of non-redundant mRNAs ($N = 1240$) by determining the mean normalized signal intensities across the individual arrays, using computational algorithms as described previously [1,30,62].

A restricted platelet proteomic database was generated by analysis of proteins in cytosolic fraction of human platelets. Cytosolic fractions were extracted using 0.015% digitonin followed by centrifugation to remove platelet membrane proteins. Samples from four different donors were pooled and analyzed in duplicate using liquid chromatography coupled to tandem mass spectrometry (μLC-MS/MS). Automated protein identifications were obtained using Pro ID Software 1.0 (Applied Biosystems) linked to the Swiss-Prot database. Spectral (peptide) counts were used as a semiquantitative means of establishing protein abundance among normalized datasets [63,64], with excellent concordance between platelet runs (Spearman rank correlation $r = 0.87$, $p < 0.0001$).

Two comprehensive bioinformatics approaches (Fig. 10.3) were used to dissect relationships between transcriptome and proteome in platelets as a unique cellular model system devoid of active transcriptional activity. The first approach was based on traditional proteomic strategy, where spectral counts are generally used to identify individual proteins by searching through the entire universe of known proteins represented in Swiss-Prot database. A pool of proteins identified this way included many duplicates and proteins from other species and thus needed additional processing. The second approach was novel in the sense that it used platelet transcriptome to predict all possible platelet-expressed proteins; then spectral counts database was used to (1) confirm or validate these predictions and (2) evaluate protein abundance. This approach searches for proteins that can be expressed only from transcripts detected in platelets (platelet transcriptome). The resulting list of proteins by definition included human proteins only, although some additional effort was needed to remove duplicates.

### 10.7.1 Two Approaches to Dissection of Platelet Proteome

***10.7.1.1 GenomeWide Approach*** The target nucleotide sequences for each Affymetrix probe set were downloaded from the Affymetrix analysis Web database; these "non-full-length" sequences were then used to download full-length platelet nucleotide sequences from RefSeq, a curated and nonredundant collection of sequences representing genomic data, transcripts, and proteins [65]. Full-length sequences were available for 1603 of the 1640 Affymetrix accessions, 1240 of which represented unique, nonredundant sequences; this 1240-member subset was used for all subsequent platelet transcript analyses. Amino acid sequences for each accession number identified by μLC-MS/MS were downloaded from the NCBI database [65], and two methods were applied in parallel to further develop the platelet protein database. In the first case, the amino acid sequences in FASTA format were queried against the human RefSeq database using blastp [protein–protein BLAST (basic alignment search tool)] [66], thereby providing a unique RefSeq identification number (RefSeq NP ID) for subsequent comparison to platelet RefSeq mRNA transcripts. Additionally, a computer program (BlastClust, NCBI) was used to cluster the platelet protein (and platelet nucleotide sequences) into nonredundant sequence sets. This was particularly important in the platelet protein dataset in which abundant proteins (e.g., actin) were represented by hundreds of NCBI accessions. BlastClust utilizes pairwise matches and places a sequence in a cluster if the sequence matches at least one sequence already in the cluster. For proteins, the BlastP algorithm is used to compute the pairwise matches, whereas the Megablast algorithm is

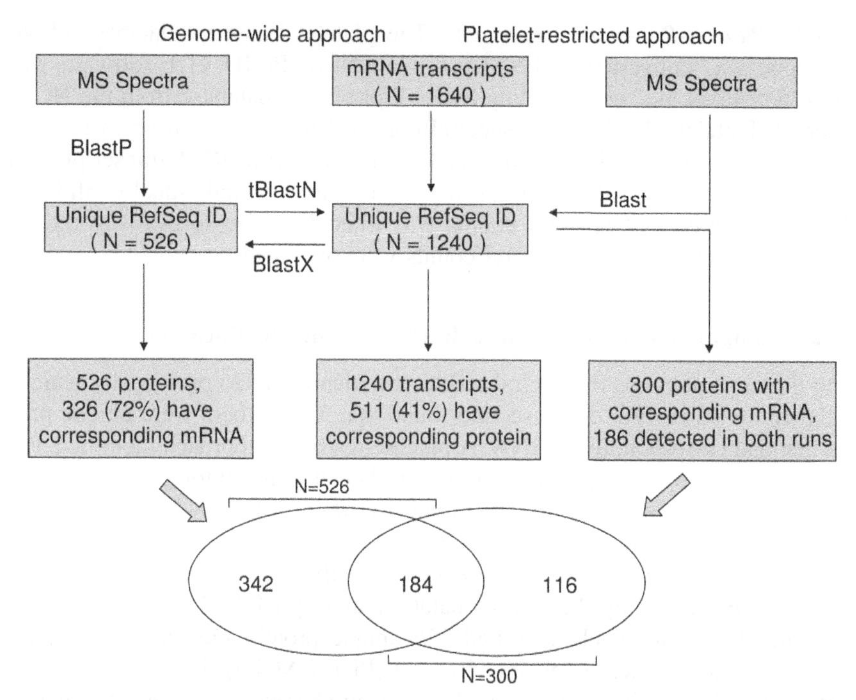


**Figure 10.3** Overall schema for integrated bioinformatics approach used to analyze proteomic and genomic datasets. Platelet protein and platelet transcripts were identified by 2D LC-MS/MS and microarray technology, respectively. The amino acid sequences of the identified proteins and the nucleotide sequences of the detectable transcripts were used to generate individual databases containing full-length sequences. These platelet protein (526 sequences) and platelet nucleotide (1240 sequences) databases were queried against one another using appropriate "blast" algorithms. Blast results were filtered for $E < 0.001$. Below—Venn diagram characterizing two approaches to protein identification.

used for nucleotide sequences. In order to control the stringency of clustering, parameters of the BlastClust program were set to require pairwise matches be 50% identical over an area covering 50% of the length of at least one sequence.

Direct (i.e. platelet-RNA) database comparisons were completed using tBlastN and BlastX. BlastX compares translated products of the platelet nucleotide query against the platelet protein database, thereby identifying proteins that are homologous to a nucleotide coding region [67]. The reported $E$ values provide an estimate of the statistical significance of the match between protein and nucleotide or nucleotide and protein. An $E$ value of $<0.001$ was considered statistically significant. In addition, protein sequences were then queried against the platelet nucleotide sequence database using tBlastN (NCBI), thereby allowing comparison of platelet protein amino acid sequences to the six-frame translations of the platelet nucleotide database. All relational database analyses are derived using $E$ values $<0.001$.

***10.7.1.2 Platelet-Restricted Approach*** The platelet transcript database (1240 transcripts) was converted to FASTA format. Next, ProID v1.1 software was used to download the corresponding Swiss-Prot protein database from the NCBI ftp-server. Platelet MS data (the spectral counts database) was used to identify proteins that can potentially be expressed from given set of RNA transcripts and corresponding proteins. The resulting list of proteins with individual confidence values and numbers of spectral counts was filtered by the software to reduce multiple representation, and spectral counts were averaged.

### 10.7.2 Comparison of Two Approaches to Protein Identification

Using the same database of spectral counts, we identified 526 proteins by searching the entire Swiss-Prot database, and found that 326 of them were unique proteins that had corresponding mRNA. A platelet-restricted approach identified 300 proteins, using the same spectral counts and platelet transcriptome database as a transcript database. In this study, 184 proteins were identified by both approaches (Fig. 10.3), 118 of which were identified in both mass spectrometry runs. Each of these proteins had a unique corresponding mRNA expressed in platelets, as verified independently by the RefSeq database and by BLAST analyses.

The traditional approach identified 326 unique protein sequences, including various chemokines such as platelet factor 4 (PF4, CXCL4), RANTES (CCL5), proplatelet basic protein (CXCL7), and other well-characterized platelet proteins. The majority of platelet transcript/protein match sets in our database are represented in the previously elucidated platelet microparticle proteome [68]. These included proteins expressed at high levels, such as those involved in cytoskeletal regulation and protein synthesis. Presumably, the latter proteins are derived from megakaryocytes and indeed, mRNA was detected for many of these proteins at both the platelet and megakaryocyte levels.

Unlike the traditional approach, our platelet-restricted approach used the same set of spectra counts but matched it to a very limited set of proteins that can be expressed from platelet-expressed mRNAs. Using ProID v1.0 software, we searched to determine which proteins translated from platelet transcriptome database (1240 transcripts) can give peptide sets that match peptides actually detected by MS. This approach generated list of 300 protein/gene pairs. Among those, 136 proteins were not detected by traditional techniques. The union of these two lists constitutes 642 proteins, with 184 proteins identified by both techniques. 342 proteins were identified only by the traditional approach, only 116 were identified whereas using the platelet transcriptome database as a source of proteins.

These results demonstrate that an integrated analytical platform incorporating both transcriptomic and proteomic databases considerably enhances efficiency of platelet protein detection. Indeed, integrating the results of genomic and proteomic studies may help to elucidate functional complexes and signaling pathways involved in platelet activation responses, as well as facilitate the identification of novel platelet therapeutic targets [69].

The establishment of a comprehensive database of platelet transcripts and proteins would help platelet researchers, and ensure a comprehensive representation of the platelet proteome [70]. To date, platelet proteomic data have become increasingly available in various databases. For example, the Swiss-Prot Webpage has published the platelet proteome on a 2D gel (`http://ca.expasy.org`). The Reactome Website (`http://www.reactome.org`) includes information about the complex pathways that operate in platelets [41]. The Human Protein Reference Database (`http://www.hprd.org`) includes features such as PhosphoMotif Finder and Human Proteinpedia [71]. Dittrich et al. [72] created a central resource for the platelet proteome, interactome, and phosphorylation state known as PlateletWeb (`http://plateletweb.bioapps.biozentrum.uni-wuerzburg.de`). This "virtual platelet" knowledge base also includes a characterization of the platelet protein kinase repertoire (kinome).

## 10.8 PROTEINS LACKING A CORRESPONDING mRNA

The degree of correlation between platelet proteomic and transcriptomic data in the endpoints of detection and identification varies by study [72–74]. Some investigators have found that about a third of platelet proteins identified by proteomic methods are not reflected in the transcriptome [37,73]. The discordance may be due to (1) the limited mRNA stability of these genes; (2) failure of microarray analysis to detect very low levels of RNA; (3) the occurrence of proteins that may be synthesized in megakaryocytes, after which mRNA is degraded; and (4) the fact that some proteins may be taken up from plasma or from other cells rather than synthesized in megakaryocytes or platelets [75].

Traditional genomewide proteomic analysis of our proteomic and transcriptomic data allowed identification of 43 platelet proteins without corresponding mRNA. Probes for the majority of their mRNAs (41 of 43) were represented on the Affymetrix microarray chip; however, the fluorescent signal was below the threshold level. As a result, Affymetrix software flagged these mRNA as absent or below detection level. Proteins in this group include albumin (likely explained as a high-abundance plasma "contaminant"), and fibrinogen—previously known to be endocytosed through megakaryocyte/platelet GPIIb/IIIa ($\alpha$IIb$\beta$3) receptors [76]. Interestingly, no $\alpha$- or $\gamma$-fibrinogen transcripts were detected, although mRNAs were present for $\beta$-fibrinogen. Finally, this group also encompassed a wide variety of enzymes (catalase, superoxide dismutase, ATPases, etc.) and other proteins (clathrin, cyclin C, SNAREs, etc.) that presumably identify a discrete subset of platelet proteins with short mRNA half-lives. Interestingly, more recent data suggest that platelets contain mRNAs with intronic sequences that retain the unusual capacity for extranuclear (cytoplasmic) pre-mRNA splicing, providing an additional level of diversity in controlling platelet protein expression [6]. In some instances, RNA may have degraded after protein synthesis in the megakaryocytes or the circulating platelet. Differences in the half-lives of proteins and mRNAs may have influenced this relationship. Other platelet proteins such as fibrinogen are largely scavenged from plasma, and thus might not have a corresponding transcript at the platelet level.

## 10.9 CONCLUDING REMARKS

Lacking ongoing genomic DNA transcription, the platelet represents a unique model system for studying transcriptome–proteome interactions. Translation of platelet mRNAs into proteins is a tightly regulated process, which allows the platelet to respond effectively to different internal stimuli. This process is regulated at several levels: (1) it is regulated by selection of megakaryocyte-derived mRNAs and pre-mRNAs to be packaged into platelets; (2) it is regulated by splicing of selected pre-mRNAs and biosynthesis of specific proteins in response to different stimuli—platelet proteome changes during platelet aging, with 117 proteins changing their relative abundance during 5-day storage [77]; and (3) proteome of normal platelets and platelets of patients with type 2 diabetes is significantly different—the quantitative proteomic approach identified 122 proteins that were either up- or downregulated in type 2 diabetics relative to nondiabetic controls. The question as to whether these differences are reflected in platelet transcriptome remains to be answered.

Identification of novel proteins in platelets can benefit from integration of transcriptomic and proteomic studies. Macaulay et al. [78] have demonstrated feasibility of discovery of novel platelet receptors using transcriptome analysis and protein prediction. Comparison of transcriptomes of *in vitro* differentiated megakaryocytes and erythroblasts identified 151 megakaryocyte-overexpressed transcripts that encode transmembrane-domain-containing proteins, including many proteins not previously characterized in human platelets. Flow cytometry analysis confirmed expression of *G6b, G6f*, and *LRRC32* receptors on the surface of platelets. Expression of these receptors is restricted to platelets. Immunological studies confirmed expression of two other receptors in platelets: *LAT2* and *SUCNR1*. Their expression was not restricted to platelet lineage. An important role of *LRRC32* in promotion of clot formation was demonstrated using functional genomics studies in *Danio rerio* (zebrafish) [79]. Antisense morpholinooligonucleotide—based knockdown in zebrafish represents convenient, fast, and powerful model system to study the function of novel proteins in human platelets, as shown in Chapter 11.

In the future, integration of proteomic and transcriptomic studies can be used to overcome existing limitations of proteomic technologies, such as identification of hydrophobic proteins and proteins with significant levels of posttranslational modifications. As already demonstrated, comparative transcriptomic studies can be used to predict novel proteins and pathways in platelets. Transcriptomic analysis of pre-mRNA splicing in platelets can shed light on novel molecular mechanisms of platelet function.

## REFERENCES


1. Gnatenko DV, Dunn JJ, McCorkle SR, Weissmann D, Perrotta PL, Bahou WF, Transcript profiling of human platelets using microarray and serial analysis of gene expression, *Blood* **101**(6):2285–2293 (2003).
2. Bugert P, Dugrillon A, Gunaydin A, Eichler H, Kluter H, Messenger RNA profiling of human platelets by microarray hybridization, *Thromb. Haemost.* **90**(4):738–748 (2003).
3. McRedmond JP, Park SD, Reilly DF, Coppinger JA, Maguire PB, Shields DC, et al., Integration of proteomics and genomics in platelets: A profile of platelet proteins and platelet-specific genes, *Mol. Cell. Proteom.* **3**(2):133–144 (2004).
4. Dittrich M, Birschmann I, Pfrang J, Herterich S, Smolenski A, Walter U, et al., Analysis of SAGE data in human platelets: Features of the transcriptome in an anucleate cell, *Thromb. Haemost.* **95**(4):643–651 (2006).
5. Sauer S, Lange BM, Gobom J, Nyarsik L, Seitz H, Lehrach H, Miniaturization in functional genomics and proteomics, *Nat. Rev. Genet.* **6**(6):465–476 (2005).
6. Denis MM, Tolley ND, Bunting M, Schwertz H, Jiang H, Lindemann S, et al., Escaping the nuclear confines: Signal-dependent pre-mRNA splicing in anucleate platelets, *Cell* **122**(3):379–391 (2005).
7. Stenberg PE, Hill RL, Platelets and megakaryocytes, in Lee G, Foerster J, Lukens J, Paraskevas F, Greer J, Rodgers G, eds., *Wintrobe's Clinical Hematology*, Lippincott William & Wilkins, Philadelphia, 1999, pp. 615–660.
8. Newman P, Gorski J, White G, Gidwitz S, Cretney C, Aster R, Enzymatic amplification of platelet-specific messenger RNA using the polymerase chain reaction, *J. Clin. Invest.* **82**:739–743 (1988).
9. Fink L, Holschermann H, Kwapiszewska G, Muyal JP, Lengemann B, Bohle RM, et al., Characterization of platelet-specific mRNA by real-time PCR after laser-assisted microdissection, *Thromb. Haemost.* **90**(4):749–756 (2003).
10. Rinder H, Schuster J, Rinder C, Wang C, Schwseidler H, Smith B, Correlation of thrombosis with increased platelet turnover in thrombocytosis, *Blood* **91**:1288–1294 (1998).
11. Kieffer N, Guichard J, Farcet JP, Vainchenker W, Breton-Gorius J, Biosynthesis of major platelet proteins in human blood platelets, *Eur. J. Biochem.* **164**(1):189–195 (1987).
12. Weyrich AS, Zimmerman GA, Platelets: Signaling cells in the immune continuum, *Trends Immunol.* **25**(9):489–495 (2004).
13. Weyrich A, Dixon D, Pabla R, Elstad M, McIntyre T, Prescott S, et al., Signal-dependent translation of a regulatory protein, Bcl-2, in activated human platelets, *Proc. Natl. Acad. Sci. USA* **95**:5556–5561 (1998).
14. Hasegawa S, Pawankar R, Suzuki K, Nakahata T, Furukawa S, Okumura K, et al., Functional expression of the high affinity receptor for IgE (FcepsilonRI) in human platelets and its' intracellular expression in human megakaryocytes, *Blood* **93**(8):2543–2551 (1999).

15. Power CA, Clemetson JM, Clemetson KJ, Wells TN, Chemokine and chemokine receptor mRNA expression in human platelets, *Cytokine* **7**(6):479–482 (1995).
16. Hsu TC, Shore SK, Seshsmma T, Bagasra O, Walsh PN, Molecular cloning of platelet factor XI, an alternative splicing product of the plasma factor XI gene, *J. Biol. Chem.* **273**(22):13787–13793 (1998).
17. Weyrich AS, Dixon DA, Pabla R, Elstad MR, McIntyre TM, Prescott SM, et al., Signal-dependent translation of a regulatory protein, Bcl-3, in activated human platelets, *Proc. Natl. Acad. Sci. USA* **95**(10):5556–5561 (1998).
18. Hoheisel JD, Microarray technology: Beyond transcript profiling and genotype analysis, *Nat. Rev. Genet.* **7**(3):200–210 (2006).
19. Sausville EA, Holbeck SL, Transcription profiling of gene expression in drug discovery and development: the NCI experience, *Eur. J. Cancer.* **40**(17):2544–2549 (2004).
20. Butte A, The use and analysis of microarray data, *Nat. Rev. Drug Discov.* **1**(12):951–960 (2002).
21. Velculescu V, Zhang L, Vogelstein B, Kinzler K, Serial analysis of gene expression, *Science* **270**:484–487 (1995).
22. Zhang L, Zhou W, Velculescu V, Kern S, Hruban R, Hamilton S, et al., Gene expression profiles in normal and cancer cells, *Science* **276**:1268–1272 (1997).
23. Dunn JJ, McCorkle SR, Praissman LA, Hind G, Van Der Lelie D, Bahou WF, et al., Genomic signature tags (GSTs): A system for profiling genomic DNA, *Genome Res.* **12**(11):1756–1765 (2002).
24. Wang E, Miller L, Ohnmacht G, Liu E, Marincola F, High-fidelity mRNA amplification for gene profiling, *Nat. Biotechnol.* **18**:457–459 (2000).
25. Peters D, Kassam A, Feingold E, Heidrich-O'Hare E, Yonas H, Ferrell R, et al., Comprehensive transcript analysis in small quantities of mRNA by SAGE-Lite, *Nucl. Acid Res.* **27**(24):e39 (1999).
26. Canales RD, Luo Y, Willey JC, Austermiller B, Barbacioru CC, Boysen C, et al., Evaluation of DNA microarray results with quantitative gene expression platforms, Nat. *Biotechnol.* **24**(9):1115–1122 (2006).
27. Zheng Z, Luo Y, McMaster GK, Sensitive and quantitative measurement of gene expression directly from a small amount of whole blood, *Clin. Chem.* **52**(7):1294–1302 (2006).
28. Flagella M, Bui S, Zheng Z, Nguyen CT, Zhang A, Pastor L, et al., A multiplex branched DNA assay for parallel quantitative gene expression profiling, *Anal. Biochem.* **352**(1):50–60 (2006).
29. Gnatenko DV, Zhu W, Bahou WF, Multiplexed genetic profiling of human blood platelets using fluorescent microspheres, *Thromb. Haemost.* **100**(5):929–936 (2008).
30. Gnatenko DV, Cupit LD, Huang EC, Dhundale A, Perrotta PL, Bahou WF, Platelets express steroidogenic 17beta-hydroxysteroid dehydrogenases. Distinct profiles predict the essential thrombocythemic phenotype, *Thromb. Haemost.* **94**(2):412–421 (2005).
31. Wicki AN, Walz A, Gerber-Huber SN, Wenger RH, Vornhagen R, Clemetson KJ, Isolation and characterization of human blood platelet mRNA and construction of a cDNA library in lambda gt11. Confirmation of the platelet derivation by identification of GPIb coding mRNA and cloning of a GPIb coding cDNA insert, *Thromb. Haemost.* **61**(3):448–453 (1989).

32. Newman PJ, Gorski J, White GC 2nd, Gidwitz S, Cretney CJ, Aster RH, Enzymatic amplification of platelet-specific messenger RNA using the polymerase chain reaction, *J. Clin. Invest*. **82**(2):739–743 (1988).
33. Rox JM, Bugert P, Muller J, Schorr A, Hanfland P, Madlener K, et al., Gene expression analysis in platelets from a single donor: Evaluation of a PCR-based amplification technique, *Clin. Chem*. **50**(12):2271–2278 (2004).
34. Dittrich M, Birschmann I, Stuhlfelder C, Sickmann A, Herterich S, Nieswandt B, et al., Understanding platelets. Lessons from proteomics, genomics and promises from network analysis, *Thromb. Haemost*. **94**(5):916–925 (2005).
35. Welle S, Bhatt K, Thornton C, Inventory of high-abundance mRNAs in skeletal muscle of normal men, *Genome Res*. **9**:506–513 (1999).
36. Gnatenko DV, Bahou WF, Recent advances in platelet transcriptomics, *Transfus. Med. Hemother*. **33**:217–226 (2006).
37. Gnatenko DV, Perrotta PL, Bahou WF, Proteomic approaches to dissect platelet function: One-half of the story, *Blood* **108**(13):3983–3991 (2006).
38. Macaulay IC, Carr P, Gusnanto A, Ouwehand WH, Fitzgerald D, Watkins NA, Platelet genomics and proteomics in human health and disease, *J. Clin. Invest*. **115**(12):3370–3377 (2005).
39. Hyman T, Huizing M, Blumberg PM, Falik-Zaccai TC, Anikster Y, Gahl WA, Use of a cDNA microarray to determine molecular mechanisms involved in grey platelet syndrome, *Br. J. Haematol*. **122**(1):142–149 (2003).
40. Goerttler PS, Kreutz C, Donauer J, Faller D, Maiwald T, Marz E, et al., Gene expression profiling in polycythaemia vera: Overexpression of transcription factor NF-E2, *Br. J. Haematol*. **129**(1):138–150 (2005).
41. Kralovics R, Teo SS, Buser AS, Brutsche M, Tiedt R, Tichelli A, et al., Altered gene expression in myeloproliferative disorders correlates with activation of signaling by the V617F mutation of Jak2, *Blood* **106**(10):3374–3376 (2005).
42. Briere JB, Essential thrombocythemia, *Orphanet J. Rare Dis*. **2**:3 (2007).
43. Geissler WM, Davis DL, Wu L, Bradshaw KD, Patel S, Mendonca BB, et al., Male pseudohermaphroditism caused by mutations of testicular 17 beta-hydroxysteroid dehydrogenase 3, *Nat. Genet*. **7**(1):34–39 (1994).
44. Bray JE, Marsden BD, Oppermann U, The human short-chain dehydrogenase/reductase (SDR) superfamily: A bioinformatics summary, *Chemicobiol. Interact*. **178**(1–3):99–109 (2009).
45. Johnson M, Ramey E, Ramwell PW, Sex and age differences in human platelet aggregation, *Nature* **253**(5490):355–357 (1975).
46. Jones SB, Bylund DB, Rieser CA, Shekim WO, Byer JA, Carr GW, Alpha 2-adrenergic receptor binding in human platelets: Alterations during the menstrual cycle, *Clin. Pharmacol. Ther*. **34**(1):90–96 (1983).
47. Pilo R, Aharony D, Raz A, Testosterone potentiation of ionophore and ADP induced platelet aggregation: Relationship to arachidonic acid metabolism, *Thromb. Haemost*. **46**(2):538–542 (1981).
48. Castellsague J, Perez Gutthann S, García Rodriguez LA, Recent epidemiological studies of the association between hormone replacement therapy and venous thromboembolism. A review, *Drug Saf*. **18**(2):117–123 (1998).

49. Daly E, Vessey MP, Hawkins MM, Carson JL, Gough P, Marsh S, Risk of venous thromboembolism in users of hormone replacement therapy, *Lancet* **348**(9033):977–980 (1996).
50. Wang HY, Malek RL, Kwitek AE, Greene AS, Luu TV, Behbahani B, et al., Assessing unmodified 70-mer oligonucleotide probe performance on glass-slide microarrays, *Genome Biol*. **4**(1):R5 (2003).
51. Cawley GC, Talbot NL, Fast exact leave-one-out cross-validation of sparse least-squares support vector machines, *Neural Netw*. **17**(10):1467–1475 (2004).
52. Gnatenko DV, Zhu W, Xu X, Samuel ET, Monaghan M, Zarrabi MH, et al., Class prediction models of thrombocytosis using genetic biomarkers, *Blood* **115**(1):7–14 (2010).
53. Griffin TJ, Gygi SP, Ideker T, Rist B, Eng J, Hood L, et al., Complementary profiling of gene expression at the transcriptome and proteome levels in Saccharomyces cerevisiae, *Mol. Cell. Proteom*. **1**(4):323–333 (2002).
54. Gygi SP, Rochon Y, Franza BR, Aebersold R, Correlation between protein and mRNA abundance in yeast, *Mol. Cell. Biol*. **19**(3):1720–1730 (1999).
55. Futcher B, Latter GI, Monardo P, McLaughlin CS, Garrels JI, A sampling of the yeast proteome, *Mol. Cell. Biol*. **19**(11):7357–7368 (1999).
56. Anderson L, Seilhamer J, A comparison of selected mRNA and protein abundances in human liver, *Electrophoresis* **18**(3–4):533–537 (1997).
57. McCarthy JE, Posttranscriptional control of gene expression in yeast, *Microbiol. Mol. Biol. Rev*. **62**(4):1492–1553 (1998).
58. Hack CJ, Integrated transcriptome and proteome data: the challenges ahead, *Brief Funct. Genom. Proteom*. **3**(3):212–219 (2004).
59. Hatizmanikatis V, Choe L, Lee K, Proteomics: Theoretical and experimental considerations, *Biotechnol. Prog*. **15**:312–318 (1999).
60. Hatizmanikatis V, Lee K, Dynamical analysis of gene networks requires both mRNA and protein expression information, *Metabol. Eng*. **1**:275–281 (1999).
61. Macaulay IC, Carr P, Gusnanto A, Ouwehand WH, Fitzgerald D, Watkins NA, Platelet genomics and proteomics in human health and disease, *J. Clin. Invest*. **115**(12):3370–3377 (2005).
62. Gnatenko D, Perrotta P, Ji C, Zhu W, Bahou W, Platelet gene/protein expression analyses using an integrated platform, *Blood* **110**(11):1065A (2007).
63. Sandhu C, Connor M, Kislinger T, Slingerland J, Emili A, Global protein shotgun expression profiling of proliferating mcf-7 breast cancer cells, *J. Proteome Res*. **4**(3):674–689 (2005).
64. Liu H, Sadygov RG, Yates JR 3rd, A model for random sampling and estimation of relative protein abundance in shotgun proteomics, *Anal. Chem*. **76**(14):4193–4201 (2004).
65. Pruitt KD, Tatusova T, Maglott DR, NCBI Reference Sequence (RefSeq): A curated non-redundant sequence database of genomes, transcripts and proteins, *Nucleic Acids Res*. **33**(database issue):D501–D504 (2005).
66. Altschul SF, Gish W, Miller W, Myers EW, Lipman DJ, Basic local alignment search tool, *J. Mol. Biol*. **215**(3):403–410 (1990).

67. Altschul SF, Madden TL, Schaffer AA, Zhang J, Zhang Z, Miller W, et al., Gapped BLAST and PSI-BLAST: A new generation of protein database search programs, *Nucleic Acids Res*. **25**(17):3389–3402 (1997).

68. García BA, Smalley DM, Cho H, Shabanowitz J, Ley K, Hunt DF, The platelet microparticle proteome, *J. Proteome Res*. **4**(5):1516–1521 (2005).

69. Maguire PB, Platelet proteomics: Identification of potential therapeutic targets, *Pathophysiol. Haemost. Thromb*. **33**(5–6):481–486 (2004).

70. Watson SP, Bahou WF, Fitzgerald D, Ouwehand W, Rao AK, Leavitt AD, Mapping the platelet proteome: A report of the ISTH Platelet Physiology Subcommittee, *J. Thromb. Haemost*. **3**(9):2098–2101 (2005).

71. Keshava Prasad TS, Goel R, Kandasamy K, Keerthikumar S, Kumar S, Mathivanan S, et al., Human Protein Reference Database—2009 update, *Nucleic Acids Res*. **37**(database issue):D767–D772 (2009).

72. Dittrich M, Birschmann I, Mietner S, Sickmann A, Walter U, Dandekar T, Platelet protein interactions: Map, signaling components, and phosphorylation groundstate, *Arterioscler. Thromb. Vasc. Biol*. **28**(7):1326–1331 (2008).

73. Greening DW, Glenister KM, Kapp EA, Moritz RL, R.L. S, G.W. L, et al., Comparison of human platelet membrane-cytoskeletal proteins with the plasma proteome: Towards understanding the platelet-plasma nexus, *Proteom. Clin. Appl*. **2**(1):63–77 (2008).

74. Hillmann AG, Harmon S, Park SD, O'Brien J, Shields DC, Kenny D, Comparative RNA expression analyses from small-scale, single-donor platelet samples, *J. Thromb. Haemost*. **4**(2):349–356 (2006).

75. Bugert P, Ficht M, Kluter H, Towards the identification of novel platelet receptors: comparing RNA and proteome approaches, *Transfus. Med. Hemother*. **33**(3):236–243 (2006).

76. Handagama PJ, Amrani DL, Shuman MA, Endocytosis of fibrinogen into hamster megakaryocyte alpha granules is dependent on a dimeric gamma A configuration, *Blood* **85**(7):1790–1795 (1995).

77. Springer DL, Miller JH, Spinelli SL, Pasa-Tolic L, Purvine SO, Daly DS, et al., Platelet proteome changes associated with diabetes and during platelet storage for transfusion, *J. Proteome Res*. **8**(5):2261–2272 (2009).

78. Macaulay IC, Tijssen MR, Thijssen-Timmer DC, Gusnanto A, Steward M, Burns P, et al., Comparative gene expression profiling of *in vitro* differentiated megakaryocytes and erythroblasts identifies novel activatory and inhibitory platelet membrane proteins, *Blood* **109**(8):3260–3269 (2007).

79. O'Connor MN, Salles, II, Cvejic A, Watkins NA, Walker A, Garner SF, et al., Functional genomics in zebrafish permits rapid characterization of novel platelet membrane proteins, *Blood* **113**(19):4754–4752 (2009).

# 11

# PLATELET FUNCTIONAL GENOMICS

Isabelle I. Salles, Marie N. O'Connor, Daphne C. Thijssen-Timmer, Katleen Broos, and Hans Deckmyn

**Abstract**

During the last few decades, major advances in our understanding of platelet generation and function have been achieved through rapid development of molecular and genetic technologies. Nevertheless, it remains a challenge to assign the functional roles of candidate genes and proteins identified by genomic and proteomic studies in platelets and megakaryocytes, as direct molecular biology approaches are hampered by the anucleated nature of platelets. This chapter summarizes different approaches that can be applied for platelet functional genomics.

## 11.1 INTRODUCTION

The central role that platelets play in hemostasis and thrombotic events is due to their capacity to adhere to damaged or activated endothelium and to aggregate to form a thrombus. The role of quite a number of ligand–receptor interactions and signal transduction events in these processes is known; however, more recent transcriptome and proteome studies of platelets and the megakaryocyte (MK), their precursor cell, have dramatically increased our knowledge of the existence of the proteins present in human platelets [1]. Analysis of the data revealed that 10,314 genes were expressed in MK, 279 of which were exclusively expressed in MK compared to all other blood cells [2]. Of these, a substantial number encode

*Platelet Proteomics: Principles, Analysis and Applications*, First Edition.
Edited by Ángel García and Yotis A. Senis.

proteins whose roles in hemostasis and thrombosis are currently unknown. Unfortunately, most of the approaches that can classically be applied in functional genomics to characterize the function of novel proteins are of limited use when studying anucleated human platelets or investigating large sets of candidate genes. In this chapter we review and summarize the experimental options that are currently (as of 2010) being investigated for platelet functional genomics, most of which are still very much in development and require proof of their feasibility and applicability (Fig. 11.1).

## 11.2 NONMAMMALIAN VERTEBRATE MODEL SYSTEMS

### 11.2.1 General Characteristics

Nonvertebrate model organisms such as *Drosophila* or *Caenorhabditis elegans* are powerful genetic tools, yet their hematologic systems share very little similarity with that of humans. At the genetic level in *Drosophila*, hematologically important transcription factor families and signaling pathways are present [3]; however, they lack a cell equivalent of the human platelet, and thus are unsuitable for use as a model for platelet function. Nonmammalian vertebrates, on the other hand, are well suited to modeling of hemostasis. There is evidence for a coagulation system in all vertebrates, from the "primitive" one present in jawless vertebrates (e.g., hagfish and lamprey) involving tissue factor, prothrombin, fibrinogen, and a subset of the vitamin K–dependent serine proteases, to the more complex and remarkably well-conserved system found in jawed vertebrates and mammals [4]. Nonmammalian vertebrates such as birds [5], frogs [6], and fish [7,8] have nucleated cells equivalent to platelets, called thrombocytes. Since the 1990s, the zebrafish (*Danio rerio*) has come to the fore as a valid model for thrombosis and hemostasis [9,10], and a number of tools have been developed to facilitate functional genomics in this model. In this section, the relevance of zebrafish as a model, key studies, and its potential will be discussed.

### 11.2.2 Zebrafish Hematopoiesis and Hemostasis

The zebrafish is a freshwater tropical cyprinid fish. It was first introduced as a genetic model organism in the 1980s [11] and is now the premier nonmammalian vertebrate model of development and disease [12]. In particular, the zebrafish is an attractive model organism for studying vertebrate hemostasis, as it uniquely combines the advantages of genetic tractability with biological relevance [9,10]. Blood cell morphology and number are similar [7,13], and the coagulation system is highly conserved [14,15]. Zebrafish have erythrocytes, monocytes, granulocytes, and thrombocytes, and a fully functioning adaptive immune system including T and B cells, antigen presenting cells, and natural killer (NK)-like cells (reviewed in Ref. 16).

Conserved genetic programmes regulate hematopoiesis in humans and zebrafish. Large-scale forward genetic screens have identified zebrafish mutants

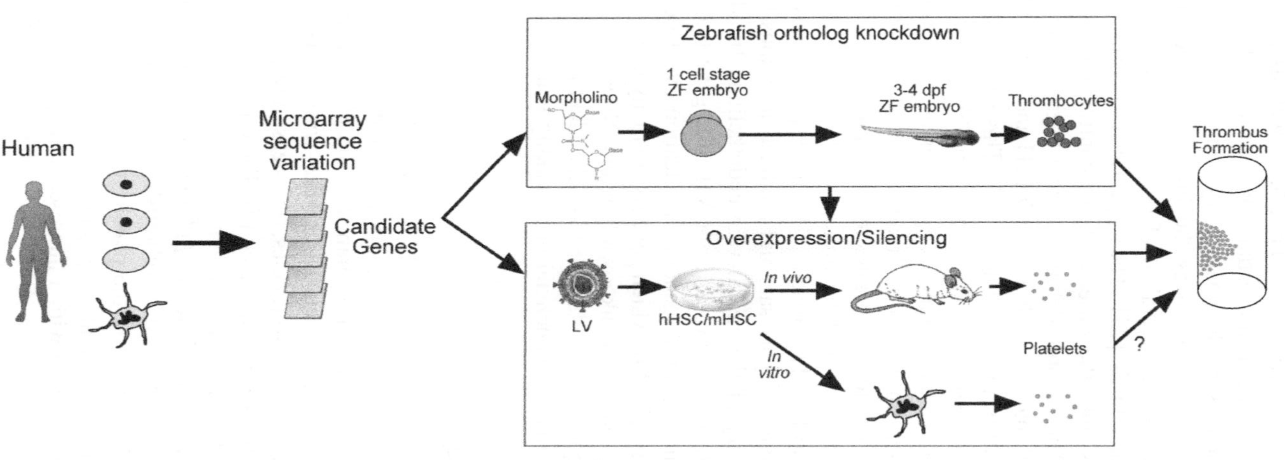


**Figure 11.1** Overview of platelet functional genomic approaches. Candidate genes identified from various gene expression studies can be further characterized in a zebrafish thrombosis model using morpholino technology or using *in vivo* or *in vitro* systems to produce genetically modified platelets from hematopoietic progenitor cells.

with hematopoietic defects that model human hematological disorders, and show that transcription factors such as *scl* [17,18], *gata1, gata2* [19], *runx1* [20], and *lmo2* [21] have conserved function. In zebrafish, as well as in mammals, hematopoiesis occurs in two sequential and spatially distinct waves [10,13,22,23]. The first, transient wave generates predominantly erythroid cells and occurs in an area of the posterior mesoderm known as the intermediate cell mass [4]. This is the functional and structural equivalent of the extraembryonic yolk sac blood islands in mammals and birds [24]. The appearance of primitive blood cells in circulation coincides with the onset of heart contractions, at 26–28 h postfertilization (hpf) [19]. The second, definitive wave begins around 30 hpf, when hematopoietic stem cell markers such as *c-myb* [21] and *runx1* [20,25,26] are detected in the ventral wall of the dorsal aorta [27]. This site corresponds to the aorta–gonad–mesonephros region of other vertebrates [28].

Molecular [14] and biochemical [15] studies suggest that all components of the coagulation network are present in zebrafish. Treatment of adult zebrafish with warfarin leads to a bleeding phenotype, suggesting a requirement for vitamin K–dependent γ-carboxylation of hemostatic proteins. Knockdown of the prothrombin gene by a morpholino-based approach [29] caused spontaneous bleeding around eye and brain and along the yolk sac, and inhibited thrombus formation induced by laser-induced injury of the caudal vein endothelium. In addition to a gene for factor VII, zebrafish also have a factor VII–like gene that plays an inhibitory role in coagulation [30,31]. A genome duplication event occurred in teleosts ~350 million years ago [32,33], which gave rise to many gene duplicates, or ohnologs, some of which persist with unknown function. Now that we have an almost complete genome sequence for zebrafish [34], the identification of these ohnologs is more robust and may be beneficial in unpicking gene function.

### 11.2.3 Zebrafish Thrombocytes

Unlike their mammalian counterparts, zebrafish thrombocytes and erythrocytes are nucleated. Thrombocytes are also larger, ~5 μm in diameter, and similar in size than lymphocytes [13]. When examined at the ultrastructural level, however, thrombocytes share many structural features with human platelets, including an open canalicular system, α-granules, and dense granules, and they form pseudopodia-like projections when activated [7]. The ontogeny of zebrafish thrombocytes is unknown, although genes encoding transcription factors important for MK development in mammals, such as *fli1* [35], *fog1* [36], *gata1* [19], *nfe2* [37], and *runx1* [20], are expressed in zebrafish, as well as the thrombocytic/MK marker genes *itga2b*, encoding the platelet glycoprotein GPIIb (also known as CD41 and integrin αIIb), and *c-mpl*, encoding the thrombopoietin receptor [8]. Labeling of thrombocytes with a fluorescent lipophilic dye DiI-C18 [31], or with the green fluorescent protein (GFP) under the control of the CD41 promotor by a transgenic approach [8], shows that thrombocytes enter circulation as early as 36 hpf. There is also evidence for two populations of thrombocytes: "young" thrombocytes, which initiate formation of arterial thrombi; and older, "mature" thrombocytes [38]. Further investigation is needed to determine the source of these embryonic thrombocytes, their lifespan, and later sites of thrombopoiesis [4].

Importantly, thrombocytes function in a manner similar to that for human platelets [7,38,39]. *In vitro*, they adhere or aggregate in response to ristocetin, collagen, thrombin, ADP, and arachidonic acid [7], and this response can be inhibited by antiplatelet drugs such as the GPIIb/IIIa antagonist Reopro® and acetylsalicylic acid [7]. *In vivo*, thrombus formation can be induced by chemical or laser injury of blood vessel endothelium of the dorsal aorta or caudal vein [2,31], and these thrombi are composed primarily of thrombocytes [2]. As yet it is unclear what influence the nucleus has on thrombocyte function, and to what extent this will cause differences in behavior to humans. More recent studies have shown that splicing and translation can occur *de novo* in human platelets [40]; therefore, this may not be such an important concern, in particular when studying acute responses such as thrombus formation.

### 11.2.4 Zebrafish as a Powerful Tool for Functional Genomics

The mouse is the gold standard vertebrate model because it is a mammal, small in size, with a rapid breeding time and numerous genetic and immunologic tools available. Gene targeting techniques in pluripotent embryonic stem (ES) cells in combination with more recent advances using site-specific recombinase systems now allow full manipulation of the mouse genome. Despite the invaluable knowledge gained from these methodologies in modeling human diseases, mouse studies still require extensive laboratory work, are time-consuming and expensive, especially for *in vivo* genomewide genetic studies. Over the years, the zebrafish has emerged as an attractive vertebrate model organism for studying human hematopoiesis [10] and also many aspects of human biology and diseases [12,41,42]. The external fertilization and the optical clarity of the embryo facilitate phenotypic analysis of developmental processes and organ formation *in vivo* in real time, without invasive manipulation, and better visualization of early blood-related phenotypes compared to mice, where development occurs *in utero*. Moreover, zebrafish gain sufficient oxygenation via passive diffusion to develop normally for several days without blood circulation [43,44], thus allowing mutations of genes perturbing the vascular system to be investigated, which would cause very early embryonic lethality in mammals. Zebrafish reach sexual maturity in only 3 or 4 months, and adult females can produce up to 200–300 eggs per week. Because of their small size, thousands of animals can be maintained in a fish facility and require much less space than mice and therefore are very cost-effective for large-scale genetic screens. Zebrafish maintain diploidy during development as opposed to other fish that can be tri- or tetraploid, facilitating genetic analysis. Additionally, considerable resources of the zebrafish genome are readily available and constantly evolving (for information on the *Danio rerio* sequencing project, see Ref. 45), facilitating identification and characterization of mutations or orthologs of human candidate genes.

Forward genetic approaches have traditionally been used for the identification of mutant phenotypes. The first large-scale mutagenesis screenings in the zebrafish were performed using the chemical mutagen ethylnitrosourea (ENU), where about 2000 zebrafish mutants were generated [46,47], followed by other studies that allowed the identification of zebrafish hematopoietic mutants [10]. To perform these screens, adult male fish are exposed to ENU, which induces point mutations in the germ cells, and then are mated to normal female fish. Heterozygous F1 offspring are then mated to wild-type fish again, and the resulting F2 families are intercrossed in order to identify the mutation carrier. Forward genetic screens are limited by redundancy, however, hindering straightforward identification of genes whose mutations are responsible for the particular phenotype. An alternative approach is insertional mutagenesis where exogenous DNA serves as the mutagen and also as a molecular tag for identifying the mutated gene responsible for the phenotype [48]. The retroviral murine leukemia virus (MLV) [49] and *Tol2* transposon [50] have been used successfully in zebrafish.

As opposed to forward genetics, reverse genetics involves studying the phenotypic consequences of altering the function of a gene of interest. Potential candidate genes issued from human gene expression or genome wide association studies can be studied in zebrafish by mutating or knocking down the ortholog gene. In the zebrafish, mutants in orthologs of selected human genes are currently generated by TILLING (targeting-induced local lesions in genomes) [51–53]. Chemical (e.g., ENU), but also insertional, mutagenesis using transposons and retroviruses can be used to achieve target-selected mutagenesis [54]. Random mutagenesis followed by PCR is used to amplify the regions of gene of interest to identify point mutations leading to potential hypomorphs, antimorphs, and neomorphs [54]. The success of identifying mutations depends however on the size and the high frequency of mutations in the generated DNA library and of the target gene size. Random mutagenesis is performed on male zebrafish that are subsequently mated to wild-type females. The F1 males are sacrificed for genotyping and their sperm samples cryopreserved to recover the line of interest, or alternatively, DNA from live F1 male fish is obtained from tail clipping [55]. This method allows identification of not only mutations in an arbitrary gene, but also allelic series of mutations that can provide insights into the molecular mechanism of disease [55]. As both forward and reverse genetic screens are generated from the same starting material, they can be combined and offer complementary valuable information [52].

A most powerful technique in zebrafish genomics has been the development of knockdown technology by morpholino oligonucleotides (MOs) [56]. These are antisense oligonucleotide analogs with a morpholine ring that prevents their degradation in the cell. As such, they can persist for several days when microinjected into fertilized one-cell embryos. MOs are designed to bind a specific mRNA at either the translational start site, thus blocking translation [57], or at the splice junction, thereby preventing the correct splicing of the mRNA [58]. They are technically easy to use and induce a phenotype dependent on the dose, thus allowing the identification of phenotypes that may be masked in a null genetic background.

Finally, the generation of transgenic zebrafish by microinjection of DNA constructs into one-cell stage fertilized embryos has also facilitated efficient large-scale genetic screening. For example, tissue-specific expression of GFP placed under control of the *gata1* [59] or *fli1* [60] promoters has facilitated the study of erythropoiesis and angiogenesis, respectively. Because of the particular interest to use the zebrafish for the study of thrombocytes, as mentioned above, Lin et al. have engineered a transgenic line where the zebrafish CD41 promoter was fused to GFP and injected into one-cell embryos [8]. In these transgenic fish GFP is expressed in thrombocytes and thrombocyte precursors, and they have proved invaluable in understanding the fate of hematopoietic stem cells (HSCs) [27,61].

### 11.2.5 Characterization of Novel Platelet Membrane Proteins by Reverse Genetics in Zebrafish

In 2009 we demonstrated the suitability of zebrafish for functional analysis of novel platelet genes by using a laser-induced thrombosis model in combination with MO technology [2]. Knockdown of the genes encoding GPIIb or coagulation factor (F)VIII replicated the well-characterized Glanzmann thrombasthenia and hemophilia A phenotypes, respectively. MO-injected fish did not display any developmental defects, nor was spontaneous bleeding observed. However, when the system was challenged, for instance, by laser injury of artery endothelium or by knocking both genes down in the same individual, impaired thrombus formation or, in the latter case, spontaneous bleeding was observed.

In the thrombosis model, laser injury was performed on the endothelial cell wall of the caudal artery of 3–4 days postfertilization larvae as described previously [29,31]. The transparency of the larvae allows direct visualization by intravital light microscopy of the subsequent thrombus formation without intervention. In this study, the time to attachment (TTA), defined as the time required for the first cell to adhere at the site of injury, and thrombus size area (TSA) at a fixed timepoint postinjury, were measured in control compared to MO-injected sibling larvae. As expected, GPIIb MO-injected fish had a normal TTA compared to control siblings, but displayed a significant reduction in TSA ($>50\%$, $p < 0.001$). Similarly, a normal TTA accompanied by a decreased TSA ($>70\%$, $p < 0.05$) were observed in FVIII MO-injected embryos compared to controls. To assess the specific contribution of thrombocytes in the thrombosis model, we used the CD41-GFP transgenic zebrafish line that has fluorescent thrombocytes [8]. Analysis of thrombus formation in the CD41-GFP transgenic zebrafish showed that thrombi were most often (but not always) initiated by thrombocytes, that GPIIb and FVIII MO-injected transgenic fish displayed the same defects in thrombus formation as did MO-injected wild types, and most importantly that thrombi were composed primarily of thrombocytes.

Previous studies from the Bloodomics consortium revealed that of the 10,314 genes that are expressed in MK, 279 are MK-specific, and of these, 75 encode putative transmembrane domains. Of those, we choose four for functional analysis in the zebrafish: *BAMBI* (bone morphogenetic protein and activin membrane-bound inhibitor), *DCBLD2* (discoidin, CUB, and LCCL domain), *ESAM* (endothelial cell-selective adhesion molecule), and *LRRC32* (leucine-rich repeat C32). A control gene, encoding for one of the two anthrax receptors (ANTXR2), which had a more ubiquitous expression, was also included [2].

After confirming the expression of all genes in human platelets by Western blotting using purpose-generated polyclonal antibodies, and identification of the zebrafish ortholog for each gene, we designed two MOs per gene to act as mutual controls for assessment of the knockdown phenotype. Overall, 42–70 larvae were tested for each gene and combined analysis of data for both MO for each gene revealed significant alterations in cell attachment and/or thrombus growth for all four candidate genes, but not for the ANTXR2 MO-injected fish. BAMBI and LRRC32 MO-injected fish had a significant slower adhesion (TTA) and an impaired thrombus development (TSA). In contrast, both DCBLD2 and ESAM MO-injected fish displayed a normal TTA but increased TSA [2].

This was the first example of reverse genetics using the zebrafish for platelet functional genomics. With the increasing number of candidate genes selected by transcriptional studies of MK/platelets and also in genomewide association studies (e.g., in patients with myocardial infarction) [62–64], this model will undoubtedly lead to better understanding of platelet biology and its role in cardiovascular diseases by identifying candidate genes playing an important role in platelet function.

### 11.2.6 Perspectives

The two best-established approaches to test gene function in zebrafish are knockdown and overexpression of the gene of interest by microinjection of MO [56] or of nucleic acids [65–68], respectively. So far, only the former of these is possible for studying thrombocyte function: DNA or mRNA used in overexpression studies are reliable for only the first 24 hpf, due to degradation and dilution by cell division [65–67], and thrombocytes do not enter blood circulation until 36 hpf [8,31]. Transgenic approaches have been developed to overcome these problems, allowing spatiotemporal control of expression using inducible heat-shock promotor [69–71] or tissue-specific promotors [8]; however, they require a significant time investment for generation of stable transgenic lines. New methodologies include use of temperature-sensitive inteins (protein introns [72,73]) and small molecule/protein tags [74]. MO electroporation [75] or ribozyme-mediated knockdown [76] could also be used for tissue-specific block of gene function, but these tools are still in the early stages of development; selectivity and reproducibility issues are still to be addressed.

Technology for permanently knocking out a gene in zebrafish will allow phenotype analysis at later stages of development and in adults, and with 100% loss of gene function. Creation of disease models by gene knockout could facilitate screens for genes with subtle phenotypes [2], or could be used in drug discovery. The zebrafish is very well suited to high-throughput small-molecule screens.

Zebrafish embryos are small, are fertilized externally, and develop in a simple aqueous solution; thus treatment with any low-molecular-weight pharmacological reagent is very simple as long as it can be dissolved in embryo medium. Large screens have been conducted to identify compounds that influence cell cycle [77], hematopoietic stem cell homeostasis [78], angiogenesis [79], or heart rate [80]. Notably, Peterson et al. screened a small molecule library to discover compounds capable of reversing the phenotype of gridlock mutants (suffering from aortic occlusion similar to aortic coarctation in humans) [81].

Knockout technology is thus far available only in mice, but there has been progress in zebrafish. Cells with ES-like properties have been isolated [82], and nuclear transfer of genetically modified culture cells has been achieved using cultured embryonic fibroblast cells [83]. Still, rates of germline transmission are at present prohibitively low. An exciting alternative is the use of zinc-finger nucleases, which have been used to significantly enhance homologous recombination and the insertion of DNA sequences at specific sites in the genome [84]. Moreover, two more recently published studies describe successful use of designer zinc-finger nucleases to introduce targeted mutations into the zebrafish genome [85,86], thus creating custom knockout fish without the collateral genetic damage associated with TILLING. If this technology lives up to its potential, then the zebrafish will be greatly enhanced as an experimental model for understanding human disease.

## 11.3 GENETIC MANIPULATION OF MAMMALIAN MEGAKARYOCYTES

Knockout and other transgenic mice (knockin, Cre-Lox) have allowed us to study many target proteins in murine platelets; however, these approaches are low-throughput, technically challenging, and time-consuming. The following section gives an overview of how to generate MK and platelets by *in vitro* and *in vivo* approaches and how to genetically modify these cells from hematopoietic and progenitor cells using viral transduction technology.

### 11.3.1 Megakaryopoiesis and Generation of Platelets from Progenitors

***11.3.1.1 Megakaryopoiesis*** Platelet assembly and release are the final events of a multistep process involving commitment, proliferation, and differentiation of hematopoietic progenitors to MK. Maturation of MK occurs in the bone marrow (BM), where polyploid MK extend their proplatelets into microvessels where they are sheared by flowing blood [87]. It is not clear whether platelets are released from proplatelet ends or whether proplatelets detach from MK in bulk and fragment further into platelets [87,88]. Identification of the growth factor thrombopoietin (Tpo) and its receptor c-MPL which is expressed on MK, has allowed for the development of *in vitro* culture techniques to study megakaryopoiesis and platelet biogenesis in more detail and perform genetic manipulations [89,90].

Multipotent HSCs are the only cells with self-renewal capacity, and during commitment decisions, HSCs produce a progeny of cells that first loose their ability to self-renew and then their multipotential properties. Multipotent progenitor cells (MPPs) also named hematopoietic progenitor cells, are defined by the expression of CD34 and CD38, or in the case of mouse MPP cells, by absence or low expression of lineage (Lin) markers and the expression of c-kit, Sca1, and $Flk2^+$ [91,92]. MPP cells give rise to the common lymphoid progenitors (CLPs) and the common myeloid progenitors (CMPs). The common bipotent progenitor for the erythroid and MK lineage is called either MEP or BFU-E/MK and derives from the CMP or, alternatively, directly from the HSC. All fractions—HSC, CMP, CLP, and MEP—can be distinguished phenotypically and purified from BM or cord blood (CB) after selection for $CD34^+$ cells, lineage depletion, and subsequent staining for CD38, Il3RA, and CD45RA [93]. The MK-committed progenitors can be distinguished based on their proliferative capacities in clonal assays [94]: BFU-MK are the most primitive MK progenitors with the highest proliferation, followed by the CFU-MK with a lower proliferative capacity, and finally by more mature progenitors that have been described only in the mouse. During *in vitro* culture, different MK stages can be distinguished by focusing on platelet membrane proteins. The expression of CD34 disappears, overlapping with the appearance of CD41 (GPIIb)/CD61 (GPIIIa), and followed by expression of GPIb, GPIX, GPVI, PF4, VWF, and GPV and at final stage, $\beta_1$-tubulin. In the first stage of megakaryopoiesis, MK cytoplasm volume increases and cells become polyploid by a process called endomitosis. Although polyploidization reflects the MK maturation stage, proplatelets and platelets can also be formed by $2N$ MK. During the second stage, proplatelets are formed that have the appearance of beads linked by thin cytoplasmic bridges. Proplatelets are filled with microtubule bundles originating in the MK cell body, which slide along the shaft and form loops at the tips, before returning to the cell body [88]. Along the microtubules, granules are transported and become trapped in the proplatelet ends [95]. The ends of the proplatelets are then separated from the shafts by sliding movements of the microtubules, mediating platelet release [88]. Although there is strong evidence that platelets are formed primarily at the end of proplatelets, even when proplatelets are dissociated from the MK body [88,96], the precise mechanism on how platelets are released from the proplatelet ends is not known. In the *in vivo* situation, mature MKs protrude their proplatelets into sinusoidal microvessels [87], which are then sheared by the bloodflow, causing mechanical forces that eventually also mediate platelet formation. *In vitro*, MK fragment into many proplatelets and larger cytoplasmic fragments of which the nuclei are extruded. Platelets as well as other (micro) particles are formed, although not as many as one would expect from the estimation that each MK can give rise to 1000–3000 platelets, based on the fact that each day $1 \times 10^{11}$ platelets are produced.

***11.3.1.2 In vitro Platelet Production*** For *in vitro* culturing of human MK and platelets, a single selection for $CD34^+$ cells is performed to obtain the hematopoietic progenitor cells from various sources such as CB, peripheral blood (PB), or

BM. Although differences exist between the proliferation rate, the expression of differentiation markers, and the extent of polyploidization between CB and PB, *in vitro* expanded MK from each source can give rise to functional platelets when transplanted into irradiated NOD/SCID mice (see the following section). No direct comparison has been made between *in vitro*–derived platelets from the different $CD34^+$ sources, and most studies have used CB cells because they are readily available. Studies on the production of murine platelets make use of progenitors isolated from mouse fetal livers [97], BM [98], or murine ES cells [99]. MK expansion protocols vary considerably, using only Tpo [100] or combinations of Tpo with interleukin-$1_\beta$(IL$\beta$) [101] or Tpo with a cocktail of various combinations of cytokines such as IL3, IL6, IL9, IL11, Flit3-L, GM-CSF, or SCF [98,102–104]. Multistep culturing systems using CB $CD34^+$ cells have been developed by several groups in which the MK are first expanded and then differentiated using media containing different cytokines [105–107]. The Proulx group developed a relatively short two-phase culture system with the expansion of $CD34^+$ cells and differentiation of MK in the following culture step, leading to a high MK purity after 14 days [105]. A positive effect on MK progenitor expansion was reported when $CD34^+$ cells were maintained at 39°C during the expansion phase [108]. Culturing MK at high oxygen tension (20%) stimulates the expansion, while a low oxygen tension (5%) stimulates the differentiation [109]. In general, the yield of cultured platelets is very low, which is a major drawback. Since the last step of platelet shedding is the most difficult part of the *in vitro* culturing process, many approaches were undertaken to mimic the natural BM environment by growing mature MK on endothelial cells or mesenchymal stem cells (MSCs). These feeder layers provide an extracellular matrix and may produce soluble chemokines or cytokines necessary for growth and differentiation cues. Indeed, there is evidence for a better expansion of $CD34^+$ cells on feeder cells such as MSC [110] or increased MK proliferation and polyploidization on bone marrow endothelial cells (BMECs) [111], but no supporting effects have been reported on the production of platelets. Only one publication has successfully described the production of platelets on a feeder layer [106]. In the three-step culture system developed by Matsunaga et al., MKs were initially expanded on telomerase gene-transduced human stromal cells (hTERT stroma) and were subsequently transferred to a liquid culture system for the final stage of platelet formation [106]. This result is the highest overall production of platelets observed until now ($2 \times 10^{11}$ platelets per CB unit) but has a relatively low MK purity in the expansion phase requiring considerable material resources.

Instead of a feeder layer, two- or three-dimensional scaffolds have been used to produce platelets [107]. It seems that the 3D milieu in the bioreactor allows for a prolonged continuous platelet production (32 days vs. 24 days in 3D scaffolds in a well or 10 days in a 2D situation), although further improvements are necessary to optimize the yield. Other approaches to improve platelet production without the use of feeder cells include the addition of various chemokines or other substrates putatively provided *in vivo*, directly to the MK culture. Good candidates for which MK and platelets have receptors on their surface include bone morphogenetic proteins (BMPs), peroxisome proliferator-activated receptor gamma (PPARγ) ligands, and c-MPL agonists. It has been shown in an *in vitro* culturing assay that BMP4 can efficiently regulate MK differentiation as well as platelet production in the absence of TPO [112]. The PPARγ ligand 15-deoxy-$\Delta^{12,14}$ prostaglandin $J_2$(15d-$PGJ_2$) can support platelet formation in culture derived from mouse and human MK by induction of reactive oxygen species (ROS) [113]. Finally, regulation of metalloproteinases and Src kinases in the culture medium could prevent shedding of important platelet receptors and increase platelet production. Indeed, addition of the SRC kinase inhibitor SU6656 significantly enhanced platelet production from a MK cell line [114], whereas addition of the metalloproteinase inhibitor GM6001 improved *in vitro* generation of platelets from mouse ES cells [115].

Megakaryocytes can also be derived from murine or human ES cells, which have the advantage of rapid proliferation and the readily availability of ES cell lines. Gaur et al. [116] established an OP9 stromal cell coculture system to generate MKs from human ES (hES) cells, but the yield of MK was low. Takayama et al. [117] modified this system by addition of vascular endothelial growth factor to promote hES cell-derived sacs that provide a suitable environment for hematopoietic progenitors. Indeed, mature MK could be cultured from these sacs as well as functional platelets, on average $4.8 \times 10^6$ platelets from $10^5$ hES cells. The generation of platelets from murine ES cells is more efficient, yielding $1 \times 10^8$ platelets from $1 \times 10^4$ ES cells in two waves of platelet production [99].

*In vitro*–derived platelets have a size (3–5 μM) and volume (6–10 μL) similar to those of plasma-derived platelets [118]. Flow cytometry is used for phenotypic analysis of platelets using fluorochrome-conjugated antibodies as well as DNA staining, and fresh platelets are always used for proper gate setting [119]. The function of cultured-derived platelets can be assessed by fibrinogen and annexin V binding, or P-selectin expression after activation by known platelet agonists by flow cytometry [99,104,106,117]. One should be aware that many *in vitro*–generated platelets are already preactivated by either compounds in the medium or the isolation process [104]. Microparticles might interfere with the analyses since these particles also express CD41 and CD42b, and the majority also phosphatidylserine on their surface. These microparticles are not shed from the proplatelet ends but emanate from the MK surface [97]. They can be distinguished from activated platelet-derived microparticles by the expression of filamin A. Additional functional tests with cultured-derived platelets include aggregation studies performed with both murine and human platelets [98,120], and a thrombosis model using culture-derived murine platelets [120], but these are still hampered by the low numbers of platelets produced.

***11.3.1.3 In vivo Platelet Production*** As illustrated in the previous section, there have been considerable efforts in developing *in vitro* culture systems to generate MK/platelets; however, the low yield of platelets prohibits easy assessment of platelet functionality. Therefore, alternative *in vivo* models of MK and platelet formation have been established. This has been especially successful in mouse transplantation studies, where murine thrombopoiesis is reconstituted in lethally irradiated recipient mice by transplantation of BM cells from donor mice. Examples of successful genetic modifications of platelets using this approach will be given in Section 11.3.2.3. Nonobese diabetic/severe combined immunodeficient (NOD/SCID) mice, on the other hand, provide an excellent model for study of human hematopoiesis after transplantation of human HSC [121]. These animals present multiple defects in both innate and adaptive immunity: T- and B-cell deficiencies, defective natural killer (NK) cells, macrophage dysfunction, and absence of circulating complement [122]. Transplantation of various sources (e.g., CB, mobilized PB) of human HSC and progenitors results in high levels of engraftment and differentiation of multiple blood cell lineages in the BM of NOD/SCID mice [123]. The generation of human platelets after transplantation of human HSC into sublethally irradiated NOD/SCID mice were initially described by Ueda et al. [124]. An antiasialo GM1 antiserum to further deplete NK cells was used to improve engraftment capability. Circulating human platelets representing up to 35.5% of the total platelet population were observed in NOD/SCID PB after transplantation of $5 \times 10^4$CD34$^+$ cells [124]. Perez et al. used human CD34$^+$ cells isolated from PB as a source of platelet progenitors and showed that human platelets could be detected 1 week after injection of $0.6 \times 10^6$ cells with a maximal production of human platelets ($20 \times 10^6$/mL) obtained at 3 weeks [125]. In addition, functionality of human platelets was demonstrated by flow cytometry with an increased expression of human P-selectin after activation of platelets with thrombin [125]. Since then, several other groups have demonstrated that injection of human CD34$^+$ cells isolated from PB [126–128] and CB [100,129–137] into NOD/SCID mice can lead to the production of human platelets with nevertheless variable yield. It is generally accepted that, in comparison to BM or PB, CB transplantation, despite better long-term engraftment capacity, is characterized by slower engraftment kinetics, in particular by a delayed platelet recovery [138]. Indeed, differences in engraftment potential between CB and PB progenitors/HSC are reflected in the megakaryocytic lineage, with several CB studies showing a peak production of human platelets in NOC/SCID mice toward 4–9 weeks posttransplantation [100,131,133,134], while, as observed by Perez et al. [125], the maximal human platelet production in NOD/SCID mice injected with human PB-CD34$^+$ cells occurred 3 weeks postinjection [126,127].

Thrombocytopenia is an unavoidable consequence of high-dose chemotherapy, and the NOD/SCID model has been used to provide knowledge of the optimization of the transplant in order to accelerate platelet recovery. Several groups have therefore investigated the potential advantages of culturing CB- or PB-CD34$^+$ cells to improve platelet recovery by increasing numbers of injected progenitor cells. While a clear benefit was observed in CB studies up to a few days posttransplantation [100,129,133], this was not the case for later stages [100,133], and the numbers of circulating human platelets were modest. In a very promising study, Tijssen et al. demonstrated that MK generated from PB-CD34$^+$ cells *in vitro* are able to significantly improve human platelet production in NOD/SCID mice during the earlier stages of transplantation ($<$10 days) [128]. A combination of increased numbers of injected PB-CD34$^+$ cells ($> 2 \times 10^6$ cells) compared to previous studies [125–127], with an optimized culture protocol of 7 days, led to the detection of human platelets as early as 3 days posttransplant. Here also a peak production was reached 3 weeks posttransplantation with absolute numbers of circulating human platelets ($27 \times 10^6$ mL$^{-1}$), similar to previously published data [125]. Functionality of human platelets produced in mice after CB or PB transplantation of human progenitors has been demonstrated by flow cytometry, where an increase in P-selectin expression on the platelet surface was observed after *ex vivo* stimulation with thrombin [125,127,129,131,133] or thrombin receptor activating peptide (TRAP) [128,135].

On the basis of the studies conducted with NOD/SCID mice, and in an attempt to determine the optimal conditions for functional assessment of human platelets obtained from transduced progenitors, we transplanted high numbers of CB-CD34$^+$ cells in sub-lethally irradiated NOD/SCID mice. Production of human platelets after transplantation of $3 \times 10^6$ unexpanded CD34$^+$ cells was detected within 10 days posttransplantation with mean levels of human platelets of $2.2 \pm 0.43 \times 10^6$ mL$^{-1}$. High levels of human platelets were still detected 8–9 weeks posttransplantation, with human platelets representing 2–9% of the total platelet population [139]. Human platelets issued from human HSC were functional, as demonstrated by an increased CD62P expression and binding of monoclonal antibody PAC1 after *ex vivo* stimulation with TRAP and collagen-related peptide (CRP) [139]. By specifically labeling murine and human platelets with monoclonal antibodies before perfusion over a collagen coated surface, we could visualize human platelets incorporating into murine thrombi, which could be abrogated by the addition of inhibitory antibodies against human GPIb and GPIIb/IIIa [139].

Thus, it currently is possible to produce human platelets from different sources of human HSC and progenitors at levels that allow functional studies. However, for assessment of the function of genetically modified platelets, one needs to factor in different parameters, such as the source of HSC and the different kinetics of platelet production associated with each, the genetic background of the NOD/SCID mice (e.g., NOD/SCID, NOD/SCID treated with GM1 antiasialo serum, NOD/SCID/IL2rg$^{-/-}$), reviewed elsewhere [123], and the time after the transplantation.

### 11.3.2 Current Approaches in Genetic Modification of Platelets

***11.3.2.1 Overview of Technology*** Nowadays lentiviral transduction is the method of choice for achieving stable transgene expression in HSC and their progeny. Lentiviruses (LV) are, in contrast to onco/$\gamma$-retroviruses, able to transduce nondividing cells without altering cell multipotency and self-renewing capacity [140]. Despite all efforts to increase biosafety of lentiviral vectors, concerns about the generation of replication competent viruses and insertional mutagenesis in human gene therapy still exist. Feline or simian LV, for which analogous vectors are developed, may serve as an alternative, as these forms of

HIV are not transmittable to humans [141]. Nonintegrating vectors are currently being developed to avoid nonspecific integration of a transgene, however, restraining the use in terminally differentiated cells [142].

$CD34^+$ cells from different sources (umbilical CB, mobilized PB, and BM) have different susceptibility to LV transduction and therefore require slightly adapted transduction protocols [143]; however, a general feature is their low permissiveness for the vector. Short-term stimulation of HSC with early-acting cytokines, incubation with a proteasome inhibitor [144], transduction with high vector titers, a long transduction period [145], or repeated transductions [146–148] appear to be key issues in achieving efficient transduction. In addition, transgene expression is highly dependent on its promoter [143]. Indeed, using a GFP reporter gene, Salmon et al. showed that the transcriptional activity of the CMV promoter in HSC was limited [140]. Moreover, whereas a self-inactivating (SIN) LV vector design did not influence the expression from the EF1α promoter, it dramatically decreased expression from the murine phosphoglycerate kinase (PGK) (424–930) promoter [140]. This problem could be overcome by integration of a woodchuck hepatitis virus posttranscriptional regulatory element (WPRE) upstream of the 3′-LTR; however, the WPRE element was shown to negatively influence the expression of the EF1α promoter [140]. Transfer vectors are being further optimized to enhance trangene expression by incorporation of several *cis*-acting elements to enhance transgene expression [147]. In addition, bicistronic, multiple gene, tetracycline, or doxycycline inducible vectors and miRNA vectors have been developed to allow coexpression of reporter genes or gene knockdown, respectively [149].

The optimization of lentiviral vectors allowing high and platelet-specific expression by the incorporation of tissue-specific promoters will be discussed further in the next section.

***11.3.2.2 Genetic Modification of Megakaryocytes/Platelets in vitro*** Wilcox et al. were the first to investigate whether a murine leukemia virus (MuLV)-derived SIN vector carrying the −813/+33 regulatory element of the GPIIb promoter could drive megakaryocyte-specific transgene expression [150]. Using a β-galactosidase reporter gene, they demonstrated that the CMV and GPIIb(−889) promoter equally directed gene expression in the megakaryocytic progeny of retrovirally transduced PB $CD34^+$ cells. Next, a MuLV construct expressing the human alloantigen 2 form of the GPIIIa subunit (HPA1b, $Pl^{A2}$) under the control of the GPIIb(−889) promoter was used to transduce $CD34^+$ cells from two Glanzmann thrombasthenia (GT) patients with defects in the GPIIIa gene [151]. After differentiation, 19% of the MK progeny expressed GPIIb/IIIa on the surface at 34% of the normal receptor levels. These MK cells could be activated by agonists as assessed by increased PAC1 binding at 10% of the mean fluorescence values of normal control MK. Additionally, −889/GPIIIa $Pl^{A2}$-transduced MK could retract a fibrin clot *in vitro* at levels comparable to those extracted from nonthrombasthenic individuals [151] demonstrating that the GT phenotype can be corrected despite suboptimal expression levels of the receptor.

In order to achieve blood-lineage-restricted expression from lentiviral vectors, Shi et al. [146] cloned the GPIIb(−889) promoter fragment in a LV SIN vector together with the *GPIBA* gene to correct *in vitro* the platelet GPIbα deficiency associated with Bernard–Soulier syndrome (BSS). After repeated transduction of PB-CD34$^+$ cells with this vector, HA-tagged GPIbα was expressed specifically in the MK lineage and incorporated in the GPIb/IX complex at the surface of 50% of the differentiated MK. Although the GPIIb promoter is the best characterized MK-specific promoter, it is also partially transcriptionally active in HSC. To strictly limit transgene expression to MK lineage, the potency of a short GPIbα promoter fragment (−322/+19) was also investigated as this promoter is active only later in MK differentiation. The GFP expression pattern after LV transduction of human CD34$^+$ cells of different sources (CB, PB, BM) under control of this GPIbα promoter fragment was compared to that under transcriptional control of the EF1α and GPIIb(−889) promoters [152]. Whereas GPIbα promoter activity is strictly targeted to mature MK with a percentage of CD41a$^-$/CD42b$^-$/GFP$^+$ cells never exceeding 10% of the total population, this percentage reaches 60% in GPIIb-GFP- and EF1α-GFP-transduced cells [152]. These results suggest that the GPIIb-driven expression is not as MK-restricted as previously believed, at least for the early steps of MK differentiation *in vitro*. However, the expression driven from the GPIbα promoter resulted in significantly lower transgene levels, most likely due to its late activation during megakaryocytic differentiation. It is interesting to note that PB-CD34$^+$ are not as efficient in transgene expression compared to other sources of CD34$^+$ cells, and this held with all three promoters studied [152].

Raslova et al. [145] used the PGK/WPRE combination in a bicistronic IRES-eGFP construct to overexpress the transcription factor FLI1 in CD34$^+$ cells from Paris–Trousseau syndrome (PTS) patients. PTS is an autosomal dominant disease caused by a chromosomal deletion of a DNA fragment harboring the transcription factors ETS1 and FLI1. Lentivirus-mediated expression of *FLI1* in CD34$^+$ cells successfully restored megakaryopoiesis *in vitro* as demonstrated by the correction of CD41, CD42, and von Willebrand (vWF) expression, and the increased ploidization and MK differentiation. It is noteworthy that eGFP was poorly translated from the IRES sequence and therefore may not be useful as a reporter construct.

Besides *in vitro* correction of deficiencies associated with platelet disorders, few laboratories have obtained genetically modified platelets *in vitro*. Ungerer et al. successfully transduced CD34$^+$ cells by adenoviral infection leading to GFP expressing culture-derived platelets [98]. However, interestingly, they later showed that these platelets were not functional [120]. In contrast, retroviral transduction of CD34$^+$ cells had no impact on platelets generated thereafter [120]. Sufficient genetically modified murine platelets were produced to perform aggregation studies and to be injected back into mice, allowing *in vivo* studies to be carried out by intravital microscopy (e.g., thrombosis model) [120]. More recently, Gillitzer et al. investigated the molecular mechanism in which the syntactin and synoptome-associated protein (SNAP-23) contributes to granule release

[153]. *In vitro* differentiated platelets, derived from RV-transduced murine BM cells, expressing a dominant negative SNAP-23 mutant, showed an impaired agonist-induced degranulation and aggregation response [153].

Finally, the trafficking of platelet factor 4 (PF4) to α-granules was investigated by PF4 to GPF. In MK differentiated *in vitro* from transduced CB-CD34$^+$ cells, PF4-GFP similar to PF4 alone, collocalized with vWF in α-granules, and was secreted on thrombin stimulation [154]. The fusion of the signal peptide to GFP directed the entry into the endoplasmic reticulum (ER)–Golgi secretory pathway, but was not sufficient to target GFP to the granular compartment, suggesting that storage in α-granules of MK results from a specific sorting mechanism [154].

***11.3.2.3 Genetic Modification of Megakaryocytes/Platelets in vivo*** One of the first examples describing a successful retroviral transduction of mouse marrow cells was carried out by Yan et al. [155]. Although the ectopic overexpression of murine c-mpl was not directly demonstrated in hematopoietic cells (including platelets) of the reconstituted mice, mice displayed a similar phenotype to that reported in c-mpl$^{-/-}$ and Tpo$^{-/-}$ mice with decreased numbers of MK and platelets [155]. In another retroviral transduction study, an increased thrombopoiesis up to 10 days was observed in mice reconstituted with enriched MK progenitor cells overexpressing p45-NF-E2 [156]. Although Li et al. showed successful transgene expression through serial transplantations in lymphoid, myeloid, erythroid, and megakaryocytic (platelets) lineages [157], transgene silencing associated with retroviral transduction has been observed on serial BM transplantation in other studies [158].

To date, the most successful means of producing genetically modified platelets *in vivo* has been through gene therapy approaches aiming to correct expression of proteins essential to platelet functions or to target the expression of the transgene in the MK, which has been possible largely with the advancement of LV. After demonstrating the feasibility of lineage-specific gene expression of GPIIIa in MK using the platelet-specific promoter GPIIb [150], Wilcox et al. aimed to restore platelet function in GPIIIa-deficient mice by transducing their BM with a HIV1 self-inactivating vector carrying the human GPIIb promoter/GPIIIa transgene construct [148]. Human GPIIIa formed a stable complex with murine GPIIb that could restore platelet function and improved primary hemostasis with a minimum of only 7% of the normal levels of the GPIIb/IIIa complex needed [148]. This chimeric mouse model was also used to study the inhibition of immune-mediated HPA1a platelet destruction, which occurs in, for instance, fetomaternal alloimmune thrombocytopenia, by using blocking antibodies lacking a destructive constant region *in vivo* [159]. Very promising results in the gene therapy field came with an elegant study showing successful transplantation of *ex vivo* transduced murine HSC with lentiviral vector expressing the human Wiskott–Aldrich syndrome protein (WASP) in *WAS*$^{-/-}$ mice [160]. *WAS* is normally classified as a hereditary thrombocytopenia, but is not restricted to platelets, as B and T lymphocytes are also affected [161]. Expression of the human WASP protein was successfully restored in platelets as well as in other blood cell lineages [160]. It is noteworthy to mention that nonlethally irradiated mice were used in this study, and that the expression of human WASP was obtained with both the WASP or PGK promoters [160]. Another promising study for the gene therapy field and demonstrating the feasibility of genetic modification of platelets was initiated by Horn et al. [162], who used a clinically relevant large-animal model to test efficient gene transfer to repopulating HSC. In this study, canine CD34$^+$ cells were successfully transduced with LV carrying the GFP transgene under the PGK promoter and transplanted into irradiated dogs. Transgene expression was detected in B and T cells, granulocytes, red blood cells, and platelets; remarkably, 8% of GFP$^+$ platelets were detected in the PB of a representative transplanted dog ($n = 3$), and this one year posttransplantation, attesting for the long-term expression of the transgene [162].

The most successful approach in both gene therapy (targeting expression to platelets) and the genetic modification of platelets *in vivo* by means of viral transduction has been with attempts of correcting hemophilia A. After moderate success in detection of FVIII expression *in vivo* in gene therapy studies with LV, due to development of inhibitory antibodies and/or inefficacy of promoters [163,164], Wilcox et al. [165] developed a strategy aiming to target FVIII expression to platelets to sequester it from potential inhibitors until its release upon platelet activation at sites of injury. They initially showed the collocalization of FVIII with VWF in platelet α-granules of NOD/SCID mice transplanted with human CD34$^+$ cells transduced with a retrovirus expressing the human FVIII under a constitutive promoter [165]. Later, using a transgenic approach, they demonstrated the feasibility of targeting the expression of FVIII to platelets to successfully correct the bleeding phenotype in hemophilia A mice [166–168]. Lentiviruses were also used to express the FVIII gene in combination with the GPIIb(−889) promoter in FVIII$^{-/-}$ mice by transplantation of transduced BM [147]. FVIII activity levels were comparable to the ones in mice heterozygous for the human FVIII transgene 4 weeks posttransplantation, and were sustained for several months as assessed in secondary transplanted mice. More importantly, phenotypic correction of murine hemophilia A was observed in the tail clip survival test [147]. Using similar LV vectors targeting the expression of FVIII to platelets, canine BM cells were transduced and transplanted into sublethally irradiated NOD/SCID mice. About 30–40% of the platelet population was of canine origin 4 weeks posttransplantation, and FVIII could be successfully expressed in NOD/SCID mice [169]. Finally, Sakata et al. used similar approaches to express human FVIII in human platelets using, however, a simian LV system. Although using a CMV promoter, they could successfully detect FVIII expression for at least 60 days in plasma but only very low amounts in the human platelets [170]. Increased platelet transgene expression was obtained by using LV containing the eGFP gene under the control of the platelet specific GPIb promoter as demonstrated by expression of eGFP in 16–20% of platelets from mice transplanted with transduced murine HSC [171]. Despite very low levels of FVIII detected in plasma of hemophilia A mice transplanted with HSC transduced with simian

LV carrying the GPIb-FVIII transgene, a partial correction of the phenotype was observed [171]. More recently, eGFP expression in platelets was further increased by eliminating BM cell sorting before transduction [172]. Interestingly, correction of the bleeding phenotype in haemophilia A mice was achieved by expression of murine FVIIa in platelets, although with scant levels [172].

***11.3.2.4 RNA Interference*** Lentiviruses can also be used to silence the expression of a specific protein in hematopoietic progenitor cells prior to differentiation into MK/platelets or transplantation in NOD/SCID mice. Lentiviral vectors containing short hairpin RNA (shRNA) sequences, directed against target genes, are transduced into hematopoetic progenitor cells, resulting in a reduction of target protein expression by a process called RNA interference (RNAi). The majority of the studies reported so far describe the effect of silencing a gene of interest on MK proliferation and differentiation *in vitro* using siRNA-based transfection [173–175] or shRNA-RV/LV transduction of human $CD34^+$ cells or murine BM and ES cells [176–180]. Berthebaud et al. used LV vectors expressing a shRNA specific to RGS16, a member of the cell signaling RGS protein family, to knock down RGS16 in human $CD34^+$ cells cultured toward MK. They showed that CXCR4 signaling in MK was upregulated, indicating that RGS16 is a negative regulator of CXCR4 [181]. A role for WAVE2 and Abi1 in proplatelet production was demonstrated by Eto et al. using similar approaches via shRNA expressing LV vectors in mouse ES cells [182].

The most promising study thus far in successfully silencing the expression of a selected gene in platelets has been achieved by Ohmori et al. [183]: transplantation of transduced murine BM cells with a LV vector expressing GFP and a shRNA targeting $\alpha IIb\beta 3$ under the GPIb$\alpha$ and U6 promoters, respectively, which caused a significant reduction in $\alpha IIb\beta 3$ expression in $GFP^+$ platelets [183]. Platelet activation by convulxin and ADP was affected in platelets from recipient mice transplanted with cells transduced with the $\alpha IIb\beta 3$-shRNA LV vector. In addition, they also used this method to knock down talin in platelets. A reduction of talin expression was confirmed in $GFP^+$ murine platelets and was accompanied by impaired $\alpha IIb\beta 3$ activation, P-selectin expression, and spreading on fibrinogen matrix after agonist stimulation. Unfortunately, a low transduction efficiency of $\sim$20% prevented additional functional tests to be performed (e.g., aggregation). However, the role of talin *in vivo* identified in this study was confirmed by several groups using Cre-Lox technology to delete talin in platelets [184,185].

## 11.4 CONCLUDING REMARKS

The promise of the ability to generate genetically modified platelets is exciting, indeed. Clearly, this would allow numerous fundamental research questions aiming to delineate the mechanisms involved in megakaryopoiesis/thrombopoiesis to be addressed and to further extend our understanding of platelet activation and inhibition. Obviously, the promise of producing human platelets in experimental animals also introduces new perspectives as this will enable researchers to perform preclinical tests in small animals of new human-platelet-specific compounds (antibodies) designed to, for example, stimulate human thrombopoiesis to reverse thrombocytopenia from different ethiologies, or to define the *in vivo* action of new antithrombotics. However, it should be clearly kept in mind that such model systems will suffer from species incompatibilities as human platelets do not necessarily interact correctly with murine ligands, for instance. The most rewarding application nevertheless remains in cell-based gene therapy with the production of genetically manipulated autologous platelets in patients where inherited deficits in platelet proteins can be corrected. In addition, platelets can be used to deliver a specific foreign cargo to a bleeding or thrombotic site, as has been exemplified with FVIII, but which clearly could also include antithrombotics. It nevertheless is clear that before all these possibilities can be realized, there is still a long way to go to overcome all the current technical hurdles that stand in the way of the ability to produce safely and efficiently transduced genetically manipulated platelets with good yields.

## REFERENCES

1. Watkins NA, Gusnanto A, de Bono B, De S, Miranda-Saavedra D, Hardie DL, et al., A HaemAtlas: Characterizing gene expression in differentiated human blood cells, *Blood* **113**(19):e1-9 (2009).
2. O'Connor MN, Salles, II, Cvejic A, Watkins NA, Walker A, Garner SF, et al., Functional genomics in zebrafish permits rapid characterization of novel platelet membrane proteins, *Blood* **113**(19):4754–4762 (2009).
3. Crozatier M, Meister M, Drosophila haematopoiesis, *Cell. Microbiol*. **9**(5): 1117–1126 (2007).
4. Davidson CJ, Tuddenham EG, McVey JH, 450 million years of hemostasis, *J. Thromb. Haemost*. **1**(7):1487–1494 (2003).
5. Grecchi R, Saliba AM, Mariano M, Morphological changes, surface receptors and phagocytic potential of fowl mono-nuclear phagocytes and thrombocytes *in vivo* and *in vitro*, *J. Pathol*. **130**(1):23–31 (1980).
6. Daimon T, Mizuhira V, Uchida K, Fine structural distribution of the surface-connected canalicular system in frog thrombocytes, *Cell Tissue Res*. **201**(3):431–439 (1979).
7. Jagadeeswaran P, Sheehan JP, Craig FE, Troyer D, Identification and characterization of zebrafish thrombocytes, *Br. J. Haematol*. **107**(4):731–738 (1999).
8. Lin HF, Traver D, Zhu H, Dooley K, Paw BH, Zon LI, et al., Analysis of thrombocyte development in CD41-GFP transgenic zebrafish, *Blood* **106**(12):3803–3810 (2005).
9. Jagadeeswaran P, Gregory M, Day K, Cykowski M, Thattaliyath B, Zebrafish: A genetic model for hemostasis and thrombosis, *J. Thromb. Haemost*. **3**(1):46–53 (2005).
10. Carradice D, Lieschke GJ, Zebrafish in hematology: Sushi or science? *Blood* **111**(7):3331–3342 (2008).

11. Streisinger G, Walker C, Dower N, Knauber D, Singer F, Production of clones of homozygous diploid zebra fish (Brachydanio rerio), *Nature* **291**(5813):293–296 (1981).

12. Lieschke GJ, Currie PD, Animal models of human disease: Zebrafish swim into view, *Nat. Rev. Genet.* **8**(5):353–367 (2007).

13. Davidson AJ, Zon LI, The "definitive" (and "primitive") guide to zebrafish hematopoiesis, *Oncogene* **23**(43):7233–7246 (2004).

14. Hanumanthaiah R, Day K, Jagadeeswaran P, Comprehensive analysis of blood coagulation pathways in teleostei: Evolution of coagulation factor genes and identification of zebrafish factor VIIi, *Blood Cells Mol. Dis.* **29**(1):57–68 (2002).

15. Jagadeeswaran P, Sheehan JP, Analysis of blood coagulation in the zebrafish, *Blood Cells Mol. Dis.* **25**(3–4):239–249 (1999).

16. Traver D, Herbomel P, Patton EE, Murphey RD, Yoder JA, Litman GW, et al., The zebrafish as a model organism to study development of the immune system, *Adv. Immunol.* **81**:253–330 (2003).

17. Gering M, Rodaway AR, Gottgens B, Patient RK, Green AR, The SCL gene specifies haemangioblast development from early mesoderm, *EMBO J.* **17**(14):4029–4045 (1998).

18. Gottgens B, Barton LM, Chapman MA, Sinclair AM, Knudsen B, Grafham D, et al., Transcriptional regulation of the stem cell leukemia gene (SCL)—comparative analysis of five vertebrate SCL loci, *Genome Res.* **12**(5):749–759 (2002).

19. Detrich HW 3rd, Kieran MW, Chan FY, Barone LM, Yee K, Rundstadler JA, et al., Intraembryonic hematopoietic cell migration during vertebrate development, *Proc. Natl. Acad. Sci. USA* **92**(23):10713–10717 (1995).

20. Kalev-Zylinska ML, Horsfield JA, Flores MV, Postlethwait JH, Vitas MR, Baas AM, et al., Runx1 is required for zebrafish blood and vessel development and expression of a human RUNX1-CBF2T1 transgene advances a model for studies of leukemogenesis, *Development* **129**(8):2015–2030 (2002).

21. Thompson MA, Ransom DG, Pratt SJ, MacLennan H, Kieran MW, Detrich HW 3rd, et al., The cloche and spadetail genes differentially affect hematopoiesis and vasculogenesis, *Dev. Biol.* **197**(2):248–269 (1998).

22. Berman J, Hsu K, Look AT, Zebrafish as a model organism for blood diseases, *Br. J. Haematol.* **123**(4):568–576 (2003).

23. de Jong JL, Zon LI, Use of the zebrafish system to study primitive and definitive hematopoiesis, *Annu. Rev. Genet.* **39**:481–501 (2005).

24. Willett CE, Cortes A, Zuasti A, Zapata AG, Early hematopoiesis and developing lymphoid organs in the zebrafish, *Dev. Dyn.* **214**(4):323–336 (1999).

25. Kataoka H, Ochi M, Enomoto K, Yamaguchi A, Cloning and embryonic expression patterns of the zebrafish Runt domain genes, runxa and runxb, *Mech. Dev.* **98**(1–2):139–143 (2000).

26. Burns CE, DeBlasio T, Zhou Y, Zhang J, Zon L, Nimer SD, Isolation and characterization of runxa and runxb, zebrafish members of the runt family of transcriptional regulators, *Exp. Hematol.* **30**(12):1381–1389 (2002).

27. Bertrand JY, Kim AD, Teng S, Traver D, CD41 + cmyb + precursors colonize the zebrafish pronephros by a novel migration route to initiate adult hematopoiesis, *Development* **135**(10):1853–1862 (2008).

28. Palis J, Yoder MC, Yolk-sac hematopoiesis: The first blood cells of mouse and man, *Exp. Hematol*. **29**(8):927–936 (2001).
29. Day K, Krishnegowda N, Jagadeeswaran P, Knockdown of prothrombin in zebrafish, *Blood Cells Mol. Dis*. **32**(1):191–198 (2004).
30. Sheehan J, Templer M, Gregory M, Hanumanthaiah R, Troyer D, Phan T, et al., Demonstration of the extrinsic coagulation pathway in teleostei: Identification of zebrafish coagulation factor VII, *Proc. Natl. Acad. Sci. USA* **98**(15):8768–8773 (2001).
31. Gregory M, Hanumanthaiah R, Jagadeeswaran P, Genetic analysis of hemostasis and thrombosis using vascular occlusion, *Blood Cells Mol. Dis*. **29**(3):286–295 (2002).
32. Amores A, Force A, Yan YL, Joly L, Amemiya C, Fritz A, et al., Zebrafish hox clusters and vertebrate genome evolution, *Science* **282**(5394):1711–1714 (1998).
33. Postlethwait JH, Yan YL, Gates MA, Horne S, Amores A, Brownlie A, et al., Vertebrate genome evolution and the zebrafish gene map, *Nat. Genet*. **18**(4):345–349 (1998).
34. Wilming LG, Gilbert JG, Howe K, Trevanion S, Hubbard T, Harrow JL, The vertebrate genome annotation (Vega) database, *Nucleic Acids Res*. **36**(database issue):D753–D760 (2008).
35. Brown LA, Rodaway AR, Schilling TF, Jowett T, Ingham PW, Patient RK, et al., Insights into early vasculogenesis revealed by expression of the ETS-domain transcription factor Fli-1 in wild-type and mutant zebrafish embryos, *Mech. Dev*. **90**(2):237–252 (2000).
36. Nishikawa K, Kobayashi M, Masumi A, Lyons SE, Weinstein BM, Liu PP, et al., Self-association of Gata1 enhances transcriptional activity *in vivo* in zebra fish embryos, *Mol. Cell. Biol*. **23**(22):8295–8305 (2003).
37. Pratt SJ, Drejer A, Foott H, Barut B, Brownlie A, Postlethwait J, et al., Isolation and characterization of zebrafish NFE2, *Physiol. Genom*. **11**(2):91–98 (2002).
38. Thattaliyath B, Cykowski M, Jagadeeswaran P, Young thrombocytes initiate the formation of arterial thrombi in zebrafish, *Blood* **106**(1):118–124 (2005).
39. Jagadeeswaran P, Liu YC, Sheehan JP, Analysis of hemostasis in the zebrafish, *Meth. Cell. Biol*. **59**:337–357 (1999).
40. Denis MM, Tolley ND, Bunting M, Schwertz H, Jiang H, Lindemann S, et al., Escaping the nuclear confines: Signal-dependent pre-mRNA splicing in anucleate platelets, *Cell* **122**(3):379–391 (2005).
41. Goessling W, North TE, Zon LI, New waves of discovery: Modeling cancer in zebrafish, *J. Clin. Oncol*. **25**(17):2473–2479 (2007).
42. Chico TJ, Ingham PW, Crossman DC, Modeling cardiovascular disease in the zebrafish, *Trends Cardiovasc. Med*. **18**(4):150–155 (2008).
43. Pelster B, Burggren WW, Disruption of hemoglobin oxygen transport does not impact oxygen-dependent physiological processes in developing embryos of zebra fish (Danio rerio), *Circ. Res*. **79**(2):358–362 (1996).
44. Grillitsch S, Medgyesy N, Schwerte T, Pelster B, The influence of environmental P(O(2)) on hemoglobin oxygen saturation in developing zebrafish Danio rerio, *J. Exp. Biol*. **208**(Pt. 2):309–316 (2005).

45. Ekker SC, Stemple DL, Clark M, Chien CB, Rasooly RS, Javois LC, Zebrafish genome project: Bringing new biology to the vertebrate genome field, *Zebrafish* **4**(4):239–251 (2007).
46. Driever W, Solnica-Krezel L, Schier AF, Neuhauss SC, Malicki J, Stemple DL, et al., A genetic screen for mutations affecting embryogenesis in zebrafish, *Development* **123**:37–46 (1996).
47. Haffter P, Granato M, Brand M, Mullins MC, Hammerschmidt M, Kane DA, et al., The identification of genes with unique and essential functions in the development of the zebrafish, Danio rerio, *Development* **123**:1–36 (1996).
48. Sivasubbu S, Balciunas D, Amsterdam A, Ekker SC, Insertional mutagenesis strategies in zebrafish, *Genome Biol*. **8**(Suppl. 1):S9 (2007).
49. Wang D, Jao LE, Zheng N, Dolan K, Ivey J, Zonies S, et al., Efficient genome-wide mutagenesis of zebrafish genes by retroviral insertions, *Proc. Natl. Acad. Sci. USA* **104**(30):12428–12433 (2007).
50. Kawakami K, Takeda H, Kawakami N, Kobayashi M, Matsuda N, Mishina M, A transposon-mediated gene trap approach identifies developmentally regulated genes in zebrafish, *Dev. Cell* **7**(1):133–144 (2004).
51. McCallum CM, Comai L, Greene EA, Henikoff S, Targeted screening for induced mutations, *Nat. Biotechnol*. **18**(4):455–457 (2000).
52. Wienholds E, Schulte-Merker S, Walderich B, Plasterk RH, Target-selected inactivation of the zebrafish rag1 gene, *Science* **297**(5578):99–102 (2002).
53. Wienholds E, van Eeden F, Kosters M, Mudde J, Plasterk RH, Cuppen E, Efficient target-selected mutagenesis in zebrafish, *Genome Res*. **13**(12):2700–2707 (2003).
54. Skromne I, Prince VE, Current perspectives in zebrafish reverse genetics: Moving forward, *Dev. Dyn*. **237**(4):861–882 (2008).
55. Stemple DL, TILLING—a high-throughput harvest for functional genomics, *Nat. Rev. Genet*. **5**(2):145–150 (2004).
56. Nasevicius A, Ekker SC, Effective targeted gene "knockdown" in zebrafish, *Nat. Genet*. **26**(2):216–220 (2000).
57. Summerton J, Weller D, Morpholino antisense oligomers: Design, preparation, and properties, *Antisense Nucleic Acid Drug Dev*. **7**(3):187–195 (1997).
58. Heasman J, Morpholino oligos: Making sense of antisense? *Dev. Biol*. **243**(2): 209–214 (2002).
59. Long Q, Meng A, Wang H, Jessen JR, Farrell MJ, Lin S, GATA-1 expression pattern can be recapitulated in living transgenic zebrafish using GFP reporter gene, *Development* **124**(20):4105–4111 (1997).
60. Lawson ND, Weinstein BM, *In vivo* imaging of embryonic vascular development using transgenic zebrafish, *Dev. Biol*. **248**(2):307–318 (2002).
61. Kissa K, Murayama E, Zapata A, Cortes A, Perret E, Machu C, et al., Live imaging of emerging hematopoietic stem cells and early thymus colonization, *Blood* **111**(3):1147–1156 (2008).
62. Welcome Trust Case Control Consortium et al., Genome-wide association study of 14,000 cases of seven common diseases and 3,000 shared controls, *Nature* **447**(7145):661–678 (2007).
63. Ouwehand WH, Platelet genomics and the risk of atherothrombosis, *J. Thromb. Haemost*. **5**(Suppl. 1):188–195 (2007).

64. Samani NJ, Erdmann J, Hall AS, Hengstenberg C, Mangino M, Mayer B, et al., Genomewide association analysis of coronary artery disease, *N. Engl. J. Med.* **357**(5):443–453 (2007).
65. Kelly GM, Erezyilmaz DF, Moon RT, Induction of a secondary embryonic axis in zebrafish occurs following the overexpression of beta-catenin, *Mech. Dev.* **53**(2):261–273 (1995).
66. Toyama R, O'Connell ML, Wright CV, Kuehn MR, Dawid IB, Nodal induces ectopic goosecoid and lim1 expression and axis duplication in zebrafish, *Development* **121**(2):383–391 (1995).
67. Nikaido M, Tada M, Saji T, Ueno N, Conservation of BMP signaling in zebrafish mesoderm patterning, *Mech. Dev.* **61**(1–2):75–88 (1997).
68. Koos DS, Ho RK, The nieuwkoid gene characterizes and mediates a Nieuwkoop-center-like activity in the zebrafish, *Curr. Biol.* **8**(22):1199–1206 (1998).
69. Halloran MC, Sato-Maeda M, Warren JT, Su F, Lele Z, Krone PH, et al., Laser-induced gene expression in specific cells of transgenic zebrafish, *Development* **127**(9):1953–1960 (2000).
70. Hardy ME, Ross LV, Chien CB, Focal gene misexpression in zebrafish embryos induced by local heat shock using a modified soldering iron, *Dev. Dyn.* **236**(11):3071–3076 (2007).
71. Skromne I, Thorsen D, Hale M, Prince VE, Ho RK, Repression of the hindbrain developmental program by Cdx factors is required for the specification of the vertebrate spinal cord, *Development* **134**(11):2147–2158 (2007).
72. Paulus H, Protein splicing and related forms of protein autoprocessing, *Annu. Rev. Biochem.* **69**:447–496 (2000).
73. Zeidler MP, Tan C, Bellaiche Y, Cherry S, Hader S, Gayko U, et al., Temperature-sensitive control of protein activity by conditionally splicing inteins, *Nat. Biotechnol.* **22**(7):871–876 (2004).
74. Liu KJ, Arron JR, Stankunas K, Crabtree GR, Longaker MT, Chemical rescue of cleft palate and midline defects in conditional GSK-3beta mice, *Nature* **446**(7131):79–82 (2007).
75. Taneyhill LA, Coles EG, Bronner-Fraser M, Snail2 directly represses cadherin6B during epithelial-to-mesenchymal transitions of the neural crest, *Development* **134**(8):1481–1490 (2007).
76. Pei DS, Sun YH, Long Y, Zhu ZY, Inhibition of no tail (ntl) gene expression in zebrafish by external guide sequence (EGS) technique, *Mol. Biol. Rep.* **35**(2):139–143 (2008).
77. Murphey RD, Stern HM, Straub CT, Zon LI, A chemical genetic screen for cell cycle inhibitors in zebrafish embryos, *Chem. Biol. Drug Des.* **68**(4):213–219 (2006).
78. North TE, Goessling W, Walkley CR, Lengerke C, Kopani KR, Lord AM, et al., Prostaglandin E2 regulates vertebrate haematopoietic stem cell homeostasis, *Nature* **447**(7147):1007–1011 (2007).
79. Tran TC, Sneed B, Haider J, Blavo D, White A, Aiyejorun T, et al., Automated, quantitative screening assay for antiangiogenic compounds using transgenic zebrafish, *Cancer Res.* **67**(23):11386–11392 (2007).
80. Burns CG, Milan DJ, Grande EJ, Rottbauer W, MacRae CA, Fishman MC, High-throughput assay for small molecules that modulate zebrafish embryonic heart rate, *Nat. Chem. Biol.* **1**(5):263–264 (2005).

81. Peterson RT, Shaw SY, Peterson TA, Milan DJ, Zhong TP, Schreiber SL, et al., Chemical suppression of a genetic mutation in a zebrafish model of aortic coarctation, *Nat. Biotechnol*. **22**(5):595–599 (2004).

82. Sun L, Bradford CS, Ghosh C, Collodi P, Barnes DW, ES-like cell cultures derived from early zebrafish embryos, *Mol. Mar. Biol. Biotechnol*. **4**(3):193–199 (1995).

83. Lee KY, Huang H, Ju B, Yang Z, Lin S, Cloned zebrafish by nuclear transfer from long-term-cultured cells, *Nat. Biotechnol*. **20**(8):795–799 (2002).

84. Moehle EA, Rock JM, Lee YL, Jouvenot Y, DeKelver RC, Gregory PD, et al., Targeted gene addition into a specified location in the human genome using designed zinc finger nucleases, *Proc. Natl. Acad. Sci. USA* **104**(9):3055–3060 (2007).

85. Ekker SC, Zinc finger-based knockout punches for zebrafish genes, *Zebrafish* **5**(2):121–123 (2008).

86. Foley JE, Yeh JR, Maeder ML, Reyon D, Sander JD, Peterson RT, et al., Rapid mutation of endogenous zebrafish genes using zinc finger nucleases made by Oligomerized Pool ENgineering (OPEN), *PLoS ONE* **4**(2):e4348 (2009).

87. Junt T, Schulze H, Chen Z, Massberg S, Goerge T, Krueger A, et al., Dynamic visualization of thrombopoiesis within bone marrow, *Science* **317**(5845):1767–1770 (2007).

88. Italiano JE Jr., Patel-Hett S, Hartwig JH, Mechanics of proplatelet elaboration, *J. Thromb. Haemost*. **5**(Suppl. 1):18–23 (2007).

89. Guerriero R, Testa U, Gabbianelli M, Mattia G, Montesoro E, Macioce G, et al., Unilineage megakaryocytic proliferation and differentiation of purified hematopoietic progenitors in serum-free liquid culture, *Blood* **86**(10):3725–3736 (1995).

90. Debili N, Wendling F, Cosman D, Titeux M, Florindo C, Dusanter-Fourt I, et al., The Mpl receptor is expressed in the megakaryocytic lineage from late progenitors to platelets, *Blood* **85**(2):391–401 (1995).

91. Christensen JL, Weissman IL, Flk-2 is a marker in hematopoietic stem cell differentiation: A simple method to isolate long-term stem cells, *Proc. Natl. Acad. Sci. USA* **98**(25):14541–14546 (2001).

92. Uchida N, Weissman IL, Searching for hematopoietic stem cells: evidence that Thy-1.1lo Lin- Sca-1+ cells are the only stem cells in C57BL/Ka-Thy-1.1 bone marrow, *J. Exp. Med*. **175**(1):175–184 (1992).

93. Manz MG, Miyamoto T, Akashi K, Weissman IL, Prospective isolation of human clonogenic common myeloid progenitors, *Proc. Natl. Acad. Sci. USA* **99**(18):11872–11877 (2002).

94. Chang Y, Bluteau D, Debili N, Vainchenker W, From hematopoietic stem cells to platelets, *J. Thromb. Haemost*. **5**(Suppl. 1):318–327 (2007).

95. Richardson JL, Shivdasani RA, Boers C, Hartwig JH, Italiano JE Jr., Mechanisms of organelle transport and capture along proplatelets during platelet production, *Blood* **106**(13):4066–4075 (2005).

96. Cramer EM, Norol F, Guichard J, Breton-Gorius J, Vainchenker W, Masse JM, et al., Ultrastructure of platelet formation by human megakaryocytes cultured with the Mpl ligand, *Blood* **89**(7):2336–2346 (1997).

97. Flaumenhaft R, Dilks JR, Richardson J, Alden E, Patel-Hett SR, Battinelli E, et al., Megakaryocyte-derived microparticles: direct visualization and distinction from platelet-derived microparticles, *Blood* **113**(5):1112–1121 (2009).

98. Ungerer M, Peluso M, Gillitzer A, Massberg S, Heinzmann U, Schulz C, et al., Generation of functional culture-derived platelets from CD34+ progenitor cells to study transgenes in the platelet environment, *Circ. Res.* **95**(5):e36–e44 (2004).
99. Fujimoto TT, Kohata S, Suzuki H, Miyazaki H, Fujimura K, Production of functional platelets by differentiated embryonic stem (ES) cells *in vitro*, *Blood* **102**(12):4044–4051 (2003).
100. van Hensbergen Y, Schipper LF, Brand A, Slot MC, Welling M, Nauta AJ, et al., *Ex vivo* culture of human CD34+ cord blood cells with thrombopoietin (TPO) accelerates platelet engraftment in a NOD/SCID mouse model, *Exp. Hematol.* **34**(7):943–950 (2006).
101. van den Oudenrijn S, de Haas M, Calafat J, van der Schoot CE, von dem Borne AE, A combination of megakaryocyte growth and development factor and interleukin-1 is sufficient to culture large numbers of megakaryocytic progenitors and megakaryocytes for transfusion purposes, *Br. J. Haematol.* **106**(2):553–563 (1999).
102. Cortin V, Garnier A, Pineault N, Lemieux R, Boyer L, Proulx C, Efficient *in vitro* megakaryocyte maturation using cytokine cocktails optimized by statistical experimental design, *Exp. Hematol.* **33**(10):1182–1191 (2005).
103. Balduini A, d'Apolito M, Arcelli D, Conti V, Pecci A, Pietra D, et al., Cord blood *in vitro* expanded CD41 cells: Identification of novel components of megakaryocytopoiesis, *J. Thromb. Haemost.* **4**(4):848–860 (2006).
104. Chen TW, Yao CL, Chu IM, Chuang TL, Hsieh TB, Hwang SM, Large generation of megakaryocytes from serum-free expanded human CD34+ cells, *Biochem. Biophys. Res. Commun.* **378**(1):112–117 (2009).
105. Boyer L, Robert A, Proulx C, Pineault N, Increased production of megakaryocytes near purity from cord blood CD34+ cells using a short two-phase culture system, *J. Immunol. Meth.* **332**(1–2):82–91 (2008).
106. Matsunaga T, Tanaka I, Kobune M, Kawano Y, Tanaka M, Kuribayashi K, et al., *Ex vivo* large-scale generation of human platelets from cord blood CD34+ cells, *Stem Cells* **24**(12):2877–2887 (2006).
107. Sullenbarger B, Bahng JH, Gruner R, Kotov N, Lasky LC, Prolonged continuous *in vitro* human platelet production using three-dimensional scaffolds, *Exp. Hematol.* **37**(1):101–110 (2009).
108. Proulx C, Dupuis N, St-Amour I, Boyer L, Lemieux R, Increased megakaryopoiesis in cultures of CD34-enriched cord blood cells maintained at 39 degrees C, *Biotechnol. Bioeng.* **88**(6):675–680 (2004).
109. Mostafa SS, Miller WM, Papoutsakis ET, Oxygen tension influences the differentiation, maturation and apoptosis of human megakaryocytes, *Br. J. Haematol.* **111**(3):879–889 (2000).
110. Cheng L, Qasba P, Vanguri P, Thiede MA, Human mesenchymal stem cells support megakaryocyte and pro-platelet formation from CD34(+) hematopoietic progenitor cells, *J. Cell. Physiol.* **184**(1):58–69 (2000).
111. Avecilla ST, Hattori K, Heissig B, Tejada R, Liao F, Shido K, et al., Chemokine-mediated interaction of hematopoietic progenitors with the bone marrow vascular niche is required for thrombopoiesis, *Nat. Med.* **10**(1):64–71 (2004).
112. Jeanpierre S, Nicolini FE, Kaniewski B, Dumontet C, Rimokh R, Puisieux A, et al., BMP4 regulation of human megakaryocytic differentiation is involved in thrombopoietin signaling, *Blood* **112**(8):3154–3163 (2008).

113. O'Brien JJ, Spinelli SL, Tober J, Blumberg N, Francis CW, Taubman MB, et al., 15-Deoxy-delta12,14-PGJ2 enhances platelet production from megakaryocytes, *Blood* **112**(10):4051–4060 (2008).

114. Gandhi MJ, Drachman JG, Reems JA, Thorning D, Lannutti BJ, A novel strategy for generating platelet-like fragments from megakaryocytic cell lines and human progenitor cells, *Blood Cells Mol. Dis*. **35**(1):70–73 (2005).

115. Nishikii H, Eto K, Tamura N, Hattori K, Heissig B, Kanaji T, et al., Metalloproteinase regulation improves *in vitro* generation of efficacious platelets from mouse embryonic stem cells, *J. Exp. Med*. **205**(8):1917–1927 (2008).

116. Gaur M, Kamata T, Wang S, Moran B, Shattil SJ, Leavitt AD, Megakaryocytes derived from human embryonic stem cells: a genetically tractable system to study megakaryocytopoiesis and integrin function, *J. Thromb. Haemost*. **4**(2):436–442 (2006).

117. Takayama N, Nishikii H, Usui J, Tsukui H, Sawaguchi A, Hiroyama T, et al., Generation of functional platelets from human embryonic stem cells *in vitro* via ES-sacs, VEGF-promoted structures that concentrate hematopoietic progenitors, *Blood* **111**(11):5298–5306 (2008).

118. Choi ES, Nichol JL, Hokom MM, Hornkohl AC, Hunt P, Platelets generated *in vitro* from proplatelet-displaying human megakaryocytes are functional, *Blood* **85**(2):402–413 (1995).

119. Cortin V, Pineault N, Garnier A, *Ex vivo* megakaryocyte expansion and platelet production from human cord blood stem cells, *Meth. Mol. Biol*. **482**:109–126 (2009).

120. Gillitzer A, Peluso M, Laugwitz KL, Munch G, Massberg S, Konrad I, et al., Retroviral infection and selection of culture-derived platelets allows study of the effect of transgenes on platelet physiology *ex vivo* and on thrombus formation *in vivo*, *Arterioscler. Thromb. Vasc. Biol*. **25**(8):1750–1755 (2005).

121. Lapidot T, Fajerman Y, Kollet O, Immune-deficient SCID and NOD/SCID mice models as functional assays for studying normal and malignant human hematopoiesis, *J. Mol. Med*. **75**(9):664–673 (1997).

122. Shultz LD, Schweitzer PA, Christianson SW, Gott B, Schweitzer IB, Tennent B, et al., Multiple defects in innate and adaptive immunologic function in NOD/LtSz-scid mice, *J. Immunol*. **154**(1):180–191 (1995).

123. Shultz LD, Ishikawa F, Greiner DL, Humanized mice in translational biomedical research, *Nat. Rev. Immunol*. **7**(2):118–130 (2007).

124. Ueda T, Yoshino H, Kobayashi K, Kawahata M, Ebihara Y, Ito M, et al., Hematopoietic repopulating ability of cord blood CD34(+) cells in NOD/Shi-scid mice, *Stem Cells* **18**(3):204–213 (2000).

125. Perez LE, Rinder HM, Wang C, Tracey JB, Maun N, Krause DS, Xenotransplantation of immunodeficient mice with mobilized human blood CD34+ cells provides an *in vivo* model for human megakaryocytopoiesis and platelet production, *Blood* **97**(6):1635–1643 (2001).

126. Angelopoulou M, Novelli E, Grove JE, Rinder HM, Civin C, Cheng L, et al., Cotransplantation of human mesenchymal stem cells enhances human myelopoiesis and megakaryocytopoiesis in NOD/SCID mice, *Exp. Hematol*. **31**(5):413–420 (2003).

127. Angelopoulou MK, Rinder H, Wang C, Burtness B, Cooper DL, Krause DS, A preclinical xenotransplantation animal model to assess human hematopoietic stem cell engraftment, *Transfusion* **44**(4):555–566 (2004).
128. Tijssen MR, van Hennik PB, di Summa F, Zwaginga JJ, van der Schoot CE, Voermans C, Transplantation of human peripheral blood CD34-positive cells in combination with *ex vivo* generated megakaryocytes results in fast platelet formation in NOD/SCID mice, *Leukemia* **22**(1):203–208 (2008).
129. Feng Y, Zhang L, Xiao ZJ, Li B, Liu B, Fan CG, et al., An effective and simple expansion system for megakaryocyte progenitor cells using a combination of heparin with thrombopoietin and interleukin-11, *Exp. Hematol.* **33**(12):1537–1543 (2005).
130. Pick M, Perry C, Lapidot T, Guimaraes-Sternberg C, Naparstek E, Deutsch V, et al., Stress-induced cholinergic signaling promotes inflammation-associated thrombopoiesis, *Blood* **107**(8):3397–3406 (2006).
131. Mattia G, Milazzo L, Vulcano F, Pascuccio M, Macioce G, Hassan HJ, et al., Long-term platelet production assessed in NOD/SCID mice injected with cord blood CD34(+) cells, thrombopoietin-amplified in clinical grade serum-free culture, *Exp. Hematol.* **36**(2):244–252 (2008).
132. Bruno S, Gunetti M, Gammaitoni L, Dane A, Cavalloni G, Sanavio F, et al., *In vitro* and *in vivo* megakaryocyte differentiation of fresh and *ex-vivo* expanded cord blood cells: Rapid and transient megakaryocyte reconstitution, *Haematologica* **88**(4):379–387 (2003).
133. Bruno S, Gunetti M, Gammaitoni L, Perissinotto E, Caione L, Sanavio F, et al., Fast but durable megakaryocyte repopulation and platelet production in NOD/SCID mice transplanted with *ex-vivo* expanded human cord blood CD34+ cells, *Stem Cells* **22**(2):135–143 (2004).
134. Perez LE, Alpdogan O, Shieh JH, Wong D, Merzouk A, Salari H, et al., Increased plasma levels of stromal-derived factor-1 (SDF-1/CXCL12) enhance human thrombopoiesis and mobilize human colony-forming cells (CFC) in NOD/SCID mice, *Exp Hematol.* **32**(3):300–307 (2004).
135. Suzuki K, Hiramatsu H, Fukushima-Shintani M, Heike T, Nakahata T, Efficient assay for evaluating human thrombopoiesis using NOD/SCID mice transplanted with cord blood CD34+ cells, *Eur. J. Haematol.* **78**(2):123–130 (2006).
136. Schipper LF, van Hensbergen Y, Fibbe WE, Brand A, A sensitive quantitative single-platform flow cytometry protocol to measure human platelets in mouse peripheral blood, *Transfusion* **47**(12):2305–2314 (2007).
137. Nakamura T, Miyakawa Y, Miyamura A, Yamane A, Suzuki H, Ito M, et al., A novel nonpeptidyl human c-Mpl activator stimulates human megakaryopoiesis and thrombopoiesis, *Blood* **107**(11):4300–4307 (2006).
138. Tse WW, Zang SL, Bunting KD, Laughlin MJ, Umbilical cord blood transplantation in adult myeloid leukemia, *Bone Marrow Transpl.* **41**(5):465–472 (2008).
139. Salles II, Thijs T, Brunaud C, De Meyer SF, Thys J, Vanhoorelbeke K, Deckmyn H, Human platelets produced in nonobese diabetic/severe combined immunodeficient (NOD/SCID) mice upon transplantation of human cord blood CD34(+) cells are functionally active in an ex vivo flow model of thrombosis, *Blood* **114**(24):5044–5051 (2009).

140. Salmon P, Kindler V, Ducrey O, Chapuis B, Zubler RH, Trono D, High-level transgene expression in human hematopoietic progenitors and differentiated blood lineages after transduction with improved lentiviral vectors, *Blood* **96**(10):3392–3398 (2000).

141. Barraza RA, Poeschla EM, Human gene therapy vectors derived from feline lentiviruses, *Vet. Immunol. Immunopathol*. **123**(1–2):23–31 (2008).

142. Philpott NJ, Thrasher AJ, Use of nonintegrating lentiviral vectors for gene therapy, *Hum Gene Ther*. **18**(6):483–489 (2007).

143. Santoni de Sio F, Naldini L, Short-term culture of human CD34+ cells for lentiviral gene transfer, *Meth. Mol. Biol*. **506**:59–70 (2009).

144. Santoni de Sio FR, Gritti A, Cascio P, Neri M, Sampaolesi M, Galli C, et al., Lentiviral vector gene transfer is limited by the proteasome at postentry steps in various types of stem cells, *Stem Cells* **26**(8):2142–2152 (2008).

145. Raslova H, Komura E, Le Couedic JP, Larbret F, Debili N, Feunteun J, et al., FLI1 monoallelic expression combined with its hemizygous loss underlies Paris-Trousseau/Jacobsen thrombopenia, *J. Clin. Invest*. **114**(1):77–84 (2004).

146. Shi Q, Wilcox DA, Morateck PA, Fahs SA, Kenny D, Montgomery RR, Targeting platelet GPIbalpha transgene expression to human megakaryocytes and forming a complete complex with endogenous GPIbbeta and GPIX, *J. Thromb. Haemost*. **2**(11):1989–1997 (2004).

147. Shi Q, Wilcox DA, Fahs SA, Fang J, Johnson BD, Du LM, et al., Lentivirus-mediated platelet-derived factor VIII gene therapy in murine haemophilia A, *J. Thromb. Haemost*. **5**(2):352–361 (2007).

148. Fang J, Hodivala-Dilke K, Johnson BD, Du LM, Hynes RO, White GC 2nd, Wilcox DA, et al., Therapeutic expression of the platelet-specific integrin, alphaIIbbeta3, in a murine model for Glanzmann thrombasthenia, *Blood* **106**(8):2671–2679 (2005).

149. Reiser J, Lai Z, Zhang XY, Brady RO, Development of multigene and regulated lentivirus vectors, *J. Virol*. **74**(22):10589–10599 (2000).

150. Wilcox DA, Olsen JC, Ishizawa L, Griffith M, White GC 2nd, Integrin alphaIIb promoter-targeted expression of gene products in megakaryocytes derived from retrovirus-transduced human hematopoietic cells, *Proc. Natl. Acad. Sci. USA* **96**(17):9654–9659 (1999).

151. Wilcox DA, Olsen JC, Ishizawa L, Bray PF, French DL, Steeber DA, et al., Megakaryocyte-targeted synthesis of the integrin beta(3)-subunit results in the phenotypic correction of Glanzmann thrombasthenia, *Blood* **95**(12):3645–3651 (2000).

152. Lavenu-Bombled C, Izac B, Legrand F, Cambot M, Vigier A, Masse JM, et al., Glycoprotein Ibalpha promoter drives megakaryocytic lineage-restricted expression after hematopoietic stem cell transduction using a self-inactivating lentiviral vector, *Stem Cells* **25**(6):1571–1577 (2007).

153. Gillitzer A, Peluso M, Bultmann A, Munch G, Gawaz M, Ungerer M, Effect of dominant negative SNAP-23 expression on platelet function, *J. Thromb. Haemost*. **6**(10):1757–1763 (2008).

154. Briquet-Laugier V, Lavenu-Bombled C, Schmitt A, Leboeuf M, Uzan G, Dubart-Kupperschmitt A, et al., Probing platelet factor 4 alpha-granule targeting, *J. Thromb. Haemost*. **2**(12):2231–2240 (2004).

155. Yan XQ, Lacey DL, Saris C, Mu S, Hill D, Hawley RG, et al., Ectopic overexpression of c-mpl by retroviral-mediated gene transfer suppressed megakaryopoiesis but enhanced erythropoiesis in mice, *Exp. Hematol*. **27**(9):1409–1417 (1999).

156. Fock EL, Yan F, Pan S, Chong BH, NF-E2-mediated enhancement of megakaryocytic differentiation and platelet production *in vitro* and *in vivo*, *Exp. Hematol*. **36**(1):78–92 (2008).

157. Li Z, Fehse B, Schiedlmeier B, Dullmann J, Frank O, Zander AR, et al., Persisting multilineage transgene expression in the clonal progeny of a hematopoietic stem cell, *Leukemia* **16**(9):1655–1663 (2002).

158. Klug CA, Cheshier S, Weissman IL, Inactivation of a GFP retrovirus occurs at multiple levels in long-term repopulating stem cells and their differentiated progeny, *Blood* **96**(3):894–901 (2000).

159. Ghevaert C, Wilcox DA, Fang J, Armour KL, Clark MR, Ouwehand WH, et al., Developing recombinant HPA-1a-specific antibodies with abrogated Fcgamma receptor binding for the treatment of fetomaternal alloimmune thrombocytopenia, *J. Clin. Invest*. **118**(8):2929–2938 (2008).

160. Dupre L, Marangoni F, Scaramuzza S, Trifari S, Hernandez RJ, Aiuti A, et al., Efficacy of gene therapy for Wiskott-Aldrich syndrome using a WAS promoter/cDNA-containing lentiviral vector and nonlethal irradiation, *Hum Gene Ther*. **17**(3):303–313 (2006).

161. Orange JS, Stone KD, Turvey SE, Krzewski K, The Wiskott-Aldrich syndrome, *Cell. Mol. Life Sci*. **61**(18):2361–2385 (2004).

162. Horn PA, Keyser KA, Peterson LJ, Neff T, Thomasson BM, Thompson J, et al., Efficient lentiviral gene transfer to canine repopulating cells using an overnight transduction protocol, *Blood* **103**(10):3710–3716 (2004).

163. Kootstra NA, Matsumura R, Verma IM, Efficient production of human FVIII in hemophilic mice using lentiviral vectors, *Mol. Ther*. **7**(5Pt 1):623–631 (2003).

164. Tiede A, Eder M, von Depka M, Battmer K, Luther S, Kiem HP, et al., Recombinant factor VIII expression in hematopoietic cells following lentiviral transduction, *Gene Ther*. **10**(22):1917–1925 (2003).

165. Wilcox DA, Shi Q, Nurden P, Haberichter SL, Rosenberg JB, Johnson BD, et al., Induction of megakaryocytes to synthesize and store a releasable pool of human factor VIII, *J. Thromb. Haemost*. **1**(12):2477–2489 (2003).

166. Yarovoi HV, Kufrin D, Eslin DE, Thornton MA, Haberichter SL, Shi Q, et al., Factor VIII ectopically expressed in platelets: Efficacy in hemophilia A treatment, *Blood* **102**(12):4006–4013 (2003).

167. Shi Q, Wilcox DA, Fahs SA, Weiler H, Wells CW, Cooley BC, et al., Factor VIII ectopically targeted to platelets is therapeutic in hemophilia A with high-titer inhibitory antibodies, *J. Clin. Invest*. **116**(7):1974–1982 (2006).

168. Shi Q, Fahs SA, Wilcox DA, Kuether EL, Morateck PA, Mareno N, et al., Syngeneic transplantation of hematopoietic stem cells that are genetically modified to express factor VIII in platelets restores hemostasis to hemophilia A mice with preexisting FVIII immunity, *Blood* **112**(7):2713–2721 (2008).

169. Wilcox DA, Du LM, Haberichter SL, Jacobi PM, Fang J, Jensen ES, et al., Platelet-targeted expression of coagulation factor VIII (FVIII) shows efficacy for using the dog as a large animal model for gene therapy of hemophilia A, *Proc. 50th Meeting American Society of Hematology*, San Francisco, Dec. 6–9, 2008.

170. Kikuchi J, Mimuro J, Ogata K, Tabata T, Ueda Y, Ishiwata A, Sakata Y, et al., Sustained transgene expression by human cord blood derived CD34+ cells transduced with simian immunodeficiency virus agmTYO1-based vectors carrying the human coagulation factor VIII gene in NOD/SCID mice, *J. Gene Med.* **6**(10):1049–1060 (2004).

171. Ohmori T, Mimuro J, Takano K, Madoiwa S, Kashiwakura Y, Ishiwata A, et al., Efficient expression of a transgene in platelets using simian immunodeficiency virus-based vector harboring glycoprotein Ibalpha promoter: *In vivo* model for platelet-targeting gene therapy, *FASEB J*. **20**(9):1522–1524 (2006).

172. Ohmori T, Ishiwata A, Kashiwakura Y, Madoiwa S, Mitomo K, Suzuki H, et al., Phenotypic correction of hemophilia A by ectopic expression of activated factor VII in platelets, *Mol. Ther*. **16**(8):1359–1365 (2008).

173. Labbaye C, Spinello I, Quaranta MT, Pelosi E, Pasquini L, Petrucci E, et al., A three-step pathway comprising PLZF/miR-146a/CXCR4 controls megakaryopoiesis, *Nat. Cell. Biol* **10**(7):788–801 (2008).

174. Gushiken FC, Patel V, Liu Y, Pradhan S, Bergeron AL, Peng Y, et al., Protein phosphatase 2A negatively regulates integrin alpha(IIb)beta(3) signaling, *J. Biol. Chem*. **283**(19):12862–12869 (2008).

175. Jung YJ, Chae HC, Seoh JY, Ryu KH, Park HK, Kim YJ, et al., Pim-1 induced polyploidy but did not affect megakaryocytic differentiation of K562 cells and CD34+ cells from cord blood, *Eur. J. Haematol*. **78**(2):131–138 (2007).

176. Schulze H, Dose M, Korpal M, Meyer I, Italiano JE, Jr., Shivdasani RA, RanBP10 is a cytoplasmic guanine nucleotide exchange factor that modulates noncentrosomal microtubules, *J. Biol. Chem*. **283**(20):14109–14119 (2008).

177. Bouilloux F, Juban G, Cohet N, Buet D, Guyot B, Vainchenker W, et al., EKLF restricts megakaryocytic differentiation at the benefit of erythrocytic differentiation, *Blood* **112**(3):576–584 (2008).

178. Levay K, Slepak VZ, Tescalcin is an essential factor in megakaryocytic differentiation associated with Ets family gene expression, *J. Clin. Invest*. **117**(9):2672–2683 (2007).

179. Gurbuxani S, Xu Y, Keerthivasan G, Wickrema A, Crispino JD, Differential requirements for survivin in hematopoietic cell development, *Proc. Natl. Acad. Sci. USA* **102**(32):11480–11485 (2005).

180. Olthof SG, Fatrai S, Drayer AL, Tyl MR, Vellenga E, Schuringa JJ, Downregulation of signal transducer and activator of transcription 5 (STAT5) in CD34+ cells promotes megakaryocytic development, whereas activation of STAT5 drives erythropoiesis, *Stem Cells* **26**(7):1732–1742 (2008).

181. Berthebaud M, Riviere C, Jarrier P, Foudi A, Zhang Y, Compagno D, et al., RGS16 is a negative regulator of SDF-1-CXCR4 signaling in megakaryocytes, *Blood* **106**(9):2962–2968 (2005).

182. Eto K, Nishikii H, Ogaeri T, Suetsugu S, Kamiya A, Kobayashi T, et al., The WAVE2/Abi1 complex differentially regulates megakaryocyte development and spreading: Implications for platelet biogenesis and spreading machinery, *Blood* **110**(10):3637–3647 (2007).

183. Ohmori T, Kashiwakura Y, Ishiwata A, Madoiwa S, Mimuro J, Sakata Y, Silencing of a targeted protein in *in vivo* platelets using a lentiviral vector delivering short hairpin RNA sequence, *Arterioscler. Thromb. Vasc. Biol.* **27**(10):2266–2272 (2007).
184. Petrich BG, Marchese P, Ruggeri ZM, Spiess S, Weichert RA, Ye F, et al., Talin is required for integrin-mediated platelet function in hemostasis and thrombosis, *J. Exp. Med.* **204**(13):3103–3111 (2007).
185. Nieswandt B, Moser M, Pleines I, Varga-Szabo D, Monkley S, Critchley D, et al., Loss of talin1 in platelets abrogates integrin activation, platelet aggregation, and thrombus formation *in vitro* and *in vivo*, *J. Exp. Med.* **204**(13):3113–3118 (2007).

# 12

# SYSTEMS BIOLOGY TO STUDY PLATELET-RELATED BLEEDING DISORDERS

JAN-WILLEM N. AKKERMAN AND BERNARD DE BONO

**Abstract**

Systems biology provides a crucial platform for knowledge accumulation of signaling pathways, which control platelet functions or their inhibition. It assists in interpreting the impact of congenital and acquired defects that result in an increased tendency to bleed.

## 12.1 INTRODUCTION: INITIATION, INHIBITION, AND TERMINATION OF PLATELET FUNCTIONS

The realization that cells and organs can be understood only as an integrated system of linked processes is now taking over from the era of reductionist simplification whereby biological processes were dissected at the molecular and atomic levels. Such a network of connectivities provides an overview of, as well as a crucial platform for, knowledge accumulation. It also facilitates comparisons with other cell types as a way to discover new centers for signal regulation. Systems biology aims to identify the rules that govern the emergent behavior of biological systems by generating models that are able to distinguish the normal from the pathological [1–3]. Pathway complexity is now a key issue, requiring a level of knowledge integration that is often deemed unsustainable.

*Platelet Proteomics: Principles, Analysis and Applications*, First Edition.
Edited by Ángel García and Yotis A. Senis.

The platelet field is undergoing a radical transformation from reductionist simplification to large-scale integration. Considerable amounts of information now originate from high-throughput analyses of transcription and translation, as well as insights into the processes that control proteins through translocation, glycosylation, methylation, and phosphorylation of Tyr, Ser, and Thr residues. This bulk of information paves the way for what is now the biggest challenge in this field: to translate it into a better understanding of the mechanisms that regulate initiation and suppression of platelet function.

This chapter focuses on human platelets. Information of their murine counterparts or from other cell types is used only when connectivity shows gaps because of lack of information from the human platelet/megakaryocyte field. Knowledge accumulated so far is presented as connectivity schemes. For reasons of clarity, the figures do not show the translocation of signaling components between subcellular compartments and the role of anchoring proteins therein. Furthermore, affinity changes of rate-limiting enzymes during platelet activation as well as the complexity introduced by isozyme expression are discussed elsewhere [4], and are beyond the scope of this overview. The focus of this work is on the activation of pathways either triggering functions or neutralizing their induction, and the mechanisms that turn off these signals, such as through dephosphorylation reactions; our insight into this part of the signaling machinery is still somewhat limited.

Systems biology is a platform for generating new hypotheses about integration of signals. Therefore, following definition of a framework of signaling routes in the platelet, future work will be facilitated through comparisons between networks within platelets and between different cell types, other blood cells, and neurons in particular. Important nodes identified in other cells might well serve a role in platelets. A better insight in these integrated networks will help to understand the mechanisms that drive platelet functions and the disturbances that interfere with their regulation, resulting in an increased tendency to bleed.

With a few notable exceptions, all agents that initiate or inhibit platelet functions are unable to cross the plasma membrane but transfer information through a sequence known as *stimulus–response coupling*. This mechanism involves the transfer of messages through transitions between hydrophilic and hydrophobic compounds, amplification through internal circuits, and via the merging of pathways that allow integration of information.

A 2007 review separates the surface-expressed receptors (Rs) for activation and inhibition of platelet functions into five categories [5]: (1) seven-transmembrane-Rs, (2) member of the leucine-rich repeat family, (3) integrins, (4) immunoglobulin-Rs, and (5) Tyr-kinase-Rs. Although platelet activation at the level of surface receptors can be separated into different mechanisms, downstream signaling sees a strong integration of activating and inhibitory routes that is at odds with the concept of agonist-specific platelet activation. Thus, a first challenge for systems biology is to decipher intertwining signalingy mechanisms downstream of receptor activation. In addition to intracellular crosstalk, there is the extracellular feedback of information loops through thromboxane $A_2(TxA_2)$ release and ADP secretion, as well as the induction of a new wave of signaling events initiated by ligand-occupied integrin αIIbβ3. However, platelet functions initiated by thrombin differ from those induced by ADP, indicating that not only the type of signaling sequence but also the strength of its activation determines the biological response.

The highly simplified scheme in Figure 12.1 illustrates the main flows of information induced by platelet stimulating agents. An arbitrary division in fields of signaling is as follows:

1. *Signal induction*—the generation of signals through ligand–receptor interaction and their distribution over multiple pathways immediately downstream of the receptor
2. $Ca^{2+}$ *homeostasis*—the induction of $Ca^{2+}$ mobilization from intracellular storage sites, influx of extracellular $Ca^{2+}$, and restoration of preactivation levels by reuptake in the storage sites and efflux
3. *Mobility, spreading, and contraction*—changes in cell shape, spreading to an adhesive surface and contraction supported by the actin cytoskeleton and actomyosin contractile activity
4. *Granule secretion*—secretion responses that liberate low-molecular-weight substances and many different proteins
5. *Procoagulant activity*—the generation of a procoagulant surface through phosphatidylserine (PS) transfer
6. *Aggregation*—platelet–platelet interaction through fibrinogen and von Willebrand factor binding to integrin αIIbβ3
7. *Signal amplification*—I acceleration of signal induction through release of $TxA_2$ and extracellular feedback
8. *Signal amplification*—II acceleration of signal induction through secretion of ADP stored in δ-granules, which, similarly to $TxA_2$, accelerates primary signal induction and in addition suppresses the formation of the endogenous platelet inhibitor cAMP
9. *Signal inhibition*—prevention of signal induction through a rise in cAMP and cGMP and their kinases by phosphorylating intermediates in activating pathways
10. *Signal termination*—the arrest of signal induction by inhibitory receptors, thereby setting timeframes for transmission of signals that initiate platelet functions

A number of excellent reviews on this subject are available [6–9].

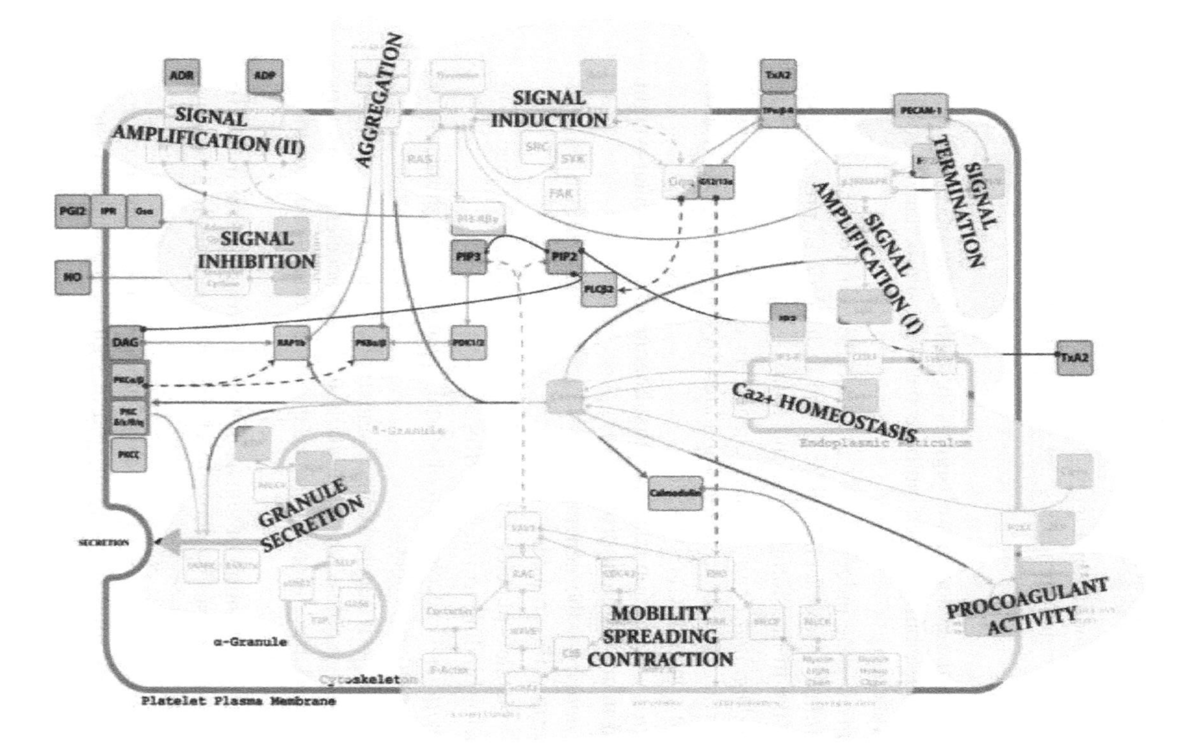


**Figure 12.1** An arbitrary separation of fields of signal transduction in platelets. Following initial ligand–receptor contact, intracellular messengers are generated (signal induction). They transduce signaling to $Ca^{2+}$ mobilization and $Ca^{2+}$ influx ($Ca^{2+}$ homeostasis), the mechanisms that control the cytoskeleton and contractility (mobility of megakaryocytes, spreading on an adhesive surface and contraction of platelets), the extrusion of the contents of δ−, α−, and lysosomal granule contents (granule secretion) and the surface exposure of negatively charged phosphatidylserine, which assists in coagulation (procoagulant activity). These responses are strongly enhanced by extracellular feedback signaling through released thromboxane A2 and secreted ADP (signal amplification type (I) and (II), respectively; see item 7 in list at end of section 12.1). The activation pathways are blocked by ligands that raise the level of the intracellular inhibitor cAMP and by nitricmonooxide, which increases the level of the inhibitor cGMP (signal inhibition). Following induction of activating pathways, receptors turn the signaling pathways off (signal termination). A further description is given in the text.

## 12.2 PLATELET ACTIVATION THROUGH SEVEN-TRANSMEMBRANE RECEPTORS

### 12.2.1 Signal Induction

The seven-transmembrane receptor family is an important class of receptors characterized by their coupling to GTP binding proteins, G-proteins in short (Fig. 12.2). For the receptors for thrombin (PAR1/4), ADP (P2Y1-R), serotonin, or 5-hydroxytryptamine (5HT2A-R), platelet activating factor (PAF1-R), and $TxA_2$($TP\alpha/\beta$-R), the G involved is Gq, which, through its $\alpha$-subunit, activates PLC$\beta$2/3 [10]. This enzyme splits $PIP_2$ into $IP_3$ and DAG. $PIP_2$ serves as a docking site for the GTPase CDC42, a participant in filapodia formation, and is a substrate for PI3−K that generates PIP3. This phospholipid is a docking site for the inactive, cytosolic PKB$\alpha/\beta$(AKT1/2) isoforms as well as for PDK1/2, which, when brought together, activate PKB through phosphorylation of Ser473 (PKB$\alpha$) and Ser474 (PKB$\beta$) and Thr308 (PKB$\alpha/\beta$) [11,12]. PKB controls a wide range of biological activities whose participation in integrin $\alpha$IIb$\beta$3 activation [13], glucose transport through the transporter GLUT3 [14], and induction of prosurvival signals [15] have been established in platelets and/or megakaryocytes.

Because IP3 is water-soluble, it easily transfers to the endoplasmatic reticulum, where it initiates $Ca^{2+}$ mobilization by activating the IP3-R. DAG and $Ca^{2+}$ together activate PKC$\alpha/\beta$. These PKC isozymes are major steps in the induction of secretion, liberating ADP and serotonin from $\delta$-granules, numerous soluble proteins from the $\alpha$-granule, and inducing surface expression of membrane-bound $\alpha$-granule proteins, including GLUT3, which mediates glucose uptake; P-selectin, which binds counterreceptors on leukocytes and endothelial cells; and a second population of $\alpha$IIb$\beta$3 molecules that bind fibrinogen and von Willebrand facor (vWF), thereby strengthening the aggregate.

In addition to its role in Gq-mediated signaling, the $TP\alpha/\beta$-Rs share with PAR1/4 the ability to activate G12/13. Ligand occupancy starts signaling to the GTPase Rho and control of stress fiber formation by activating PAK and MLC (myosin light chain) phosphorylation by inhibiting its phosphatase. Also, the receptors for LPA, denoted EDG1/2/3, are coupled to G13 and further signaling to Rho [16].

Activated PAR1/4 also triggers signaling to the GTPase Ras through the adapter protein Grb2 and the exchange factor SOS. Activated Ras (Ras $\alpha$-GTP), in turn, activates Raf, which in most cells directly activates MEK and ERK2. In platelets, MEK induction is by PKC and $Ca^{2+}$ [17,18], and downstream signaling by Ras remains to be clarified.

A second class of inducers of platelet functions are the seven-transmembrane receptors that are associated with the G-protein Gi, an inhibitor of adenylylcyclase. Binding of ADP to the $P2Y_{12}$-R and epinephrine to the $\alpha_2$-adrenergic-R releases $\alpha_i$ and $\alpha_z$ from their $\beta\gamma$ complexes, respectively. This starts inhibition of adenylyl cyclase-7 and activation of PI3-kinase-$\beta\gamma$, thereby suppressing cAMP formation ($G\alpha_i$, $G\alpha_z$) and inducing PKB activation ($\beta\gamma$).

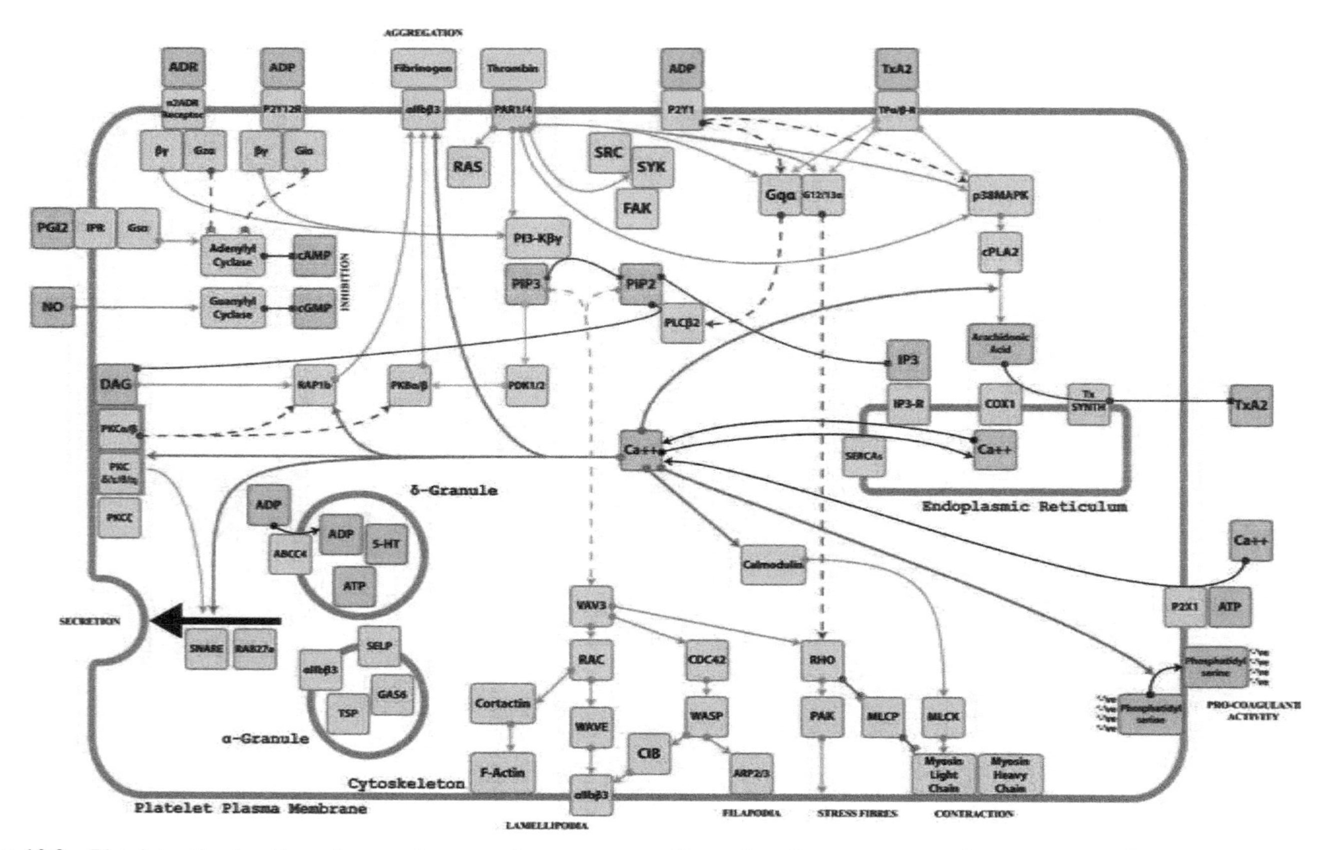


**Figure 12.2** Platelet activation through seven transmembrane receptors. Examples of seven-transmembrane receptors that start platelet functions include the receptors for thrombin, the P2Y1 receptor for ADP, and the thromboxane A2 receptor. Other receptors stimulate signaling by primary receptors by lowering the level of the inhibitor cAMP (e.g., the α2-adrenergic receptor, the P2Y12 receptor for ADP). Boxes in blue represent proteins; those in green, metabolites; “-”ve: negative charge; → activation;—<< inhibition.

### 12.2.2 $Ca^{2+}$ Homeostasis

The IP3-R consists of a ligand recognition site and a $Ca^{2+}$ channel. Ligand occupancy opens the channels, enabling the flux of $Ca^{2+}$ ions through a gradient induced by a high $Ca^{2+}$ concentration in the stores (100–800 μM) and a low $Ca^{2+}$ concentration in the cytosol (70–100 nM) of the resting platelet. The fall in stored $Ca^{2+}$ is sensed by STIM1, which, in turn, activates the CRACs in the plasma membrane. Their pore forming unit ORAI1 then facilitates $Ca^{2+}$ influx through a gradient established by a high extracellular $Ca^{2+}$ concentration (about 1 mM) and a low cytosolic $Ca^{2+}$ concentration (70–100 nM) [19,20]. The consequences of the increase in cytosolic $Ca^{2+}$ are diverse:

1. Disappearance of microtubules facilitating shape change driven by actin reassembly
2. Formation of a $Ca^{2+}$-calmodulin complex, which activates MLCK and phosphorylates myosin light chain
3. Induction of actomyosin contractility; activation of $Ca^{2+}$ dependent PKC isoforms that signal to secretion and aggregation, activation of cPLA2, which generates $TxA_2$
4. Activation of integrin αIIbβ3 through $Ca^{2+}$ and calcium and integrin binding protein (CIB)
5. Induction of granule secretion and generation of a procoagulant surface that supports the clotting
6. Activation of SERCA 2b and 3, which pump back $Ca^{2+}$ ions until prestimulation levels are restored [21]

### 12.2.3 Mobility, Spreading, and Contraction

The cytoskeleton consists of microtubules, intermediate filaments, and actin filaments. Microtubules are polymers of tubulin-α/β and maintain the discoid shape of the resting platelet. Intermediate filaments organize the three-dimensional structure of the cell, anchoring organelles and serving in intracellular transport. Actin filaments are composed of two intertwined actin chains and are concentrated beneath the plasma membrane of the resting platelet in a structure called the *membrane skeleton*. They preserve the discoid shape of the resting platelet and on activation drive shape change and spreading to an adhesive surface by forming cytoplasmatic protuberances such as filapodia and lamellipodia. Together with myosin, actin polymers form the actomyosin complex that drives motility of megakaryocytes and contraction of the platelet aggregate. Stress fibers are bundles of actin–myosin–actin that serve as the major mediator of cell contractility [22].

In resting platelets, actin polymers are severed by the $Ca^{2+}$-induced activation of gelsolin through binding to the barbed ends, which are covered by CapZ and adducin, thereby protecting against binding of new actin monomer. Following the formation of PIP3, other polyphosphoinositides and LPA then inactivate gelsolin, triggering its release together with adducin and CapZ and exposing nucleation sites for binding of actin monomers in a process mediated by thymosin β4 and profilin. The result is polymer elongation and formation of new actin filaments [22].

Platelets contain myosin type IIa, which consists of two heavy and four light chains. Phosphorylation enables the light chains to bind to the heavy chains. Formation of an actomyosin complex enables contraction by mechanisms analogous to those found in striated muscle. Following adhesion to a surface coated with fibrinogen or vWF, platelets spread, thereby strengthening their interaction with an adhesive surface. An important step in this process is formation of lamellipodia. These are cytoskeletal actin projections formed by a two-dimensional actin mesh starting at the plasma membrane. Microspikes that spread beyond the lamellipodia border are filopodia. Actin polymers grow through crosslinkage enforced by filamin A and α-actinin. Arp2/3 complexes are found at microfilament–microfilament junctions together with proteins that sustain the actin mesh such as WASP.

Three members of the GTPase family are key intermediates in the mechanisms that drive cytoskeleton reassembly. These are Rac, Cdc42, and Rho. In reactions under control of the guanine nucleotide exchange factor Vav, the active, GTP-bound form controls formation of lamellipodia, filapodia, and stress fibers. Mainstreams of signaling are the phosphorylation of WAVE by Rac, of WASP by Cdc42 and the activation of PAK and inhibition of MLC phosphatase by Rho/Rho kinase. WAVE and WASP are scaffolds that link upstream signals to activation of the ARP2/3 complex leading to a burst in actin polymerization [23]. Numerous membrane binding proteins interact with WAVE and WASP, thereby controlling membrane-dependent actin polymerization. Human platelets contain WAVE1/2/3, which are substrates for calpain and couple to the cytoskeleton on activation by thrombin [24]. WAVE contains binding sites for PIP3, ARP2/3, Rac, and monomeric actin, among others. Tyr-phosphorylated WASP induces a reorganization of the actin cytoskeleton, formation of podosomes involved in mobility, and induces migration and cell trafficking. WASP is activated by phosphoinositides and Cdc42 in a synergistic manner and by the Src family Tyr-kinases. Phosphorylated WASP activates ARP2/3 [24].

### 12.2.4 Granule Secretion

Secretion of α-granules, δ-granules, and lysosomes involves fusion of the lipid bilayers of the granules and the open canalicular system and release of soluble granule contents. In the resting platelet, fusion is prevented by electrostatic repulsive and hydration forces [25]. SNARE proteins on both membranes bring the two bilayers together, inducing fusion of the outer layers, formation of a fusion pore, and liberation of granule contents. This process is under the control of PA and PIP2 that recruit cytoskeletal proteins (e.g., α-actinin, myosin, N-WASP), proteins involved in lipid metabolism (PLC, PI3-K), SNARE chaperone molecules (synaptotagmin, MUNC18), and regulators of small GTPases

(GTPase activating proteins, guanine nucleotide exchange factors). The vesicle SNARE proteins or VAMP3/8 bind to target SNARE proteins syntaxin-2/4/7 and SNAP-23 family proteins. Several chaperone molecules direct the function of the SNARE proteins, such as the small GTPase Rab family, of which the members Rab1a/1b/3B/4/6c/8/11/27a/27b and −31 are present in platelets [26]. Interestingly, the Rab GDP dissociation inhibitor suppresses α-granule but not δ-granule secretion.

Secretion is constrained by the actin cytoskeleton but requires actomysin contractility for granule centralization and extrusion. A rise in cytosolic $Ca^{2+}$ and activation of PKC are key players in initiation of secretion. $Ca^{2+}$-mediated activation includes the EF-hand proteins and $Ca^{2+}$/phospholipid-binding proteins. EF-hand proteins are calmodulin and calcyclin. Calmodulin binds to VAMP and contributes to pore formation. PKC initiates secretion in a $Ca^{2+}$-dependent and independent manner by phosphorylating SNARE proteins and certain chaperone molecules, the integral membrane proteins (MARCs) in particular.

The α-granules contain several proteins that, when released, begin activating receptors that contribute to further platelet stimulation. Examples are fibrinogen and von Willebrand factor (vWF), which are ligands for activated αIIbβ3 and when attached to a surface bind αIIbβ3 in the resting conformation, Gas6, a ligand for certain Tyr-kinase-receptors (see text below) and thrombospondin, which, through one of its receptors, suppresses activation of guanylylcyclase by NO [27].

Secretion of δ-granule contents contributes with extrusion of ADP and ATP, serotonin, and possibly histamine. Concentrations of δ-granule ATP and ADP are 2.02 and 1.74 μmol $10^{-11}$ platelets, respectively [28]. The presence of ADP in the δ-granules might reflect the activity of the nucleotide transporter ABCC4 [29]. ADP is a ligand for the P2Y1-R coupled to Gq and the P2Y12-R coupled to Gi. ATP is ligand for the P2X1 $Ca^{2+}$ channel and serotonin for the 5HT2A-R coupled to Gq.

### 12.2.5 Procoagulant Activity

Resting platelets maintain an asymmetric phospholipid distribution in the plasmamembrane through the activity of ATP-dependent flippase (aminophospholipid translocase), which catalyzes inward lipid transport and floppase [ATP binding cassette transporter 1, previously known as MRP1 (multidrug resistance protein 1)], which catalyzes outward lipid transport. Their combined activities preserve the exclusive localization of PS at the inner leaflet of the plasma membrane [30]. Platelet agonists that induce a strong and persistent increase in cytosolic $Ca^{2+}$ start a third enzyme called *scramblase* that rapidly moves phospholipids from inner to outer leaflet and vice versa (flip-flop) and within minutes collapses membrane asymmetry. The result is expression of negatively charged PS on the platelet surface, which binds the coagulation factors II (prothrombin), VII, IX, and X, inducing a manyfold acceleration in the reactions that lead to formation of fibrin fibers. PS exposure, together with $Ca^{2+}$-induced calpain activation, also triggers the outward blebbing of the plasma membrane followed by the shedding of microvesicles. Their enrichment in tissue factor, the starter of the coagulation cascade, is thought to be a main factor contributing to intravascular coagulation initiation (bloodborne tissue factor) [31]. PS is also a cofactor for several PKC isozymes and promotes the $Ca^{2+}$-induced membrane fusion that precedes exocytosis.

### 12.2.6 Aggregation

Integrin αIIbβ3 (glycoprotein IIb-IIIa) is the fibrinogen receptor that couples one platelet to another in a process known as *aggregation*. It is also a receptor for vWF, which also sustains aggregation, and for fibronectin. In resting platelets, the complex is inaccessible to soluble fibrinogen. Platelet activation by ligands that induce a rise in $Ca^{2+}$ and activate PKC start *inside-out activation of* the receptor, initiating a conformation change that exposes ligand binding sites. Following ligand binding, clusters form that not only mediate platelet–platelet interaction but also are a source for a second wave of signal generation known as *outside-in signaling* (discussed below). Integrin αIIbβ3 in the resting state is capable of binding surface-bound fibrinogen, but metabolic arrest induced by agents that raise cAMP or cGMP completely closes αIIbβ3.

Gene deletion experiments in mice make clear that multiple signaling routes control the affinity state of αIIbβ3. A major step in αIIbβ3 activation is binding of the 270-kDa protein talin to the β3 cytosolic tail [32,33]. Talin consists of a 50-kDa FERM domain with the β3 binding site and a 220-kDa Rod domain that binds actin. FERM and Rod domains are separated by a calpain cleavage site and thereby are sensitive to changes in $Ca^{2+}$. Recruitment of talin is under control of the small GTPase Rap1b. Also PKC contributes to Rap1b activation. Activation of Rap1b is promoted by the guanine nucleotide exchange factor CalDAG-GEF1 and therefore is induced by a rise in $Ca^{2+}$ and formation of DAG, both results of PLCβ activation. A Rap1 GAP converts the GTPase back to the inactive conformation and inhibits αIIbβ3 activation. These reactions are controlled by of RIAM. Further control may come from binding to talin of PIP-K1γ, PIP2, and calpain and the phosphorylation of Ser and Thr residues by PKC. The constitutative association of αIIbβ3 and PP1c, Rack1, and Csk with αIIbβ3 in resting platelets and binding of kindlin (to β3), CIB (to αIIb), and β3-endonexin (to β3) further adds to the complexity of the mechanisms that expose ligand binding sites of αIIbβ3.

### 12.2.7 Signal Amplification

***12.2.7.1 Signal Amplification I: TxA*2** This pathway is known for its sensitivity to aspirin, a successful antithrombotic agent that reduces chances for a second cardiovascular event by 25% [32]. How seven-transmembrane receptors start this pathway is not precisely known [7]. Following activation of upstream MKK3, p38MAP-kinase is activated, which, together with a rise in $Ca^{2+}$, activates cPLA2. This enzyme liberates arachidonate from position 2 of membrane phospholipids. The next step is the conversion into prostaglandin endoperoxides by COX1 and a peroxidation step resulting in formation of TxA2 in a reaction catalyzed by Tx-synthase. Both COX1 and the synthase are bound to the endoplasmatic reticulum, which implies that a transport mechanism must bring arachidonate to these enzymes. TxA2 is released through facilitated diffusion and activates the same or adjacent platelets via the TP$\alpha$/$\beta$-R.

***12.2.7.2 Signal Amplification II: ADP*** Adenosine diphosphate is a ligand for the seven-transmembrane receptor P2Y12-R, the target for the antithrombotic agents clopidogrel and prasugrel. This member of the purinergic receptor family is present as oligomers in lipid-rich patches in the plasma membrane known as *rafts* and coupled to the inhibitory G protein of adenylyl cyclase-7, Gi [34]. Other seven-transmembrane receptors coupled to $G_i$ are EP3-R (for PGE2) and the $\alpha_2$-adrenergic-R (for adrenaline and noradrenaline), which is coupled to Gz, a member of the Gi family. Receptor occupancy triggers the dissociation of GTP-bound G$\alpha$i-2, which inhibits adenylyl cyclase and the formation of cAMP. Released $\beta\gamma$ subunits drive activation of PI3-K.

### 12.2.8 Signal Inhibition

The countermechanism of ADP-driven adenylylcyclase suppression consists of the seven-transmembrane receptors coupled to the stimulatory G-protein of adenylylcyclase, Gs. PGI2 is released from endothelial cells and binds to the PGI2-R (IPR), thereby inducing G$\alpha$s activation. Other members are the EP2-R (for PGE1/2), EP4-R (for PGE1/2), the DP-R (for PGD2), and the A2A-R (for adenosine) Thus, platelets have two opposing mechanisms to stabilize the basal level of cAMP, and slight disturbances in this mechanism have a direct impact on aggregation (see Ref 35 and references cited therein).

Cyclic AMP inhibits the induction of platelet function through PKA. This Ser/Thr kinase consists of a tetramer of two catalytic and two regulatory domains. Each regulatory domain is noncovalently bound to a catalytic subunit, and binding of two molecules of cAMP to each regulatory subunit, releases the catalytic subunits which then become active [36]. PKA interferes with activating pathways at the level of receptors (PARs, TP-Rs, IP3-R), enzymes (PLC$\beta$), and SERCAs (see Abbreviations list at end of this chapter), among others. A major substrate of PKA is VASP, an inhibitor of $\alpha$IIb$\beta$3 activation.

Breakdown of cAMP to AMP is mediated through PDE2/3/5 (phosphodiesterase). Stimulation of the $\alpha$2-adrenergic-R activates Gz, inducing a 20% fall in cAMP, thereby enhancing functional responses [37]. A similar fall is induced by a high thrombin concentration through PAR1/4 and Gi by released ADP [38]. Pharmacological blockade of PDE induces a gradual increase in cAMP in the absence of ligands for receptors that modulate the activity of Gs or Gi. Thus, there is a rapid turnover of cAMP in the resting platelet in the absence of occupied receptors coupled to Gs [39,40]. Overstimulation by seven transmembrane receptors is preventend by PKA-induced receptor phosphorylation and arrestin-mediated endocytosis [41]. A closely related mechanism is the control of cGMP.

The mechanism is unique in the sense that it is activated by receptor-independent transport of NO, which directly activates the haem group of guanylylcyclase, thereby generating cGMP. Cyclic GMP activates PKG, which also inhibits signal transduction to aggregation and secretion.

## 12.3 PLATELET ACTIVATION THROUGH THE LEUCINE-RICH REPEAT FAMILY: GLYCOPROTEIN Ibα

### 12.3.1 Signal Induction

Platelet GPIbα (glycoprotein Ibα) is a component of the $[GPIb\alpha]_2[GPIb\beta]_4GPV[GPIX]_2$ complex and serves as a receptor for multiple ligands such as vWF, P-selectin, α-thrombin, coagulation factors XI and XIIa, high-molecular-weight kininogen, thrombospondin 1, and $\beta_2$-glycoprotein I (Fig. 12.3). The highly glycosylated GPIbα is a class 1 transmembrane receptor containing an extracellular region with an *N*-terminal flank, seven leucine-rich repeats, a *C*-terminal flank, a sulfated region, and a *C*-terminal macroglycopeptide region. The ligand binding sites for vWF and thrombin are located in the *N*-terminal 282 residues. A major function of GPIbα is binding to the A1 domain of vWF, which in the disturbed vessel wall binds to collagen fibers. Binding induces a conformational change, and active vWF becomes a docking site for platelets in flowing blood. The rapid and reversible GPIbα–vWF interaction slows down the platelets and enables firm attachment and further platelet activation through the collagen receptors integrin α2β1 and GPVI and the fibrinogen receptor integrin αIIbβ3.

Signal generation by GPIbα is weak and strongly enhanced by receptor clustering and its capacity to activate integrin αIIbβ3, which following ligand binding, positively feeds back to pathways activated by GPIbα. Because it is a class 1 receptor, GPIbα is unable to activate G proteins and instead activates Src family members and the Tyr-kinase Syk for downstream signal induction [42–46].

There is activation of PKC in platelet suspensions and a weak, oscillating $Ca^{2+}$ signal in platelets adhered to coated vWF as a result of Src/Syk-mediated activation of SLP76 and PLCγ2 [47]. A parallel pathway is the activation of PI3-kinase resulting in PIP3 formation and PKB activation, as described above. GPIbα is constitutively associated with 14-3-3ζ. Changes in 14-3-3ζ association affect signaling to PI3-K only in part and have a profound effect on the binding of vWF.

A major pathway in signaling induced by GPIbα leads to the formation of TxA2. Upstream effectors again include Src family members and Syk, which signal to Raf, MEK, and ERK2. Interference with pharmacological inhibitors reveals that these steps are upstream of p38MAPK.

The signal induction phase in platelets activated by GPIbα-vWF binding is weak. Output is activation of PKC, $Ca^{2+}$ increases, activation of PKB through

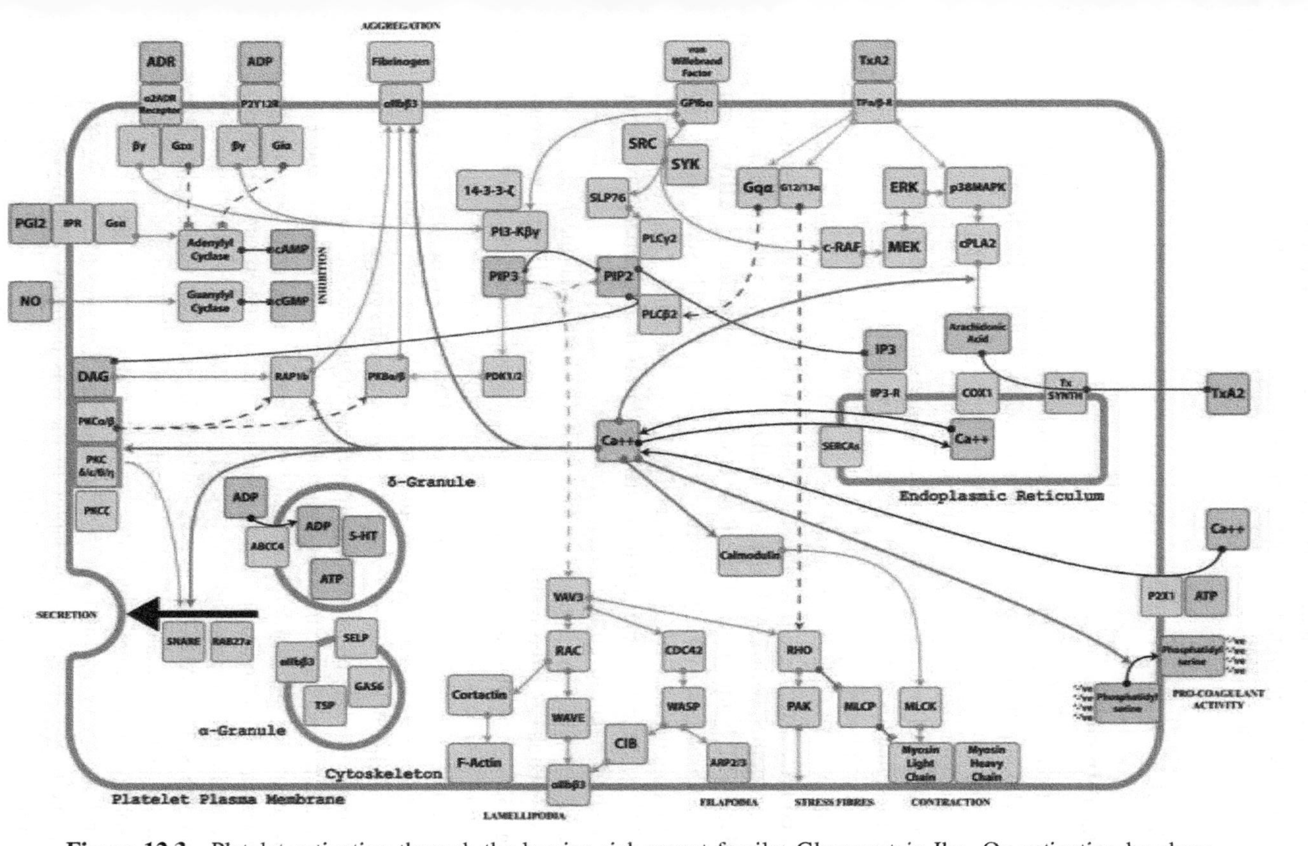


**Figure 12.3** Platelet activation through the leucine-rich repeat family: Glycoprotein Ibα. On activation by shear in solution and adhesion at a collagen surface, the von Willebrand factor activates glycoprotein Ibα. However, it is a weak platelet activator. Although connectivity is present, signaling pathways slow down before activation is complete.

formation of PIP3, and especially formation of TxA2. Theoretically, these signaling elements have the capacity to activate effectors participating in seven-transmembrane receptor signaling, but their activation is too weak to induce detectable downstream signaling.

## 12.4 PLATELET ACTIVATION THROUGH INTEGRINS

### 12.4.1 Signal Induction

Platelet activation by collagen involves two receptors, the integrin α2β1 and GPVI (Fig. 12.4). The current concept is that in mice α2β1 is the prime receptor that mediates attachment of platelets to the collagen fibers in the wound, giving GPVI a chance to generate signals that induce spreading, aggregate formation, secretion, and the generation of a procoagulant surface. In human platelets both receptors contribute to signal induction [48,49]. Receptor occupancy of α2β1 initiates a sequence of Tyr phosphorylations that involve Src, SLP76, and FAK, leading to the activation of PLCγ2. The result is induction of platelet functions through IP3-mediated release of $Ca^{2+}$ from the endoplasmatic reticulum and formation of DAG. $Ca^{2+}$, together with DAG, triggers CalDAG-GEF1, which drives Rap1b in the active, GTP-bound form. GTP-Rap1b supports activation of αIIbβ3 and spreading on fibrinogen/vWF and surface exposure of PS. These reactions are facilitated by concurrent activation of the p38MAPK route and TxA2 formation. Integrin α2β1 is subject to inside-out control by mechanisms initiated by other receptors such as ligand-occupied αIIbβ3 and GPIb [50–52].

Outside-in signaling by ligand occupied αIIbβ3 initates a mechanism for anchorage of platelets to the endothelial matrix, thereby stabilizing the hemostatic plug [32]. This mechanism starts actin polymerization and cytoskeletal reorganization, which sustain spreading under high shear stress, aggregate stabilization, and clot retraction. Prolonged platelet attachment to surface-coated collagen also intiates splicing of pre-mature mRNA and the translation of message to proteins such as *tissue factor*, which initiates the coagulation cascade.

The binding of fibrinogen to αIIbβ3 activated through inside-out signaling (see text above) and the receptor clustering that follows intitiate activation of the Tyr-kinases Src and Syk [33]. In the resting platelet, Src is constitutively bound to β3 in an inactive state. Fibrinogen binding and integrin clustering activate Src, which, in turn, recruits and activates Syk. Active Src and Syk then phosphorylate the adapter proteins SLP76, ABAP, and c-Cbl which facilitate activation of the Rac GTPase Vav, a major regulator of cytoskeletal reorganization. In a second wave of Tyr phosphorylations induced by αIIbβ3 and associated proteins, the β3 subunit is phosphorylated on Tyr747 and Tyr759, inducing binding sites for Shc, an actin bundling protein and an upstream adapter in the Ras pathway; and myosin, a motor protein involved in aggregate stabilization and clot retraction. A second target is the phosphorylation by Src and Btk of PLCγ2, which generates DAG and IP3, a trigger for $Ca^{2+}$ mobilization. A third target is α-actinin, an actin binding protein that localizes the integrin to adhesion sites in a reaction supported by FAK.

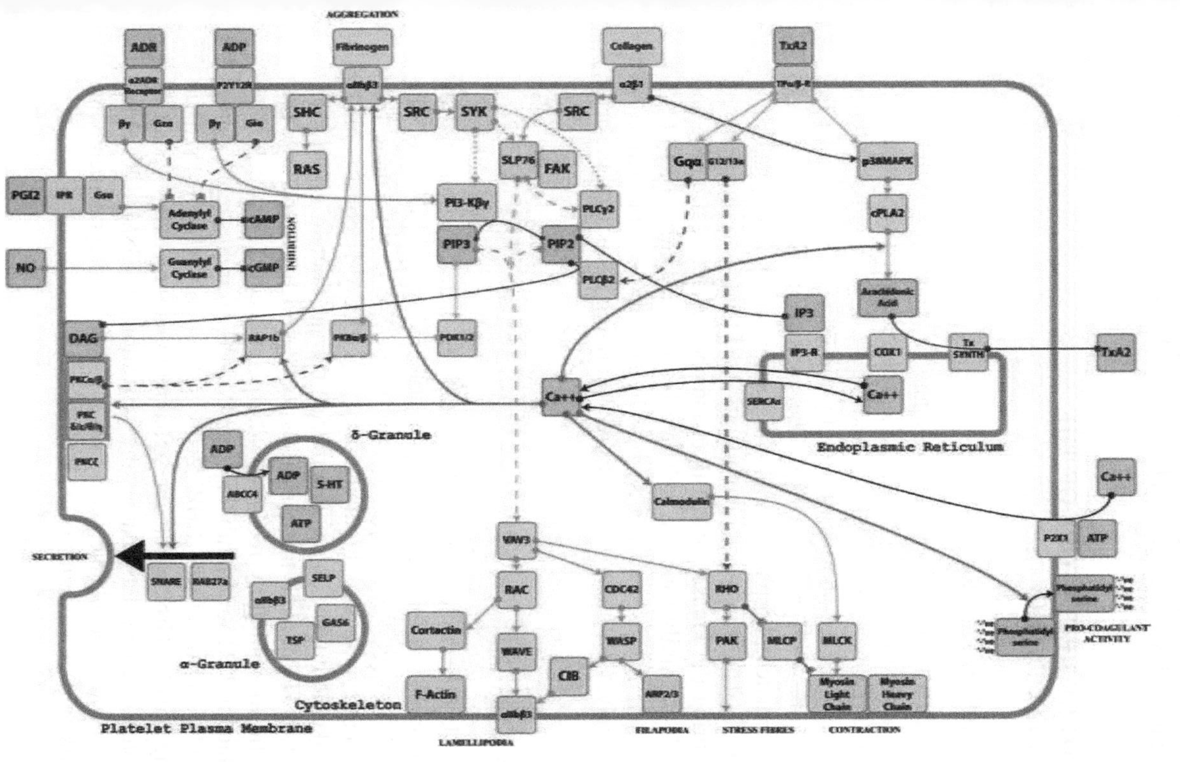


**Figure 12.4** Platelet activation through integrins. Contact with collagen activates integrin α2β1; contact of fibrinogen activates integrin αIIbβ3. The regulation of integrin αIIbβ3 is complicated by the conformational change that exposes ligand binding sites induced by platelet stimulation with thrombin and ADP (inside-out signaling) and the second wave of signalling initiated by the binding of fibrinogen (outside-in signaling). For reasons of clarity, outside-in signaling to the Tyr-kinase receptors that serve in stabilization of the thrombus has been omitted.

Tyr-phosphorylated α-actinin binds vinculin, zyxin, and membrane proximal parts of β3. Other proteins involved in signaling by ligand-occupied, clustered αIIbβ3 include RACK1, ILK, PTP-1B, kindlin, talin, and CIB, some of which also serve in inside-out regulation of αIIbβ3. Together, these complexes provide the support for firm adhesion under flow, aggregate stability, and clot retraction.

## 12.5 SIGNALING BY RECEPTORS OF THE IMMUNOGLOBULIN FAMILY

### 12.5.1 Signal Induction

Immunoglobulin-R GPVI contains two IgG domains, a mucin-rich stalk and a cytosolic tail of 51 amino acids [53] (Fig. 12.5). The receptor is constitutively associated with the Tyr-kinases Fyn and Lyn and the Fc-R γ-chain homodimer. Ligand occupancy triggers the phosphorylation by Src of an ITAM on each Fc-R γ-chain and recruits the Tyr kinase Syk. The autophosphorylation of Syk, together with Src-mediated phosphorylation, initiates formation of a signalosome that activates PLCγ2. The transmembrane adapter LAT and cytosolic adapters SLP76 and Gads bring the kinases Syk, Btk, a member of the Tec family, and members of the PI3-K family together, inducing optimal activation of PLCγ2. The result is formation of DAG and IP3, such as by analogy with the activation of PLCβ2. Interestingly, GPVI also activates VAV3, forming a link with the actin cytoskeleton.

### 12.5.2 Signal Termination

PECAM1 is a 130-kDa transmembrane glycoprotein. In platelets stimulated by thrombin, ligand-occupied αIIbβ3 [54], collagen [55], and native LDL [56], the protein is activated through Tyr phosphorylation by the Src family kinase Fgr, thereby inhibiting aggregation and thrombus formation. PECAM1 consists of six extracellular IgG-like homology domains, a transmembrane domain, and a cytoplasmic tail with two ITIMs with the consensus sequence L/I/V/S-x-Y-x-x-L/V that is characteristic for inhibitory receptors. Phosphorylated ITIMs recruit the SH2 domain-containing PP2A, an inhibitor of p38MAPK or an upstream kinase. Phosphorylated ITIMs also activate SHP1/2 [57]. Phosphorylation of SHP1/2 introduces docking sites for Grb2 regulating signaling to integrin αIIbβ3. SHP1/2 inhibition is accomplished by terminating the activation of p38MAPK. Other ITIM-containing receptors on platelets are CD72, a type 2 integral membrane protein, TLT1 expressed in α-granules, and G6b-B.

## 12.6 Tyr-KINASE RECEPTORS

Platelets have a number of receptors that are Tyr-phosphorylated on ligand binding. Binding of TPO to the c-Mpl receptor induces receptor dimerization and

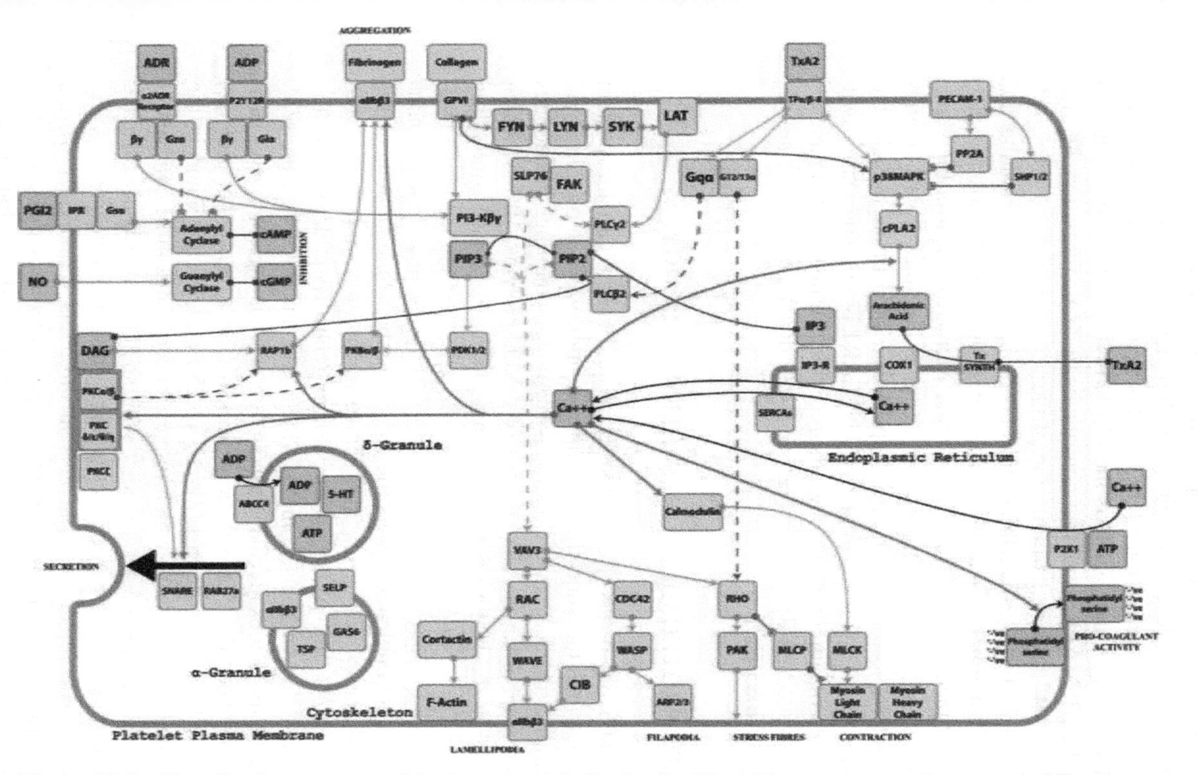


**Figure 12.5** Signaling by receptors of the immunoglobulin family. The collagen receptor glycoprotein VI activates platelets through its associated Fc-receptor γ-chains, which contain immunoreceptor Tyr-based activation (ITAM) motifs. The inhibitory receptor PECAM1 terminates platelet activation through its immunoreceptor Tyr-based inhibition (ITIM) motifs.

signaling. The receptor is vital for the differentiation of hematopoietic stem cells to megakaryocytes, the progenitors of platelets. Also platelets express this receptor, and stimulation by TPO enhances the formation of TxA2 via ERK2 and p38MAPK [58]. A platelet expresses approximately 600 receptors for insulin, and receptor occupancy induces Tyr phosphorylation of the insulin receptor. This leads to downstream activation of IRS1. This is an important protein that controls several signaling routes, such as the PI3-K/PKB pathway that regulates glucose uptake, the p38MAPK route that contributes to TxA2 formation, and the accumulation of cAMP through its inhibition of G$\alpha$i-2.

Ligand occupancy of integrin $\alpha$IIb$\beta$3 and outside-in signaling start the Tyr phosphorylation of a number of receptors that help stabilize the hemostatic plug and mediate clot retraction [59]. Axl, Tyro-3, and Mer are receptors for Gas6, stored in $\alpha$-granules [60]. The ephrin receptors A and B bind their ephrin ligands on adjacent platelets. These receptors are coupled to $\alpha$IIb$\beta$3 through PKB, but insight into these connections is incomplete [61]. PEAR1 is an epidermal growth factor (EGR)-repeat-containing transmembrane receptor that in activated platelets becomes phosphorylated on Tyr925 and Ser953/1029 downstream of outside-in signaling through $\alpha$IIb$\beta$3 [62].

## 12.7 PLATELET PROTEOMICS AND SYSTEMS BIOLOGY

Research since 2006 or so has greatly increased our insight into the protein composition of platelets, the subcellular compartmentalization of proteins, and their role in complex mechanisms of stimulus–response coupling. Proteomics [63–66] reveals that a human platelet consists of ~3000 proteins, 10% of which are phosphorylated in the resting state. There are about 130 kinases, including 90 Ser/Thr-, 30 Tyr-, and 5 dual-specificity kinases [67,68]. The membrane fraction consists of about 600 proteins with 30 with seven transmembrane domains indicative of G-protein-coupled receptors [69,70]. Thrombin stimulation increases expression of membrane proteins, reflecting the translocation from the cytosol to the platelet surface [71]. Stimulation with collagen-related peptide triggers the Tyr phopshorylation of 96 proteins, some of which have unknown function [72]. Pulldown of specific molecules reveals the induction of signaling complexes such as the association between the PKC substrate pleckstrin with eight other proteins, some mediating the binding to actin [73]. About 250 proteins are found in the releasate of activated platelets [74], which in part reflects the soluble contents of $\delta$-granules [75] and $\alpha$-granules [76], together with shedded microparticles [77]. These are the constituents that participate not only in maintenance of the vessel wall [78] and tissue repair [79] but also in thrombus formation [80] and arthritis [81]. The rapidly emerging field of phosphoproteomics links Tyr, Ser, and Thr phosphorylations with agonist induction of platelet shape change, adhesion, aggregation, secretion, and procoagulant activity. These classical functions now merge with new fields of signal transduction that control protein synthesis from pre-mRNA [82,83], initiate apoptosis [84], and control generation of surface markers that initiate platelet destruction by macrophages and hepatocytes

[85]. It is only a matter of time before this technology has mapped the signaling sequences that link growth factor stimulation to transcription factors that initiate megakaryocyte differentiation in hematopoietic stem cells, proplatelet formation, and platelet shedding. The HaemAtlas lists the genes that are specifically transcribed for megakaryocytes [86]. Only a systems biology approach can combine all this information in a comprehensive manner. State-of-the-art databases now combine the knowledge of protein–protein interactions discovered in other cell types with identified platelet proteins [87] and those restricted to platelet membranes [88]. Functional genomics couple SNPs in receptors and downstream targets to signaling molecules and platelet hyperactivity *in vitro* [89]. Reactome is an open-source and manually curated pathway database that provides pathway analysis tools for human platelets on the basis of literature data and provides detailed information of individual reactions [90].

## 12.8 INCREASED BLEEDING TENDENCY

A few excellent reviews have appeared [91–95] and are the basis of this section (Table 12.1).

### 12.8.1 Seven Transmembrane Receptors

A study on the aggregometer is often an early step in the diagnosis of a platelet function defect. It is clear that a lowered aggregation response can be related to a defect in primary signaling only when signal amplification type I (formation of $TxA_2$ and signal transduction through TPα/β-R; see item 7, type I, in list at end of Section 12.1) and signal amplification type II (δ-granule ADP and signal transduction through P2Y12-R) function normally.

Congenital abnormalities in seven-transmembrane-Rs are rare and generally lead to a mild bleeding phenotype. There are variable mucocutaneous bleeding manifestations and excess bleeding after trauma and surgery. A majority of these patients manifest prolonged bleeding times [92].

For PAR1/4 and P2Y1-R no abnormalities have been reported so far. One patient showed a selective impairement in aggregation induced by PAF [96]. Different types of TPα-R defects have been described [97]. An Arg60-Leu replacement in the first cytoplasmatic loop shows normal agonist binding but reduced GTPase activity, PLCβ activation, $Ca^{2+}$ mobilization, IP3 formation, and TxA2-induced aggregation [98]. An Asp304-Asn replacement disturbed ligand binding [99]. Endogenous TxA2 formation induced by thrombin is normal. Since the TPα/β-R is a key component of signal amplification type a (see list item 7, type I, in Section 12.1), aggregation responses by other agonists are also disturbed. Platelets with a reduced Gq expression caused by impaired mRNA transcription or stability show a lowered GTPase activity and $^{35}$S-GTPγS binding to a platelet membrane preparation. The result is reduced $Ca^{2+}$ mobilization, activation of the fibrinogen receptor αIIbβ3, and phosphorylation of pleckstrin, a substrate for PKCs.

**TABLE 12.1 Congenital Defects in Signaling Elements in Human Platelets**

| Protein | Mutation | Site | Affected Response | Reference(s) |
|---|---|---|---|---|
| PAF-R | — | — | Aggregation | 96 |
| TPα–R | Arg60-Leu | IL1 | G-protein activation | 98 |
| | Asp304-Asn | TMD7 | Ligand binding | 99 |
| Gα protein | — | — | $^{35}$S-GTP binding, GTPase activity<br>Aggregation by STM-R agonists | 100 |
| PLCβ | — | — | PI hydrolysis, PA formation | 101 |
| WASP | — | — | WASP expression or αIIb binding, thrombocytopenia, aggregation defect | 105–107 |
| Scramblase | — | — | Procoagulant activity, microparticle shedding | 30<br>109 |
| αIIbβ3 | 58 mutations | αIIb | Expression | 110 |
| | 40 mutations | β3 | Inside-out activation, ligand binding | 110 |
| | Arg214-Trp | β3 | Activation | 111,112 |
| | Ser752-Pro | β3 | Activation | 113,114 |
| | Point mutation | β3 | Locked high-affinity state | 115 |
| Kinlin3 | TGG(Trp)-TGA(stop) | αIIbβ3 | activation | 116 |
| cPLA2/$Ca^{2+}$ | — | | Arachidonate release from membranes | 117 |
| COX1 | — | | Expression, catalytic function | 118,119 |
| Tx-synthase | — | | Signaling | 120–122 |
| ABCC4 | — | | Feedback activation by secreted ADP | 123 |
| P2Y12–R | Arg256-Glu/Trp | TM6/EL3 | Expression, $G_i$ activation 97 | 126–128 |
| | 2-bp frameshift, stopcodon | | Expression 96 | 126 |
| | Pro258-Thr, | EL3 | — | 128 |
| | 378delC | | Expression | 127 |
| Gi | | | Expression | 129 |
| XL-Gαs | | | Expression (increased), hyperactive IP-R | 133 |
| Gαs | — | | Expression (decreased), hypoactive EP-R2 | 134 |
| PACAP | — | | Expression (increased), basal cAMP (increased)<br>Aggregation, secretion | 135 |
| GPIbα | 20 mutations | | Expression, signaling | 136–138 |

**TABLE 12.1 (*Continued*)**

| Protein | Mutation | Site | Affected Response | Reference(s) |
|---|---|---|---|---|
| GPIbα | Val239-Met,Gly233-Ser | | Gain of function, spontaneous vWF Binding | 139–141 |
| GPIbβ | 16 mutations | | Expression, signaling | 142 |
| GPIX | 11 mutations | | Expression | 139 |
| α2β1 | — | | Adhesion to, aggregation by collagen | 143,144 |
| GPVI | — | | Expression, signaling | 145–149 |

*Notation*: EL—extracellular loop; IL—intracellular loop; TMD—transmembrane domain; STM-R—seven-transmembrane receptor. See also Abbreviation list at end of chapter (preceding References list).

As expected, the abnormality affects aggregation responses by all agonists whose receptors signal through Gq [100].

A patient with impaired aggregation responses induced by ADP, epinephrine, and PAF showed a lowered hydrolysis of PI with concurrent reduction in PA formation and pleckstrin phosphorylation, together pointing at a defect in PLC [101]. A heterogenous group of platelet signaling defects is characterized by an impaired increase in cytosolic $Ca^{2+}$ due to defects in $Ca^{2+}$ mobilization and influx on stimulation with different agonists. This abnormality is often accompanied by reduced formation of IP3 and DAG and therefore is probabably caused by a defect in PLCβ2. Treatment with a mixture of $Ca^{2+}$ ionophore and $Ca^{2+}$-rich buffer to induce a $Ca^{2+}$ increase, while bypassing receptor activation, restores the aggregation response, again pointing at a defect upstream of $Ca^{2+}$ rises.

So far, platelets with a defect in the mechanisms that control IP3-mediated release of $Ca^{2+}$ from the endoplasmic reticulum, the reaccumulation by SERCAs, and in/efflux through the plasma membrane have not been reported. The more recent characterization of STIM1 and store-operated $Ca^{2+}$ interplay paves the way for a search for primary defects in $Ca^{2+}$ homeostasis. Platelets from mice with a mutated, inactive ORAI show reduced agonist-stimulated $Ca^{2+}$ increase, and impaired αIIbβ3 activation and secretion. Strikingly, there is a strong reduction in exposure of PS and generation of procoagulant activity [102].

Mutations in the X-chromosome gene (Location Xp11.22) lead to the Wiskott–Aldrich syndrome, a disorder affecting T-lymphocytes and platelets. The bleeding tendency varies from mild to severe. There is thrombocytopenia; platelets are small with decreased survival and abnormalities in δ-granule content, GPIb, integrin α2β1, impaired aggregation, and abnormalities in energy metabolism [103]. WASP binds to CIB attached to the cytosolic tail of the αIIb part of integrin αIIbβ3 [104]. There are patients with reduced WASP expression and those with normal expression combined with a reduced binding affinity for αIIb, which explains the aggregation defect [105–107]. Murine WASP-deficient megakaryocytes show a loss in the negative signaling by the collagen receptor

$\alpha 2\beta 1$ to proplatelet formation, loss of podosomes, and reduced chemotactic migration [108].

The Scott syndrome is characterized by a moderate to severe bleeding tendency and a severe impairment in the generation of a procoagulant surface. The cause is not precisely known but must reside in a dysfunctional scramblase [30]. Agonist-induced Tyr-phosphorylation of FAK, Src family kinases, and Syk is impaired, but whether these are causes or consequences of the defective scramblase is uncertain. There is impaired shedding of PS-containing microvesicles [109].

An absent or severely reduced aggregation response is characteristic for Glanzmann thrombasthenia [91,93,95]. The bleeding time is strongly prolonged, and there is a severe hemorrhagic diathesis. Mutations, splicing errors, or folding defects induce defects in surface expression of $\alpha IIb\beta 3$, its activation through inside-out signaling, or the binding of soluble fibrinogen or vWF [110]. Defects in $\alpha IIb\beta 3$ activation have been attributed to Arg214-Trp and Ser753-Pro replacements and a deletion of 39 amino acids at R724, both in $\beta 3$ [111–115]. In terms of signaling, the importance of an absent or defective $\alpha IIb\beta 3$ lies in the failure to mediate crosstalk with other surface receptors, such as $\alpha 2\beta 1$. An important contributor to cytoskeleton anchorage of integrin $\alpha IIb\beta 3$ is the talin-like protein kindlin. A point mutation creating a premature stopcodon TGG (Trp)-TGA (stop) in Kinlin3 causes failure to open the integrin through inside-out signaling and severe bleeding [116].

Signal amplification type I (item 7, type I, in list at end of Section 12.1) I involves release of arachidonate from membrane phospholipids by cPLA2, its conversion to TxA2 through COX1, and Tx-synthase and activation of TP$\alpha$/$\beta$-R initiating Gq- and G12/13-mediated activation routes. This receptor also takes part in primary signal induction when occupied by TxA2 from adjacent platelets.

A few patients with reduced release of [$^3$H]-labeled arachidonate from membrane phospholipids and formation of TxA2 in thrombin-activated platelets have been described [117]. The activity of cPLA2 was normal, suggesting an impaired capacity to raise $Ca^{2+}$, which, together with p38MAPK, activates cPLA2. Inhibition of TxA2 formation by targeting COX1 with aspirin is an important antithrombotic treatment.

Medication-free individuals with reduced COX1 activity have been described, with some showing a decreased expression of COX1 protein and others a dysfunctional protein [118,119]. Patients with a Tx-synthase deficiency have the phenotype of platelets treated with aspirin: TxA2 formation and arachidonate-induced aggregation are absent, but they respond normally to a stable TxA2 mimetic [120–122]. This contrasts with patients whose platelets lack a normal TP$\alpha$/$\beta$-R. Here, formation of TxA2 is normal, but the cells do not respond to endogenous TxA2 or a stable mimetic. Signal amplification type II (see item 7, type II in list at end of Section 12.1) involves the extrusion of $\delta$-granule ADP followed by P2Y12-R-mediated signaling to Gi, cAMP suppression, and PI3-K activation. A defect in granule ATP-ADP is known as $\delta$-*storage pool deficiency* (SPD). Levels of these constituents vary highly between affected individuals, and granule ADP shows the strongest correlation with prolongation of the bleeding time [28]. There is evidence for the existence of "empty granules," indicating that the defect is in the package of granules with adenine nucleotides. Transport of ADP from the cytosol to the intragranule space is mediated through ABCC4, suggesting that the low content of ADP and possibly ATP reflects a defect in this transporter [123]. Congenital diseases are known that combine δ-storage pool deficiency with oculocutaneous albinism (Hermansky–Pudlak syndrome), albinism, and defects in T cells (Chediak–Higaski syndrome) and α-granule deficiency (α-storage pool disease). Although evidence in humans is lacking, studies in mice and rats indicate a role for Rab proteins in control of protein trafficking and granule formation. A Rab38 mutation causes a specifc δ-granule deficiency [124]. A defect in Rab geranyl–geranyl transferease α-subunit in mice caused a macrothrombocytopenia with abnormal δ- and α-granules [125].

Deletions in the P2Y12-R gene introduce a frameshift and premature stop-codon and a fall in receptor surface expression [126–128]. Arg256-Glu and Arg265-Trp replacements in the third extracelular loop lead to reduced suppression of cAMP, reflecting impaired receptor activation of Gi. ADP aggregation is reduced. Since the P2Y12-R is a key component of signal amplification type II (see item 7, type II in list at end of Section 12.1), aggregation responses by other agonists are also disturbed. Depending on conditions, the impact of P2Y12-R deficiency in patients mimicks the effect of the receptor antagonist clopidogrel in healthy individuals. P2Y12 signals through the inhibitory G protein of adenylyl cyclase Gi. Reduced Gi expression is accompanied by an increased tendency to bleed [129].

Decreased surface expression of α2-adrenergic-R leads to the expected reduction in epinephrine-induced aggregation. Unexpectedly, suppression of adenylyl-cyclase through Gz is normal. Inhibition of the P2Y12-R pathway is also seen when normal platelets bind insulin, which triggers the IRS1-mediated inactivation of Gαi–2 [130,131]. Type 2 diabetes mellitus platelets are often resistant to signaling by insulin.

In normal platelets, secretion of δ-granule contents and α-granule contents generally go hand in hand. Although α-granules contain several components with signaling properties (see discussion above), a defect in this pathway does not change the platelet activating machinery. The prolonged bleeding time observed in these individuals is probably caused by degradation of coagulation factor V, which normally stimulates thrombin formation on the platelet surface.

Cyclic AMP is a very potent platelet inhibitor, and a rise from 1.5 to 4.0 nmol/$10^{11}$ platelets suffices to lock αIIbβ3 in the closed conformation [132]. Freson and Van Geet [133] described a patient with a normal expression of the extralarge Gαs protein and increased expression of Gαs protein. There is an increase in trauma-related bleeding tendency, possibly caused by the hypersensitivity of the Gs-coupled IP-R to ligands such as prostacyclin. Despite this abnormality, the level of cAMP in resting platelets is normal. The opposite condition, a decrease in Gαs expression, is seen in patients with pseudohypoparathyroidism [134]. Basal cAMP is normal, and there is no increased bleeding tendency, but a 400-fold higher dose of PGE1 is required to induce the same cAMP increase as seen in normal platelets. A patient with an increased expression of PACAP (pituitary adenylylcyclase activating polypeptide) shows a strongly prolonged bleeding time. The basal cAMP concentration is increased, leading to reduced aggregation and secretion [135].

### 12.8.2 Membership in the Leucine-Rich Repeat Family

Patients with a deficiency of GPIb suffer from thrombocytopenia with giant platelets caused by quantitative or qualitative abnormalities of the GPIb-V-IX complex [136–138]. The bleeding time is prolonged. Mutations have been reported in GPIbα, GPIbβ, and GPIX but not in GPV. There are missence mutations interfering with intracellular trafficking and nonsence mutations leading to a truncated GPIbα devoid of a transmembrane domain. Patients with platelet-type von Willebrand disease have a gain of function in GPIbα due to a Gly233-Val/Ser or Met239-Val mutation [139–141]. These receptors spontaneously bind vWF. This is a life-threatening disorder caused by the removal of activated platelets coupled to active, large vWF multimers [134]. A mutation in GPIbβ interfered with GPIb signaling, suggesting that the GPIbβ subunit contributes to signal induction by the GPIb-V-IX complex [142].

### 12.8.3 Integrins

Very few patients have been reported with a defect in the expression of α2β1 [143,144]. There is a mild bleeding tendency, a prolonged bleeding time, and absent adhesion and aggregation following contact with collagen. For obvious reasons, platelets with absent or strongly reduced αIIbβ3 expression lack normal aggregation and ligand-induced outside-in signaling. Of specific interest are those variations of Glanzmann thrombasthenia that combine normal expression of αIIbβ3 with a failure to expose fibrinogen binding sites or with a constitutively active conformation. Outside-in signaling is strongly impaired. The signaling defects in these patients await exploration.

### 12.8.4 Immunoglobulin Receptors

Patients have been reported with a more than 90% reduction in the expression of GPVI [145–149]. This abnormality leads to a mild bleeding disorder with weak or absent adhesion to a collagen-coated surface and collagen-induced shape change and aggregation in suspensions.

### 12.8.5 Tyr-Kinase Receptors

So far, no cases of patients with a bleeding tendency caused by a defect in Tyr-kinasse receptors have been reported.

## 12.9 CONCLUDING REMARKS

Connectivity schemes of the platelet signaling machinery help us understand the complexity of activating and inhibitory pathways. They form a scaffold for later addition of new information in a stepwise process leading to completeness. They link with genetic abnormalities that affect platelet function. Furthermore, they allow comparisons with megakaryocytes at different stages of maturation, and other blood cells as well as neurons with which platelets share many signaling steps. Crucially, they also provide insights in the congenital and acquired abnormalities that cause hemorrhagic diathesis.

## ACKNOWLEDGMENT

JWNA gratefully acknowledges support by the Netherlands Thrombosis Foundation.

## ABBREVIATIONS

| | |
|---|---|
| ABCC | ATP binding cassette transporter |
| ARP | Actin-related protein |
| BTK | Bruton Tyr-kinase |
| CBL | Casitas *b*-lineage lymphoma protooncogene |
| CDC | Cell division control (protein) |
| CIB | Calcium and integrin binding (protein) |
| c-MPL | Myeloproliferative leukemia protein, cyclic |
| COX | Cyclooxygenase |
| CRAC | $Ca^{2+}$ release–activated $Ca^{2+}$ channels |
| CSK | c-src kinase |
| DAG | Diacylglycerol |
| EDG | Endothelial differentiation gene |
| ERK | Extracellular kinase |
| FAK | Focal adhesion kinase |
| GAP | GTPase activating protein |
| GEF | Guanylate exchange factor |
| GLUT | Glucose transporter |
| GP | Glycoprotein |
| GRB | Growth factor receptor binding (protein) |
| ILK | Integrin-linked kinase |
| IP | Inositol P |
| IRS | Insulin-R (receptor) substrate |

| | |
|---|---|
| ITAM | Immunoreceptor Tyr-based activation motif |
| ITIM | Immunoreceptor Tyr-based inhibitory motif |
| LAT | Linker for activation of T cells |
| LPA | Lysophosphatidic acid |
| MARCS | Integral membrane protein |
| MEK | Mitogen extracellular kinase |
| MKK | Mitogen-activated protein kinase–kinase |
| MLCK | Myosin light-chain kinase |
| MUNC | Regulatory protein munc |
| NO | Nitric monoxide |
| ORAI | Store-operated calcium channel |
| P2Y12-R | P2Y purinoreceptor 12 |
| P2Y1-R | P2Y purinoreceptor 1 |
| MAPK | Mitogen-activated protein kinase |
| PACAP | Pituitary adenylylcyclase activating polypeptide |
| PAF | Platelet activating factor |
| PAK | p21-activated kinase |
| PAR | Proteolytic activated receptor |
| PDE | Phosphodiesterase |
| PDK | PI-dependent kinase |
| PEAR | Platelet endothelial aggregation receptor |
| PECAM | Platelet endothelial cell adhesion molecule |
| PG | Prostaglandin |
| PI | Phosphatidylinositol |
| PK | Protein kinase |
| PL | Phospholipase |
| PLA | Phospholipase A |
| PP | Protein phosphatase |
| PS | Phosphatidylserine |
| PTB | Phosphobutyryltransferase |
| PTP | Protein Tyr phosphatase |
| RAB | ras-related protein rab |
| RAC | res-related protein rac |
| RACK | Receptor for activated C kinase |
| RHO | GTP binding protein Rho |
| RIAM | Molecular adapter and Rap1 effector |
| SERCA | Sarcoplasmic/endoplasmic reticulum calcium (channels) |
| SHC | src homology 2 domain containing transforming protein |
| SHP | src homology domain containing tyrosine phosphatase |
| SNAP | Synaptosomel-associated protein |
| SNARE | Soluble NSF (*N*-ethylmaleimide-sensitive factor) attachment (protein) receptors |
| SOS | Son of sevenless |
| SRC | Protooncogene tyrosine protein kinase |
| STIM | Stromal interaction molecule |

| | |
|---|---|
| SYK | Spleen tyrosine kinase |
| TLT | TREM-like transcript |
| TPO | Thrombopoietin |
| TxA2 | Thromboxaan A2 |
| VAMP | Vesicle-associated membrane protein |
| VASP | Vasodilator-stimulated phosphoprotein |
| VAV | Guanine nucleotide exchange factor |
| vWF | Von Willebrand factor |
| WASP | Wiskott–Aldrich syndrome protein |
| WAVE | Verprolin homology-domain-containing protein |

*Source:* UniProt, `http://www.uniprot.org`.

## REFERENCES


1. Noble D, Systems biology and the heart, *Biosystems* **83**:75–80 (2006).
2. Bader S, Kühner S, Gavin AC, Interaction networks for systems biology, *FEBS Lett*. **582**:1220–1224 (2008).
3. Harrington ED, Jensen LJ, Bork P, Predicting biological networks from genomic data, *FEBS Lett*. **582**:1251–1258 (2008).
4. Purvis JE, Chatterjee MS, Brass LF, Diamond SL, A molecular signaling model of platelet phosphoinositide and calcium regulation during homeostasis and P2Y1 activation, *Blood* **112**:4069–4079 (2008).
5. Clemetson KJ, Clemetson JM, Platelet receptors, in Michelson AD, ed., *Platelets*, Academic Press, Amsterdam, 2007, pp. 117–144.
6. Gibbins JM, Platelet adhesion signalling and the regulation of thrombus formation, *J. Cell Sci*. **117**:3415–3425 (2004).
7. Adam F, Kauskot A, Rosa JP, Bryckaert M, Mitogen-activated protein kinases in hemostasis and thrombosis, *J. Thromb. Haemost*. **6**:2007–2016 (2008).
8. Brass LF, Zhu L, Stalker TJ, Novel therapeutic targets at the platelet vascular interface, *Arterioscler. Thromb. Vasc. Biol*. **28**:43–50 (2008).
9. Nieswandt B, How do platelets prevent bleeding? *Blood* **111**:4835 (2008).
10. Offermanns S, Activation of platelet function through G protein-coupled receptors, *Circ. Res*. **99**:1293–1304 (2006).
11. Kim S, Jin J, Kunapuli SP, Relative contribution of G-protein-coupled pathways to protease-activated receptor-mediated Akt phosphorylation in platelets, *Blood* **107**:947–954 (2006).
12. Kroner C, Eybrechts K, Akkerman JWN, Dual regulation of platelet protein kinase B, *J. Biol. Chem*. **275**:27790–27798 (2000).
13. Woulfe D, Jiang H, Morgans A, Monks R, Birnbaum M, Brass LF, Defects in secretion, aggregation, and thrombus formation in platelets from mice lacking Akt2, *J. Clin. Invest*. **113**:441–450 (2004).
14. Ferreira IA, Mocking AI, Urbanus RT, Varlack S, Wnuk M, Akkerman JWN, Glucose uptake via glucose transporter 3 by human platelets is regulated by protein kinase B, *J. Biol. Chem*. **280**:32625–32633 (2005).

15. Kozuma Y, Kojima H, Yuki S, Suzuki H, Nagasawa T, Continuous expression of Bcl-xL protein during megakaryopoiesis is post-translationally regulated by thrombopoietin-mediated Akt activation, which prevents the cleavage of Bcl-xL, *J. Thromb. Haemost*. **5**:1274–1282 (2007).
16. Siess W, Platelet interaction with bioactive lipids formed by mild oxidation of low-density lipoprotein, *Pathophysiol. Haemost. Thromb*. **35**:292–304 (2006).
17. Nadal-Wollbold F, Pawlowski M, Lévy-Toledano S, Berrou E, Rosa JP, Bryckaert M, Platelet ERK2 activation by thrombin is dependent on calcium and conventional protein kinases C but not Raf-1 or B-Raf, *FEBS Lett*. **531**:475–482 (2002).
18. Shock DD, He K, Wencel-Drake JD, Parise LV, Ras activation in platelets after stimulation of the thrombin receptor, thromboxane A2 receptor or protein kinase C, *Biochem. J*. **321**:525–530 (1997).
19. Braun A, Varga-Szabo D, Kleinschnitz C, Pleines I, Bender M, Austinat M, Bosl M, Stoll G, Nieswandt B, Orai1 (CRACM1) is the platelet SOC channel and essential for pathological thrombus formation, *Blood* **113**:1097–1104 (2009).
20. Feske S, Calcium signalling in lymphocyte activation and disease, *Nat. Rev. Immunol*. **7**:690–702 (2007).
21. Redondo PC, Salido GM, Pariente JA, Sage SO, Rosado JA, SERCA2b and 3 play a regulatory role in store-operated calcium entry in human platelets, *Cell Signal*. **20**:337–346 (2008).
22. Hartwig JH, Barkalow K, Azim A, Italiano J, The elegant platelet: Signals controlling actin assembly, *Thromb. Haemost*. **82**:392–398 (1999).
23. Rohatgi R, Ma L, Miki H, Lopez M, Kirchhausen T, Takenawa T, Kirschner MW, The interaction between N-WASP and the Arp2/3 complex links Cdc42-dependent signals to actin assembly, *Cell* **97**:221–231 (1999).
24. Takenawa T, Suetsugu S, The WASP-WAVE protein network: connecting the membrane to the cytoskeleton, *Nat. Rev. Mol. Cell Biol*. **8**:37–48 (2007).
25. Flaumenhaft R, Molecular basis of platelet granule secretion, *Arterioscler. Thromb. Vasc. Biol*. **23**:1152–1160 (2003).
26. Tolmachova T, Abrink M, Futter CE, Authi KS, Seabra MC, Rab27b regulates number and secretion of platelet dense granules, *Proc. Natl. Acad. Sci. USA* **104**:5872–5877 (2007).
27. Isenberg JS, Frazier WA, Roberts DD, Thrombospondin-1: a physiological regulator of nitric oxide signaling, *Cell. Mol. Life Sci*. **65**:728–742 (2008).
28. Akkerman JW, Nieuwenhuis HK., Mommersteeg-Leautaud ME, Gorter G, Sixma JJ, ATP-ADP compartmentation in storage pool deficient platelets: Correlation between granule-bound ADP and the bleeding time, *Br. J. Haematol*. **55**:135–143 (1983).
29. Jedlitschky G, Tirschmann K, Lubenow LE, Nieuwenhuis HK, Akkerman JW, Greinacher A, Kroemer HK, The nucleotide transporter MRP4 (ABCC4) is highly expressed in human platelets and present in dense granules, indicating a role in mediator storage, *Blood* **104**:3603–3610 (2004).
30. Zwaal RF, Comfurius P, Bevers EM, Surface exposure of phosphatidylserine in pathological cells, *Cell. Mol. Life Sci*. **62**:971–988 (2005).
31. Morel O, Toti F, Hugel B, Freyssinet JM, Cellular microparticles: A disseminated storage pool of bioactive vascular effectors, *Curr. Opin. Hematol*. **11**:156–164 (2004).

32. Coller BS, Shattil SJ, The GPIIb/IIIa (integrin αIIbβ3) odyssey: A technology-driven saga of a receptor with twists, turns, and even a bend, *Blood* **112**:3011–3025 (2008).

33. Shattil SJ, Newman PJ, Integrins: Dynamic scaffolds for adhesion and signaling in platelets, *Blood* **104**:1606–1615 (2004).

34. Savi P, Zachayus JL, Delesque-Touchard N, Labouret C, Hervé C, Uzabiaga MF, Pereillo JM, Culouscou JM, Bono F, Ferrara P, Herbert JM, The active metabolite of Clopidogrel disrupts P2Y12 receptor oligomers and partitions them out of lipid rafts, *Proc. Natl. Acad. Sci. USA* **103**:11069–11074 (2006).

35. Daniel JL, Ashby B, Pulcinelli FM, Platelet signalling: cAMP and cGMP, in Gresele P, Page C, Fuster V, Vermylen J, eds., *Platelets in Thrombotic and Nonthrombotic Disorders*, Cambridge Univ. Press, 2002, pp. 290–303.

36. Francis SH, Corbin JD, Structure and function of cyclic nucleotide-dependent protein kinases, *Annu. Rev. Physiol.* **56**:237–272 (1994).

37. Keularts IM, van Gorp RM, Feijge MA, Vuist WM, Heemskerk JW, alpha(2A)-adrenergic receptor stimulation potentiates calcium release in platelets by modulating cAMP levels, *J. Biol. Chem.* **275**:1763–1772 (2000).

38. Ferreira IA, Eybrechts KL, Mocking AI, Kroner C, Akkerman JW, IRS-1 mediates inhibition of Ca2+ mobilization by insulin via the inhibitory G-protein Gi, *J. Biol. Chem.* **279**:3254–3264 (2004).

39. Feijge MA, Ansink K, Vanschoonbeek K, Heemskerk JW, Control of platelet activation by cyclic AMP turnover and cyclic nucleotide phosphodiesterase type-3, *Biochem. Pharmacol.* **67**:1559–1567 (2004).

40. Cheitlin MD, Hutter AM, Brindis RG, Ganz P, Kaul S, Russell RO, Zusman RM, Use of sildenafil (Viagra) in patients with cardiovascular disease. Technology and Practice Executive Committee, *Circulation* **99**:168–177 (1999).

41. Chen Z, Dupré DJ, Le Gouill C, Rola-Pleszczynski M, Stanková J, Agonist-induced internalization of the platelet-activating factor receptor is dependent on arrestins but independent of G-protein activation. Role of the C terminus and the (D/N)PXXY motif, *J. Biol. Chem.* **277**:7356–7362 (2002).

42. Ozaki Y, Asazuma N, Suzuki-Inoue K, Berndt MC, Platelet GPIb-IX-V-dependent signaling, *J. Thromb. Haemost.* **3**:1745–1751 (2005).

43. Du X. Signaling and regulation of the platelet glycoprotein Ib-IX-V complex, *Curr. Opin. Hematol.* **14**:262–269 (2007).

44. Yin H, Stojanovic A, Hay N, Du X, The role of Akt in the signaling pathway of the glycoprotein Ib-IX induced platelet activation, *Blood* **111**:658–665 (2008).

45. Mu FT, Andrews RK, Arthur JF, Munday AD, Cranmer SL, Jackson SP, Stomski FC, Lopez AF, Berndt MC, A functional 14-3-3zeta-independent association of PI3-kinase with glycoprotein Ib alpha, the major ligand-binding subunit of the platelet glycoprotein Ib-IX-V complex, *Blood* **111**:4580–4587 (2008).

46. Liu J, Pestina TI, Berndt MC, Steward SA, Jackson CW, Gartner TK, The roles of ADP and TXA in botrocetin/VWF-induced aggregation of washed platelets, *J. Thromb. Haemost.* **2**:2213–2222 (2004).

47. van Lier M, Verhoef S, Cauwenberghs S, Heemskerk JW, Akkerman JW, Heijnen HF, Role of membrane cholesterol in platelet calcium signalling in response to VWF and collagen under stasis and flow, *Thromb. Haemost.* **99**:1068–1078 (2008).

48. Sarratt KL, Chen H, Zutter MM, Santoro SA, Hammer DA, Kahn ML, GPVI and alpha2beta1 play independent critical roles during platelet adhesion and aggregate formation to collagen under flow, *Blood* **106**:1268–1277 (2005).
49. Siljander PR, Munnix IC, Smethurst PA, Deckmyn H, Lindhout T, Ouwehand WH, Farndale RW, Heemskerk JW, Platelet receptor interplay regulates collagen-induced thrombus formation in flowing human blood, *Blood* **103**:1333–1341 (2004).
50. Inoue O, Suzuki-Inoue K, Dean WL, Frampton J, Watson SP, Integrin alpha2beta1 mediates outside-in regulation of platelet spreading on collagen through activation of Src kinases and PLCgamma2, *J. Cell Biol.* **160**:769–780 (2003).
51. Bernardi B, Guidetti GF, Campus F, Crittenden JR, Graybiel AM, Balduini C, Torti M, The small GTPase Rap1b regulates the cross talk between platelet integrin alpha2beta1 and integrin alphaIIbbeta3, *Blood* **107**:2728–2735 (2006).
52. Atkinson BT, Jarvis GE, Watson SP, Activation of GPVI by collagen is regulated by alpha2beta1 and secondary mediators, *J. Thromb. Haemost.* **1**:1278–1287 (2003).
53. Watson SP, Auger JM, McCarty OJ, Pearce AC, GPVI and integrin alphaIIb beta3 signaling in platelets, *J. Thromb. Haemost.* **3**:1752–1762 (2005).
54. Relou IA, Gorter G, Ferreira IA, van Rijn HJ, Akkerman JW, Platelet endothelial cell adhesion molecule-1 (PECAM-1) inhibits low density lipoprotein-induced signaling in platelets, *J. Biol. Chem.* **278**:32638–32644 (2003).
55. Jones KL, Hughan SC, Dopheide SM, Farndale RW, Jackson SP, Jackson DE, Platelet endothelial cell adhesion molecule-1 is a negative regulator of platelet-collagen interactions, *Blood* **98**:1456–1463 (2001).
56. Korporaal SJ, Akkerman JW, Platelet activation by low density lipoprotein and high density lipoprotein, *Pathophysiol. Haemost. Thromb.* **35**:270–280 (2006).
57. Korporaal SJ, Koekman CA, Verhoef S, van der Wal DE, Bezemer M, Van Eck M, Akkerman JW, Downregulation of platelet responsiveness upon contact with LDL by the protein-tyrosine phosphatases SHP-1 and SHP-2, *Arterioscler. Thromb. Vasc. Biol.* **29**:372–379 (2009).
58. van Willigen G, Gorter G, Akkerman JW, Thrombopoietin increases platelet sensitivity to alpha-thrombin via activation of the ERK2-cPLA2 pathway, *Thromb. Haemost.* **83**:610–616 (2000).
59. Phillips DR, Conley PB, Sinha U, Andre P, Therapeutic approaches in arterial thrombosis, *J. Thromb. Haemost.* **3**:1577–1589 (2005).
60. Angelillo-Scherrer A, Burnier L, Flores N, Savi P, DeMol M, Schaeffer P, Herbert JM, Lemke G, Goff SP, Matsushima GK, Earp HS, Vesin C, Hoylaerts MF, Plaisance S, Collen D, Conway EM, Wehrle-Haller B, Carmeliet P, Role of Gas6 receptors in platelet signaling during thrombus stabilization and implications for antithrombotic therapy, *J. Clin. Invest.* **115**:237–246 (2005).
61. Prévost N, Woulfe DS, Jiang H, Stalker TJ, Marchese P, Ruggeri ZM., Brass LF, Eph kinases and ephrins support thrombus growth and stability by regulating integrin outside-in signaling in platelets, *Proc. Natl. Acad. Sci. USA* **102**:9820–9825 (2005).
62. Nanda N, Bao M, Lin H, Clauser K, Komuves L, Quertermous T, Conley PB, Phillips DR, Hart MJ, Platelet endothelial aggregation receptor 1 (PEAR1), a novel epidermal growth factor repeat-containing transmembrane receptor, participates in platelet contact-induced activation, *J. Biol. Chem.* **280**:24680–24689 (2005).
63. Senzel L, Gnatenko DV, Bahou WF, The platelet proteome, *Curr. Opin. Hematol.* **16**:329–333 (2009).

64. Power KA, McRedmond JP, de Stefani A, Gallagher WM, Gaora PO, High-throughput proteomics detection of novel splice isoforms in human platelets, *PLoS ONE* **4**:e5001 (2009).
65. Springer DL, Miller JH, Spinelli SL, Pasa-Tolic L, Purvine SO, Daly DS, Zangar RC, Jin S, Blumberg N, Francis CW, Taubman MB, Casey AE, Wittlin SD, Phipps RP, Platelet proteome changes associated with diabetes and during platelet storage for transfusion, *J. Proteome Res.* **8**:2261–2272 (2009).
66. Haudek VJ, Slany A, Gundacker NC, Wimmer H, Drach J, Gerner C, Proteome maps of the main human peripheral blood constituents, *J. Proteome Res.* **8**:3834–3843 (2009).
67. Dittrich M, Birschmann I, Mietner S, Sickmann A, Walter U, Dandekar T, Platelet protein interactions: Map, signaling components, and phosphorylation groundstate, *Arterioscler. Thromb. Vasc. Biol.* **28**:1326–1331 (2008).
68. Zahedi RP, Lewandrowski U, Wiesner J, Wortelkamp S, Moebius J, Schütz C, Walter U, Gambaryan S, Sickmann A, Phosphoproteome of resting human platelets, *J. Proteome Res.* **7**:526–534 (2008).
69. Lewandrowski U, Wortelkamp S, Lohrig K, Zahedi RP, Wolters DA, Walter U, Sickmann A, Platelet membrane proteomics: A novel repository for functional research, *Blood* **114**:e10–9 (2009).
70. Senis YA, Tomlinson MG, García A, Dumon S, Heath VL, Herbert J, Cobbold SP, Spalton JC, Ayman S, Antrobus R, Zitzmann N, Bicknell R, Frampton J, Authi KS, Martin A, Wakelam MJ, Watson SP, A comprehensive proteomics and genomics analysis reveals novel transmembrane proteins in human platelets and mouse megakaryocytes including G6b-B, a novel immunoreceptor tyrosine-based inhibitory motif protein, *Mol. Cell Proteom.* **6**:548–564 (2007).
71. Tucker KL, Kaiser WJ, Bergeron AL, Hu H, Dong JF, Tan TH, Gibbins JM, Proteomic analysis of resting and thrombin-stimulated platelets reveals the translocation and functional relevance of HIP-55 in platelets, *Proteomics* **9**:4340–4354 (2009).
72. García A, Senis YA, Antrobus R, Hughes CE, Dwek RA, Watson SP, Zitzmann N, A global proteomics approach identifies novel phosphorylated signaling proteins in GPVI-activated platelets: Involvement of G6f, a novel platelet Grb2-binding membrane adapter, *Proteomics* **6**:5332–5343 (2006).
73. Baig A, Bao X, Haslam RJ, Proteomic identification of pleckstrin-associated proteins in platelets: Possible interactions with actin, *Proteomics* **9**:4254–4258 (2009).
74. Piersma SR, Broxterman HJ, Kapci M, de Haas RR, Hoekman K, Verheul HM, Jiménez CR, Proteomics of the TRAP-induced platelet releasate, *J. Proteom.* **72**:91–109 (2009).
75. Hernandez-Ruiz L, Valverde F, Jimenez-Nuñez MD, Ocaña E, Sáez-Benito A, Rodríguez-Martorell J, Bohórquez JC, Serrano A, Ruiz FA, Organellar proteomics of human platelet dense granules reveals that 14-3-3zeta is a granule protein related to atherosclerosis, *J. Proteome Res.* **6**:4449–4457 (2007).
76. Maynard DM, Heijnen HF, Horne MK, White JG, Gahl WA, Proteomic analysis of platelet alpha-granules using mass spectrometry, *J. Thromb. Haemost.* **5**:1945–1955 (2007).
77. García BA, Smalley DM, Cho H, Shabanowitz J, Ley K, Hunt DF, The platelet microparticle proteome, *J. Proteome Res.* **4**:1516–1521 (2005).

78. Nachman RL, Rafii S, Platelets, petechiae, and preservation of the vascular wall, *N. Engl. J. Med.* **359**:1261–1270 (2008).

79. Smyth SS, McEver RP, Weyrich AS, Morrell CN, Hoffman MR, Arepally GM, French PA, Dauerman HL, Becker RC, 2009 Platelet Colloquium Participants, Platelet functions beyond hemostasis, *J. Thromb. Haemost.* **11**:1759–1766 (2009).

80. Furie B, Furie BC, Thrombus formation *in vivo*, *J. Clin. Invest.* **115**:3355–3362 (2005).

81. Boilard E, Nigrovic PA, Larabee K, Watts GF, Coblyn JS, Weinblatt ME, Massarotti EM, Remold-O'Donnell E, Farndale RW, Ware J, Lee DM, Platelets amplify inflammation in arthritis via collagen-dependent microparticle production, *Science* **327**:580–583 (2010).

82. Schwertz H, Tolley ND, Foulks JM, Denis MM, Risenmay BW, Buerke M, Tilley RE, Rondina MT, Harris EM, Kraiss LW, Mackman N, Zimmerman GA, Weyrich AS, Signal-dependent splicing of tissue factor pre-mRNA modulates the thrombogenicity of human platelets, *J. Exp. Med.* **203**:2433–2440 (2006).

83. Gerrits AJ, Koekman CA, van Haeften TW, Akkerman JWN, Platelet tissue factor synthesis in type 2 diabetes patients is resistant to inhibition by insulin, *Diabetes.* **59**:1487–1495 (2010).

84. Lopez JJ, Salido GM, Pariente JA, Rosado JA, Thrombin induces activation and translocation of Bid, Bax and Bak to the mitochondria in human platelets, *J. Thromb. Haemost.* **6**:1780–1788 (2008).

85. Rumjantseva V, Grewal PK, Wandall HH, Josefsson EC, Sørensen AL, Larson G, Marth JD, Hartwig JH, Hoffmeister KM, Dual roles for hepatic lectin receptors in the clearance of chilled platelets, *Nat. Med.* **15**:1273–1280 (2009).

86. Watkins NA, Gusnanto A, de Bono B, De S, Miranda-Saavedra D, Hardie DL, Angenent WG, Attwood AP, Ellis PD, Erber W, Foad NS, Garner SF, Isacke CM, Jolley J, Koch K, Macaulay IC, Morley SL, Rendon A, Rice KM, Taylor N, Thijssen-Timmer DC, Tijssen MR, van der Schoot CE, Wernisch L, Winzer T, Dudbridge F, Buckley CD, Langford CF, Teichmann S, Göttgens B, Ouwehand WH, Bloodomics Consortium, A HaemAtlas: Characterizing gene expression in differentiated human blood cells, *Blood* **113**:e1–9 (2009).

87. http://plateletweb.bioapps.biozentrum.uni-wuerzburg.de/PlateletWeb.php.

88. http://bloodjournal.hematologylibrary.org/cgi/content/full/blood-2009-02-203828/DC1.

89. Jones CI, Bray S, Garner SF, Stephens J, de Bono B, Angenent WG, Bentley D, Burns P, Coffey A, Deloukas P, Earthrowl M, Farndale RW, Hoylaerts MF, Koch K, Rankin A, Rice CM, Rogers J, Samani NJ, Steward M, Walker A, Watkins NA, Akkerman JW, Dudbridge F, Goodall AH, Ouwehand WH, Bloodomics Consortium, A functional genomics approach reveals novel quantitative trait loci associated with platelet signaling pathways, *Blood* **114**:1405–1416 (2009).

90. http://www.reactome.org/.

91. Cattaneo M, Congenital disorders of platelet secretion, in Gresele P, Page C, Fuster V, Vermylen J, eds., *Platelets in Thrombotic and Nonthrombotic Disorders*, Cambridge Univ. Press, 2002, pp. 655–673.

92. Rao AK, Inherited defects in platelet signaling mechanisms, *J. Thromb. Haemost.* **1**:671–681 (2003).

93. Nurden AT, Glanzmann thrombasthenia, *Orphanet. J. Rare Dis*. **1**:1–8 (2006).
94. Salles II, Feys HB, Iserbyt BF, De Meyer SF, Vanhoorelbeke K, Deckmyn H, Inherited traits affecting platelet function, *Blood Rev*. **22**:155–172 (2008).
95. Nurden P, Nurden AT, Congenital disorders associated with platelet dysfunctions, *Thromb. Haemost*. **99**:253–263 (2008).
96. Pelczar-Wissner CJ, McDonald EG, Sussman II, Absence of platelet activating factor (PAF) mediated platelet aggregation: A new platelet defect, *Am. J. Hematol*. **16**:419–422 (1984).
97. Higuchi W, Fuse I, Hattori A, Aizawa Y, Mutations of the platelet thromboxane A2 (TXA2) receptor in patients characterized by the absence of TXA2-induced platelet aggregation despite normal TXA2 binding activity, *Thromb. Haemost*. **82**:1528–1531 (1999).
98. Hirata T, Kakizuka A, Ushikubi F, Fuse I, Okuma M, Narumiya S, Arg60 to Leu mutation of the human thromboxane A2 receptor in a dominantly inherited bleeding disorder, *J. Clin. Invest*. **94**:1662–1667 (1994).
99. Mumford AD, Dawood BB, Daly ME, Murden SL, Williams MD, Protty MB, Spalton JC, Wheatley M, Mundell SJ, Watson SP, A novel thromboxane A2 receptor D304N variant that abrogates ligand binding in a patient with a bleeding diathesis, *Blood* **115**:363–369 (2010).
100. Gabbeta J, Yang X, Kowalska MA, Sun L, Dhanasekaran N, Rao AK, Platelet signal transduction defect with Galpha subunit dysfunction and diminished Galphaq in a patient with abnormal platelet responses, *Proc. Natl. Acad. Sci. USA* **94**:8750–8755 (1997).
101. Lages B, Weiss HJ, Impairment of phosphatidylinositol metabolism in a patient with a bleeding disorder associated with defects of initial platelet responses, *Thromb. Haemost*. **59**:175–179 (1988).
102. Bergmeier W, Oh-Hora M, McCarl CA, Roden RC, Bray PF, Feske S, R93W mutation in Orai1 causes impaired calcium influx in platelets, *Blood* **113**:675–678 (2009).
103. Notarangelo LD, Miao CH, Ochs HD, Wiskott-Aldrich syndrome, *Curr. Opin. Hematol*. **15**:30–36 (2008).
104. Tsuboi S, Nonoyama S, Ochs HD, Wiskott-Aldrich syndrome protein is involved in alphaIIb beta3-mediated cell adhesion, *EMBO Rep*. **7**:506–511 (2006).
105. Zhu Q, Watanabe C, Liu T, Hollenbaugh D, Blaese RM, Kanner SB, Aruffo A, Ochs HD, Wiskott-Aldrich syndrome/X-linked thrombocytopenia: WASP gene mutations, protein expression, and phenotype, *Blood* **90**:2680–2689 (1997).
106. Albert MH, Bittner TC, Nonoyama S, Notarangelo LD, Burns S, Imai K, Espanol T, Fasth A, Pellier I, Strauss G, Morio T, Gathmann B, Noordzij JG, Fillat C, Hoenig M, Nathrath M, Meindl A, Pagel P, Wintergerst U, Fischer A, Thrasher AJ, Belohradsky BH, Ochs HD, X-linked thrombocytopenia (XLT) due to WAS mutations: Clinical characteristics, long-term outcome, and treatment options, *Blood* **115**:3231–3238 (2010).
107. `http://bioinf.uta.fi/WASbase/`.
108. Sabri S, Foudi A, Boukour S, Franc B, Charrier S, Jandrot-Perrus M, Farndale RW, Jalil A, Blundell MP, Cramer EM, Louache F, Debili N, Thrasher AJ, Vainchenker W, Deficiency in the Wiskott-Aldrich protein induces premature

proplatelet formation and platelet production in the bone marrow compartment, *Blood* **108**:134–140 (2006).

109. Piccin A, Murphy WG, Smith OP, Circulating microparticles: pathophysiology and clinical implications, *Blood Rev.* **21**:157–171 (2007).
110. `http://sinaicentral.mssm.edu/intranet/research/glanzmann`.
111. Loftus JC, O'Toole TE, Plow EF, Glass A, Frelinger AL 3rd, Ginsberg MH, A beta 3 integrin mutation abolishes ligand binding and alters divalent cation-dependent conformation, *Science* **249**:915–918 (1990).
112. Lanza F, Stierlé A, Fournier D, Morales M, André G, Nurden AT, Cazenave JP, A new variant of Glanzmann's thrombasthenia (Strasbourg I). Platelets with functionally defective glycoprotein IIb-IIIa complexes and a glycoprotein IIIa 214Arg–214Trp mutation, *J. Clin. Invest.* **89**:1995–2004 (1992).
113. Chen YP, Djaffar I, Pidard D, Steiner B, Cieutat AM, Caen JP, Rosa JP, Ser-752–> Pro mutation in the cytoplasmic domain of integrin beta 3 subunit and defective activation of platelet integrin alpha IIb beta 3 (glycoprotein IIb-IIIa) in a variant of Glanzmann thrombasthenia, *Proc. Natl. Acad. Sci. USA* **89**:10169–10173 (1992).
114. Wang R, Shattil SJ, Ambruso DR, Newman PJ, Truncation of the cytoplasmic domain of beta3 in a variant form of Glanzmann thrombasthenia abrogates signaling through the integrin alpha(IIb)beta3 complex, *J. Clin. Invest.* **100**:2393–2403 (1997).
115. Ruiz C, Liu CY, Sun QH, Sigaud-Fiks M, Fressinaud E, Muller JY, Nurden P, Nurden AT, Newman PJ, Valentin N, A point mutation in the cysteine-rich domain of glycoprotein (GP) IIIa results in the expression of a GPIIb-IIIa (alphaIIbbeta3) integrin receptor locked in a high-affinity state and a Glanzmann thrombasthenia-like phenotype, *Blood* **98**:2432–2441 (2001).
116. Malinin NL, Zhang L, Choi J, Ciocea A, Razorenova O, Ma YQ, Podrez EA, Tosi M, Lennon DP, Caplan AI, Shurin SB, Plow EF, Byzova TV, A point mutation in KINDLIN3 ablates activation of three integrin subfamilies in humans, *Nat. Med.* **15**:313–318 (2009).
117. Rao AK, Koike K, Willis J, Daniel JL, Beckett C, Hassel B, Day HJ, Smith JB, Holmsen H, Platelet secretion defect associated with impaired liberation of arachidonic acid and normal myosin light chain phosphorylation, *Blood* **64**:914–921 (1984).
118. Pareti FI, Mannucci PM, D'Angelo A, Smith JB, Sautebin L, Galli G, Congenital deficiency of thromboxane and prostacyclin, *Lancet* **1**:898–901 (1980).
119. Rolf N, Knoefler R, Bugert P, Gehrisch S, Siegert G, Kuhlisch E, Suttorp M, Clinical and laboratory phenotypes associated with the aspirin-like defect: a study in 17 unrelated families, *Br. J. Haematol.* **144**:416–424 (2009).
120. Virgolini I, O'Grady J, Peskar BA, Sinzinger H, Defects in the prostaglandin-system—heredity, prevalence and vascular risk analysis, *Prostaglandins Leukot. Essent. Fatty Acids* **40**:227–237 (1990).
121. Dubé JN, Drouin J, Aminian M, Plant MH, Laneuville O, Characterization of a partial prostaglandin endoperoxide H synthase-1 deficiency in a patient with a bleeding disorder, *Br. J. Haematol.* **113**:878–885 (2001).
122. Defreyn G, Machin SJ, Carreras LO, Dauden MV, Chamone DA, Vermylen J, Familial bleeding tendency with partial platelet thromboxane synthetase deficiency:

reorientation of cyclic endoperoxide metabolism, *Br. J. Haematol.* **49**:29–41 (1981).

123. Jedlitschky G, Cattaneo M, Lubenow LE, Rosskopf D, Lecchi A, Artoni A, Motta G, Niessen J, Kroemer HK, Greinacher A, Role of MRP4 (ABCC4) in platelet adenine nucleotide-storage: Evidence from patients with delta-storage pool deficiencies, *Am. J. Pathol.* **176**:1097–1103 (2010).
124. Ninkovic I, White JG, Rangel-Filho A, Datta YH, The role of Rab38 in platelet dense granule defects, *J. Thromb. Haemost*. **6**:2143–2151 (2008).
125. Barral DC, Ramalho JS, Anders R, Hume AN, Knapton HJ, Tolmachova T, Collinson LM, Goulding D, Authi KS, Seabra MC, Functional redundancy of Rab27 proteins and the pathogenesis of Griscelli syndrome, *J. Clin. Invest*. **110**:247–257 (2002).
126. Cattaneo M, The P2 receptors and congenital platelet function defects, *Semin. Thromb. Hemost*. **31**:168–173 (2005).
127. Fontana G, Ware J, Cattaneo M, Haploinsufficiency of the platelet P2Y12 gene in a family with congenital bleeding diathesis, *Haematologica* **94**:581–584 (2009).
128. Remijn JA, IJsseldijk MJ, Strunk AL, Abbes AP, Engel H, Dikkeschei B, Dompeling EC, de Groot PG, Slingerland RJ, Novel molecular defect in the platelet ADP receptor P2Y12 of a patient with haemorrhagic diathesis, *Clin. Chem. Lab. Med*. **45**:187–189 (2007).
129. Patel YM, Patel K, Rahman S, Smith MP, Spooner G, Sumathipala R, Mitchell M, Flynn G, Aitken A, Savidge G, Evidence for a role for Galphai1 in mediating weak agonist-induced platelet aggregation in human platelets: reduced Galphai1 expression and defective Gi signaling in the platelets of a patient with a chronic bleeding disorder, Blood **101**:4828–4835 (2003).
130. Ferreira IA, Akkerman JWN, IRS-1 and the vascular complications in diabetes mellitus, *Vitamins Hormones* **70**:25–67 (2005).
131. Ferreira IA, Mocking AI, Feijge MA, Gorter G, van Haeften TW, Heemskerk JW, Akkerman JW, Platelet inhibition by insulin is absent in type 2 diabetes mellitus, *Arterioscler. Thromb. Vasc. Biol*. **26**:417–422 (2006).
132. Giesberts AN, van Willigen G, Lapetina EG, Akkerman JW, Regulation of platelet glycoprotein IIb/IIIa (integrin alpha IIB beta 3) function via the thrombin receptor, *Biochem. J*. **309**:613–620 (1995).
133. Freson K, Jaeken J, Van Helvoirt M, de Zegher F, Wittevrongel C, Thys C, Hoylaerts MF, Vermylen J, Van Geet C, Functional polymorphisms in the paternally expressed XLalphas and its cofactor ALEX decrease their mutual interaction and enhance receptor-mediated cAMP formation, *Hum. Mol. Genet*. **12**:1121–1130 (2003).
134. Freson K, Thys C, Wittevrongel C, Proesmans W, Hoylaerts MF, Vermylen J, Van Geet C, Pseudohypoparathyroidism type Ib with disturbed imprinting in the GNAS1 cluster and Gsalpha deficiency in platelets, *Hum. Mol. Genet*. **11**:2741–2750 (2002).
135. Freson K, Hashimoto H, Thys C, Wittevrongel C, Danloy S, Morita Y, Shintani N, Tomiyama Y, Vermylen J, Hoylaerts MF, Baba A, Van Geet C, The pituitary adenylate cyclase-activating polypeptide is a physiological inhibitor of platelet activation, *J. Clin. Invest*. **113**:905–912 (2004).

136. Lanza F, Bernard-Soulier syndrome (hemorrhagiparous thrombocytic dystrophy), *Orphanet. J. Rare Dis*. **1**:46 (2006).

137. `http://www.biomedcentral.com/content/supplementary/1750-1172-1-46-S1.doc`.

138. `http://www.bernardsoulier.org/`.

139. Takahashi H, Murata M, Moriki T, Anbo H, Furukawa T, Nikkuni K, Shibata A, Handa M, Kawai Y, Watanabe K, et al., Substitution of Val for Met at residue 239 of platelet glycoprotein Ib alpha in Japanese patients with platelet-type von Willebrand disease, *Blood* **85**:727–733 (1995).

140. Miller JL, Cunningham D, Lyle VA, Finch CN, Mutation in the gene encoding the alpha chain of platelet glycoprotein Ib in platelet-type von Willebrand disease, *Proc. Natl. Acad. Sci. USA* **88**:4761–4765 (1991).

141. Nurden P, Lanza F, Bonnafous-Faurie C, Nurden A, A second report of platelet-type von Willebrand disease with a Gly233Ser mutation in the GPIBA gene, *Thromb. Haemost*. **97**:319–321 (2007).

142. Strassel C, David T, Eckly A, Baas MJ, Moog S, Ravanat C, Trzeciak MC, Vinciguerra C, Cazenave JP, Gachet C, Lanza F, Synthesis of GPIb beta with novel transmembrane and cytoplasmic sequences in a Bernard-Soulier patient resulting in GPIb-defective signaling in CHO cells, *J. Thromb. Haemost*. **4**:217–228 (2006).

143. Nieuwenhuis HK, Akkerman JWN, Houdijk WPM, Sixma JJ, Human blood platelets showing no response to collagen fail to express surface glycoprotein Ia, *Nature* **318**:470–472 (1985).

144. Kehrel B, Balleisen L, Kokott R, Mesters R, Stenzinger W, Clemetson KJ, van de Loo J, Deficiency of intact thrombospondin and membrane glycoprotein Ia in platelets with defective collagen-induced aggregation and spontaneous loss of disorder, *Blood* **71**:1074–1078 (1988).

145. Kehrel B, Wierwille S, Clemetson KJ, Anders O, Steiner M, Knight CG, Farndale RW, Okuma M, Barnes MJ, Glycoprotein VI is a major collagen receptor for platelet activation: It recognizes the platelet-activating quaternary structure of collagen, whereas CD36, glycoprotein IIb/IIIa, and von Willebrand factor do not, *Blood* **91**:491–499 (1998).

146. Moroi M, Jung SM, Okuma M, Shinmyozu K, A patient with platelets deficient in glycoprotein VI that lack both collagen-induced aggregation and adhesion, *J. Clin. Invest*. **84**:1440–1445 (1989).

147. Dunkley S, Arthur JF, Evans S, Gardiner EE, Shen Y, Andrews RK, A familial platelet function disorder associated with abnormal signalling through the glycoprotein VI pathway, *Br. J. Haematol*. **137**:569–577 (2007).

148. Ichinohe T, Takayama H, Ezumi Y, Arai M, Yamamoto N, Takahashi H, Okuma M, Collagen-stimulated activation of Syk but not c-Src is severely compromised in human platelets lacking membrane glycoprotein VI, *J. Biol. Chem*. **272**:63–68 (1997).

149. Arthur JF, Dunkley S, Andrews RK, Platelet glycoprotein VI-related clinical defects, *Br. J. Haematol*. **139**:363–372 (2007).

# 13

# PLATELET PROTEOMICS IN TRANSFUSION MEDICINE

THOMAS THIELE, LEIF STEIL, UWE VÖLKER, AND
ANDREAS GREINACHER

**Abstract**

Human platelet concentrates are a frequently used therapeutic. Production, processing, and storage of platelet concentrates as well as platelet substitution therapy are a main issue of transfusion medicine. Proteomic technologies have the potential to provide comprehensive information about changes occurring during processing and storage of platelet concentrates. Thus proteomic applications will likely guide improvements of routine quality assessment and new developments such as pathogen inactivation procedures and will contribute to elucidation of aspects of platelet biology that are important for the prevention of platelet storage lesions.

## 13.1 INTRODUCTION

### 13.1.1 Platelets in the Context of Transfusion Medicine

Platelets provide a wide spectrum of diverse functions in the human organism. In clinical medicine, the most important aspect of platelet function is their role in hemostasis. Platelets build the cellular basis for blood coagulation. In the event of endothelial damage, platelets become activated and initiate vessel

*Platelet Proteomics: Principles, Analysis and Applications*, First Edition.
Edited by Ángel García and Yotis A. Senis.

recovery by local clot formation. Beside this well-known role, platelets also release cytokines, clotting factors, and mediators of inflammation and act as signaling cells in immunological processes. Platelets contain significant amounts of RNA, which can be utilized for *de novo* protein synthesis and are thus able to adapt to specific physiological challenges by posttranscriptional mechanisms by adjusting their proteome, phenotype, and functions [1,2]. It is therefore a growing issue whether, besides their effects in hemostasis, additional biological effects might be provoked by transfusion of platelets that are altered by storage or the process of preparation [3].

In 1950 Hirsch and colleagues published the first report on transfusion of platelet-rich plasma to overcome bleeding in a thrombocytopenic patient [4]. Nowadays platelet concentrates (PCs) are extensively used in clinical medicine. Platelet transfusions are mandatory for the prevention and treatment of bleeding in thrombocytopenic patients receiving high-dose chemotherapy or hematologic stem cell transplantation. Platelet substitution is also one of the basic treatments for restoring sufficient hemostasis in patients with bleeding during major surgery or in bleeding trauma patients. Thus, platelets with hemostatic capacity became a therapeutic with increasing importance for many modern therapies.

Transfusion medicine focuses on producing the "drug" platelet concentrate by obtaining platelets from healthy blood donors, to preserve them for storage, and to provide them in sufficient quality and quantity for transfusion to patients in need of platelets. In this manner transfusion medicine has established one of the most advanced quality management systems in clinical medicine to provide patients with blood products such as PCs. This includes the selection of blood donors according to strict criteria, standardized procedures for blood donation, processing, testing, and storage of PCs. Modern PCs maintain sufficient functionality over several days of storage and have a relatively low potential to cause adverse reactions or to transmit infectious agents like bacteria or viruses.

As do all blood components, PCs differ from "typical drugs" because of their origin from individual human donors. They are not clearly chemically defined but are highly complex mixtures of plasma proteins and even more complex cells. During the production of PCs, platelets have to be handled in an artificial environment, which may result in changes in the integrity of platelets, explicitly at the protein or even at the peptide level. In this context, protein modifications can occur during various steps of the production process, such as blood donation, processing, conservation, and the storage of PCs. The functional damage of platelets caused by any of these steps is summarized by the term *platelet storage lesions* (PSLs) [5,6]. It is a major challenge in transfusion medicine to identify the molecular mechanisms responsible for the PSL.

Despite all precautions it cannot be excluded that PCs transmit pathogens from the donor to the patient. Therefore, the efficient inactivation of pathogens within PCs is a recurrent and persistent issue. While these treatments reduce the risk of transmission of virus and bacteria with the PC, they may also alter platelet integrity [7] and induce further storage lesions. A global understanding of the PSL and the knowledge of the impact of any particular production step causing

PSL changes will help to improve platelet collection, processing, and storage and to enhance the quality of PCs.

### 13.1.2 Platelet Proteomics in Transfusion Medicine

Proteomics has developed into a mature leading discipline in life science since the mid-1990s. Contemporary proteomic strategies allow a comprehensive assessment of protein modifications with high coverage and offer capabilities for qualitative and quantitative analysis and for high-throughput protein identification [8]. In this manner proteomic applications represent a very useful tool to monitor platelet protein changes without major restrictions [9]. This unbiased view is necessary to effectively screen for changes in platelets processed for transfusion and to uncover storage lesions.

On the basis of proteomic findings, key marker proteins can be identified that indicate platelet damage. These markers may then be used in less complex test systems such as flow cytometry or enzyme immuno assays suitable for high-throughput routine monitoring of PC quality.

Proteomic applications overcome some limitations of currently applied quality monitoring systems, because they can disclose changes at the molecular level much more comprehensively than selective methods used today. In this context proteomic methods qualify for the large-scale screening of protein changes induced by PC production and storage. In this regard proteomic approaches can be applied for the evaluation of new protocols or methods for PC production before their introduction into practice. However, without additional functional analysis, proteomic technologies alone are not sufficient to define the biologic relevance of certain changes in the platelet proteome. Otherwise, proteomic experiments will help define the appropriate targets leading to the development of functional assays, which will, in turn, provide biological information.

In this chapter we review the application of platelet proteomics in transfusion medicine as an example of the opportunities of modern proteomic technologies for transfusion medical research and practice. Major achievements will be highlighted, and useful proteomic strategies will be discussed in detail. Initially, we briefly summarize the options of how platelets are transferred from the blood of a healthy blood donor into a plastic bag to become a PC.

## 13.2 BACKGROUND FOR PROTEOMIC STUDIES OF PLATELETS STORED FOR TRANSFUSION

### 13.2.1 Production, Preservation, and Storage of Platelet Concentrates

Platelet concentrates are prepared from individual whole-blood donations (usually, with subsequent pooling) or collected from a single donor by apheresis. There are two types of whole-blood-derived PCs: (1) PCs from platelet-rich plasma (PRP) and (2) PCs prepared from buffy coats (BCs). The PRP method has been the standard procedure used in North America for more than 35 years.

In Europe, the BC method has been most commonly used for approximately two decades and has also been introduced in Canada [10].

Preparation of platelets from PRP, whole blood is collected, anticoagulated with citrate, and stored at room temperature for up to 8 h. It is first centrifuged at low speed ("soft spin") to yield red blood cells (RBC) and platelet-rich plasma. PRP is extracted and then centrifuged at high speed ("hard spin"), separating plasma from platelets. The platelet pellet is left undisturbed for 1 h, allowing platelets to disaggregate. After 1 h of resting, platelets are resuspended and stored at room temperature with continuous agitation. Four to six units of these platelets are then pooled for a therapeutic platelet unit.

With the BC method, citrated whole blood is stored at room temperature (usually for $\leq$2 h) before processing. Afterward the whole blood is centrifuged at high speed, and packed red cells and the plasma are separated by top (plasma) and bottom (red cells) outlets of the blood bag. The intervening layer, which contains platelets, white cells, some red cells, and plasma, is the so-called buffy coat. Four to six BCs are pooled along with plasma from one of the whole-blood donors or an electrolyte solution as storage media. The pooled BCs are centrifuged at low speed to separate PRP from the red cells and leukocytes and to yield the final platelet unit. In contrast to the PRP method, the platelets rest on a layer of red blood cells during the hard spin and not on the non-physiological plastic surface.

Another method for obtaining PCs is single-donor apheresis. Donors are connected via an intravenous line to a blood cell separator. The venous blood is mixed with citrate at the side of the cannula and then flows into the separator chamber. The rotating centrifugation chamber separates platelets from other blood cells and stores them in a separate container. The platelet-depleted whole blood is then given back to the donor. In some apheresis systems, platelets are collected as PRP and do not require resuspension, whereas other systems produce a concentrated platelet pellet that must be resuspended. Figure 13.1 gives an overview of the three currently applied production methods of PCs.

All methods discussed yield a platelet product with a comparable mean platelet concentration of $2 - 6 \times 10^{11}$ cells in a final volume of approximately 200–250 mL.

In all three methods, PCs are depleted from leukocytes. This is achieved either by using leukoreduction filters (PRP and BC pooled platelets), or by the highly efficient platelet separation within the apheresis procedure. The introduction of leukoreduction has greatly minimized leukocyte-specific side effects such as posttransfusion fever induced by the release of cytokines by transfused leukocytes.

In addition, PCs scheduled for transfusion in immunocompromised individuals have to be irradiated using $\gamma$ irradiation, which breaks DNA strands of leucocytes [11], thereby preventing graft-versus-host disease caused by donor T-lymphocyte cell division in the recipient after transfusion. As thrombocytopenia after chemotherapy is a leading indication for platelet transfusion, and as these patients are most often immunocompromised, changes in platelets by $\gamma$ irradiation are of major clinical relevance.

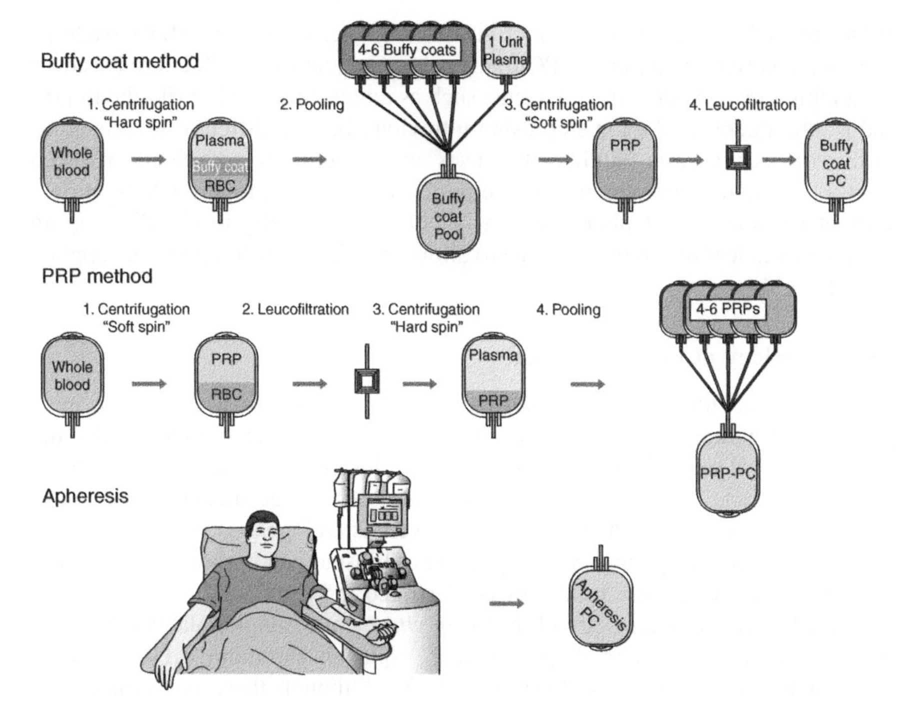


**Figure 13.1** Standard approaches for the production of platelet concentrates.

After production, PCs are stored at 22°C under continuous agitation. For optimal gas diffusion, gas-permeable storage bags are used [12]. As platelets have to be stored between 20°C and 24°C, the risk of bacterial growth is much higher than for red blood cell concentrates, which are stored at 4°C, or plasma, stored frozen below −20°C. The risk for bacterial growth in the PC increases significantly with increasing storage time. Therefore, the maximum period for platelet storage must not exceed 5 days in most jurisdictions.

Platelets can be stored in either plasma or an additive isotonic, buffered salt solution containing about 30% plasma. The latter has several advantages: (1) the plasma of the donor can be used for treatment of another patient; (2) antibodies in the plasma, especially isoagglutinines directed against blood groups A and B, are reduced by 70%, which enhances PC compatibility and simplifies PC logistics because it permits transfusion of blood group O—PCs to recipients with blood group A or B with a reduced risk of adverse effects caused by the anti-A and anti-B isoagglutinins present in the blood group O plasma; and (3) potentially pathogenic antibodies (e.g., anti-HLA or antineutrophil antibodies) or other plasma constituents causing adverse reactions are reduced in the PC [13,14].

Another important aspect of using additive solutions for PCs is the progressing introduction of pathogen reduction technologies (PRTs) for PCs. PRTs aim to reduce the risk of the transmission of viruses (e.g., hepatitis C and HIV), bacteria, and protozoa (e.g., malaria) by PC transfusion. Contemporary PRTs are based on the addition of photochemical reagents such as amotosalen [15] or riboflavin [16] and a subsequent irradiation step using ultraviolet light at different wavelengths. In this manner, a large number of viruses and bacteria are sufficiently reduced in PCs [17]. Other phototreatments are in development employing UV light only without the addition of photochemical agents. All currently used PRTs require storage media less turbid than plasma to allow sufficient penetration of the applied illuminant [18].

### 13.2.2 Biology of Platelet Storage

Platelets stored under blood bank conditions pass through a variety of morphologic, structural, and functional changes, which are summarized under the term *platelet storage lesion* (PSL) [6]. These lesions are caused by single or combined effects of exogenous conditions such as storage media, storage temperature, and method of PC preparation as well as endogenous processes within the platelets, such as platelet viability, platelet lifespan, and protein degradation. A short overview of these processes is illustrated in Figure 13.2.

Some functional defects are observed *in vitro*. For example, platelet aggregation responses to a number of agonists (e.g., collagen, thrombin, ADP) decrease significantly with ongoing storage (Fig. 13.3). Although these observable processes are used for current quality testing (see text below), the molecular events underlying these effects are not yet completely understood.

Platelets stored in PCs remain metabolically dynamic at room temperature. Some metabolites such as lactate accumulate, resulting in a decrease of the pH over time. Moreover, platelets become activated during storage and change from a resting discoid shape to an activated spherical shape [5]. Platelet activation is indicated by an increase in surface levels of the platelet activation marker P-selectin (CD62P), which is associated with the release of platelet α-granules and release of ß-thromboglobulin and platelet factor 4, which accumulate in the storage media.

Besides platelet activation, a possible role of apoptosis or apoptosis-like events throughout platelet storage is pointed out by other investigators [19], probably mediated by caspase-3 [20]. For example, procaspase-3 is processed following thrombin activation. Converted to its active form, caspase-3 is able to degrade membrane–cytoskeletal–linker proteins such as moesin [21]. In this context, additional platelet activation during preparation might induce irreversible defects in platelets and might enhance early apoptosis in PC platelets.

The introduction of new PRT might induce additional changes in the proteins of platelets in the PC. For example, a higher rate of platelet destruction, accelerated metabolic changes, and reduced agonist-induced aggregation responses have been observed in pathogen-inactivated platelets [22].

The relevance of PSLs in platelets prepared for transfusion was shown in radiolabeling studies. Platelets of autologous PCs of volunteer donors have a significantly decreased recovery and survival following storage compared to fresh

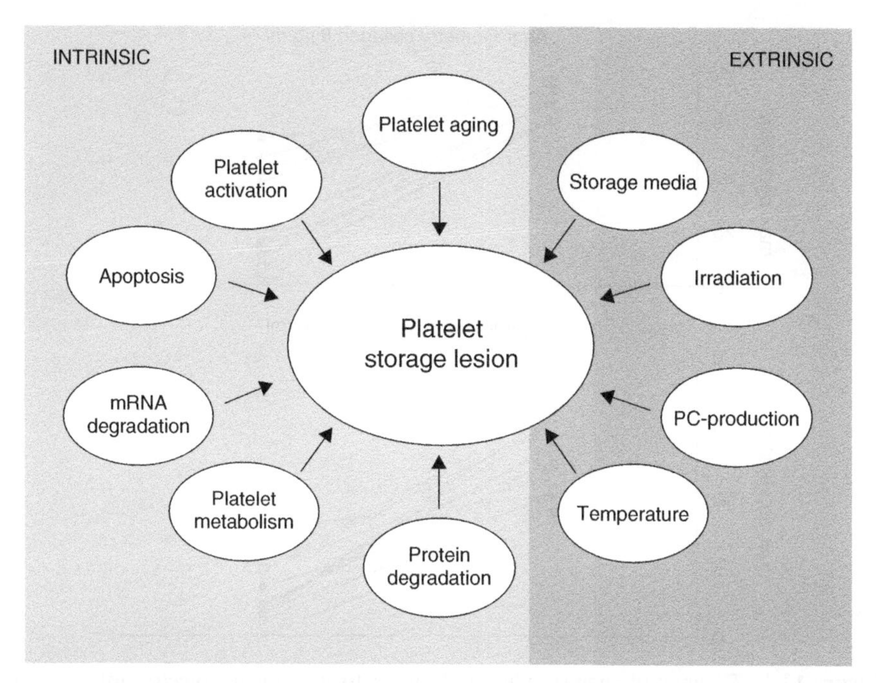


**Figure 13.2** Factors influencing the platelet storage lesion can be divided into intrinsic and extrinsic factors. Intrinsic factors, such as platelet aging and activation, are related to platelet biology; otherwise they also reflect single or combined effects of exogenous conditions such as storage media and temperature and method of PC preparation.

platelets [23,24]. Also, platelet survival and recovery after 5 and 7 days of storage [25] were found to have a mean recovery of 63% after 5 days with a mean survival of ~160 h and only a recovery of nearly 54% and a mean survival of 134 h after 7 days, respectively.

### 13.2.3 Current Quality Monitoring of Platelet Concentrates

Quality control is carried out *in vitro* by selective methods assessing platelet morphology, metabolism, or function. Morphologic changes can be classified by the Kunicki morphology score [26]; metabolism is determined by changes in pH, lactate, or glucose levels.

The function of stored platelets is evaluated by (1) testing the ability of platelets to reconstitute after their exposition to a hypotonic environment (hypotonic shock response); (2) measuring the extent of shape change after agonist-induced platelet aggregation (aggregometry); or (3) flow cytometry assessing platelet activation markers such as CD62P, CD40L, or CD63; changes in platelet surface glycoproteins such as GPIb, GPIIb, and GPIIIa; and early apoptosis markers such as annexin V binding [27].

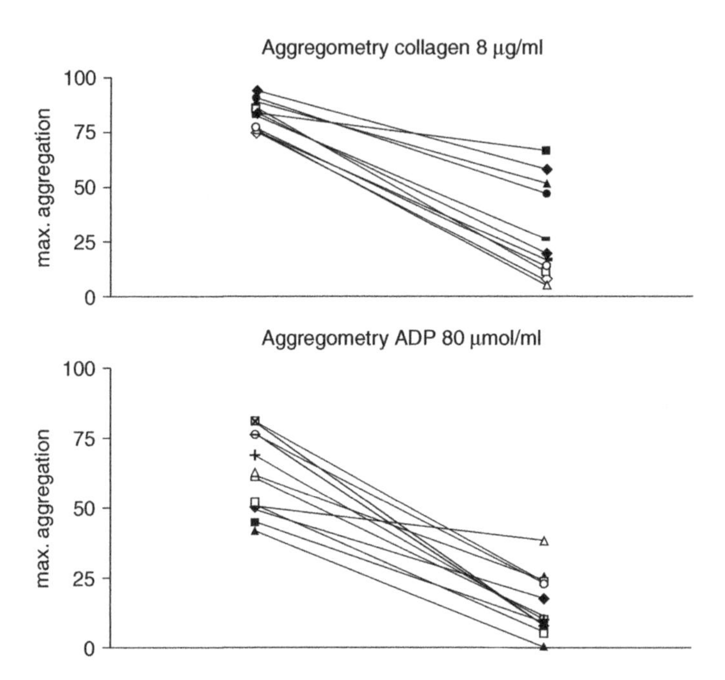


**Figure 13.3** Diagram of agonist-induced platelet activation (aggregometry) following 6 days of PC-storage. One effect of platelet storage is the reduced ability to activate after incubation with a physiologic agonist such as collagen or ADP. This can be monitored by platelet aggregometry, where the decrease in optical density that occurs in solution when platelets aggregate is measured.

The readout systems currently applied in transfusion medicine to identify PSLs are insufficient to identify the platelet proteins affected by the various steps of PC production and frequently fail to reveal the molecular nature of production-induced changes. Therefore, it is of particular importance to develop analytical tools that facilitate a broad view of the processes occurring in platelets stored for transfusion. Proteomic approaches might serve this purpose very well because they allow for a comprehensive monitoring of changes in protein levels and their post-translational modifications at a global scale.

### 13.2.4 Technical Requirements for Proteomic Studies on Platelet Concentrates

Before applying proteomics for the analysis of therapeutic blood products such as PCs, one must critically choose from a broad range of proteomic applications. The decision is determined by the specific question and should balance specific needs, efforts, and costs. For routine quality screening a cost-effective and robust approach with respect to reproducibility and interassay variability would be desirable. Moreover, the opportunity for high-throughput analysis should be included. In contrast, advanced scientific questions demand a most comprehensive proteomic overview addressing proteins and their modifications with high sensitivity and coverage. To date (as of 2010), none of the individual proteomic approaches satisfies all these requirements simultaneously.

The application of two-dimensional gel electrophoresis (2DGE) using fluorescent dyes such as *differential* (or *difference*) *in-gel electrophoresis* (DIGE) has several advantages for the study of platelets stored in PCs because they allow comprehensive and quantitative monitoring of the platelet proteome at different timepoints, so that one can easily follow up the kinetics of a whole proteome over time (Fig. 13.4). Modern fluorescent dyes combine reasonable sensitivity and a broad dynamic range for quantitation. Furthermore, protein modifications such as degradation can be monitored. In this regard these gel-based approaches are not far from being a robust quality screening tool, despite their well-known limitations such as reduced coverage of hydrophobic

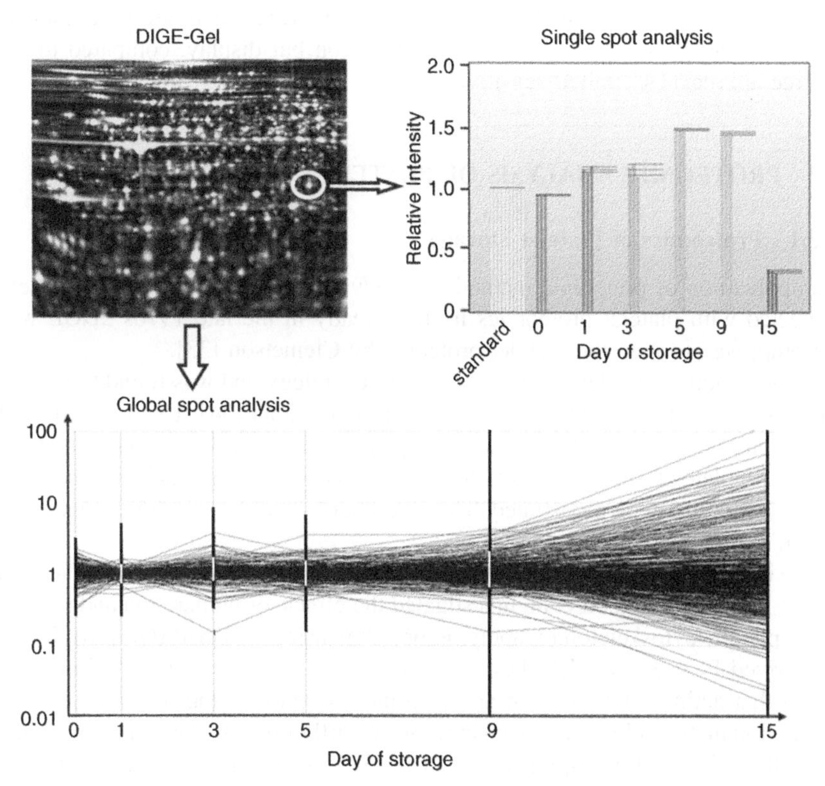


**Figure 13.4** Schematic representation of a quantitative gel-based analysis of stored platelet concentrates. Differential in gel electrophoresis enables the quantification of a large number of proteins with high reproducibility (upper left panel). Besides the single-spot analysis over time (upper right panel), the insertion of an internal standard and computer-assisted analysis allows the accurate quantification of a whole dataset (panel below). Using this approach, time-resolved changes can be monitored which makes it an effective tool for the investigation of PSL-changes.

proteins and those with a low or high molecular mass. However, gel-based methods are at present inappropriate for high-throughput routine testing.

Current gel-free proteomics approaches provide better sensitivity than do 2D gels and can also yield data on the absolute protein levels, employing external standard labeled with stable isotopes [28–32]. In addition, full automation is more likely to be achieved in the near future for gel-free, mass spectrometry-centered proteomic approaches. However, at this stage both gel-based and gel-free approaches still have their pros and cons. Gel-free approaches clearly provide greater sensitivity and better coverage, but are limited mainly to the peptide level, because in standard workflows proteins are initially digested with proteases such as trypsin, fractionated by one- or two-dimensional liquid chromatography (LC), and then analyzed online by electrospray ionization–tandem mass spectrometry (ESI-MS/MS). In such an approach information about protein isoforms and concurrence of multiple posttranslational modifications is lost or at least compromised. On the other hand, gel-based approaches facilitate separation at the protein level and thus retain such information but display, compared to the gel-free approaches, limited sensitivity.

## 13.3 PROTEOMIC ANALYSIS OF PLATELET CONCENTRATES

### 13.3.1 Proteomics of Platelet Storage

The application of proteomic methods to study storage lesions in PCs is closely connected with platelet proteomics itself. Already in the late 1970s 2DGE was first employed to map the platelet proteome by Clemetson [33].

In 1987, actin was identified by a proteomic strategy and was found to change during storage of platelets under blood bank conditions [34]. Using 2DGE and silver staining, Snyder and coworkers mapped 30 altered protein spots throughout 7 days of platelet storage. Later, for two of these spots a calcium-dependent proteolysis could be demonstrated. They were identified as actin fragments using Edman's amino acid sequencing [35].

The issue of how PC production affects platelet storage lesions also surfaced in the pre-ms era. Estebanell et al. investigated the effects of preparation and storage on the platelet cytoskeleton by applying one-dimensional SDS-PAGE [36]. Using BC-derived PCs as a model, they again identified actin as a target and demonstrated that actin polymerization and tyrosine phosphorylation differed between preparation methods [37]. These changes were partly reversible when the platelets were allowed to rest after preparation of the PC, but recurrence of these protein changes was observed after prolonged storage.

The true capabilities of modern proteomics using prefractionation of proteins and peptides in combination with identification by mass spectrometry for the investigation of platelets were demonstrated relatively recently [38–41]. A whole array of these technologies can now be applied to monitor proteome changes of the platelets in PCs. Gel-based proteomic approaches allow assessment of nearly 2300 protein spots and more than 400 different proteins [42,43], while gel-free approaches such as combined fractional diagonal chromatography (COFRADIC) led to the detection of 641 proteins, with an overlap of ~40% to 2DGE-based methods [44,45]. This allowed (1) global mapping of platelet proteins [42,43,46–48], (2) investigation of phosphorylated proteins and the impact of protein phosphorylation on signal transduction pathways/cascades [49–52], (3) exploration of *N*-glycosylation sites on human platelet proteins [53], (4) study of the membrane proteome [54,55]; (5) investigation of the platelet releasate [56], and (6) the study of the microparticle proteome platelet microparticle proteome [57]. Even identification of whole signaling cascades under different conditions will be feasible [58]. All these aspects have been covered in previous chapters of this book.

For assessment of the impact of different manipulations of PCs, it is, however, important to compare changes in protein patterns quantitatively before and after the manipulation. A major breakthrough for the investigation of storage related proteome changes was the development of techniques allowing precise, reproducible quantification of these changes within one PC over time. Our group employed DIGE to analyze protein changes in platelets during an extended storage over 15 days [59]. In this study, 401 cytosolic protein spots present in both (1) platelets of PCs derived from single-donor apheresis and (2) BC pooling were compared. By processing of the quantitative DIGE data with a gene expression analysis software suite, a comparison of all protein spots mapped for the experiment in all analyzed PCs on every timepoint investigated was feasible (Fig. 13.4). More than 99% of the protein spots did not change significantly during the first 5 days of storage, and 97% remained unchanged even after 9 days. However, major alterations of the platelet proteome were observed between days 9 and 15. β-Actin, septin-2, and gelsolin were identified as potential surrogate markers indicating storage related changes in PCs.

Another example of the application of proteomics for assessment of PCs is the study by Glenister et al., who investigated the influence of platelet storage on the PC supernatant using a cytokine antibody microarray and 2DGE [60]. In this study several cytokines, including platelet-derived growth factor or brain-derived neurotrophic factor, were shown to be released from platelets during a storage period of 7 days.

Gel-free proteomic techniques are increasingly applied to assess platelets in transfusion medicine. Thon et al. combined the gel-based DIGE system with two gel-free quantitative proteomic approaches, namely, isotope-coded affinity tagging (ICAT) and isobaric tags for relative and absolute quantification (iTRAQ) to identify storage-related platelet protein changes [61]. The application of these complementing technologies allowed identification of changes in intensity for a total of 503 proteins between days 1 and 7 of PC storage. Of these, 93 proteins were identified by 2D gel/DIGE, 355 by iTRAQ, and 139 by ICAT. However, only 5 proteins were identified by all three approaches, thus highlighting the fact that gel-based and gel-free approaches are, indeed, complementary. Other gel-free approaches identified platelet factor 4 and β-thromboglobulin by LC-MS/MS of supernatant peptides in stored apheresis and BC-derived PCs [62].

Besides assessment of storage lesions for improved PC quality, methods developed in transfusion medicine can also be used for more basic research on platelet changes. Springer and coworkers prepared PCs from diabetic patients and from healthy controls and compared changes related to platelet storage with protein alterations that are attributable to type 2 diabetes in a LC-ESI-MS/MS experiment [63]. In this study platelet storage was used as a stress model to further investigate changes in platelets of diabetic patients. In this study, 122 proteins that were either up- or down-regulated in type 2 diabetes relative to nondiabetic controls and 117 proteins whose abundances changed during a 5-day storage period were detected. The characterization of changes in the proteome of platelets obtained from diabetic patients will most probably allow further studies on the reason why diabetic patients have an enhanced risk for cardiovascular events.

### 13.3.2 Proteomics of Pathogen-Inactivated Platelet Concentrates

The impact of novel pathogen reduction technologies on the platelet proteome has been investigated using DIGE [64]. In this study, the effect of UVC treatment on cytosolic platelet proteins with other well-established phototreatments, namely, UVB and $\gamma$ irradiation, was compared. A large overlap of more than 40 cytosolic platelet proteins was found to be affected by all three modes of irradiation, with $\gamma$ irradiation having the most pronounced effect with over 80 proteins being changed after treatment. UVB caused changes in about 60 proteins, and UVC modified less than 40 proteins. UVC irradiation induced only two individual changes. However, one spot was characterized as ERp72, a protein disulfide isomerase $A_4$, which catalyses the rearrangement of S–S bonds in proteins [65]. This raises the issue as to whether UVC-irradiation affects integrin conformation. This, however, could not be addressed by the gel-based DIGE technology with its known drawbacks in resolving hydrophobic or high-molecular-weight proteins such as membrane proteins (see text above). It is therefore necessary to adopt proteomic applications feasible for the quantitative monitoring of platelet membrane proteins to study the effect of pathogen reduction technologies on the platelet surface proteome.

## 13.4 IMPACT OF PROTEOMIC DATA

### 13.4.1 Relevance for Preparation of Platelet Concentrates

Proteomic studies investigating the PSL by proteomic methods concomitantly revealed that a number of proteins such as gelsolin, septin, talin, or vinculin change during platelet storage. These proteins, identified by independent groups, could be used as surrogate markers for PSLs and employed for routine quality monitoring of PCs. For example, determination of the extent of changes generated on these proteins might be utilized to characterize a new production method, diverse storage media, or new PRTs. However, for applications under routine conditions, much cheaper and more easily applied test systems need to be established.

The possible role of apoptosis or apoptosis-like events during platelet storage has been challenged in a number of publications [66]. Proteomic studies revealed storage-related changes in septin-2 and gelsolin, which are known substrates of caspase-3. Septins and gelsolin are also supposedly relevant in apoptosis in various other tissues [67]. As these proteins are altered by apoptosis and not primarily by platelet activation, apoptosis might have a more substantial impact on the PSL than currently assumed. Thus, inhibition of apoptosis might become an additional approach to prolonging storage time of PCs [68]. If apoptosis plays a major role in platelet alteration in PCs, its potential biologic effects in the transfusion recipient should be analyzed further.

### 13.4.2 Relevance for Understanding Platelet Biology

Several new biologic mechanisms causing PSL changes have been found by proteomic approaches. One example is the involvement of the integrin GPIIb/IIIa signaling pathway in the PSL. Proteomics revealed a number of proteins such as zyxin, vinculin, talin, or α-actinin, which change in level during platelet storage [59,61]. These proteins are players in the GPIIb/IIIa pathway [69]. Consistent with the observation that the GPIIb/IIIa complex is activated during storage, as indicated by enhanced PAC1 expression in flow cytometry [70], GPIIb/IIIa signaling is likely to play a role in the PSL. It is conceivable that the observed changes in proteins involved in this pathway reflect ongoing platelet activation, which, in turn, may lead to irreversible functional defects. Monitoring of the GPIIb/IIIa signaling pathway may be instrumental in developing methods of PC production that reduce PSLs.

Another field of interest is the possible ability of *in vitro*–stored platelets to synthesize proteins. A Western blot analysis of GPIIIa revealed a twofold increase in concentration on day 7 of storage and a fourfold increase on day 10 [71]. It was further demonstrated that full-length GPIIIa mRNA is present throughout a 10-day storage period and that translation of GPIIb and GPIIIa during storage takes place for this storage interval. These findings indicate that platelets are capable of synthesizing biologically relevant proteins *ex vivo* and indicate a possible role of protein *de novo* synthesis in the PSL. Other reports demonstrated that vital platelets contain mRNA and are capable of mRNA processing and even protein *de novo* synthesis in platelets [72,73]. Quantitative proteomics should be able to identify whether new proteins are synthesized in PCs. In this context, mRNA degradation, for example, by pathogen reduction methods, and thereby impairment of *de novo* protein synthesis, might also be of relevance for PSLs.

### 13.4.3 Future Prospects for Platelet Proteomics in Transfusion Medicine

Notably, the maximum storage time of 4 (e.g., in Germany) or 5 (e.g., in the United States) days is a logistic challenge for the provision of PCs because production depends on the blood donation frequency, and consumption is often hardly predictable. Consequently, several new concepts were developed to maintain platelet integrity during prolonged storage. Unfortunately, platelets stored at temperatures below 15°C perform very disappointingly *in vivo*, which was found to be based on an elevated clearance of transfused cold-stored platelets [74]. While galactosylation was shown to prevent this effect *in vitro* [75], *in vivo* tests revealed an even less effective performance of galactosylated and cold-stored platelets [76,77].

Moreover, metalloproteinases might play an important role in protein degradation in platelets during activation. Interestingly, a mouse model analyzed by Bergmeier and coworkers revealed that the inhibition of endogenous metalloproteinases clearly improved posttransfusion recovery of *in vitro*–aged mouse platelets [78]. Therefore, it might be an attractive option to improve storage conditions in order to prolong platelet storage using inhibitors of endogenous metalloproteinases.

Proteomic applications have the potential to be implemented into the validation phase of new concepts to prepare, manipulate, and store PCs. They will at least indicate whether a new method of PC preparation has major effects on the integrity of platelets and might be instrumental to direct further research to certain candidate proteins. Despite the great potential of this technology, changes defined by proteomic techniques must be carefully interpreted in regard to their biological relevance. It is important to state that ultimately only clinical studies will reveal whether PCs provide sufficient hemostatic capacity in thrombocytopenic patients to prevent major bleeding.

## 13.5 CONCLUDING REMARKS

Contemporary proteomic strategies allow a comprehensive assessment of protein modifications with high coverage, offer capabilities for qualitative and even quantitative analysis, and enable high-throughput protein identification. For the therapeutical product platelet concentrate, the identification of mechanisms leading to PSL and avoidance of these alterations is of major clinical importance. Proteomics allows comprehensive investigations of the PSL, but the different proteomic technologies have to be used in a complementary manner. Thus, gel-based and gel-free approaches should both be applied in order to fully exploit the advantages of state of the art proteomic technologies.

Although proteomics is currently the most promising tool for comprehensive quality assessment of the production process of PCs, it provides only limited information on the functional activity of proteins. Therefore, proteomics and functional assays have to complement one another.

Despite all the potentials of proteomic technologies for transfusion medicine, their limited capability of resolving conformational changes of proteins is a major limitation. These conformational changes can cause major alterations in function and, especially if they occur in membrane proteins, may induce immune responses in the PCs recipient.

Platelet proteomics has already reached an advanced stage and will provide a much better understanding of platelets in the future. Transfusion medicine began to adopt these techniques to enable researchers to better understand and improve production of platelets as a therapeutic for the bleeding patient. Further validation of PCs should include subproteomes such as the platelet membrane proteome, the phosphoproteome, and organellar subproteomes to obtain much deeper insight into the events that occur during storage of platelets for transfusion. Proteomic techniques will probably prove to be instrumental in validating and optimizing new methods of processing and storage of PCs, especially with the new pathogen inactivation procedures.

## REFERENCES

1. Weyrich AS, Lindemann S, Tolley ND, Kraiss LW, Dixon DA, Mahoney TM, Prescott SP, McIntyre TM, Zimmerman GA, Change in protein phenotype without a nucleus: Translational control in platelets, *Semin. Thromb. Hemost.* **30**(4):491–498 (2004).
2. Zimmerman GA, Weyrich AS, Signal-dependent protein synthesis by activated platelets: New pathways to altered phenotype and function, *Arterioscler. Thromb. Vasc. Biol.* **28**(3): s17–s24 (2008).
3. Greinacher A, Warkentin TE, Transfusion medicine in the era of genomics and proteomics, *Transfus. Med. Rev.* **19**(4):288–294 (2005).
4. Hirsch EO, Favre-Gilly J, Dameshek W, Thrombopathic thrombocytopenia; successful transfusion of blood platelets, *Blood* **5**(6):568–580 (1950).
5. Kaufman RM, Platelets: Testing, dosing and the storage lesion—recent advances, in *Hematology*, American Society of Hematology Education Program, 2006, pp. 492–496.
6. Klinger MH, The storage lesion of platelets: Ultrastructural and functional aspects, *Ann. Hematol.* **73**(3):103–112 (1996).
7. Blajchmany MA, Goldman M, Baeza F, Improving the bacteriological safety of platelet transfusions, *Transfus. Med. Rev.* **18**(1):11–24 (2004).
8. Aebersold R, Mann M, Mass spectrometry-based proteomics, *Nature* **422** (6928):198–207 (2003).
9. Thiele T, Steil L, Völker U, Greinacher A, Proteomics of blood-based therapeutics: A promising tool for quality assurance in transfusion medicine, *BioDrugs* **21**(3):179–193 (2007).
10. Murphy S, Platelets from pooled buffy coats: An update, *Transfusion* **45**(4):634–639 (2005).
11. Birnboim HC, Jevcak JJ, Fluorometric method for rapid detection of DNA strand breaks in human white blood cells produced by low doses of radiation, *Cancer Res.* **41**(5):1889–1892 (1981).
12. Holme S, Vaidja K, Murphy S, Platelet storage at 22 degrees C: Effect of type of agitation on morphology, viability, and function *in vitro*, *Blood* **52**(2):425–435 (1978).
13. Gulliksson H, Defining the optimal storage conditions for the long-term storage of platelets, *Transfus. Med. Rev.* **17**(3):209–215 (2003).
14. Ringwald J, Zimmermann R, Eckstein R, The new generation of platelet additive solution for storage at 22 degrees C: Development and current experience, *Transfus. Med. Rev.* **20**(2):158–164 (2006).

15. Lin L, Cook DN, Wiesehahn GP, Alfonso R, Behrman B, Cimino GD, Corten L, Damonte PB, Dikeman R, Dupuis K, Fang YM, Hanson CV, Hearst JE, Lin CY, Londe HF, Metchette K, Nerio AT, Pu JT, Reames AA, Rheinschmidt M, Tessman J, Isaacs ST, Wollowitz S, Corash L, Photochemical inactivation of viruses and bacteria in platelet concentrates by use of a novel psoralen and long-wavelength ultraviolet light, *Transfusion* **37**(4):423–435 (1997).
16. Goodrich RP, The use of riboflavin for the inactivation of pathogens in blood products, *Vox Sang*. **78**(Suppl. 2): 211–215 (2000).
17. Klein HG, Pathogen inactivation technology: Cleansing the blood supply, *J. Intern. Med*. **257**(3):224–237 (2005).
18. Pelletier JP, Transue S, Snyder EL, Pathogen inactivation techniques, *Best. Pract. Res. Clin. Haematol*. **19**(1):205–242 (2006).
19. Bertino AM, Qi XQ, Li J, Xia Y, Kuter DJ, Apoptotic markers are increased in platelets stored at 37 degrees C, *Transfusion* **43**(7):857–866 (2003).
20. Shcherbina A, Remold-O'Donnell E, Role of caspase in a subset of human platelet activation responses, *Blood* **93**(12):4222–4231 (1999).
21. Shcherbina A, Kenney DM, Bretscher A, Remold ODE, Dynamic association of moesin with the membrane skeleton of thrombin-activated platelets, *Blood* **93**(6):2128–2129 (1999).
22. Apelseth TO, Bruserud O, Wentzel-Larsen T, Bakken AM, Bjorsvik S, Hervig T, *In vitro* evaluation of metabolic changes and residual platelet responsiveness in photochemical treated and gamma-irradiated single-donor platelet concentrates during long-term storage, *Transfusion* **47**(4):653–665 (2007).
23. AuBuchon JP, Herschel L, Roger J, Murphy S, Preliminary validation of a new standard of efficacy for stored platelets, *Transfusion* **44**(1):36–41 (2004).
24. Murphy S, What's so bad about old platelets? *Transfusion* **42**(7):809–811 (2002).
25. Dumont LJ, AuBuchon JP, Whitley P, Herschel LH, Johnson A, McNeil D, Sawyer S, Roger JC, Seven-day storage of single-donor platelets: Recovery and survival in an autologous transfusion study, *Transfusion* **42**(7):847–854 (2002).
26. Kunicki TJ, Tuccelli M, Becker GA, Aster RH, A study of variables affecting the quality of platelets stored at "room temperature," *Transfusion* **15**(5):414–421 (1975).
27. Cardigan R, Turner C, Harrison P, Current methods of assessing platelet function: Relevance to transfusion medicine, *Vox Sang*. **88**(3):153–163 (2005).
28. Thon JN, Schubert P, Duguay M, Serrano K, Lin S, Kast J, Devine DV, Comprehensive proteomic analysis of protein changes during platelet storage requires complementary proteomic approaches, *Transfusion* **48**(3):425–435 (2008).
29. Beynon RJ, Doherty MK, Pratt JM, Gaskell SJ, Multiplexed absolute quantification in proteomics using artificial QCAT proteins of concatenated signature peptides, *Nat. Meth*. **2**(8):587–589 (2005).
30. Gygi SP, Rist B, Gerber SA, Turecek F, Gelb MH, Aebersold R, Quantitative analysis of complex protein mixtures using isotope-coded affinity tags, *Nat. Biotechnol*. **17**(10):994–999 (1999).
31. Bronstrup M, Absolute quantification strategies in proteomics based on mass spectrometry, *Expert Rev. Proteom*. **1**(4):503–512 (2004).
32. Ho Y, Gruhler A, Heilbut A, Bader GD, Moore L, Adams SL, Millar A, Taylor P, Bennett K, Boutilier K, Yang L, Wolting C, Donaldson I, Schandorff S,

Shewnarane J, Vo M, Taggart J, Goudreault M, Muskat B, Alfarano C, Dewar D, Lin Z, Michalickova K, Willems AR, Sassi H, Nielsen PA, Rasmussen KJ, Andersen JR, Johansen LE, Hansen LH, Jespersen H, Podtelejnikov A, Nielsen E, Crawford J, Poulsen V, Sorensen BD, Matthiesen J, Hendrickson RC, Gleeson F, Pawson T, Moran MF, Durocher D, Mann M, Hogue CW, Figeys D, Tyers M, Systematic identification of protein complexes in Saccharomyces cerevisiae by mass spectrometry, *Nature* **415**(6868):180–183 (2002).

33. Clemetson KJ, Capitanio A, Luscher EF, High resolution two-dimensional gel electrophoresis of the proteins and glycoproteins of human blood platelets and platelet membranes, *Biochim. Biophys. Acta* **553**(1):11–24 (1979).
34. Snyder EL, Dunn BE, Giometti CS, Napychank PA, Tandon NN, Ferri PM, Hofmann JP, Protein changes occurring during storage of platelet concentrates. A two-dimensional gel electrophoretic analysis, *Transfusion* **27**(4):335–341 (1987).
35. Snyder EL, Horne WC, Napychank P, Heinemann FS, Dunn B, Calcium-dependent proteolysis of actin during storage of platelet concentrates, *Blood* **73**(5):1380–1385 (1989).
36. Estebanell E, Diaz-Ricart M, Lozano M, Mazzara R, Escolar G, Ordinas A, Cytoskeletal reorganization after preparation of platelet concentrates, using the buffy coat method, and during their storage, *Haematologica* **83**(2):112–117 (1998).
37. Estebanell E, Diaz-Ricart M, Escolar G, Lozano M, Mazzara R, Ordinas A, Alterations in cytoskeletal organization and tyrosine phosphorylation in platelet concentrates prepared by the buffy coat method, *Transfusion* **40**(5):535–542 (2000).
38. García A, Watson SP, Dwek RA, Zitzmann N, Applying proteomics technology to platelet research, *Mass Spectrom. Rev*. **24**(6):918–930 (2005).
39. Macaulay IC, Carr P, Gusnanto A, Ouwehand WH, Fitzgerald D, Watkins NA, Platelet genomics and proteomics in human health and disease, *J. Clin. Invest*. **115**(12):3370–3377 (2005).
40. Maguire PB, Moran N, Cagney G, Fitzgerald DJ, Application of proteomics to the study of platelet regulatory mechanisms, *Trends Cardiovasc. Med*. **14**(6):207–220 (2004).
41. McRedmond JP, Park SD, Reilly DF, Coppinger JA, Maguire PB, Shields DC, Fitzgerald DJ, Integration of proteomics and genomics in platelets: A profile of platelet proteins and platelet-specific genes, *Mol. Cell. Proteom*. **3**(2):133–144 (2004).
42. O'Neill EE, Brock CJ, von Kriegsheim AF, Pearce AC, Dwek RA, Watson SP, Hebestreit HF, Towards complete analysis of the platelet proteome, *Proteomics* **2**(3):288–305 (2002).
43. García A, Prabhakar S, Brock CJ, Pearce AC, Dwek RA, Watson SP, Hebestreit HF, Zitzmann N, Extensive analysis of the human platelet proteome by two-dimensional gel electrophoresis and mass spectrometry, *Proteomics* **4**(3):656–668 (2004).
44. Martens L, Van Damme P, Van Damme J, Staes A, Timmerman E, Ghesquiere B, Thomas GR, Vandekerckhove J, Gevaert K, The human platelet proteome mapped by peptide-centric proteomics: A functional protein profile, *Proteomics* **5**(12):3193–3204 (2005).
45. Staes A, Demol H, Van Damme J, Martens L, Vandekerckhove J, Gevaert K, Global differential non-gel proteomics by quantitative and stable labeling of tryptic peptides with oxygen-18, *J. Proteome Res*. **3**(4):786–791 (2004).

46. Gravel P, Sanchez JC, Walzer C, Golaz O, Hochstrasser DF, Balant LP, Hughes GJ, García-Sevilla J, Guimon J, Human blood platelet protein map established by two-dimensional polyacrylamide gel electrophoresis, *Electrophoresis* **16**(7):1152–1159 (1995).
47. Immler D, Gremm D, Kirsch D, Spengler B, Presek P, Meyer HE. Identification of phosphorylated proteins from thrombin-activated human platelets isolated by two-dimensional gel electrophoresis by electrospray ionization-tandem mass spectrometry (ESI-MS/MS) and liquid chromatography-electrospray ionization-mass spectrometry (LC-ESI-MS), *Electrophoresis* **19**(6):1015–1023 (1998).
48. Marcus K, Immler D, Sternberger J, Meyer HE, Identification of platelet proteins separated by two-dimensional gel electrophoresis and analyzed by matrix assisted laser desorption/ionization-time of flight-mass spectrometry and detection of tyrosine-phosphorylated proteins, *Electrophoresis* **21**(13):2622–2636 (2000).
49. Maguire PB, Wynne KJ, Harney DF, O'Donoghue NM, Stephens G, Fitzgerald DJ, Identification of the phosphotyrosine proteome from thrombin activated platelets, *Proteomics* **2**(6):642–648 (2002).
50. Marcus K, Moebius J, Meyer HE, Differential analysis of phosphorylated proteins in resting and thrombin-stimulated human platelets, *Anal. Bioanal. Chem.* **376**(7):973–993 (2003).
51. García A, Prabhakar S, Hughan S, Anderson TW, Brock CJ, Pearce AC, Dwek RA, Watson SP, Hebestreit HF, Zitzmann N, Differential proteome analysis of TRAP-activated platelets: Involvement of DOK-2 and phosphorylation of RGS proteins, *Blood* **103**(6):2088–2095 (2004).
52. García A, Senis YA, Antrobus R, Hughes CE, Dwek RA, Watson SP, Zitzmann N, A global proteomics approach identifies novel phosphorylated signaling proteins in GPVI-activated platelets: Involvement of G6f, a novel platelet Grb2-binding membrane adapter, *Proteomics* **6**(19):5332–5343 (2006).
53. Lewandrowski U, Moebius J, Walter U, Sickmann A, Elucidation of N-glycosylation sites on human platelet proteins: A glycoproteomic approach, *Mol. Cell. Proteom.* **5**(2):226–233 (2006).
54. Moebius J, Zahedi RP, Lewandrowski U, Berger C, Walter U, Sickmann A, The human platelet membrane proteome reveals several new potential membrane proteins, *Mol. Cell. Proteom.* **4**(11):1754–1761 (2005).
55. García A, Zitzmann N, Watson SP, Analyzing the platelet proteome, *Semin. Thromb. Hemost.* **30**(4):485–489 (2004).
56. Coppinger JA, Cagney G, Toomey S, Kislinger T, Belton O, McRedmond JP, Cahill DJ, Emili A, Fitzgerald DJ, Maguire PB, Characterization of the proteins released from activated platelets leads to localization of novel platelet proteins in human atherosclerotic lesions, *Blood* **103**(6):2096–2104 (2004).
57. García BA, Smalley DM, Cho H, Shabanowitz J, Ley K, Hunt DF, The platelet microparticle proteome, *J. Proteome Res.* **4**(5):1516–1521 (2005).
58. Dittrich M, Birschmann I, Stuhlfelder C, Sickmann A, Herterich S, Nieswandt B, Walter U, Dandekar T, Understanding platelets. Lessons from proteomics, genomics and promises from network analysis, *Thromb. Haemost.* **94**(5):916–925 (2005).
59. Thiele T, Steil L, Gebhard S, Scharf C, Hammer E, Brigulla M, Lubenow N, Clemetson KJ, Völker U, Greinacher A, Profiling of alterations in platelet proteins during storage of platelet concentrates, *Transfusion* **47**(7):1221–1233 (2002).

60. Glenister KM, Payne KA, Sparrow RL, Proteomic analysis of supernatant from pooled buffy-coat platelet concentrates throughout 7-day storage, *Transfusion* **48**(1):99–107 (2008).
61. Thon JN, Schubert P, Duguay M, Serrano K, Lin S, Kast J, Devine DV, Comprehensive proteomic analysis of protein changes during platelet storage requires complementary proteomic approaches, *Transfusion* **48**(3):425–435 (2008).
62. Wurtz V, Hechler B, Ohlmann P, Isola H, Schaeffer-Reiss C, Cazenave JP, Van Dorsselaer A, Gachet C, Identification of platelet factor 4 and beta-thromboglobulin by profiling and liquid chromatography tandem mass spectrometry of supernatant peptides in stored apheresis and buffy-coat platelet concentrates, *Transfusion* **47**(6):1099–1101 (2007).
63. Springer DL, Miller JH, Spinelli SL, Pasa-Tolic L, Purvine SO, Daly DS, Zangar RC, Jin S, Blumberg N, Francis CW, Taubman MB, Casey AE, Wittlin SD, Phipps RP, Platelet proteome changes associated with diabetes and during platelet storage for transfusion, *J. Proteome Res*. **8**(5):2261–2272 (2009).
64. Mohr H, Steil L, Gravemann U, Thiele T, Hammer E, Greinacher A, Muller TH, Völker U, A novel approach to pathogen reduction in platelet concentrates using short-wave ultraviolet light, *Transfusion* **49**(12):2612–2624 (2009).
65. Forster ML, Sivick K, Park YN, Arvan P, Lencer WI, Tsai B, Protein disulfide isomerase-like proteins play opposing roles during retrotranslocation, *J. Cell. Biol*. **173**(6):853–859 (2006).
66. Seghatchian J, Krailadsiri P, Platelet storage lesion and apoptosis: Are they related? *Transfus. Apher. Sci*. **24**(1):103–105 (2001).
67. Hall PA, Jung K, Hillan KJ, Russell SE, Expression profiling the human septin gene family, *J. Pathol*. **206**(3):269–278 (2005).
68. Li J, Xia Y, Bertino AM, Coburn JP, Kuter DJ, The mechanism of apoptosis in human platelets during storage, *Transfusion* **40**(11):1320–1329 (2000).
69. Thon JN, Schubert P, Devine DV, Platelet storage lesion: A new understanding from a proteomic perspective, *Transfus. Med. Rev*. **22**(4):268–279 (2008).
70. Vetlesen A, Mirlashari MR, Ezligini F, Kjeldsen-Kragh J, Evaluation of platelet activation and cytokine release during storage of platelet concentrates processed from buffy coats either manually or by the automated OrbiSac system, *Transfusion* **47**(1):126–132 (2007).
71. Thon JN, Devine DV, Translation of glycoprotein IIIa in stored blood platelets, *Transfusion* **47**(12):2260–2270 (2007).
72. Schwertz H, Tolley ND, Foulks JM, Denis MM, Risenmay BW, Buerke M, Tilley RE, Rondina MT, Harris EM, Kraiss LW, Mackman N, Zimmerman GA, Weyrich AS, Signal-dependent splicing of tissue factor pre-mRNA modulates the thrombogenicity of human platelets, *J. Exp. Med*. **203**(11):2433–2440 (2006).
73. Weyrich AS, Schwertz H, Kraiss LW, Zimmerman GA, Protein synthesis by platelets: Historical and new perspectives, *J. Thromb. Haemost*. **7**(2):241–246 (2009).
74. Hoffmeister KM, Felbinger TW, Falet H, Denis CV, Bergmeier W, Mayadas TN, von Andrian UH, Wagner DD, Stossel TP, Hartwig JH, The clearance mechanism of chilled blood platelets, *Cell* **112**(1):87–97 (2003).
75. Hoffmeister KM, Josefsson EC, Isaac NA, Clausen H, Hartwig JH, Stossel TP, Glycosylation restores survival of chilled blood platelets, *Science* **301**(5639):1531–1534 (2003).

76. Wandall HH, Hoffmeister KM, Sorensen AL, Rumjantseva V, Clausen H, Hartwig JH, Slichter SJ, Galactosylation does not prevent the rapid clearance of long-term, 4 degrees C-stored platelets, *Blood* **111**(6):3249–3256 (2008).
77. Hornsey VS, Drummond O, McMillan L, Morrison A, Morrison L, MacGregor IR, Prowse CV, Cold storage of pooled, buffy-coat-derived, leucoreduced platelets in plasma, *Vox Sang*. **95**(1):26–32 (2008).
78. Bergmeier W, Burger PC, Piffath CL, Hoffmeister KM, Hartwig JH, Nieswandt B, Wagner DD, Metalloproteinase inhibitors improve the recovery and hemostatic function of *in vitro*-aged or -injured mouse platelets, *Blood* **102**(12):4229–4235 (2003).

# 14

# CARDIOVASCULAR PROTEOMICS

FERNANDO VIVANCO, FERNANDO DE LA CUESTA,
MARÍA G. BARDERAS, IRENE ZUBIRI, AND GLORIA ÁLVAREZ-LLAMAS

**Abstract**

Cardiovascular diseases—where platelets play a central role—encompass a group of diseases, including hypertension, coronary heart disease, stroke, and heart failure, and remain the most common cause of death in the developed world. Cardiovascular proteomics focuses on the identification and characterization of the role of proteins in the cardiovascular system in physiological and pathological states in order to identify and outline both mechanisms and markers of disease. In this chapter we summarize the current status of different proteomic approaches and the most recent findings reported in cardiovascular proteomics. Proteomics holds tremendous promise for the diagnosis, prognosis, and treatment of cardiovascular diseases.

## 14.1 INTRODUCTION

Cardiovascular disease (CVD) encompasses a group of diseases, including hypertension, coronary heart disease, stroke, congenital heart defects, and heart failure. Traditionally, CVDs have relied on epidemiological associations to identify diagnostic and prognostic risk factors, such as hypercholesterolemia and hypertension, in diverse populations. At present, the identification and validation of biomarkers of CVD are based on proteomic technologies, which associate the disease

*Platelet Proteomics: Principles, Analysis and Applications*, First Edition.
Edited by Ángel García and Yotis A. Senis.

phenotype with individual proteins or protein profiles [1–3]. One of the main problems in clinical practice is that symptoms can occur in the later phases of a disease; thus, biomarkers for early diagnosis, prognosis or evaluation of patient recovery are still needed. In general terms, a biomarker is "a characteristic that is objectively measured and evaluated as an indicator of normal biological processes, pathogenic processes or pharmacological responses to a therapeutic intervention" [4]. In the search for biomarkers, two complementary strategies can be applied: deductive (knowledge-based) and inductive (unbiased). In a deductive approach, the biological processes underlying atherosclerosis must be understood beforehand, and preselected molecules are investigated in order to discover their potential use as biomarkers. In an *inductive* approach, study of multiple molecules is carried out at the same time to characterize the biomolecular profile of a disease at a given stage. In cardiovascular pathology, individual biomarkers typically reflect existing cardiac injury (e.g., troponins and creatin kinase), wall stress (e.g., natriuretic peptides), or activation of coagulation or inflammatory pathways (e.g., C-reactive protein, plasminogen activator inhibitor, myeloperoxidase). However, the data generated by proteomic approaches are not specific for a unique molecule (protein), but rather encompass global changes in thousands of molecules simultaneously. Thus, such approaches generate profiles or datasets that reflect the general situation of the sample (cell, tissue, biopsy, serum, plasma, urine, etc.). In this situation, which has been termed *systems biology* [5], we deal with thousands of analytical data. Therefore, one of the benefits of approaching the task by means of proteomics lies in the potential to carry out unbiased analyses without preselection of putative candidate markers to obtain a global picture of interacting proteins and molecular mechanisms associated with a particular disease. The molecular effects of drugs employed in the treatment of disease can also be investigated; this specific proteomic modality is known as *pharmacoproteomics* [6].

When approaching a comprehensive biomarker research, several decisions must be adopted beforehand. Such choices may be crucial to the final results (Fig. 14.1): (1) starting *sample material* (human or animal models; serum, plasma, urine, cerebrospinal fluid (CSF), tissue/cells secretome, complete tissue/cells extracts, isolated tissue sections/cells/microparticles), which is strongly dependent on sample availability and may imply direct collaboration between laboratory and clinics; (2) *starting hypotheses*, which may focus on particular molecules or proteins, a certain subproteome, or the whole proteome; (3) robust *analytical techniques* and developed methodologies able to provide high reproducibility and sufficient sensitivity to detect low-abundance proteins in highly complex samples; and (4) *validation strategies* by methodologies alternative to those used in the discovery phase applied to a wide cohort of patients and healthy subjects. Several published reviews include the latest proteomics studies in CVDs in general and atherosclerosis in particular [1–3,6–15].

Platelets play a central role in cardiovascular diseases and other pathologies (cancer, stroke) and thus constitute the main subject of this book. Platelets have a

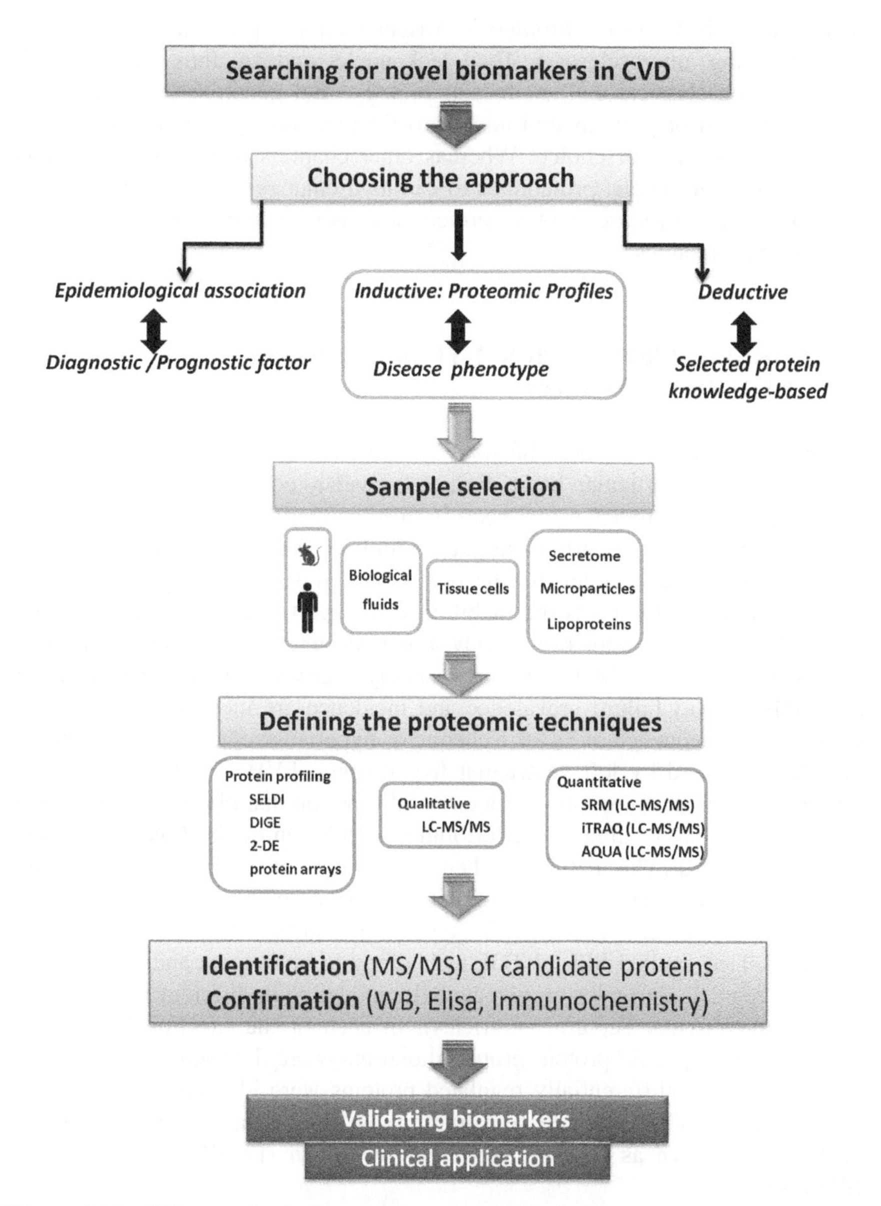


**Figure 14.1** Different alternatives that must be chosen in the search for novel biomarkers in CVD. In a proteomic approach, an inductive strategy (unbiased) is selected and applied to different types of samples using several techniques. If the identification process is successful, a confirmatory step by an independent technique is mandatory. A final validation step must be performed before using the novel biomarker in clinical practice.

pivotal role in hemostasis, thrombosis, vascular repair, and inflammatory reactions, including atherosclerosis. They lack nuclei and are thus not amenable to most of the classical cell, molecular biology, and genomic techniques. For the identification of proteins and novel protein functions, proteomic studies are therefore the method of choice. Whereas other chapters of the book highlight the application of platelet proteomics to cardiovascular research (see Chapters 1 and 4), this chapter focuses on how proteomics has been applied to other aspects of vascular proteomics.

## 14.2 CARDIAC PROTEOMICS: HYPERTENSION AND HEART DYSFUNCTION

Proteomic analysis of diseased hearts may allow for the identification of markers specific for a particular heart disease (valvular, congenital) of interest. In addition, proteomic studies of the heart in different pathological states will yield important information regarding cardiac physiology that may be useful for finding specific therapeutic targets to improve the prognosis of these patients [1–3]. Functionally, heart disease is the inability of the heart to pump sufficient blood to meet the metabolic needs of the body. It may occur suddenly, as with an acute myocardial infarction (AMI), or progress slowly over years, as with chronic heart failure (HF). Heart failure prevalence and incidence are increasing worldwide, mainly as a consequence of either ischemic or hypertensive heart disease (HDD) [16]. HDD, defined by left ventricular hypertrophy (LVH), is characterized by complex changes in myocardial structure and function that elicit the remodeling of the myocardium, leading to a deterioration of left ventricular function and the development of HF. Circulating biochemical markers of myocardial remodeling in HDD have been reviewed [17] and include cardiotrophin 1, annexin $A_5$, the *C* terminus of the propeptide of procollagen type 1, and the MMP1 : TIMP1 ratio. Melle et al. [18] applied SELDI-TOF technology to investigate and identify differentially regulated proteins on myocardial remodeling in different heart regions (atria, interventricular septum, ventricles). In each of the functionally distinct cardiac regions, specific protein profile alterations were detected on myocardial remodeling. Three differentially regulated proteins were identified: triose phosphate isomerase (TIM), the cell signaling protein Raf1 kinase inhibitory protein (RKIP) [also known as *ethanolamine binding protein* (PEBP)], and the small HSP αβ-crystallin.

Left ventricular hypertrophy is considered a reversible state in most instances, provided that blood pressure is kept under strict control [6]. It is frequently assumed that pharmacological regression of LVH normalizes myocardium physiology. In this regard, the effects of different antihypertensive treatments (ACEI, quinapril, losartan, and doxazosin plus quinapril) on proteins expressed in the left ventricle in a model of hypertension-induced LVH were studied in a spontaneous hypertensive rat (SHR) model [19–21]. Two-dimensional gel electrophoresis (2DGE) analysis showed that 36 proteins were differentially expressed in hearts from SHRs compared with normotensive Wistar–Kyoto (WKY) rats. Antihypertensive treatment normalized 15, 14, and 13 proteins after quinapril, doxazosin + quinapril, and losartan treatments, respectively. Thus, around 50% of the altered proteins were not normalized after LVH regression [19]. These results are in agreement with those obtained by Jin et al. [22], who studied the protein profile of the left ventricular myocardium in a prehypertensive and hypertensive stage in the same SHR model and the effect of treatment with losartan and enalapril. The proteins that are not affected with drug treatment may be potential candidates for the development of new antihypertensive drugs designed to prevent LVH [6]. More recently, Jülling et al. [23] performed a more focused approach and compared the cardiac mitochondrial proteome of 20-month-old SHRs and WKY controls by iTRAQ and LC-MS/MS. Many of the altered proteins corresponded with proteins identified in previous studies [19–22] that were, in fact, of mitochondrial origin, confirming the data.

Left ventricular remodeling (LVR) after AMI is a dynamic and complex process that occurs in response to myocardial damage. Pinet et al. [24,25] searched for novel biomarkers of LVR in both heart tissue and plasma. They observed that post-AMI LVR is associated with the modulation of numerous HSPs, together with proteins involved in cellular protection against oxidative stress. Interestingly, the expression of four proteins (glyceraldehyde-3-phosphate dehydrogenase, $\alpha\beta$-crystalline, peroxiredoxin 2, and isocytrate dehydrogenase) was linked to echographic parameters according to HF severity. For plasma analysis they found posttranslational modification (PTM) variants of $\alpha_1$-chain haptoglobin (Hp$\alpha$1), which was elevated in remodeling patients [25].

*Urea transporters* (UTs) are a family of small integral membrane proteins that are specifically permeable to urea. There are two families of UTs (the renal type, UT-A, and the vascular type, UT-B). Yu et al. [26] used 2DGE to study the protein expression profiles of heart tissue (constituting most conduction systems) in wild-type versus UT-B-null mice at different ages and found more than a dozen proteins altered in the myocardium of UT-B-null mice. Adaptation of the rat cardiac proteome in response to intensity-controlled endurance exercise was studied by Burniston [27]. This rodent running model replicates endurance exercises in humans and can be used to investigate molecular adaptations in the heart. Over a 6-week regime, 23 proteins were differentially expressed between exercised and control hearts. Clements et al. [28] studied changes in the myocardial protein profiles of patients undergoing cardiac surgery by applying CP/CPB (cardioplegic and cardiopulmonary bypass) in human atrial samples (pre- and post-CP/CPB). Cardiac surgery resulted in multiple changes in the human myocardial profile. CP/CPB specifically modified cytoskeletal, metabolic, and inflammatory proteins, including myosin light chain 2, glutathione-*S*-transferase $\gamma$, enoyl-CoA hydratase 1, S100A8, HSP27, superoxide dismutase, and $\alpha_1$-1 acid glycoprotein 1.

Heatstroke is a life-threatening illness characterized by multiple organ dysfunction (including cardiac dysfunction), hyperpyrexia, and central nervous system disorders [29]. Arterial hypotension may result from decreased cardiac output during heatstroke. Whole-body cooling (WBC) is the current therapy of choice for heatstroke because no pharmacologic agents are available. Cheng et al. [30] have identified the cardiac proteins that are involved in heatstroke-induced cardiac dysfunction in a rat model. Most of these proteins exhibited decreased expression levels and were associated with mitochondrial function and cytoskeletal integration.

A topic of notorious importance is the identification of altered PTMs of cardiac myofilament proteins and their relation to heart diseases, including phosphorylation, oxidation, partial proteolysis and degradation, glycosylation, and nitrosylation.

Whether these PTMs are directly correlated with cardiac pathology is a subject of active investigation [31,32]. PTMs of cardiac myofilament proteins have been described for troponins (TnI, TnT), tropomyosin (Tm), actin, desmin, nebulette (an isoform of nebulin expressed specifically in cardiac muscle), myosin light chains 1 and 2, and titin. Phosphorylation of cardiac myofilament proteins represents one of the main posttranslational mechanisms that regulate cardiac pump function. For example, TnI is phosphorylated by several kinases (PKA, PKC, PAK, and PKD), and several studies have demonstrated decreased TnI phosphorylation in HF [33–35]. Likewise, the partial proteolysis of TnI at the $N$- and $C$-terminal regions plays an important role in myocardial stunning [36]. Similar PTMs occur in TnT associated with HF. Other PTMs, particularly the oxidation and nitrosylation of cardiac myofilament proteins, play critical roles in signal reception and transduction in working myocytes [37,38].

Hypertension is a major risk factor for numerous cardiovascular diseases, including stroke, MI, HF, and end-stage renal disease. This process is associated with increased arterial wall thickening due mostly to hypertrophy, proliferation, migration, apoptosis of vascular smooth muscle cells (VSMCs), and abnormal accumulation of extracellular matrix proteins. Thus, differentially expressed proteins in aortas from 18-week-old SHRs and their normotensive counterpart, Wistar–Kyoto rats (WKY), were examined by 2DGE [39]. Among the 50 differentially expressed proteins identified, RhoGDIα [Rho (ρ) GDP dissociation inhibitor α], was prominently up-regulated. Thus, during hypertension, a high level of RhoGDIα may maintain RhoGTPases in their active state, which may result in abnormal VSMC proliferation, apoptosis, and migration [39].

Although there is no need for biomarkers to assess blood pressure, markers of vascular remodeling in response to high blood pressure may be of interest to evaluate arterial damage before the onset of clinical complications. To explore the presence of such potential biomarkers, Delbosc et al. [40] used surface-enhanced laser desorption/ionization (SELDI)-TOF to analyze the proteins released by the aortas of two rat strains with different susceptibilities to hypertension [Fischer and Brown Norway (BN) rats]. Four differentially expressed $m/z$ peaks were identified corresponding to ubiquitin, smooth muscle (SM) 22α, thymosin β4, and the $C$-terminal fragment of filamin A. Interestingly, similar variations in SM22α were observed in plasma, suggesting that this marker could be used to assess vascular damage induced by hypertension. In addition, losartan, an angiotensin II–type 1 receptor (AT1) antagonist commonly used to treat hypertension in humans, inhibited the secretion of these four biomarkers. Thus, SM22α could represent a marker of susceptibility to hypertension-induced arterial wall remodeling. In human aortas, Liao et al. [41] studied the aortic media after thoracic aortic dissection. They found decreased expression of extracellular superoxide dismutase among other proteins. Thus, an impaired antioxidative mechanism is present in the aortic wall in patients with thoracic aortic dissection.

An emerging common hallmark of many cardiovascular conditions, including hypertension, is mitochondrial dysfunction [42–44] (see discussion of ischemia and reperfusion in the next paragraph). Because the quantity of some respiratory complex subunits is decreased in hypertension, Lopez-Campistrous et al. [45] have analyzed the integrity of these complexes in the mitochondria of 12-week-old hypertensive rat brains. They found that respiratory complexes (I, III, IV, and V) exhibit assembly defects and that proteins involved in central metabolic processes had decreased abundance. In a mouse model of hypoxia-induced pulmonary hypertension, a new key protein, Fhl1, was identified as being involved in pulmonary hypertension (24 h) [46]. In cardiac hypertrophic and dilated cardiomyopathy mouse models, cardiac ventricular expression of Fhl1 (but not Fhl2 or Fhl3) is altered. The new data revealed Fhl1 as a novel protein involved in the proliferation and migration of human primary pulmonary artery SMCs (PASMCs) and confirmed the upregulation of this protein in human idiopathic pulmonary arterial hypertension (IPAH). Nath et al. [47] performed an elegant study including a proteomics-based detection screen of a protein cluster that is dysregulated during cardiovascular development. Using an LC-MS/MS proteomic approach on embryonic vascular tissue, the authors identified proteins and pathways that are disrupted during cardiovascular development. Several of these protein targets (WNT16, PCSK1, and ST14) provide the basis for a novel prenatal screen for the detection of congenital heart diseases (CHD), suggesting that dysregulation of WNT16, ST14, and PCSK1 may play a role in the etiology of human CHD.

Many studies focus on ischemia, which is one of the later effects of atherosclerotic development and induces different grades of necrosis and, in many cases, death. Even though restoration of the flow may overcome ischemic damage, reperfusion also leads to contractile dysfunction and cellular damage [48]. Effects of ischemia/reperfusion (I/R) injury on irrigated tissues can be found after MI or stroke events and also as a consequence of arterial revascularization surgery [49]. Ischemia without reperfusion (IWR) and I/R injury in *ex vivo* and *in vivo* animal models (mainly rodents) have been used frequently more recently for tissue proteomic analysis because they are very reproducible and easily implemented on a lab workflow [50]. The effect of IWR and I/R duration on the heart proteome in an *ex vivo* rat model was evaluated by Fert-Bober et al. [51]. Changes in protein levels after differential expression 2DGE analysis correlated with the duration of ischemia when this event was studied without a subsequent reperfusion step. One of the most investigated topics in heart ischemia injury is cardioprotection derived from ischemic pre- or postconditioning. Activation of two PKC isozymes, PKCε and PKCδ, has been associated with preconditioning cardioprotection. Mitochondrial and cytosolic fractions from the proteome of transgenic mouse hearts with constitutively active or dominant negative PKCε have been compared by DIGE [52]. Several cytosolic enzymes related to glucose and energy metabolism were found to be affected by PKCε activation. In addition, PKCε inhibition produced phosphorylation and mitochondrial translocation of PKCδ, thereby clarifying the synergistic effect of both PKC isoforms in cardioprotection.

Phosphorylation events in the mitochondrial proteome during pre- and post-conditioning were evaluated by isoflurane treatment (which mimics ischemic conditioning) prior to or after ischemia induction in Wistar rats [53]. Mitochondrial protein complexes from heart lysates were analyzed by two-dimensional blue native gel electrophoresis (2D BN-PAGE), followed by WB with a phosphor-Ser/Thr/Tyr antibody. From differential spots, 11 proteins were identified and correlated with oxidative phosphorylation, energy metabolism, and chaperone and carrier functions. LC-MS/MS analysis of phosphopeptides after an enrichment protocol permitted the detection of 26 potential phosphorylation sites in 19 proteins. Among these, a novel phosphorylation site in adenine nucleotide translocator 1 (ANT1) at residue $Tyr^{194}$ was identified. A more recent study reported a reduction of ischemic heart damage mediated by mitochondrial aldehyde dehydrogenase 2 (ALDH2) activation in an *ex vivo* rat model [54]. ALDH2 phosphorylation correlated with cardioprotection, as a shift in the phosphorylation state of the enzyme was observed with ethanol and PKCε activation. More importantly, the usefulness of ALDH2 activation in patients subjected to cardiac ischemia was demonstrated; an activator of the enzyme, Alda-1, reduced infarct size by 60% after ischemia in studied hearts. Thus, these relevant data may form the basis for a potential therapeutic approach to prevent or minimize myocardial ischemic injury in the clinical setting.

An alternative method to identify proteins offering cardioprotection after ischemia is to generate transgenic mice expressing the candidate molecule in a controlled manner. This methodology has been used to assay whether phosphatase 1 inhibitor 1 (PP1-I1) could protect against I/R injury [55]. A group of proteins associated with energy production and protein synthesis were altered, possibly to support the increased metabolic demands induced by PP1-I1 in transgenic hearts. Table 14.1 lists the proteins involved in the different pathologies.

## 14.3 VASCULAR PROTEOMICS

### 14.3.1 Atherosclerosis: How to Approach Biomarker Discovery

*Atherosclerosis* is a condition in which the arterial wall thickens. It is implicated in 75% of all cardiovascular-related deaths (CVDs) in the United States [56], and obesity, hypertension, diabetes mellitus, physical inactivity, and smoking are the main contributors to the development of atherosclerotic CVD. The disease starts in childhood, and atherosclerosis can be found in 80–90% of young men and women in the United States by the age of 30. Although it is the most common cause of death in Western societies and its prevalence continues to

**TABLE 14.1 Compilation of Proteomic Studies Reported to Date in the Field of Heart Disease**

HEART

| Sample Source | Pathology | Methodology | Proteins Identified | Most Relevant | Validation Method | Reference |
|---|---|---|---|---|---|---|
| Rat aorta | Hypertension | 2-DE | 50 | Rho GDI α | | 39 |
| Rat aorta | Hypertension | SELDI-TOF-MS | 3 | Ubiquitin, SM22 α, Thymosin β4, C terminal-fragment filamin | | 40 |
| Mitochondria (rat brains) | Hypertension | 2-DE MS/MS | 4 | Respiratory complexes (I,III,IV, V) | | 45 |
| Mouse heart model | Pulmonary hypertension | 2-DE | 1 | Fh1−1 ↑ | | 46 |
| Rat heart | AMI | SELDI-TOF-MS | 3 | TIM, RKIP, HSP αβ crystallin | | 18 |
| Embrionic vascular tissue | Congenital heart disease | LC-MS/MS | 3 | WNT16, ST14, PCSK1 | | 47 |
| Rat heart | Ischemia/reperfusion | 2-DE | | ATP Synthase, β chain enolase, myosin, light chain 1 | | 51 |
| Mice heart | Ischemia | 2D-DIGE | 1 | PKC ε | | 52 |
| Rat heart | Ischemia | 2D-BN-PAGE | 11 | Oxidative phosphorylation proteins, energy metabolism, chaperone proteins | WB | 53 |
| Rat heart | Ischemia | 2-DE WB | | ALDH2, PKCε | | 54 |
| Mice transgenic heart | Ischemia | 2-DE-based differential abundance and phosphoproteomic analysis | | Phosphatase-1 inhibitor-1 (PP-1-I-1) (protect function) | | 51 |

↑: up-regulated; ↓: down-regulated

increase in underdeveloped countries, the mechanisms of its etiology are not fully understood. Atherosclerosis is linked to the accumulation of inflammatory cells in the artery wall. In brief, lesions begin with endothelial dysfunction and the recruitment of monocytes and macrophages, which take up modified low-density lipoprotein (LDL) and result in the formation of foam cells. These accumulate and produce fatty streaks beneath the endothelium. The release of cytokines and chemokines induces proliferation and migration from the media of VSMCs. In advanced lesions, necrosis of macrophages and VSMCs forms a lipid-rich core covered by a fibrous cap, which protects the lesions from rupture and consists mainly of collagens and extracellular matrix (ECM) proteins synthesized by vascular cells. The rupture of the plaque leads to thrombosis due to the interaction between the lipid core and tissue factors within the blood. The clinical consequences depend on the artery where the atherotrombosis takes place (i.e., coronary, carotid, aorta, femoral) and can result in myocardial infarction (MI) or unstable angina, stroke or transient cerebral ischemia, and intermittent claudication or gangrene, which jeopardizes limb viability.

***14.3.1.1 Biological Fluids, Circulating Cells, and Microparticles*** In terms of clinical practice, biological fluids are the most accessible sources for biomarker discovery. It is assumed that protein expression patterns measured across the serum proteome have potential diagnostic utility in cardiovascular and other diseases. For this reason, proteomic studies in atherosclerosis have traditionally focused on plasma or serum, while other proteomic approaches represent a smaller portion of current atherosclerosis research (i.e., tissue proteomics [12]). Figure 14.2 shows different proteomic approaches used to study atherosclerosis and includes some representative proteins identified with each approach.

The initial 3020-protein subset of total proteins identified by the HUPO Plasma Proteome Project pilot phase (PPP) were compared with the literature for relevance to cardiovascular function and disease, resulting in 345 implicated candidates that were divided into eight different categories (markers of inflammation and CVD, vascular and coagulation, signaling, growth and differentiation, cytoskeletal, transcription factors, channel and receptor proteins, and heart failure and remodeling) [57]. Biomarkers found in the plasma/serum have the great advantage of minimally invasive quantification for diagnosis because factors released from lesion areas can accumulate in the general circulation (i.e., initial development of atherome plaque or advance progress prior to rupture). However, at the same time, plasma/serum proteomic studies are challenging becuase of (1) protein complexity; (2) highly dynamic range of protein concentrations, comprising more than 10 orders of magnitude; and (3) difficulty in detecting low-abundant proteins (the 20 most abundance proteins represent 99.9% of total plasma proteins) [58]. Moreover, some potential plasma biomarkers may be nonspecific for atherosclerosis, specifically, when reflecting an inflammatory process.

Brea et al. [59] used 2DGE to analyze serum samples from ischemic stroke patients who had previously been depleted of 12 of the most abundant plasma proteins. The results showed altered haptoglobin and serum amyloid A (SAA)

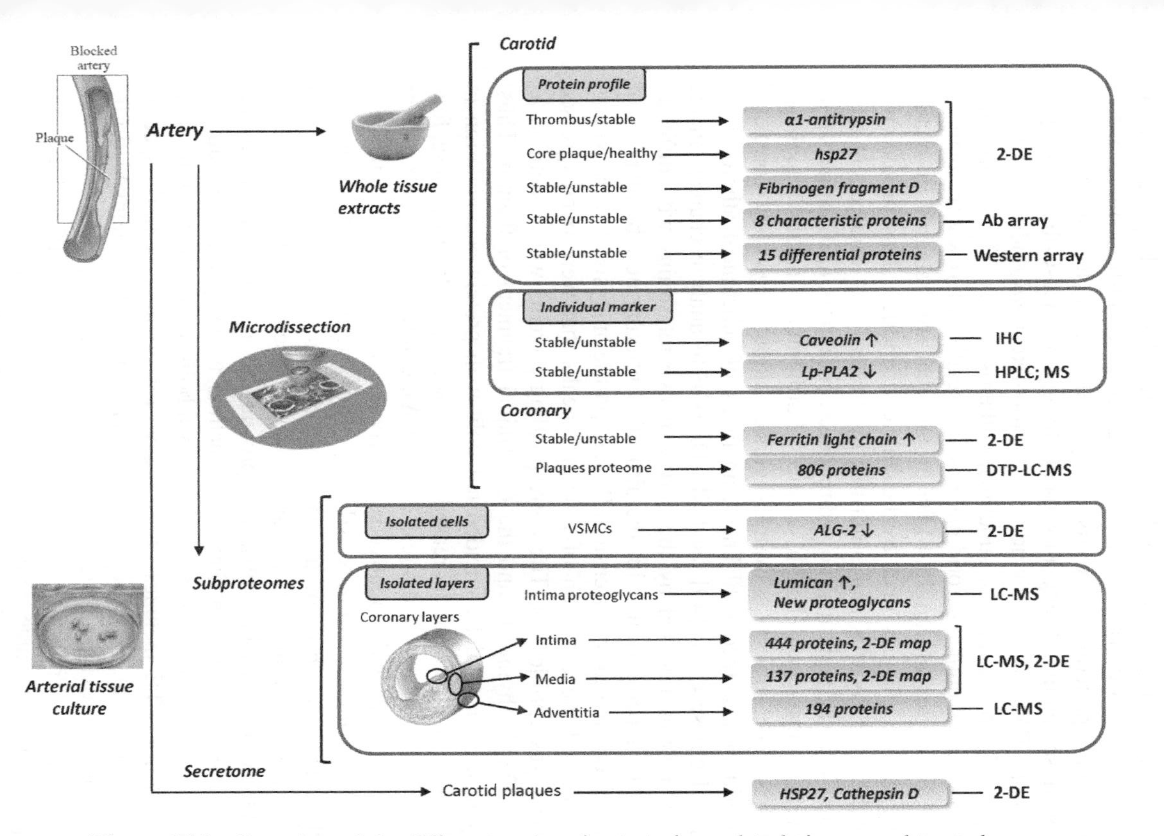


**Figure 14.2** Overview of the different proteomic strategies and techniques used to study vascular proteomes in health and disease.

expression depending on the operating atherotrombotic mechanism; higher serum levels of these proteins can predict atherotrombotic versus cardioembolic stroke [59]. SELDI-TOF-MS analysis indicated that $\beta_2$-microglobulin was elevated in the plasma of patients with peripheral arterial disease (PAD), and this elevation correlated with disease severity [60]. As initially described by Durán et al. [61], *ex vivo* incubation of vascular tissue for 24 h in serum-free culture media is compatible with proteomic analysis and provides valuable information on the vessel wall secretome. Using this methodology, Blanco-Colio et al. [62] identified soluble TWEAK (*t*umor necrosis factor–like *weak* inducer of apoptosis) as a potential biomarker for subclinical atherosclerosis. The supernatants obtained from cultured human carotid plaques and healthy arteries were analyzed by SELDI-TOF and reveald an 18.4-kDa protein that was released in smaller quantities by carotid plaques than by healthy end arteries. This protein was identified by MS as sTWEAK and confirmed by Western blot. Subsequent measurements of sTWEAK in plasma showed reduced concentrations in subjects with carotid stenosis compared with healthy controls. Furthermore, in a test population of 106 asymptomatic subjects, sTWEAK concentrations negatively correlated with the carotid intima–media thickness, an index of subclinical atherosclerosis. Similarly, Delbosc et al. [40] used SELDI-TOF to analyze changes in the abundance of proteins released by the aortas of two rat strains with different susceptibilities to hypertension, with the aim of identifying new biomarkers of arterial wall remodeling in hypertension. Protein profile analysis of secreted aortic proteins allowed the detection of four *m/z* peaks that were overexpressed in hypertension-susceptible rats. The corresponding proteins were identified as ubiquitin, SM22$\alpha$, $\beta_4$-thymosin, and the *C*-terminal fragment of filamin. Thus, hypertrophy in this rat model coincided not only with protein accumulation in the arterial wall but also with increased protein secretion. Indeed, these four proteins are linked to mechanical stress (filamin A), contractile protein turnover (SM224$\alpha$, $\beta_4$-thymosin), and oxidative stress. By the same approach, TnI forms were analyzed in patients suffering MI [63]. Thus, despite its shortcomings, SELDI-TOF analysis has become a very useful approach in the detection and identification of novel biomarkers in cardiovascular diseases. As an alternative to SELDI-TOF, which enriches subsets of proteins on a sample target surface, Ganesh et al. [64] have used a direct MS method to analyze serum samples in a cohort of patients with venous thromboembolism (VTE). Current noninvasive testing for VTE includes blood assays for D-dimer, a fibrin-specific degradation product that reflects endogenous fibrinolysis of crosslinked fibrin. Although this assay provides a high negative predictive value, it suffers from diminished specificity. Thus, the authors developed a direct MS and computational approach to determine whether protein expression profiles can predict diagnosis. The data were confirmed by gel electrophoresis of serum proteins. Proteins identified include actin, $\alpha_1$-B-glycoprotein, CD5 antigen-like, complement 4A and 9 proteins, haptoglobin, hemopexin, IgA heavy chain, leucine-rich $\alpha_2$-glycoprotein 1, myosin heavy chain, platelet coagulation factor XI, plasma kalikrein B1 precursor, and proapolipoprotein. Since these proteomic markers appear to add specificity to currently available blood assays, they could be used in combination with other methods, including radiographic and ultrasound evaluation, in order to obtain optimal sensitivity and specificity for the diagnosis of VTE. Thrombus formation is a general phenomenon that not only occurs in VTE but also mediates the development of the many of CVDs, including acute coronary syndromes and ischemic stroke. During thrombus formation, the coagulation cascade is activated, resulting in thrombin generation and fibrin formation. Many circulating cells (platelets, leukocytes, and erythrocytes) become trapped and release proteins and microparticles. Plasma and cellular proteins that contribute to thrombus formation have been analyzed by proteomic technologies and have been reviewed [65].

Liu et al. [66] have used a label-free proteomic method with LC-MS/MS (shotgun approach) to investigate the differences in protein profiles between nondiabetic ($n = 5$) and diabetic ($n = 5$) sera. They analyzed complete serum (without depletion of abundant proteins) in order to avoid protein losses and the generation of potential artifacts. After very stringent filters were applied, 942 and 1046 proteins were selected in diabetic and nondiabetic sera (888 proteins in common), respectively. To analyze this huge number of proteins, the authors developed a computing method called localized statistics of protein abundance distribution (LSPAD) to evaluate the statistical significance of protein abundance bias between diabetic and nondiabetic sera. As a result, 68 proteins were significantly overrepresented in the diabetic serum, 12 of which belonged to the complement system. This was particularly true for an upstream activator of the complement lectin pathway, ficolin 3, which was over-represented in the serum of type 2 diabetic patients. These data were validated in an independent group of patients. It remains unknown why ficolin-3, an activator of the lectin pathway of complement, is elevated in diabetic sera. In addition, the LSPAD approach could be useful for analyzing proteomic data derived for biologically complex systems, such as the plasma proteome.

Although the following studies cannot be considered "proteomic" studies, they deserve attention in terms of the importance of discovered molecules and their relationship with atherosclerosis development. These findings can be validated in further studies based on larger platforms of discovery, where a wide set of proteins are investigated simultaneously in a wide cohort of samples. Secchiero et al. [67] have assessed the relationship between the serum level of TRAIL (TNF-related apoptosis inducing ligand) and clinical outcomes in patients with AMI. Levels of TRAIL were measured in serial serum samples obtained from 60 patients admitted for AMI and 60 controls, both during hospitalization and at follow-up of 12 months. Serum levels of TRAIL were significantly decreased in patients with AMI at baseline (within 24 h from admission) compared with healthy controls and showed a significant inverse correlation with a series of negative prognostic markers, such as CK, CK-MB, and BNP. TRAIL serum levels progressively increased at hospital discharge, but normalized within 6–12 months after AMI. More importantly, low TRAIL levels at the time of patient discharge were associated with increased incidence of cardiac death and heart failure at the 12-month follow-up, even after adjusting for demographic and clinical risk parameters. Hence, circulating TRAIL might represent an important predictor of cardiovascular events, independent of conventional risk markers. A similar study was reported in 2009 describing the association between adiponectin serum levels with coronary atherosclerotic plaque burden and plaque morphology [68]. Adiponectin serum levels were measured in a cohort of 303 patients with stable typical or atypical chest pain who underwent dual-source multislice CT (DSCT) angiography to quantitatively and qualitatively assess coronary artery plaques. Low adiponectin serum levels are predictive of total coronary plaque burden and inversely correlated with the number of mixed and noncalcified plaques. The majority of the plaques included in this study were calcified; thus, adiponectin levels accounted for only 3% of the variability in total plaque number. In contrast, adiponectin accounted for 20% of the variability in mixed and noncalcified burden. Taken together, these data support the concept that adiponectin could be an important marker in the pathogenesis of atherosclerosis because adiponectin remains significantly associated with plaque morphology in a multivariate model [68]. How low adiponectin levels may contribute to coronary plaque vulnerability is unclear, but it can be considered an independent predictor of mixed and noncalcified coronary atherosclerotic plaques.

In a very different approach, Cleutjens et al. [69] applied the phage display technique to search for novel antibody markers in plasma that could distinguish unstable from stable atherosclerotic lesions. A phage display library was prepared from mRNA obtained from ruptured peripheral human atherosclerotic plaques, and phages displaying inmmunoreactive peptides were screened with serum from patients with ruptured atherosclerotic lesions. Two peptide antigens, E1 and E12, manifested 100% specificity and 76% sensitivity in discriminating between patients with ruptured atherosclerotic lesions and patients with stable plaques and healthy controls. The anti-E1 and anti-E12 antibody response preceded the increase in TnT levels in patients with AMI and, thus, was particularly sensitive for the early diagnosis of AMI. These antibodies were not autoantibodies, and the exact nature of the antigens that they recognize is unknown. The E1 peptide contains a sequence of 16 amino acids present within a potential protein termed 1NFLS, and the E12 peptide is a nonsense peptide derived from the untranslated 3′ end of a cDNA encoding PKC$\eta$. Thus, E1 and E12 probably do not exist *in vivo*, and the antibodies cross-react with unknown human targets. These intriguing data suggest that antibody profiling could be a potential approach for the noninvasive diagnosis of atherosclerotic lesions and MI [69,70]. However, identification of the antigens that they recognize is essential and could help in the understanding of atherogenesis. Figure 14.3 and Table 14.2 summarize more recent novel biomarkers measured in plasma and their association with several pathologies.

Some researchers prefer urine instead of plasma for proteomic analysis because urine is stable against proteolytic degradation, contains low concentrations of irrelevant proteins, and can be collected noninvasively. For example, Zimmerli et al. [71] examined a total of 359 urine samples from 88 patients with severe

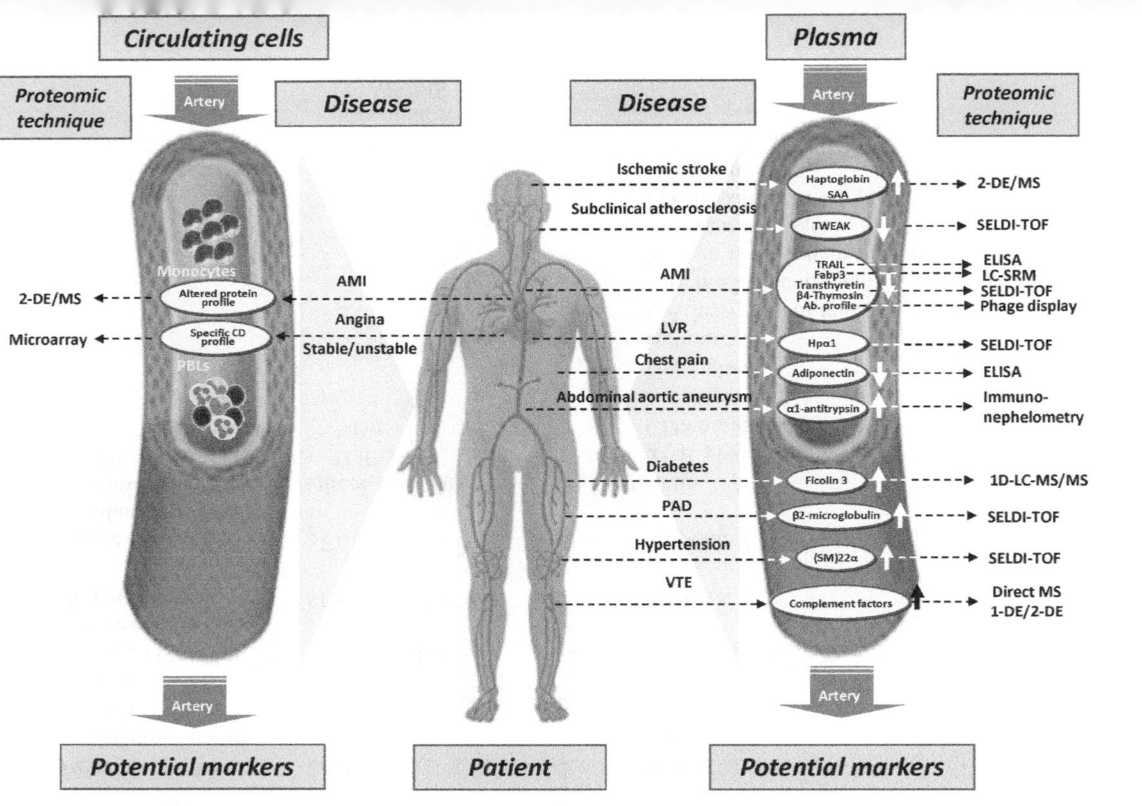


**Figure 14.3** More recent potential novel biomarkers that are associated with different cardiovascular pathologies and have been evaluated in plasma and circulating cells. Altered plasma levels in patients versus controls are shown (↓↑). The proteomic techniques used to detect the proteins are also included (right and left margins).

**TABLE 14.2 Compilation of Proteomic Studies Reported to Date in the Field of Vascular Disease**

| Sample Source | Pathology | Methodology | Proteins Identified | Most Relevant | Validation Method | Reference |
|---|---|---|---|---|---|---|
| ATHEROSCLEROSIS | | | | | | |
| Human serum/plasma | Ischemic stroke (Atherotrombotic cardioembolic stroke) | 2-DE | 2 | SAA ↑<br>haptoglobin ↑ | - | 59 |
| Human serum/plasma | Peripheral arterial disease (PAD) | SELDI-TOF-MS | 1 | $\beta_2$-microgobulin ↑ | - | 60 |
| Human carotid secretome | Atherosclerosis | SELDI-TOF-MS | 1 | TWEAK ↓ | - | 62 |
| Human serum/plasma | Acute myocardial infarction (AMI) | SELDI-TOF-MS | 1 | Troponin I | | 63 |
| Human serum/plasma | Venous thrombo-embolism (VTE) | Direct MALDI-TOF-MS<br>2-DE | 12 | Actin<br>α1-B glycoprotein<br>CD5 antigen-like complement 4A<br>Haptoglobin<br>Hemopexin<br>IgA heavy chain leucine rich-α2 glycoprotein 1<br>Myosin heavy chain platelet coagulation factor XI<br>Plasma kalikrein B1 precursor<br>Proapoliprotein | | 64 |
| Human serum/plasma | Diabetes | Label free proteomic with LC-MS/MS | 68 | Ficolin 3 ↑ | WB | 66 |

| | | | | | | |
|---|---|---|---|---|---|---|
| Human serum/plasma | AMI | Nonproteomic approach | 1 | TRAIL ↓ | - | 67 |
| Human serum/plasma | Coronary atherosclerosis | Nonproteomic approach (WB) | 2 | Adiponectin ↓ | - | 68 |
| Swine serum/plasma | Hypercholesterolemia/ Diabetes | Nonproteomic approach | 1 | Lipoprotein associated phospholipase A2 | - | 97 |
| Human serum/plasma | Atherosclerosis | Nonproteomic approach (phage display technique) | 1 | Antibody profile | - | 69 |
| Human urine | Severe coronary artery disease (CAD) | Capillary electrophoresis+ ESI-TOF/MS | Peptide profile (15 peptides) | Collagen α (I)<br>Collagen α (III) | | 71 |
| Human circulating cells (monocytes, T-cells) | AMI | Microarray | 7 | CD2, CD5, CD7, CD13, CD45, CD45RA, CD49e, CD52, CD64, CD66c | - | 72 |
| Human circulating monocytes | ACS | 2-DE | 20 | Protein profile | WB | 75,76 |
| Human circulating lipoproteins (LDL, HDL) | CAD | MS | 48 (HDL) | Calgranulin A<br>Lysozyme C<br>Complement regulatory proteins<br>Serine protease inhibitors | - | 77,78 |
| Platelets | Atherosclerosis | 2-DE LC-MS/MS | 40 | 14–3–3 ζ | - | 85 |
| ApoE mice aorta | Atherosclerosis | 2-DE | 57 | Immunoglobulins ↑ | WB | 9 |
| | | | | 1-Cys peroxiredoxin ↑ | - | |
| | | | | GSH reductase ↑ | - | |
| | | | | MOD–1 ↑ | - | |

*(continued)*

**TABLE 14.2 (*Continued*)**

| Sample Source | Pathology | Methodology | Proteins Identified | Most Relevant | Validation Method | Reference |
|---|---|---|---|---|---|---|
| ApoE mice aorta | Atherosclerosis | 2-DE | 7 | Aconitase ↑ | - | 10 |
| | | | | Fumarate | - | |
| | | | | Hydratase ↑ | - | |
| | | | | Gelsolin ↑ | - | |
| | | | | Hemoglobin ↑ | - | |
| | | | | Myosin light 3 ↑ | | |
| | | | | SOD2 ↑ | - | |
| | | | | Vesl−2 ↑ | - | |
| Human carotid | Atherosclerosis | 2-DE | 1 | $\alpha_1$-antitrypsin ↑ | WB, 2D-WB | 86 |
| Human carotid | Atherosclerosis | 2-DE | 21 | hsp27 ↓ | WB, IHC | 87 |
| Human carotid | Atherosclerosis | 2-DE | 9 | Fibrinogen fragment D ↑ | WB | 92 |
| | | | | Ferritin light subunit ↑ | - | |
| | | | | SOD2 ↑ | - | |
| | | | | Annexin A10 ↓ | - | |
| | | | | Glutathione transferase | - | |
| | | | | P1−1 ↓ | | |
| | | | | hsp20 ↓ | - | |
| | | | | hsp27 ↓ | - | |
| | | | | Rho GDI ↓ | - | |
| | | | | SOD3 ↓ | - | |
| Human aorta | Atherosclerosis | 2-DE | 27 | Annexin A5 ↑ | WB | 92 |
| | | | | Decoy receptor 1 ↑ | WB | |
| | | | | 14−3−3$\gamma$ ↓ | WB | |

| | | | | | | |
|---|---|---|---|---|---|---|
| Human carotid | Atherosclerosis | Antibody arrays | 21 | TRAF4 ↑ | WB, IHC | 93 |
| | | | | Gads ↑ | WB, IHC | |
| | | | | GIT1 ↑ | WB, IHC | |
| | | | | Caspase−9 ↑ | WB, IHC | |
| | | | | c-src ↑ | WB, IHC | |
| | | | | TOPO-I ↑I-α | WB, IHC | |
| | | | | JAM−1 ↑ | WB, IHC | |
| Human carotid | Atherosclerosis | Western arrays | 15 | TSP−2 ↑ | WB | 94 |
| | | | | MnSOD ↑ | WB | |
| | | | | apo B100 ↑ | WB | |
| | | | | PTP1C ↑ | WB | |
| | | | | ALG−2 ↓ | WB | |
| | | | | GSK−3β ↓ | WB | |
| Human coronary | Atherosclerosis | LC-MS/MS | 806 | PEDF | IHC | 99 |
| | | LMD (layers) + LC-MS/MS | | Periostin | IHC | |
| | | | | MFG-E8 | IHC | |
| | | | | Annexin I | IHC | |
| Human carotid | Atherosclerosis | Manual microdissection (intima) + proteoglycan extraction + LC-MS/MS | 8 | Lumican ↑ | IHC | 100 |
| Apo E mice aorta | Atherosclerosis | VSMCs explant culture+DIGE | 117 | PK ↑ | WB | 107 |
| | | | | LDH ↑ | WB | |
| | | | | Peroxiredoxin I ↑ | WB | |
| | | | | IGFBP-3 and IGFBP-6 ↑ | WB | |

*(continued)*

**TABLE 14.2** ***(Continued)***

| Sample Source | Pathology | Methodology | Proteins Identified | Most Relevant | Validation Method | Reference |
|---|---|---|---|---|---|---|
| Human coronary | Atherosclerosis | agLDL treated explants culture VSMCs + 2-DE | 3 | *p*-myosin RLC ↓ | WB, IHC | 108 |
| Apo E mice aorta | Atherosclerosis | Biotin perfusión (ECs surface proteins) + LC-MS/MS | 454 | Immune and inflammatory response | - | 40 |
| | | | | Cell adhesion | - | |
| | | | | Lipid metabolism | - | |
| Human carotid secretome | Atherosclerosis | 2-DE | 2 | HSP27 | WB, IHC, ELISA | 61,111 |
| | | | | Cathepsin D | | |
| Human aortic valves | Aortic stenosis | Nonproteomic approach | 7 | Apo B, apo A, apo E | | 119 |
| | | | | α actin, ACE, AT1, AT2 | | |
| Human carotid | Atherosclerosis | Western blotting | 1 | Caveolin-1 | | 96 |

↑: up-regulated; ↓: down-regulated

coronary artery disease (CAD) and 282 controls by capillary electrophoresis (CE) and ESI-TOF/MS. Although CE is not yet a widely used tool for massive separation of proteins, its utilization is progressively increasing [72]; in this work, more than 1000 polypeptides per sample were analyzed. It is important to note that the preparation procedure for urine samples must be optimized by including a delipidation step because many samples showed very high blood lipid concentrations, preventing ultrafiltration. A training set for biomarker definition was created from which multiple biomarker patterns emerged that clearly distinguished healthy controls from CAD patients. A set of 15 peptides that define a characteristic CAD signature panel was identified. Interestingly, the majority of the identified polypeptide patterns that were able to discriminate between the presence and absence of disease consisted of fragments of collagen $\alpha_1$ (type 1) chains and collagen $\alpha_1$ (type 3). Collagen types 1 and 3 are the predominant proteins in the arterial walls and appear together in the thickened intima of atherosclerotic lesions. These results show that urinary proteomics can identify CAD patients with high confidence and might also play a role in monitoring the effects of therapeutic intervention. It remains to be seen whether these fragments or the precursor proteins can be detected in the sera of these patients. It is conceivable that these fragments are the products of some arterial collagenases, such as MMP9 (which is increased in patients with atherosclerotic disease), and that only the fragments appear in blood, where they are rapidly cleared and are thus detected mainly in urine. As pointed out by the authors [71], these results illustrate an important difference between proteomic and genomic analysis. Genomics analysis identifies predisposing risk factors, whereas proteomics can identify the point in time when predisposition develops into disease. This is because the proteome is inherently dynamic and thus better reflects changes. A similar study evaluated urine proteome patterns for their potential to reflect coronary artery atherosclerosis in symptomatic patients and confirmed the previous results [73]. Thus, a comparison of polypeptide patterns obtained in these two separate studies suggests that CAD can be reflected in specific polypeptide patterns in urine. Because the peptide pattern originated mainly from collagen types 1 and 3, this finding points toward a central role of collagen in human atherosclerotic plaques.

When the analysis is focused on specific cells present in blood, information related to particular pathways involved in disease progression can be obtained. Monocytes and T cells are the predominant leukocytes involved in the development of atherosclerosis. These circulating cells were investigated in a microarray format of 82 clusters of differentiation (CD) antibodies that selectively immobilize peripheral blood mononuclear cells. The pattern of leukocyte immobilization allows monitoring of the progression from stable angina pectoris (SAP) to unstable angina pectoris (UAP). Although more clinical data are needed to further support these findings, seven antibody spots were dynamic between these two conditions. Thus, methodology may be used as an early diagnostic tool to classify cases of chest pain on admission, which ultimately avoids further complications ending in MI by prompt medical treatment [74]. In our group, proteomic profiling of circulating monocytes from acute coronary syndrome (ACS) patients was conducted by 2DGE. In a similar approach, the influence of high-dose atorvastatin (ATV) treatment compared with conventional treatments was also evaluated. Interestingly, the protein pattern observed in the monocytes of ACS patients changes with time (from time of admission up to 6 months) and results in total normalization compared with stable CAD patients at 6 months [75]. By intensive ATV treatment, the expression of 20 proteins was modified compared with the healthy pattern. Interestingly, proteins that modulate inflammation and thrombosis, such as protein disulfide isomerase ER60 (PDI), annexin I, and prohibitin, or that have other protective effects, such as HSP70, were normalized [76] (Fig. 14.3). For circulating lipoproteins, proteins truly associated with low-density lipoproteins (LDLs) and high-density lipoproteins (HDLs) were mapped and identified by MS [77,78]. In particular, identification of calgranulin A and lysozyme C in LDL suggests a role for innate immunity and inflammation. Heinecke [79] proposed that quantification of the HDL proteome, which can be performed by MS, could be a marker of CAD. Using shotgun proteomics, 48 proteins, 13 of which were not previously known to reside in HDL, were identified in HDL particles isolated from healthy control and CAD subjects [80]. Among the identified proteins, acute response proteins (23 of 48), complement regulatory proteins and serine protease inhibitors were found. These data suggest new HDL functions, including a potential role in preventing plaque rupture, protecting vascular lesions from promiscuous proteolysis and inhibiting complement activation that may prevent activation of the coagulant response in endothelia and platelets, which are two critical events in acute thrombosis [79]. A remarkable characteristic derived from the HDL proteome is that HDL is a mixture of particles containing different sets of proteins. Depending on their protein composition, some particles can be cardioprotective by inhibiting inflammation and removing toxic proteins and lipids, whereas others could have deleterious effects such as promoting cholesterol accumulation and inhibiting cardioprotective pathways. For example, the protein composition of $HDL_3$ differed between control and CAD subjects, and this composition could be modulated by lipid lowering therapies (statins) [81].

The atherome plaque and the different cell types composing it release proteins by two main routes: (1) direct protein secretion, in which proteins are shed into the bloodstream; and (2) release of membrane vesicles. By studying these secretory membrane bodies (apoptotic bodies, microparticles, exosomes, and matrix vesicles), we are also facing a *subproteome*, which again enormously simplifies the analytical strategy. Microparticles and exosomes actively participate in vascular pathophysiology and relevant results from their proteomic studies have been reviewed [82]. However, the presence of plasma contaminants and isolated particles as well as interference with the detection of low abundance proteins in the subproteome are limitations to this approach.

Platelet adhesion and secretion of its stored substances are also related to atherosclerosis [83,84]. Platelet homeostatic function controls excess bleeding, but will cause thrombus formation with potential arterial occlusions in the context of atheromatous lesion rupture. In particular, the secretion of the 14-3-3ζ protein from activated platelet dense granules has been reported [85]. The presence of 14-3-3ς in abdominal aortas with atherosclerotic plaques from patients with aneurysms was confirmed, pointing to a possible extracellular role of this protein in disease development.

***14.3.1.2 Tissue and Isolated Cells*** Tissue proteomic analysis should be considered as a complementary strategy to biological-fluids-based approaches because of their ability to identify proteins that are present in the plaque and may be shed into the circulation [12]. Such studies provide two main information datasets: (1) potential biomarkers of disease directly released from tissue lesions, which should be validated in wider cohorts before being applied to clinics; and (2) discovery of mechanisms and pathways involved in the formation and development of atherosclerosis by direct *in situ* proteomics. In this sense, three main sample sources can be approached: (1) whole-tissue proteomics, by which total proteins extracts obtained from areas of interest are studied; (2) tissue subproteomes, that is, specific arterial layers and cells isolated directly from tissue (apart from those studies based on cultured cells); and (3) tissue secretome, which is a medium enriched in proteins that derive from diseased tissue and, therefore, is one of the best sources for biomarker discovery (see Fig. 14.2).

*Arterial Whole-Tissue Proteome* Different techniques have been applied to human samples, from shotgun proteomics or 2DGE to antibody arrays and MS imaging. One of the most commonly used arterial sources is the carotid artery, particularly when biopsy material is desired, because of its accessibility and the large quantity of material that can be collected. Carotid arteries with a thrombus were compared with stable plaques by 2DGE, and an isoform from $\alpha_1$-antitrypsin was upregulated in thrombotic lesions [86]. Alfa $\alpha_1$-antitrypsin is expressed by the liver under inflammatory conditions and may enhance fibrosis processes within the plaque because it inhibits collagenases and elastases. When the atherosclerotic core regions of carotid endarterectomies were compared with nearby normal-appearing areas, 21 proteins were significantly varied [87]. HSP27 was decreased in the plaque core region with respect to normal areas, which is consistent with our earlier findings in the secretome of carotid plaques [88]. In contrast, they found lower levels of HSP27 in nonatherosclerotic reference arteries (mammary) compared with normal appearing regions of carotid endarterectomies by WB, and plasma levels were increased in patients. Subsequent studies performed by other groups showed that decreased levels of HSP27 are correlated with higher lesion areas and plaque instability [89–91]. Thus, this is an open issue that deserves further research and can be associated with a high number of HSP27 isoforms.

Comparison of a total of 48 carotid samples classified in unstable and stable plaques resulted in the identification of 33 different proteins (57 spots) [92], 70% of which were plasma-derived. Nine proteins significantly varied between groups (ferritin light subunit, fibrinogen fragment D, and SOD2 were increased; annexin A10, glutathione transferase P1-1, hsp20, hsp27, RhoGDI, and SOD3 were decreased). In particular, higher levels of fibrinogen fragment D in unstable plaques were corroborated by WB. The effects of this protein on vascular cells include increased endothelial permeability and disorganization and stimulation of the release of proinflammatory and chemotactic factors. Biopsies from affected or healthy areas of human aortas obtained during bypass surgery were also compared [92]. In this case, differential abundance analysis resulted in 39 significantly increased spots, among which annexin $A_5$, decoy receptor 1, and 14-3-3$\gamma$ were upregulated.

A different approach, antibody array technology, has undergone important advances. Commercially prearrayed platforms in diverse research areas are now available, but in only recent study were commercial antibody arrays applied to atherome plaque extracts [93]. Unstable and stable carotid plaques were compared in a 512-antibody set array. In this array, 21 proteins were overexpressed in unstable plaques, and 3 proteins were increased in stable plaques. Among the increased proteins in unstable plaques, several were previously shown to be upregulated (i.e., TNF$\alpha$, HIF1-$\alpha$, HSP40, CRP2). Eleven differentially expressed proteins were selected for WB and immunohistochemistry (IHC) validation due to their potential implication in inflammation, angiogenesis, proliferation, and apoptosis. The results showed differential expression in all 11 proteins in plaques compared with nonatherosclerotic arteries and confirmed differences between stable and unstable plaques for 8 proteins (TNF$\alpha$, TRAF4, Gads, GIT1, caspase-9, c-src, TOPO-II-$\alpha$, and JAM1). IHC experiments were carried out in parallel with antibodies to localize different cell types within the plaque: antiactin (VSMCs), anti-CD68 (macrophages/foam cells), and anti-CD105 [endothelial cells (ECs)] were used to identify different cell types within the tissue. A Western array approach (high-throughput WB analysis) was applied to study carotid plaque lysates compared with mammary artery extracts [94]. Among the 823 antibodies used, 15 proteins were considered significantly different with a fold change >5. Standard WB validation was only possible for 7 of the 15 proteins, indicating that the false-positive rate could be high with this methodology. ApoE, thrombospondin 2 (TSP2), manganese superoxide dismutase (MnSOD), apoB100, and protein-tyrosine phosphatase 1C (PTP1C) were upregulated in carotid plaques, while apoptosis-linked gene 2 (ALG2) and glycogen synthase kinase $-3\beta$ (GSK3$\beta$) were downregulated. Tissue microarray (TMA) technology has been validated for vascular tissue. A variety of vessels collected from 100 subjects undergoing autopsies were arrayed over 17 TMAs, and more than 1000 vascular tissues were evaluated in a single experiment (carotid, coronary and mammary arteries, etc.). As a proof of concept, they investigated three proteins known to be expressed in the vasculature (RAGE, CTGF, and MMP3) [95].

A correlation between plaque stability and the presence of a protein has been established for caveolin1 (Cav1) [96]. Rodriguez-Feo et al. studied the relationship between Cav1 abundance, atherosclerotic plaque characteristics, and clinical manifestations of atherosclerotic disease. The presence of Cav1 in atherosclerotic plaques was studied by western blotting in 378 subjects who underwent carotid endarterectomy. Cav1 protein expression levels were strongly reduced in atherosclerotic plaques compared with nonatherosclerotic mammary arteries. The carotid plaques showed 64% lower expression levels compared with mammary arteries. Plaques with an atheromatous phenotype showed lower Cav1 expression compared with fibrous plaques. In addition, Cav1 levels were significantly lowered in plaques with unstable characteristics, such as large lipid core, thrombus formation, macrophage infiltration, high IL6 and IL18, and elevated MMP9 activity. Recombinant peptides mimicking Cav1 scaffolding domain (Cavtratin) reduced gelatinase activity in cultured porcine arteries and impaired MMP activity and COX2 in LPS-challenged macrophages. Thus, Cav1 could be a novel potential stabilizing factor in human atherosclerosis. In addition, patients with high plaque Cav1 expression seemed to be protected from cardiovascular events within 30 days of surgery, making Cav1 the first available plaque biomarker with prognostic value. Taken together, these data indicate that low levels of Cav1 in atherosclerotic lesions contribute to plaque formation and instability. Thus, Cav1 may be considered a novel target in the prevention of atherosclerosis and a novel biomarker of vulnerable plaques with prognostic value. Contrary to the stabilizing effect of Cav1, increased levels of lipoprotein-associated phospholipase $A_2$ promote the formation of atherosclerotic lesions and dangerous, unstable atherosclerotic plaques [97]. Inhibition of Lp-$PLA_2$ with darapladib in a swine model of diabetes and hypercholesterolemia blocks the development of unstable plaques by reducing the generation of proinflammatory lipids by the enzyme.

The studies discussed so far include carotid/aortic arteries. However, incidents from coronary artery thrombosis constitute the major risk of mortality in Western societies, and this artery has been infrequently used for proteomic studies, probably due to difficulties in obtaining samples. In this sense, one can think of four main sample sources: necropsy, transplant, endarterectomy (a surgical procedure in which atheromatous plaque is removed by separating it from the arterial wall), and arterial tissue removal during a bypass surgery, which involves high risk for the patient. The first proteomic approach performed with human coronary tissue constituted a very preliminary study [98] because it was based on direct visualization of silver-stained 2DGE gels of healthy and diseased coronaries. In this way, ferritin light chain was identified as being increased in affected coronaries. WB analysis corroborated this visual evidence. However, RT-PCR assays showed decreased ferritin mRNA in diseased arteries, indicating posttranscriptional regulation of protein levels for this protein.

In 2007, an extensive study of coronary plaques based on a shotgun proteomics approach was published [99]. Efficient application of the so-called direct tissue proteomics (DTP) approach to coronary atherome plaques was shown, and the use of paraffin-embedded samples or frozen tissue sections was evaluated. A total of 225 proteins were identified from paraffin-embedded arteries, and 558 were identified from frozen sections, which imply better utility of frozen material for such proteomic analysis. Moreover, the proteome of the three layers from the coronary artery (intima, media, and adventitia) isolated by laser capture microdissection (LCM) could be identified by LC-MS/MS [see discussions of subproteomes for more details). In total, 806 unique proteins were reported, which constitutes the first enumeration of the coronary artery proteome in detail. From identified proteins, four were validated by IHC, as they had not been reported previously to be involved in atherosclerosis development and because their functions may be related to pathology: inflammation/angiogenesis [pigment epithelium-derived factor (PEDF)], extracellular matrix component (periostin) and apoptotic cells phagocytosis (milk fat globular epidermal growth factor 8 (MFG-E8), and annexin I]. The identification of known growth factors and cytokines on coronary extracts was not possible with this methodology, probably due to their low concentration levels. The novel AQUA methodology combined with multi reaction monitoring (MRM) was used to investigate transforming growth factor β(TGFβ) and stromal derived factor 1α(SDF-1α) as selected proteins. In general, this approach allows quantification of a known protein by MS through the addition of a heavy-isotope-labeled peptide of the protein of interest, so the absolute quantity of the protein of interest is calculated by comparing the area of its chromatographic peak with the peak area of the standard.

*Arterial Tissue Subproteomes* The atherome plaque represents a very complex tissue in terms of differentiated areas and cell type content. In particular, arterial layers play different roles in atherosclerosis development. In fact, atherosclerosis starts with LDL deposits at the intima, which triggers an inflammatory response that is carried out by infiltrated circulating monocytes that have differentiated into macrophages within the intima. Migration of medial VSMCs to the intima is due to a disorganization of internal elastic lamella, where they synthesize ECM. This ECM leads to intimal thickening and, subsequently, a fibrous cap that surrounds the lipid core in advanced lesions. The study of certain regions or cell types from atherome plaques gives more specific information than does global analysis of the whole tissue. Concerning proteomic analysis, proteins expressed in a certain region or by a specific cell type may be different or expressed at different levels, and these possibilities cannot be distinguished in whole-plaque studies. Different methodologies have been developed to isolate regions or cell types from tissues (i.e., microdissection, explant cultures, and cell sorting), but those techniques are not easily applied to proteomic analysis because a certain amount of cells and/or tissue is necessary. Becuase of the specific information that can be obtained, particular attention will be paid to those approaches in the current chapter.

TISSUE REGIONS ISOLATED BY MICRODISSECTION Lower levels of ALG2 in VSMCs microdissected from the human media layer of a carotid plaque with respect to those isolated from adjacent normal media were detected by Western blot array and validated by WB [94]. The proteoglycan subproteome was extracted from microdissected thickened intimas of mammary and carotid arteries and analyzed by LC-MS/MS [100]. Differential abundance analysis was performed by relative quantification of specific peptides from each proteoglycan present in most or all samples, based on the averaged MS intensities of peptides belonging to the selected proteoglycan. In particular, lumican was significantly increased in atherosclerosis-prone arteries (carotid) compared with atherosclerosis-resistant arteries (mammary), and this finding was confirmed by IHC assays. A subset of proteoglycans not previously detected in human intimas with hyperplasia was identified by this methodology, corresponding to decorin, aggrecan, fibromodulin, and prolargin/PRELP.

These studies were performed by manual microdissection, although cross-contamination with undesired tissue is common, due to challenges in properly isolating regions of interest in a manual way. The more recently developed laser microdissection (LMD) technique [101], which allows high-resolution laser-based cell/tissue isolation, overcomes this limitation and allows higher protein yields in fewer microdissection periods. This fact is strongly advantageous in proteomics studies because it minimizes protease action. Although LMD publications have increased exponentially since the technique was first described in 1996 [102], its application to proteomic analysis has been much slower due to the still small amount of samples obtained. In an initial nonproteomic approach, the human caspase-2-positive foam cell subpopulation was laser-microdissected and analyzed by WB in order to determine which isoform is expressed by these cells in atherosclerotic plaques. Indeed, the short isoform of capsase-2 was overexpressed in foam cells [103]. The first proteomics study to use tissue isolated by LMD in cardiovascular research used 2DGE gels of myocytes and blood vessels [104]. However, the first application of LMD combined with proteomics in the study of atherosclerosis was spublished in 2007 [99]. Intima, media, and adventitia from atherosclerotic coronary arteries were isolated by LMD and analyzed by LC-MS/MS, resulting in the identification of 444, 137, and 194 proteins, respectively. In total, 495 unique proteins were identified, and some of them appeared in several layers.

The benefits of LMD in the analysis of tissue subproteomes have been clearly demonstrated in terms of the compatibility of specific isolation with the detection of a wide set of proteins. However, its use with conventional 2DGE for arterial tissue analysis has not been reported in the atherosclerosis literature. This approach is undoubtedly of great interest because it will provide in-depth analysis of differences in protein expression levels in a particular region or cell type. The main disadvantage is that the small protein amounts extracted from microdissected tissue may lead to an extension of silver stain developing times for proper visualization of 2DGE maps, which may increase background signals [104]. However, this limitation can be overcome with the use of saturation labeling DIGE technology, which allows the creation of high-resolution 2DGE maps with less than 5 μg of total protein [105]. In our group, we have developed an optimum methodology for the analysis of LMD-isolated human arterial layers by 2DGE using saturation-labeling DIGE. The intima and media of human coronary and aorta arteries were efficiently isolated, labeled with saturation-labeling CyeDyes, and analyzed by 2DGE. In this way, the first 2DGE maps from human arterial layers were obtained [106]. A schematic representation of the workflow from this type of study can be observed in Figure 14.4.

Isolated Cells One of the best ways of studying molecular mechanisms in pathology development is to approach the contribution of each implicated cell

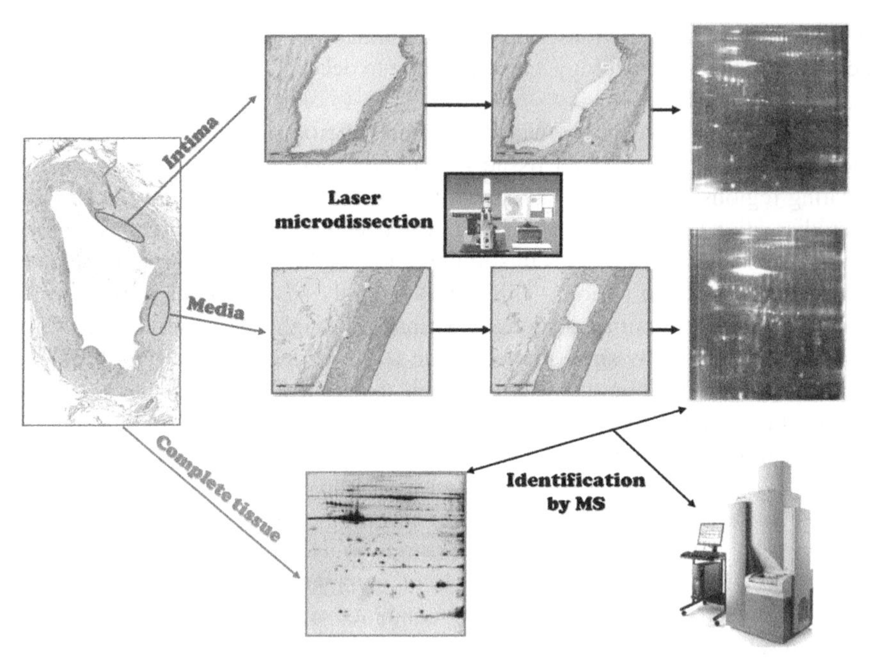


**Figure 14.4** Intima- and media-layer isolation by laser microdissection combined with saturation labeling 2D DIGE. Coronary and aorta artery layers can be isolated by LMD and analyzed by saturation labeling 2D DIGE, which allows the formation of high-resolution 2DGE maps with small amounts (<5 μg) of protein.

type, isolated from others, which better simulates the *in vivo* context. An intermediate approach consists of explant cultures of cells isolated from vascular tissue, which introduces more variability than direct analysis of isolated cells but allows greater cell yields, which is mandatory for proteomic studies. In this way, the first 2DGE maps from proteome and secretome of human arterial VSMCs resulted in the identification of 83 intracellular proteins and 18 secreted proteins [107]. Another possibility includes isolation of cells from nonatherosclerotic arteries, which can be induced to express an atherosclerotic phenotype by certain stimuli. Aggregated LDLs (agLDL) were used as such a stimulus for isolated nonatherosclerotic coronary VSMCs in a wound repair model system [108]. Proteomic 2DGE analysis revealed downregulation of phosphorylated myosin regulatory light chain in agLDL-exposed VSMCs.

The isolation of cells directly from affected arteries for proteomic analysis remains challenging because a large number of cells is required to recover enough protein for experimentation. LMD methodology is adequate for such a purpose. Although macrophages from human plaques and apoE-deficient mice have been isolated in this way, downstream applications applied to these comprise only WB [103] and RT-PCR analysis [109,110].

SECRETOME The secretome includes all proteins that are actively secreted or released by cells or tissues in the extracellular compartment as a consequence of normal metabolism or in response to certain stimuli. Apart from those cardiovascular studies focusing on the secretome of cells in culture [107], particular attention must be paid to the culture of tissue explants directly obtained from pathological and healthy arteries. This sample source simulates the best *in vivo* situation and allows detection of proteins and molecules released into the plasma, where biomarkers will be measured in the clinical setting. The use of the atherosclerotic plaque secretome for the identification of potential biomarkers of atherosclerosis has been reported [2,6,7,12,61,88]. In particular, carotid endarterectomies (pathological) and sections from radial arteries (control) were cultured in a protein-free culture medium and analyzed by 2DGE [61]. Also, secretomes from atherosclerotic carotid endarterectomies and mammary segments were cultured and differentially analyzed by 2DGE, resulting in the finding that HSP-27 release is decreased in atherosclerotic plaques and is barely detectable in complicated plaque supernatants. Circulating concentrations of HSP27 were decreased in subjects with atherosclerosis compared with healthy subjects, which confirmed the hypothesis that plasma content can reflect arterial wall secretion [98]. In complementary studies, the effects of atorvastatin on atherosclerotic plaque secretion were investigated. Interestingly, 66% of proteins differentially released by atherosclerotic plaques reverted to control values after administration of atorvastatin, including cathepsin D, which may be a potential target for therapeutic treatment [111]. The effect of atorvastatin alone or combined with amlodipine was also investigated in atherosclerotic plaque secretion. Again, either treatment normalized the levels of the different released proteins [111]. In some cases, dual treatment was superior to that observed with atorvastatin alone (e.g., β-galactoside-soluble lectin in the secretome of atherosclerotic plaques reached control values). Other examples are retinol binding protein, protein disulfide isomerase, and antioxidant protein thioredoxin peroxidase B2.

*Aortic Stenosis* Aortic stenosis (AS) is a pathology closely related to atherosclerosis, as events leading to AS and atherosclerosis development may be associated. Calcified aortic valve disease is a slow progressive disorder with a disease continuum that ranges from mild valve thickening without obstruction of bloodflow, termed aortic sclerosis, to severe calcification with impaired leaflet motion or AS. Microscopically, these areas contain evidence of chronic inflammation, with infiltration of macrophages and T lymphocytes. Accumulation of plasma lipoproteins, including LDL and lipoprotein (a), is present, and several articles have proved that these lipoproteins are oxidatively modified [112,113]. For many years, AS was thought to be the passive accumulation of calcium on the aortic surface of the valve leaflet. At present, the biological pathway leading to severe AS is considered an active process, similar to atherosclerosis, in which macrophage infiltration and neovascularization play a major role [114,115]. Furthermore, AS presents many similarities with CAD, and both share the same clinical risk factors [116–118]. *Ex vivo* studies have defined important cellular markers and potential signaling pathways in the progression of this disease. Observations in early

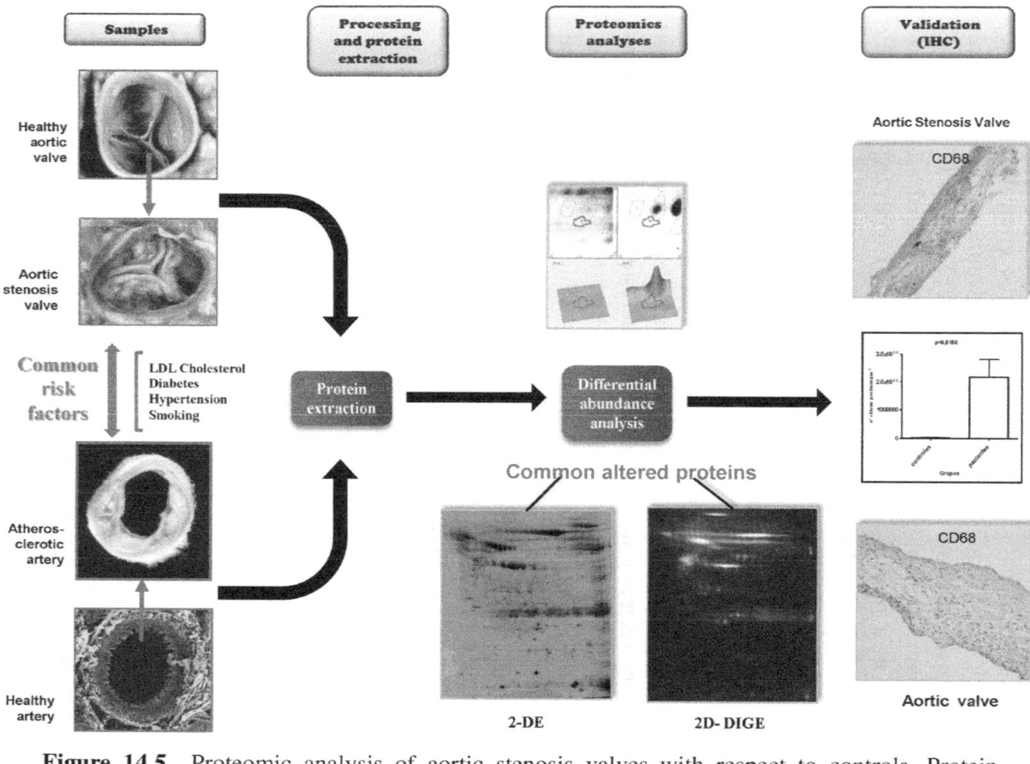


**Figure 14.5** Proteomic analysis of aortic stenosis valves with respect to controls. Protein extracts were analyzed both by silver-staining 2DGE and by minimal labeling DIGE.

valvular sclerotic lesions have shown that chronic inflammatory cells infiltrate (macrophages and T lymphocytes), lipid accumulation occurs (apoB, apoA, and apoE), and α-actin-expressing cells accumulate in the lesion and adjacent fibrosa [119]. End-stage calcified valves contain mature lamellar bone accompanied by the expression of specific bone markers that are important in the development of osteoblast bone formation [120]. In addition, angiotensin converting enzyme (ACE) and angiotensin II type 1 (AT1) and type 2 (AT2) receptors are present in stenotic aortic valves, implicating that this signaling pathway is involved in the disease process [121]. Current studies in our laboratory have shown higher levels of CD68 (a macrophage marker) in the fibrosa layer of AS valves compared with normal aortic valves using proteomic and immunochemical techniques (Fig. 14.5). Although there are similarities with the process of atherogenesis and some common risk factors, not all patients with CAD or atherosclerosis develop calcific aortic stenosis [122]. Thus, there must be other differentiating processes that may be identified by proteomic analysis.

## 14.4 CONCLUDING REMARKS

More recently, cardiovascular proteomics has experienced an impressive development, and proof of that is this book. At present hundreds of novel proteins have been associated with cardiovascular diseases using proteomic approaches; many have been identified in platelets (as seen in previous chapters), and others in the different elements of the cardiovascular system as described in this chapter. A significant number of these novel proteins are in the process of clinical validation in large cohorts of patients. Although no individual proteins have been approved to date as a new biomarker in clinical practice, the high number of candidate proteins under scrutiny is very promising. During more recent years a vast amount of information (on proteins, peptides, metabolites, signaling pathways) has accumulated from proteomic studies of cardiovascular diseases, which probably will crystallize into novel biomarkers and therapeutic targets in the coming years. This optimistic view is reinforced by the more recent application of systems biology approaches (network analysis) to cardiovascular diseases, which is allowing the identification of potential novel therapeutic targets and diagnostic markers [123–125].

## REFERENCES


1. Edwards AVG, White MY, Cordwell, The role of proteomics in clinical cardiovascular biomarker discovery, *Mol. Cell. Proteom*. **7**:1824–1837 (2008).
2. Vivanco F, Darde V, De la Cuesta F, Barderas MG, Cardiovascular proteomics, *Curr. Proteom*. **3**:147–170 (2006).
3. Arab S, Gramolini AO, Ping P, Kislinger T, Stanley B, Van Eyk J, Ouzounian M, MacLennan D, Emili A, Liu P, Cardiovascular genomic medicine. Tools to develop novel biomarkers and potential applications, *J. Am. Coll. Cardiol*. **48**:1733–1741 (2006).

4. Biomarkers Definitions Working Group, Biomarkers and surrogate endpoints: Preferred definition and conceptual framework, *Clin. Pharmacol. Ther.* **69**:89–95 (2001).
5. Drake T, Ping P, Proteomics approaches to the systems biology of cardiovascular diseases, *J. Lipid Res.* **48**:1–8 (2007).
6. Lazaro A, Alvarez-Llamas G, Gallego-Delgado J, De la Cuesta F, Osende J, Barderas MG, Vivanco F, Pharmacoproteomics in cardiac hypertrophy and atherosclerosis, *Cardiovasc. Haematol. Disord. Drug Targets* **9**:141–148 (2009).
7. Alvarez-Llamas G, De la Cuesta F, Barderas MG, Darde V, Padial LR, Vivanco F, Recent advances in atherosclerosis-based proteomics: New biomarkers and future perspectives, *Expert Rev. Proteom.* **5**:679–691 (2008).
8. Rather DJ, Daugherty A, Translating molecular discoveries into new therapies for atherosclerosis, *Nature* **45**:904–913 (2008).
9. Wang XL, Fu A, Spiro C, Lee HC, Clinical application of proteomics approaches in vascular diseases, *Proteom. Clin. Appl.* **2**:238–250 (2008).
10. Didangelos A, Simper D, Monaco C, Mayr M, Proteomics of acute coronary syndromes, *Curr. Atheroscler. Rep.* **11**:188–195 (2009).
11. Vivanco F, Padial LR, Darde V, De la Cuesta F, Alvarez-Llamas G, Diaz-Prieto N, Barderas MG, Proteomic biomarkers of atherosclerosis, *Biomarker Insights* **3**:101–113 (2008).
12. De la Cuesta F, Alvarez-Llamas G, Gil-Dones F, Martin-Rojas T, Zubiri I, Pastor C, Barderas MG, Vivanco F, Tissue proteomics in atherosclerosis: Elucidating the molecular mechanism of cardiovascular diseases, *Expert Rev. Proteom.* **6**:395–409 (2009).
13. Van Eyk J, Dunn MJ, eds, Cardiac disease, in *Clinical Proteomics. From Diagnosis to Therapy*, Wiley-VCH, 2008, pp. 203–294.
14. Schoenhoff FS, Fu Q, Van Eyk J, Cardiovascular proteomics: Implications for clinical applications, *Clin. Lab. Med.* **29**:87–99 (2009).
15. Martin-Ventura JL, Blanco-Colio L, Tuñon J, Gomez-Guerrero C, Michel JB, Meilhac O, Egido J, Proteomics in atherotrombosis: A future perspective, *Expert Rev. Proteom.* **4**:249–260 (2007).
16. Mann DL, Pathophysiology of heart failure, in Libby P, Bonow RO, Mann DL, Zipes DP, eds., *Braunwald's Heart Disease. A Textbook of Cardiovascular Medicine*, 8th ed., Saunders Elsevier, Philadelphuia, 2008, pp. 541–560.
17. Gonzalez A, Lopez B, Ravassa S, Beaumont J, Arias T, Hermida N, Zudaire A, Diez J, Biochemical markers of myocardial remodeling in hypertensive heart disease, *Cardiovasc. Res.* **81**:509–518 (2009).
18. Melle C, Camacho J, Surber R, Betge S, Von Eggeling, Zimmer T, Region-specific alterations of global protein expression in the remodeled rat myocardium, *Int. J. Mol. Med.* **18**:1207–1215 (2006).
19. Gallego-Delgado J, Lazaro A, Osende J, Barderas MG, Duran MC, Vivanco F, Egido J, Comparison of the protein profile of established and regressed hypertension-induced left ventricular hypertrophy, *J. Proteome Res.* **5**:404–413 (2006).
20. Lazaro A, Gallego-Delgado J, Osende J, Egido J, Vivanco F, Analysis of antihypertensive drugs in the heart of animal models, *Meth. Mol. Biol.* **357**:45–58 (2007).

21. Gallego-Delgado J, Lazaro A, Osende J, Esteban V, Barderas MG, Gomez-Guerrero C, Vega R, Vivanco F, Egido J, Proteomic analysis of early left ventricular hypertrophy secondary to hypertension: Modulation by hypertensive therapies, *J. Am. Soc. Nephrol*. **17**: S159–S162 (2006).
22. Jin X, Xia L, Wang L, Shi JZ, Zheng Y, Chen WL, Zhang L, Liu ZG et al., Differential protein expression in hypertrophic heart with and without hypertension in spontaneously hypertensive rats, *Proteomics* **6**:1948–1956 (2006).
23. Jülling M, Hickey AJR, Chai CC, Skea GL, Middleditch MJ, Costa S, Choong SY, Philips ARJ, Cooper GJS, Is the failing heart out of fuel or a worn engine running rich? A study of mitochondria in old spontaneously hypertensive rats, *Proteomics* **8**:2556–2572 (2008).
24. Cieniewski-Bernard C, Mulder P, Henry JP, Drobecq H, Dubois E, Pottiez G, Thuillez C, Amouyel P, Richard V, Pinet F, Proteomic analysis of left ventricular remodelin in an experimental model of heart failure, *J. Proteome Res*. **7**:5004–5016 (2008).
25. Pinet F, Beseme O, Cieniewski-Bernard C, Drovecq H, Jourdain S, Lamblin N, Amouyel P, Bauters C, Predicting left ventricular remodeling after a first myocardial infarction by plasma proteome analysis, *Proteomics* **8**:1798–1808 (2008).
26. Yu H, Meng Y, Wang LS, Jin X, Gao LF, Zhou L, Ji K, Zhao LJ, Chen GQ, Zhao X, Yang B, Differential protein expression in heart in UTB-B null mice with cardiac conduction defects, *Proteomics* **9**:504–511 (2009).
27. Burniston J, Adaptation of the rat cardiac proteome in response to intensity-controlled endurance exercise, *Proteomics* **9**:106–115 (2009).
28. Clements RT, Smejkal G, Sodha NR, Ivanov AR, Asara JM, Feng J, Lazarev A, Gautan S, Senthilnathan V, Khabbaz R, Bianchi C, Sellke W, Pilot proteomic profile of differentially regulated proteins in right atrial appendage before and after cardiac surgery using cardioplegic and cardiopulmonary bypass, *Circulation* **118**: S24–S31 (2008).
29. Bouchama A, Knochel JP, Heatstroke, *N. Engl. J. Med*. **346**:1978–1988 (2002).
30. Cheng B, Chang C, Tsay Y, Wu T, Hsu C, Lin M, Body cooling causes normalization of cardiac protein expression and function in a rat heatstroke model, *J. Proteome Res*. **7**:4935–4945 (2008).
31. Jin W, Brown AT, Murphy AM, Cardiac myofilaments: From proteome to pathophysiology, *Proteom. Clin. Appl*. **2**:800–810 (2008).
32. Yuan C, Solaro RJ, Myofilament proteins: From cardiac disorders to proteomic changes, *Proteom. Clin. Appl*. **2**:788–799 (2008).
33. Sakthivel S, Finley NL, Rosevear PR, Lorenz JN, et al., *In vivo* and *in vitro* analysis of cardiac troponin I phosphorylation, *J. Biol. Chem*. **280**:703–714 (2005).
34. Agnetti G, Kane LA, Guarnieri C, Caldarera C, Van Eyk J, Proteomic technologies in the study of kinases: Novel tools for the investigation of PKC in the heart, *Pharmacol. Res*. **55**:511–522 (2007).
35. Zaremba R, Merkus D, Hamdani N, Lamers JM, Paulus W, Dos Remedios C, Duncker DJ, Stienen GJM, Van der Velden J, Quantitative analysis of myofilament protein phosphorylation in small cardiac biopsies, *Proteom. Clin. Appl*. **1**:1285–1290 (2007).
36. Gao W, Atar D, Liu Y, Perez NG, et al., Role of troponin I proteolysis in the pathogenesis of stunned myocardium, *Circ. Res*. **80**:393–399 (1997).

37. Charles RL, Eaton P, Redox signaling in cardiovascular disease, *Proteom. Clin. Appl*. **2**:823–836 (2008).
38. Gödecke A, Schrader J, Reinartz M, Nitric oxide-mediated protein modification in cardiovascular physiology and pathology, *Proteom. Clin. Appl*. **2**:811–822 (2008).
39. Bian Y, Qi Y, Yan Z, Long D, Shen B, Jiang Z, A proteomic analysis of aorta from spontaneously hypertensive rat. RhoGDI alpha regulation by angiotensin II via $AT_1$ receptor, *Eur. J. Cell. Biol*. **87**:101–110 (2008).
40. Delbosc S, Haloui M, Loudec l, Dupuis, Cubizolles M, Podust V, Fung E, Michel JB, Meilhac O, Proteomic analysis permits the identification of new biomarkers of arterial wall remodeling in hypertension, *Mol. Med*. **14**:383–394 (2008).
41. Liao M, Liu Z, Bao J, Zhao Z, Hu J, Feng X, Feng R, Lu Q, Mei Z, Liu Y, Wu Q, Jing Z, A proteomic study of the aortic media in human thoracic aortic dissection: Implication for oxidative stress, *J. Thorac. Cardiovasc Surg*. **136**:65–72 (2008).
42. Foster B, O'Rourke B, Van Eyk, J, What can mitochondrial proteomics tell us about cardioprotection afforded by preconditioning? *Expert Rev. Proteom*. **5**:633–636 (2008).
43. White MY, Edwards AVG, Cordwell S, Van Eyk J, Mitochondria: A mirror into cellular dysfunction in heart diseases, *Proteom. Clin. Appl*. **2**:845–861 (2008).
44. Foster B, Van Eyk J, Marban E, O'Rourke B, Redox signalling and protein phosphorylation in mitochondria: progress and prospects, *J. Bioenerg. Biomembr*. **41**:159–168 (2009).
45. Lopez-Campistrous A, Hao L, Xiang W, Ton D, Semchuck P, Sander J, Ellison M, Fernandez-Patron C, Mitochondrial dysfunction in the hypertensive rat brain, *Hypertension* **51**:412–419 (2008).
46. Kwapiszewska G, Wygrecka M, Marsh LM, Schmitt S, Tröser R, Wihelm J, Helmus K, et al., Fhl-1, a new key protein in pulmonary hypertension, *Circulation* **118**:1183–1194 (2008).
47. Nath AK, Krauthhammer M, Li P, Davidov E, Butler L, Copel J, Katajamaa M, Oresic M, Buhimschi C, Snyder M, Madri JA, Proteomic-based detection of a protein cluster dysregulated during cardiovascular development identifies biomarkers of congenital heart defects, *PLoS ONE* **4**: e4221 (2009).
48. Liu B, Tewari AK, Zhang L, et al., Proteomic analysis of protein tyrosine nitration after ischemia reperfusion injury: Mitochondria as the major target, *Biochim Biophys Acta* **1794**:476–485 (2009).
49. Sung JH, Cho EH, Kim MO, Koh PO, Identification of proteins differentially expressed by melatonin treatment in cerebral ischemic injury—a proteomics approach, *J. Pineal Res*. **46**:300–306 (2009).
50. Xiong X, Liang Q, Chen J, Fan R, Cheng T, Proteomics profiling of pituitary, adrenal gland, and splenic lymphocytes in rats with middle cerebral artery occlusion, *Biosci. Biotechnol. Biochem*. **73**:657–664 (2009).
51. Fert-Bober J, Basran RS, Sawicka J, Sawicki G, Effect of duration of ischemia on myocardial proteome in ischemia/reperfusion injury, *Proteomics* **8**:2543–2555 (2008).
52. Mayr M, Liem D, Zhang J, et al., Proteomic and metabolomic analysis of cardioprotection: Interplay between protein kinase C epsilon and delta in regulating glucose metabolism of murine hearts, *J. Cell. Mol. Med*. **46**:268–277 (2009).

53. Feng J, Zhu M, Schaub MC, et al., Phosphoproteome analysis of isoflurane-protected heart mitochondria: Phosphorylation of adenine nucleotide translocator-1 on Tyr194 regulates mitochondrial function, *Cardiovasc. Res*. **80**:20–29 (2008).

54. Chen CH, Budas GR, Churchill EN, Disatnik MH, Hurley TD, Mochly-Rosen D, Activation of aldehyde dehydrogenase-2 reduces ischemic damage to the heart, *Science* **321**:1493–1495 (2008).

55. Nicolaou P, Rodriguez P, Ren X, Zhou X, Quian J, Sadayappan S, Mitton B, Pathak A, Robbins J, Hajjar R, Jones K, Kranias E, Inducible expression of active protein phosphatase-1 inhibitor-1 enhances basal cardiac function and protects against ischemia/reperfusion injury, *Circ. Res*. **104**:1012–1020 (2009).

56. Lewis SJ, Prevention and treatment of atherosclerosis: A practitioner's guide for 2008, *Am. J. Med*. **122**: S38–S50 (2009).

57. Berhane BT, Zong C, Liem DA, Huang A, Le S, Edmondson RD, Jones RC, Qiao X, Whitelegge JP, Ping P, Vondriska TM, Cardiovascular-related proteins identified in human plasma by the HUPO Plasma Proteome Project Pilot Phase, *Proteomics* **5**:3520–3530 (2005).

58. Anderson NL, and Anderson NG, The human plasma proteome: History, character, and diagnostic prospects, *Mol. Cell. Proteom*. **1**:845–867 (2002).

59. Brea D, Sobrino T, Blanco M, Fraga M, Agulla J, Rodriguez-Yañez M, Rodriguez-Gonzalez R, Perez de la Ossa N, Leira R, Forteza J, Davalos A, Castillo J, Usefulness of haptoglobin and serum amyloid A proteins as biomarkers for atherotrombotic ischemic stroke diagnoses confirmation, *Atherosclerosis* **205**:561–567 (2009).

60. Wilson AM, Kimura E, Harada RK, Nair N, Narasimhan B, Meng X, Zhang F, Beck KR, Olin JW, Fung ET, Cooke JP, β2–Microglobulin as a biomarker in peripheral arterial disease, *Circulation* **116**:1396–1403 (2007).

61. Durán MC, Mas S-, Martin-Ventura JL, Meilhac O, Michel JB, Gallego-Delgado J, Lázaro A, Tuñón J, et al., Proteomic analysis of human vessels: Application to atherosclerotic plaques, *Proteomics* **3**:973–978 (2003).

62. Blanco-Colio L, Martin-Ventura JL, Muñoz-García B, Orbe J, Paramo JA, Michel JB, Ortiz A, Meilhac O, Egido J, Identification of soluble tumour necrosis factor-like weak inducer of apoptosis (sTWEAK) as a possible biomarker of subclinical atherosclerosis, *Arterioscler. Thromb. Vasc. Biol*. **27**:916–922 (2007).

63. Peronnet E, Becquart L, Poirier F, Cubizolles M, Choquet-Kastylevsky G, Jolivet-Reynaud C, SELDI-TOF MS analysis of the cardiac troponin I forms present in plasma from patients with myocardial infarction, *Proteomics* **6**:6288–6299 (2006).

64. Ganesh SK, Sharma Y, Dayhoff J, Fales HM, Van Eyk J, Kickler TS, Billings EM, Nabel EG, Detection of venous thromboembolism by proteomic serum biomarkers, *PLoS ONE* **2**(6):e544 (2007).

65. Howes JM, Keen JN, Findlay BC, Carter AM, The application of proteomics technology to thrombosis research: the identification of potential therapeutic targets in cardiovascular diseases, *Diabetes Vasc. Dis. Res*. **5**:205–212 (2008).

66. Liu RX, Chen HB, Tu K, Zhao SH, Li SJ, Dai J, Li QR, Nie S, Li YX, Jia WP, Wu JR, Localized-statistical quantification of human serum proteome associated with type 2 diabetes, *PLoS ONE* **3**:e3224 (2008).

67. Secchiero P, Corallini F, Ceconi C, Parrinello G, Volpato S, Ferrari R, Zauli G, Potential prognostic significance of decreased serum levels of TRAIL after acute myocardial infarction, *PLoS ONE* **4**(2):e4442 (2008).

68. Broedi UC, Lebherz C, Lerhke M, Stark R, Greif M, Becker A, Von Ziegler F, Tittus J, et al., Low adiponectin levels are an independent predictor of mixed and non-calcified coronary atherosclerotic plaques, *PLoS ONE* **4**(3):e4733 (2009).

69. Cleutjens K, Faber B, Rousch M, Van Doorn R, Hackeng TM, Vink C, Geusens P, Cate H, Walterbenger J, Tchaiskovski V, Lobbes M, Somers V, Sijbers A, Black D, Kitslaar J, Daemen M, Noninvasive diagnosis of ruptures peripheral atherosclerotic lesions and myocardial infarction by antibody profiling, *J. Clin. Invest.* **118**:2979–2985 (2008).

70. Bloch DB, Wang T, Gerszten R, Novel antibody markers of unstable atherosclerotic lesions, *J. Clin. Invest.* **118**:2675–2677 (2008).

71. Zimmerli LU, Schiffer E, Zürbig P, Good DM, Kellmann M, Mouls L, Pitt AR, Coon JJ, Schmieder RE, Peter KH, Mischak H, Kolch W, Delles C, Dominiczak AF, Urinary proteomic biomarkers in coronary artery disease, *Mol. Cell. Proteom.* **7**:290–298 (2008).

72. Coon J, Zurbig P, Dakna M, Dominiczak A, Decramer et al., CE-MS analysis of the human urinary proteome for biomarker discovery and disease diagnostics, *Proteom. Clin. Appl.* **2**:964–973 (2008).

73. Von zur Muhlen C, Schiffer E, Zuerbig P, Kellmann M, Brasse M, Meert N, Vanholder RC, Dominiczak AF, Chen YC, Mischak H, Bode C, Peter K, Evaluation of urine proteome pattern analysis for its potential to reflect coronary artery atherosclerosis in symptomatic patients, *J. Proteome Res.* **8**:335–345 (2009).

74. Brown A, Lattimore J, McGrady M, et al., Stable and unstable angina: identifying novel markers on circulating leukocytes, *Proteom. Clin. Appl.* **2**:90–98 (2008).

75. Barderas MG, Tuñon J, Darde VM, et al., Circulating human monocytes in the acute coronary syndrome express a characteristic proteomic profile, *J. Proteome Res.* **6**:876–886 (2007).

76. Barderas MG, Tuñon J, Darde VM, de la Cuesta F, Jimenez-Nacher JJ, Tarin N, Lopez-Bescos L, Egido J, Vivanco F, Atorvastatin modifies the protein profile of circulating human monocytes after an acute coronary syndrome, *Proteomics* **9**:1982–1993 (2009).

77. Karlsson H, Leanderson P, Tagesson C, Lindahl M, Lipoproteomics I: mapping of proteins in low-density lipoprotein using two-dimensional gel electrophoresis and mass spectrometry, *Proteomics* **5**:551–565 (2005).

78. Karlsson H, Leanderson P, Tagesson C, Lindahl M, Lipoproteomics II: Mapping of proteins in high-density lipoprotein using two-dimensional gel electrophoresis and mass spectrometry, *Proteomics* **5**:1431–1445 (2005).

79. Heinecke JW, The HDL proteome: A marker and perhaps mediator of coronary artery disease, *J. Lipid Res.* **50**:S167–S171 (2009).

80. Vaisar T, Pennathur S, Gree PS, Gharib SA, Hoofnagle AN, Cheung MC, Byun J, Vuletic S, Kassim S, Dingh P, et al., Shotgun proteomics implicates protease inhibition and complement activation in the anti-inflammatory properties of HDL, *J. Clin. Invest.* **117**:746–756 (2007).

81. Green PS, Vaisar T, Pennhatur S, Kulstad JJ, Moore AB, Marcovina S, Brunzel J, Knopp RH, Zhao QZ, Heinecke JW, Combined statin and niacin therapy remodels the high density lipoprotein proteome, *Circulation* **118**:1259–1267 (2008).

82. Pula G, Perera S, Prokopi M, Sidibe A, Boulanger CM, Mayr M, Proteomic analysis of secretory proteins and vesicles in vascular research, *Proteom. Clin. Appl.* **2**:882–891 (2008).
83. Huo Y, Ley KF, Role of platelets in the development of atherosclerosis, *Trends Cardiovasc. Med.* **14**:18–22 (2004).
84. Jennings LK. Role of platelets in atherothrombosis, *Am. J. Cardiol.* **103**: 4A–10A (2009).
85. Hernandez-Ruiz L, Valverde F, Jimenez-Nuñez MD, et al., Organellar proteomics of human platelet dense granules reveals that 14–3–3ζ is a granule protein related to atherosclerosis, *J. Proteome Res.* **6**:4449–4457 (2007).
86. Donners M, Verluyten MJ, Bouwman FG, et al., Proteomic analysis of differential protein expression in human atherosclerotic plaque progression, *J. Pathol.* **206**:39–45 (2005).
87. Park HK, Park EC, Bae SW, et al., Expression of heat shock protein 27 in human atherosclerotic plaques and increased plasma level of heat shock protein 27 in patients with acute coronary syndrome, *Circulation* **114**:886–893 (2006).
88. Martin-Ventura JL, Duran MC, Blanco-Colio LM, et al., Identification by a differential proteomic approach of heat shock protein 27 as a potential marker of atherosclerosis, *Circulation* **110**:2216–2219 (2004).
89. Rayner K, Chen K, Mc Nulty M, et al., Extracellular release of the atheroprotective heat shock protein 27 is mediated by estrogen and competitively inhibits acLDL binding to scavenger receptor A, *Circ. Res.* **103**:133–141 (2008).
90. Miller H, Poon S, Hibbert B, et al., Modulation of estrogen signaling by the novel interaction of heat shock protein 27, a biomarker for atherosclerosis, and estrogen receptor beta: Mechanistic insight into the vascular effects of estrogens, *Arterioscler. Thromb. Vasc. Biol.* **25**:e10–4 (2005).
91. Sung HJ, Ryang YS, Jang SW, et al., Proteomic analysis of differential protein expression in atherosclerosis, *Biomarkers* **11**:279–290 (2006).
92. Leppeda AJ, Cigliano A, Cherchi GM, et al., A proteomic approach to differentiate histologically classified stable and unstable plaques from human carotid arteries, *Atherosclerosis* **203**:112–118 (2009).
93. Slevin M, Elasbali AB, Turu MM, et al., Identification of differential protein expression associated with development of unstable human carotid plaques, *Am. J. Pathol.* **168**:1004–1021 (2006).
94. Martinet W, Schrijvers DM, De Meyer GRY, et al., Western array analysis of human atherosclerotic plaques: Downregulation of apoptosis-linked gene 2, *Cardiovasc. Res.* **60**:259–267 (2003).
95. Halushka MK, Cornish TC, Lu J, Selvin S, Selvin E, Creation, validation and quantitative analysis of protein expression in vascular tissue microarrays, *Cardiovasc. Pathol.* **19**:136–146 (2010).
96. Rodriguez-Feo JA, Hellings WE, Moll FL, De Vries JP, Van Middelaar BJ, Algra A, Sluijter J, Velema E, Van der Broek T, Sessa WC, De Kleijn D, Pasterkamp G, Caveolin-1 influences vascular protease activity and is a potential stabilizing factor in human atherosclerotic disease, *PLoS ONE* **3**(7):e2612 (2009).
97. Wilensky R, Shi Y, Mohler ER, Hamamdzic D, Burger ME, Li J, Postle A, Fenning RS, Bollinger J, et al., Inhibition of lipoprotein-associated phospholipase A2 reduces

complex coronary atherosclerotic plaque development, *Nat. Med.* **14**:1059–1066 (2008).

98. You SA, Archacki SR, Angheliou G, et al, Proteomic approach to coronary atherosclerosis shows ferritin light chain as a significant marker: Evidence consistent with iron hypothesis in atherosclerosis, *Physiol. Genom.* **13**:25–30 (2003).
99. Bagnato C, Thumar J, Mayya V, et al., Proteomic analysis of human coronary atherosclerotic plaque, *Mol. Cell. Proteom.* **6**:1088–1102 (2007).
100. Talusan P, Bedri S, Yang S, et al., Analysis of intimal proteoglycans in atherosclerosis-prone and atherosclerosis-resistant human arteries by mass spectrometry, *Mol. Cell. Proteom.* **4**:1350–1357 (2005).
101. Murray GI, Curran S, eds., *Laser Capture Microdissection. Methods and Protocols*, Vol. **293**, Humana Press, Totowa, NJ.
102. Emmert-Buck MR, Bonner RF, Smith PD, et al., Laser capture microdissection, *Science* **274**:998–1001 (1996).
103. Martinet W, Knaapen MWM, De Meyer GRY, et al., Overexpression of the anti-apoptotic caspase-2 short isoform in macrophage-derived foam cells of human atherosclerotic plaques, *Am. J. Pathol.* **162**:731–736 (2003).
104. De Souza AI, McGregor E, Dunn MJ, et al., Preparation of human heart laser microdissection and proteomics, *Proteomics* **4**:578–586 (2004).
105. Kondo T, Hirohashi S, Application of highly sensitive fluorescent dyes (CyDye DIGE Fluor saturation dyes) to laser microdissection and two-dimensional difference gel electrophoresis (2D-DIGE) for cancer proteomics, *Nat. Protoc.* **1**:2940–2956 (2006).
106. De la Cuesta F, Alvarez-Llamas G, Maroto AS, et al., An optimum method designed for 2D-DIGE analysis of human arterial intima and media layers isolated by laser microdissection, *Proteom. Clin. Appl.* **3**:1174–1184 (2009).
107. Dupont A, Corseaux D, Dekeyzer O, et al., The proteome and secretome of human arterial smooth muscle cells, *Proteomics* **5**:585–596 (2005).
108. Padro T, Peña E, García-Arguinzonis M, et al., Low-density lipoproteins impair migration of human coronary vascular smooth muscle cells and induce changes in the proteomic profile of myosin light chain, *Cardiovasc. Res.* **771**:211–220 (2008).
109. Trogan E, Choudhury RP, Dansky HM, Rong JX, Breslow JL, Fisher EA, Laser capture microdissection analysis of gene expression in macrophages from atherosclerotic lesions of apolipoprotein E-deficient mice, *Proc. Natl. Acad. Sci. USA* **99**:2234–2239 (2002).
110. Trogan E, Feig JE, Dogan S, et al., Gene expression changes in foam cells and the role of chemokine receptor CCR7 during atherosclerosis regression in ApoE-deficient mice, *Proc. Natl. Acad. Sci. USA* **103**:3781–3786 (2006).
111. Duran MC, Martin-Ventura JL, Mohammed S, et al., Atorvastatin modulates the profile of proteins released by human atherosclerotic plaques, *Eur. J. Pharm.* **562**:119–129 (2007).
112. O'Brien KD, Reichenbach DD, Marcovina SM, Kuusisto J, Alpers CE, Otto CM, Apolipoproteins B, (a), and E accumulate in the morphologically early lesion of "degenerative" valvular aortic stenosis, *Arterioscler. Thromb. Vasc. Biol.* **16**:523–532 (1996).

113. Olsson M, Thyberg J, Nilsson J, Presence of oxidized low density lipoprotein in nonrheumatic stenotic aortic valves, *Arterioscler. Thromb. Vasc. Biol*. **19**:1218–1222 (1999).

114. Otto CM, Kuusisto J, Reichenbach DD, Gown AM, O'Brien KD, Characterization of the early lesion of "degenerative" valvular aortic stenosis. Histological and immunohistochemical studies, *Circulation* **90**:844–853 (1994).

115. O'Brien KD, Pathogenesis of calcific aortic valve disease: A disease process comes of age (and a good deal more), *Arterioscler. Thromb. Vasc. Biol*. **26**:1721–1728 (2006).

116. Goldbarg SH, Elmariah S, Miller MA, Fuster V, Insights into degenerative aortic valve disease, *J. Am. Coll. Cardiol*. **50**:1205–1213 (2007).

117. Stewart BF, Siscovick D, Lind BK, Gardin JM, Gottdiener JS, Smith VE, Kitzman DW, Otto CM, Clinical factors associated with calcific aortic valve disease. Cardiovascular Health Study, *J. Am. Coll. Cardiol*. **29**:630–634 (1997).

118. Mohler ER III, Mechanisms of aortic valve calcification, *Am. J. Cardiol*. **94**:1396–1402 (2004).

119. Nalini M. Rajamannan, MD; Catherine M. Otto, MD, Targeted therapy to prevent progression of calcific aortic stenosis, *Circulation* **110**:1180–1182 (2004).

120. Rajamannan NM, Subramaniam M, Rickard D, et al., Human aortic valve calcification is associated with an osteoblast phenotype, *Circulation* **107**:2181–2184 (2003).

121. O'Brien KD, Shavelle DM, Caulfield MT, et al., Association of angiotensin converting enzyme with low-density lipoprotein in aortic valvular lesions and in human plasma, *Circulation* **106**:2224–2230 (2002).

122. Yetkin E, Walteriberger J, Molecular and cellular mechanism of aortic stenosis, *Int. J. Cardiol*. **135**:4–13 (2009).

123. Ramsey SA, Gold ES, Aderem A, A systems biology approach to understanding atherosclerosis, *EMBO Mol. Med*. **2**:79–89 (2010).

124. Wheelock C, Wheelock AM, Kawashima S, Diez D, Kanehisa M, Van Erk M, Kleemann R, Haeggström JZ, Goto S, Systems biology approaches and pathways tools for investigating cardiovascular disease, *Mol. Biosyst*. **5**:588–602 (2009).

125. Lusis AJ, Weiss JN, Cardiovascular networks: Systems based approaches to cardiovascular disease, *Circulation* **121**:157–170 (2010).

# INDEX



*Platelet Proteomics: Principles, Analysis and Applications*, First Edition.
Edited by Ángel García and Yotis A. Senis.

Printed and bound by CPI Group (UK) Ltd, Croydon, CR0 4YY

16/04/2025

14658354-0002